ENVIRONMENTAL SCIENCE

ENVIRONMENTAL SCIENCE

Copyright Notice

All rights reserved. No part of this publication may be reproduced, stored in a retrieval system, or transmitted in any form or by any means, electrical, mechanical, photocopying, recording or otherwise, without the permission of BIOZONE International Ltd. This book may not be re-sold. The conditions of sale specifically prohibit the photocopying of exercises, worksheets, and diagrams from this book for any reason.

Fourth edition

ISBN 978-1-99-101409-2

Copyright ©2025 BIOZONE International Ltd

First printing

Printed by Thomson Press
using paper from responsible sources.

Acknowledgements

BIOZONE wishes to thank and acknowledge the team for their efforts and contributions to the production of this title.

Cover Photograph

Photo: iStock by Getty Images

Dubai city as seen at night from the Burj Khalifa tower – the world's tallest structure. Light pollution from artificial lights brightens the night sky. It can affect wildlife and ecosystems as well as human health.

About the Authors

Jillian Mellanby *Editor*

Jill began her science career with a degree in biochemistry and, after some time working in research institutes, became a science teacher, working in the UK and New Zealand. She spent many years as managing editor of a suite of science journals and has also written science articles for a public audience. She joined BIOZONE in late 2021.

Kent Pryor *Author*

Kent has a BSc from Massey University majoring in zoology and ecology and taught secondary school biology and chemistry for 9 years before joining BIOZONE as an author in 2009.

Sarah Gaze *Author*

Sarah has 16 years of experience as a Science and Chemistry teacher, recently completing M.Ed. (1st class hons) with a focus on curriculum, science, and climate change education. She has a background in educational resource development, academic writing, and art. Sarah joined the BIOZONE team at the start of 2022.

Lissa Bainbridge-Smith *Author*

Lissa graduated with a Masters in Science (hons) from the University of Waikato. After graduation she worked in industry in a research and development capacity for eight years. Lissa joined BIOZONE in 2006 and is hands-on developing new curricula. Lissa has also taught science theory and practical skills to international and ESL students.

North & South America	sales@biozone.com
UK & Rest of World	sales@biozone.co.uk
Australia	sales@biozone.com.au
New Zealand	sales@biozone.co.nz

www.BIOZONE.com

Contents

Using This Worktext vi
Using BIOZONE's Resource Hub ix

Course Concept Map for Environmental Science x

The Earth's Systems

Concept Map: The Earth's Systems 1
Section Focus 2

Chapter 1: The Earth's Systems

Learning Outcomes 4

- [ ] 1 Dating the Earth 5
- [ ] 2 The Earth's History 6
- [ ] 3 Structure of the Earth 8
- [ ] 4 Fossil Formation 9
- [ ] 5 Earth and Sun Cycles 11
- [ ] 6 The Earth's Crust 13
- [ ] 7 Plate Boundaries 14
- [ ] 8 The Lithosphere and Asthenosphere 16
- [ ] 9 Mechanism of Plate Movement 17
- [ ] 10 Continental Drift 18
- [ ] 11 Volcanoes and Volcanism 19
- [ ] 12 Earthquakes 21
- [ ] 13 The Rock Cycle 22
- [ ] 14 Soil Textures 24
- [ ] 15 Soil and Soil Dynamics 25
- [ ] 16 The Atmosphere 27
- [ ] 17 Atmospheric Circulation 28
- [ ] 18 El Niño and La Niña 30
- [ ] 19 Water 31
- [ ] 20 Ocean Circulation and Currents 32
- [ ] 21 Earth's Past Climate 34
- [ ] 22 Did You Get It? 36

The Living World

Concept Map: The Living World 37
Section Focus 38

Chapter 2: Ecosystems

Learning Outcomes 40

- [ ] 23 Components of an Ecosystem 41
- [ ] 24 Scales of Ecosystems 42
- [ ] 25 Factors Affecting Biome Distribution 43
- [ ] 26 The World's Terrestrial Biomes 44
- [ ] 27 The Effect of Temperature on Biomes 46
- [ ] 28 Aquatic Biomes 47
- [ ] 29 Physical Factors and Gradients 48
- [ ] 30 Physical Factors in a Forest 49
- [ ] 31 Stratification in a Forest 50
- [ ] 32 Physical Factors on a Rocky Shore 51
- [ ] 33 Physical Factors in a Small Lake 52
- [ ] 34 Habitat 53
- [ ] 35 Ecological Niche 54
- [ ] 36 Energy Inputs and Outputs 56
- [ ] 37 Plants as Producers 57
- [ ] 38 Measuring Primary Productivity 59
- [ ] 39 Cellular Respiration 60
- [ ] 40 Food Chains 61
- [ ] 41 Food Webs 62
- [ ] 42 Energy Flow in an Ecosystem 64
- [ ] 43 Production and Trophic Efficiency 66
- [ ] 44 Ecological Pyramids 69
- [ ] 45 Species Interactions in Communities 71
- [ ] 46 Ecosystem Stability 73
- [ ] 47 The Scale of Environmental Change 75
- [ ] 48 Cycles of Matter 76
- [ ] 49 The Carbon Cycle 77
- [ ] 50 The Nitrogen Cycle 79
- [ ] 51 The Oxygen Cycle 81
- [ ] 52 The Water Cycle 82
- [ ] 53 The Phosphorus Cycle 83
- [ ] 54 Primary Succession 84
- [ ] 55 Primary Succession: Surtsey Island 85
- [ ] 56 Secondary Succession 86
- [ ] 57 Wetland Succession 88
- [ ] 58 Did You Get It? 89

Chapter 3: Populations

Learning Outcomes 90

- [ ] 59 Features of Populations 91
- [ ] 60 Density and Distribution 92
- [ ] 61 Population Regulation 93
- [ ] 62 Population Growth 94
- [ ] 63 Survivorship Curves 95
- [ ] 64 Life Expectancy and Survivorship in Humans 96
- [ ] 65 Population Growth Curves 98
- [ ] 66 Modeling Population Growth 100
- [ ] 67 *r* and K Selection 101
- [ ] 68 Population Age Structure 102
- [ ] 69 World Population Distribution 104
- [ ] 70 Population Growth Rate 107

CODING Activity is marked: ⊡ to be done   ☑ when completed

- [ ] 71 Human Demography 109
- [ ] 72 Humans and Resources 111
- [ ] 73 Did You Get It? 113

Chapter 4: Investigating Ecosystems

Learning Outcomes 114

- [ ] 74 Why Do We Sample? 115
- [ ] 75 Sampling Populations 116
- [ ] 76 Sampling and Sensors 118
- [ ] 77 Monitoring Water Quality 119
- [ ] 78 Sampling Animal Populations 120
- [ ] 79 Introducing Sampling Techniques 121
- [ ] 80 Quadrat Sampling 122
- [ ] 81 Sampling a Leaf Litter Population 124
- [ ] 82 Transect Sampling 126
- [ ] 83 Mark and Recapture Sampling 128
- [ ] 84 Indirect Sampling 130
- [ ] 85 Radio, Satellite, and GPS Animal Tracking 131
- [ ] 86 Environmental DNA Sampling 132
- [ ] 87 Measuring Diversity in an Ecosystem 133
- [ ] 88 Classification Keys 135
- [ ] 89 Did You Get It? 138

Global Resources

Concept Map: Global Resources 139
Section Focus 140

Chapter 5: Land and Water

Learning Outcomes 142

- [ ] 90 Land for Agriculture 143
- [ ] 91 The Importance of Plants 144
- [ ] 92 The Green Revolution 145
- [ ] 93 Impacts of Farming 147
- [ ] 94 Intensive Farming Practices 148
- [ ] 95 Sustainable Farming 150
- [ ] 96 Cereal Crop Production 152
- [ ] 97 Meat Production 154
- [ ] 98 Food Security 155
- [ ] 99 Pest Control 156
- [ ] 100 Pesticide Resistance 158
- [ ] 101 Integrated Pest Management 159
- [ ] 102 Soil Degradation 161
- [ ] 103 Reducing Soil Erosion 163
- [ ] 104 Forestry 164
- [ ] 105 Managing Rangelands 166
- [ ] 106 Reserve Lands 168
- [ ] 107 City Planning 170
- [ ] 108 Transportation 172

- [ ] 109 Mining and Minerals 173
- [ ] 110 Globalization 175
- [ ] 111 Global Water Resources 176
- [ ] 112 Water and People 179
- [ ] 113 Water and Industry 181
- [ ] 114 Ecological Impacts of Fishing 183
- [ ] 115 Fisheries Management 185
- [ ] 116 Did You Get It? 188

Chapter 6: Energy

Learning Outcomes 189

- [ ] 117 Using Energy Transformations 190
- [ ] 118 Global Energy Consumption 191
- [ ] 119 Non-Renewable Resources 192
- [ ] 120 Coal 193
- [ ] 121 Oil and Natural Gas 195
- [ ] 122 Oil Extraction 198
- [ ] 123 Environmental Issues of Oil Extraction 201
- [ ] 124 Nuclear Power 202
- [ ] 125 Renewable Energy 205
- [ ] 126 Wind Power 206
- [ ] 127 Hydroelectricity 208
- [ ] 128 Solar Power 210
- [ ] 129 Geothermal Power 212
- [ ] 130 Ocean Power 214
- [ ] 131 Energy from Biomass 215
- [ ] 132 Hydrogen Fuel Cells 217
- [ ] 133 Comparing Fuel Choices 218
- [ ] 134 Energy Conservation 219
- [ ] 135 Energy Security 221
- [ ] 136 Energy Storage 223
- [ ] 137 Rechargeable Batteries and Energy Storage ... 225
- [ ] 138 Did You Get It? 226

Global Change

Concept Map: Global Change 227
Section Focus 228

Chapter 7: Pollution

Learning Outcomes 230

- [ ] 139 Types of Pollution 231
- [ ] 140 Water Pollution 232
- [ ] 141 Nitrogen Pollution 234
- [ ] 142 Eutrophication and Water Quality 236
- [ ] 143 Biomagnification 238
- [ ] 144 Sewage Treatment 241
- [ ] 145 Waste Management 242
- [ ] 146 Reducing Waste 243

CODING: Activity is marked: ☐ to be done   ☑ when completed

	147	Plastics in the Environment................................ 246
	148	Microplastic and Nanoplastic Pollution............ 250
	149	Air Pollution.. 252
	150	Cities and Air Pollution 254
	151	Acid Rain 256
	152	Reducing Air Pollution................................... 257
	153	Stratospheric Ozone Depletion 258
	154	Noise Pollution.. 261
	155	Pollution in the Home..................................... 262
	156	Light Pollution.. 264
	157	Health Effects of Pollution............................. 265
	158	Effect of Oil Spills.. 266
	159	Cleaning Up Oil Spills.................................... 269
	160	Fossil Fuels and Health................................... 271
	161	The Effects of Nuclear Accidents.................... 272
	162	Bhopal Disaster 274
	163	Mining Disasters.. 275
	164	Environmental Remediation 276
	165	The Economic Impact of Pollution................. 278
	166	The Role of Environmental Legislation.......... 279
	167	Did You Get It?... 281

Chapter 8: Conservation

Learning Outcomes.. 282

	168	Biodiversity Hotspots 283
	169	Loss of Biodiversity.. 284
	170	Where Have All the Insects Gone?.................. 286
	171	Tropical Deforestation 288
	172	Habitat Fragmentation.................................... 290
	173	The Impact of Introduced Species 291
	174	Control of Introduced Species........................ 292
	175	The Impact of New Diseases........................... 293
	176	Endangered Species.. 295
	177	The Sixth Mass Extinction............................... 298
	178	Managing Environmental Resources.............. 300
	179	Ecotourism in the Galápagos Islands 302
	180	*In Situ* Conservation 304
	181	*Ex Situ* Conservation 306
	182	Rewilding.. 308
	183	Conservation and Legislation 309
	184	Conservation and Sustainability 310
	185	Did You Get It?..311

Chapter 9: Climate Change

Learning Outcomes.. 312

	186	What's the Concern with Climate Change?...... 313
	187	Finding the Evidence for Climate Change..........315
	188	Climate Change Legislation 316
	189	Models of Climate Change..............................318

	190	The Enhanced Greenhouse Effect 320
	191	Warming Oceans... 322
	192	Disappearing Islands 323
	193	Ocean Acidification.. 324
	194	Albedo Effect.. 325
	195	Extreme Weather Events................................. 326
	196	Wildfires... 328
	197	Megadroughts... 329
	198	Climate Change and Range Shift.................... 330
	199	Biodiversity and Climate Change 332
	200	Positive Feedback Cycles 334
	201	Tipping Points.. 336
	202	Tipping Point: Greenland Ice Sheet................. 338
	203	Tipping Point: West Antarctic Ice Sheet 340
	204	Tipping Point: Boreal Permafrost..................... 341
	205	Tipping Point: Boreal Forests.......................... 342
	206	Tipping Point: Amazon Rainforest 343
	207	Tipping Point: Warm Water Coral Reefs.......... 344
	208	Tipping Point: AMOC and the Subpolar Gyres. 346
	209	Climate Risk 348
	210	Climate Change and Agriculture..................... 349
	211	Mitigation and Adaptation............................. 350
	212	The Climate Action Movement 352
	213	Carbon Trading 353
	214	Carbon Capture and Storage.......................... 354
	215	Carbon Sequestration 356
	216	Carbon Footprints 357
	217	Moving to Net Zero Carbon 358
	218	Possibilities of Solar Radiation Modification 360
	219	Did You Get It?... 362

Chapter 10: Science Practices

Learning Outcomes.. 363

	220	Models and Modeling 364
	221	Types of Data 365
	222	Mean, Median, Mode..................................... 366
	223	Which Graph to Use?..................................... 367
	224	Analysing and Interpreting Data.................... 368
	225	Working with Numbers 370
	226	Calculations, Conversions, and Multiples......... 372
	227	Correlation or Causation?............................... 373

Glossary 374
Questioning Terms... 379
Image Credits... 379
Index 380

CODING Activity is marked: ⊡ to be done ☑ when completed

Using This Worktext

This worktext is designed to increase your understanding of the content and skill requirements of your environmental science course, and reinforce and extend the ideas developed by your teacher. The information on the next few pages will help you navigate the content and utilize the features of the worktext. Environmental Science is divided into four sections, and each section begins with a concept map. The structure of a section is provided below.

Structure of a Section and Chapter

Section concept map
A concept map begins each section and provides an overview of content.

Section focus
A graphical double page spread highlights key ideas in the section.

Chapter introduction
Short statements summarize what you need to know for each chapter.

Did You Get It?
An assessment task allowing you to demonstrate your knowledge of the chapter's content.

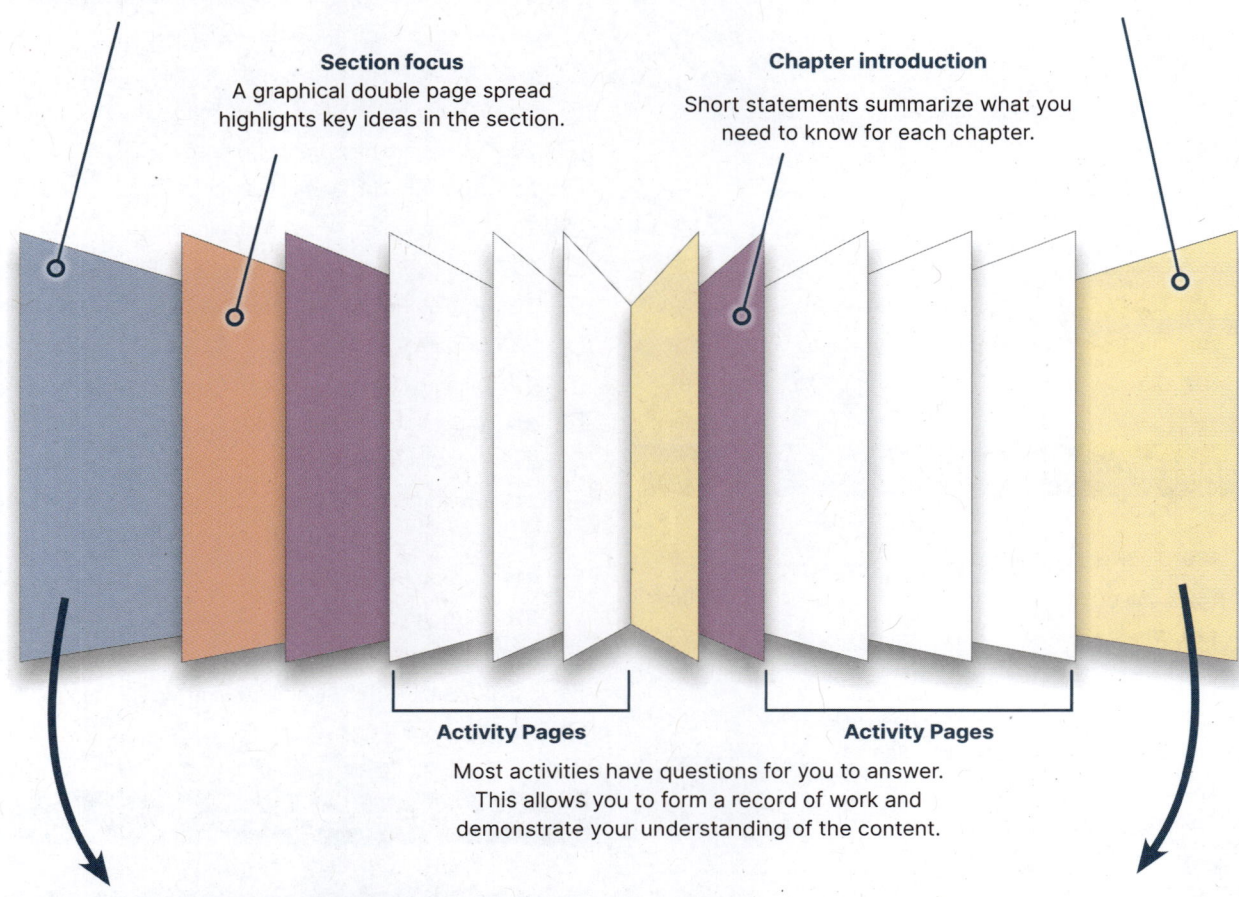

Activity Pages

Activity Pages

Most activities have questions for you to answer. This allows you to form a record of work and demonstrate your understanding of the content.

Section concept map

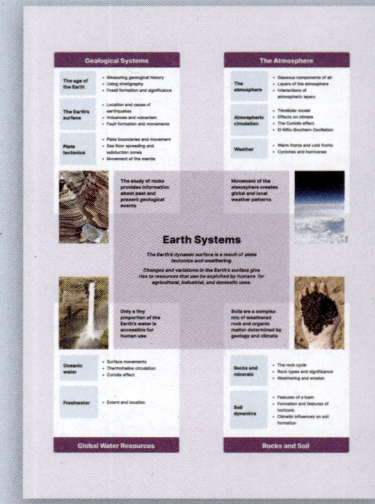

The concept maps divide the worktext in four sections:

- Earth's Systems
- The Living World
- Global Resources
- Global Change

They provide an overview or big picture approach to what you will cover in the course.

Assessment: did you get it?

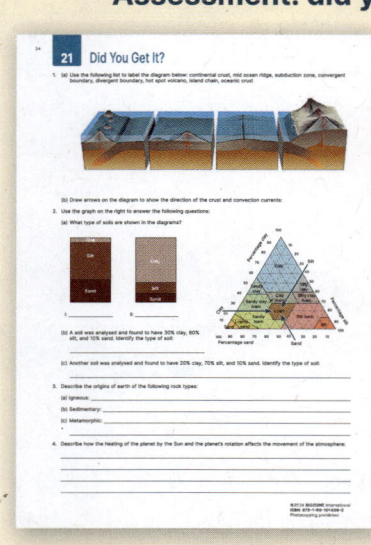

Each chapter has an end of chapter test.

It allows you to test your understanding of the content covered.

Chapter introductions

The chapter introductions contain useful information to help you navigate through the course and identify the learning outcomes (what you need to know). Use the information provided to help you learn vocabulary, identify key concepts and learning outcomes, and quickly navigate to supporting resources on BIOZONE's **Resource Hub**. The key features are explained below.

The section of the worktext you are in is identified for easy navigation.

QR codes and bitly tags allow you to quickly navigate to helpful content (e.g. videos and models) on BIOZONE's **Resource Hub**.

Chapter number and chapter title are identified for quick navigation.

Key terms
Important vocabulary you should understand and use in your course. Definitions are provided in the glossary at the back of the book.

Check boxes
Use the check boxes to keep a record of which activities you need to complete and tick them off as you work through them.

Key concepts
These are the important key ideas for the chapter. Make sure you understand the concepts summarized here.

Learning outcomes
These provide a point by point summary of what you need to know or do by the end of the chapter.

Activity numbers
The activity number for each learning outcome is identified.

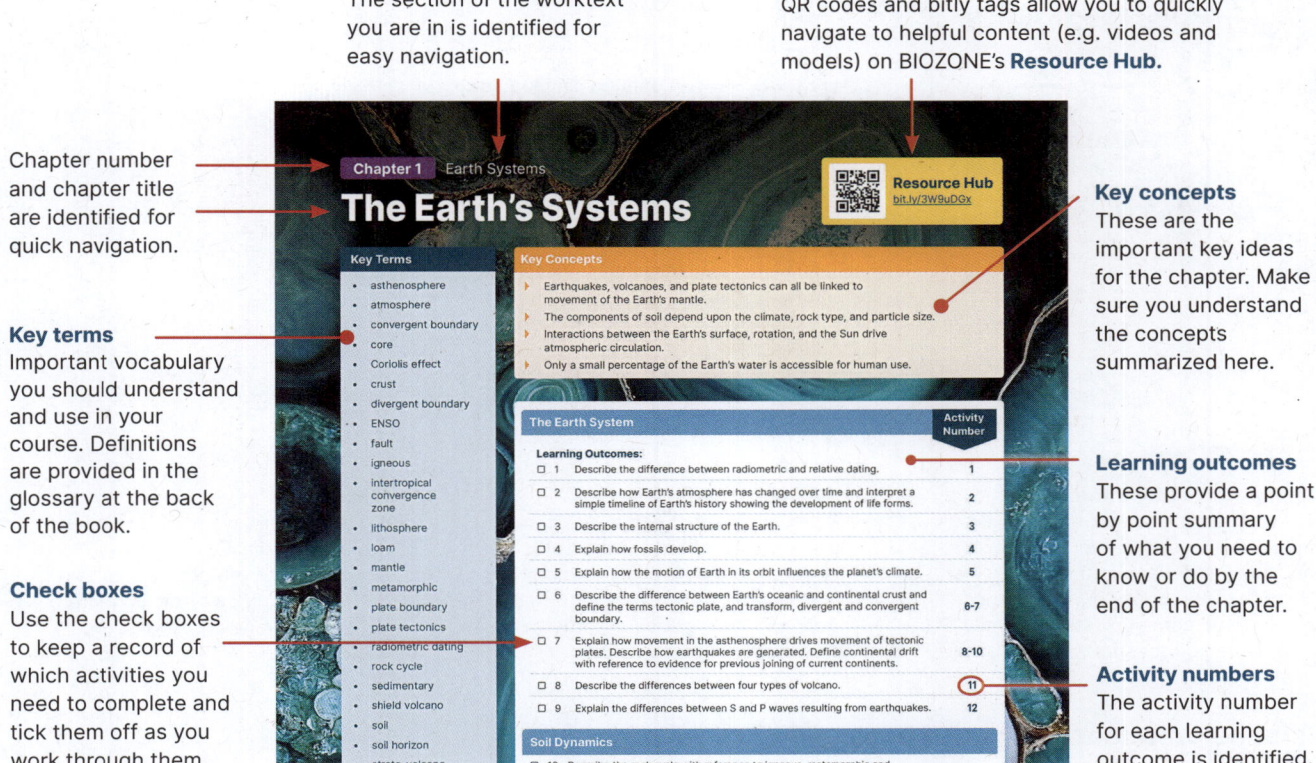

Structure of the activity pages

Activities make up most of the worktext. Be sure to interact with all the elements on the page so you don't miss any valuable information. As you work through the material, answer the questions and complete the tasks provided. Inputting your answers will form a record of work which helps to consolidate your understanding. It can also be used for revision at a later date.

Key Idea:
This provides a focus for the activity and can be used as a summary take-home point of the activity.

More information about the topic is provided through explanatory text, images, diagrams, case studies, and data.

Activity number:
Identifies the activity number to help you navigate between activities.

Introductory paragraph:
This provides background or introductory information to the topic.

Yellow QR codes:
These provide a quick link to interactive 3D models.

Blue QR codes:
These link to live data sets.

Tab system:
The tabs provide information about which section you are in and if support material is available on the **BIOZONE Resource Hub**.

Activity based questions:
Answering the questions helps reinforce your learning. Use your answers to review for tests and other assessments.

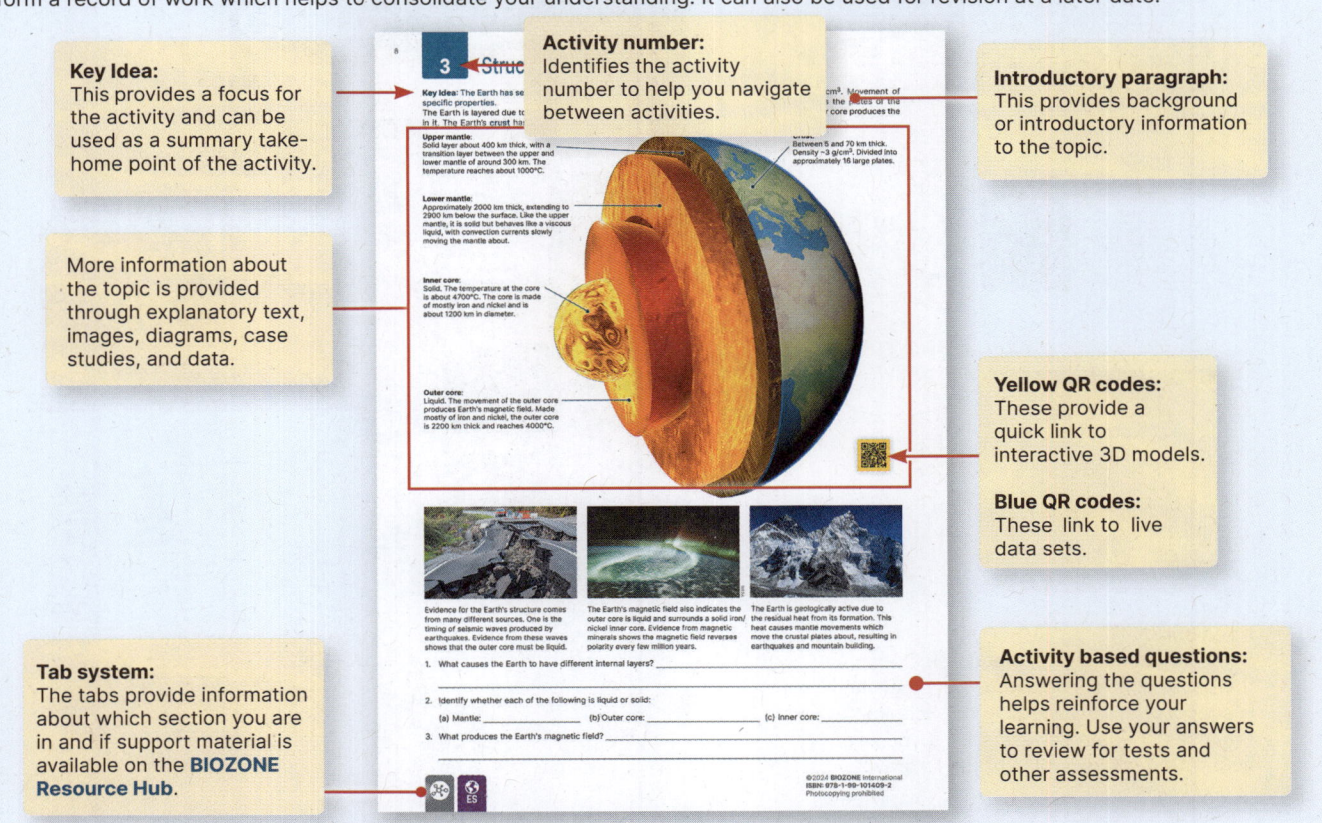

The glossary: Helping you build your science vocabulary

Building your environmental sciences vocabulary is important to help you understand ideas and communicate information about what you know. Your BIOZONE worktext has several tools to help you with this. They are explained below.

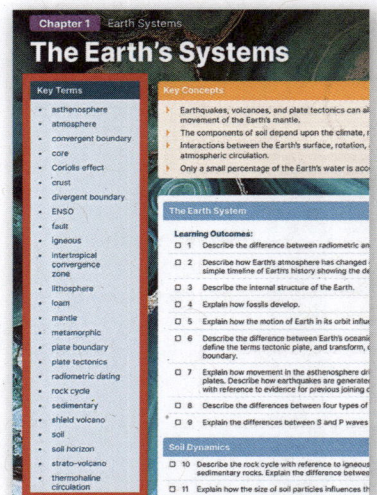

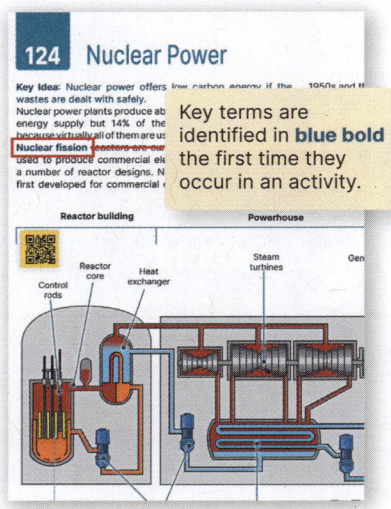

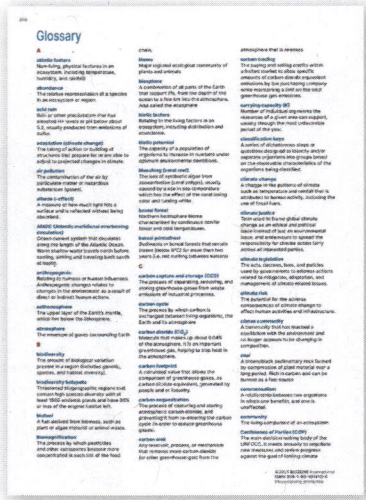

Key terms: chapter introduction
Each chapter introduction has a list of key terms. These are important words and they are defined in the glossary at the back of the book. Try to use these terms as you communicate with your classmates and teacher.

Key terms: activities
You may see that some words in an activity are written in **blue bold**. This is because they are key terms. If you are unsure of their meaning, they are defined in the glossary at the back of the book.

Glossary
A glossary of key terms is located at the back of this Environmental Science worktext. Use the glossary to find definitions for key terms and improve your understanding of what the terms mean.

Science practices: help with basic skills

The last chapter in this worktext provides information and support to help you with some basic skills you will encounter in your environmental science course. It will help you with graphing, doing simple calculations (e.g. calculating mean, median, and mode, and rates) and conversions between units. Your teacher may ask you to do certain science practice activities, but you can refer to the chapter at any time if you need help with math skills.

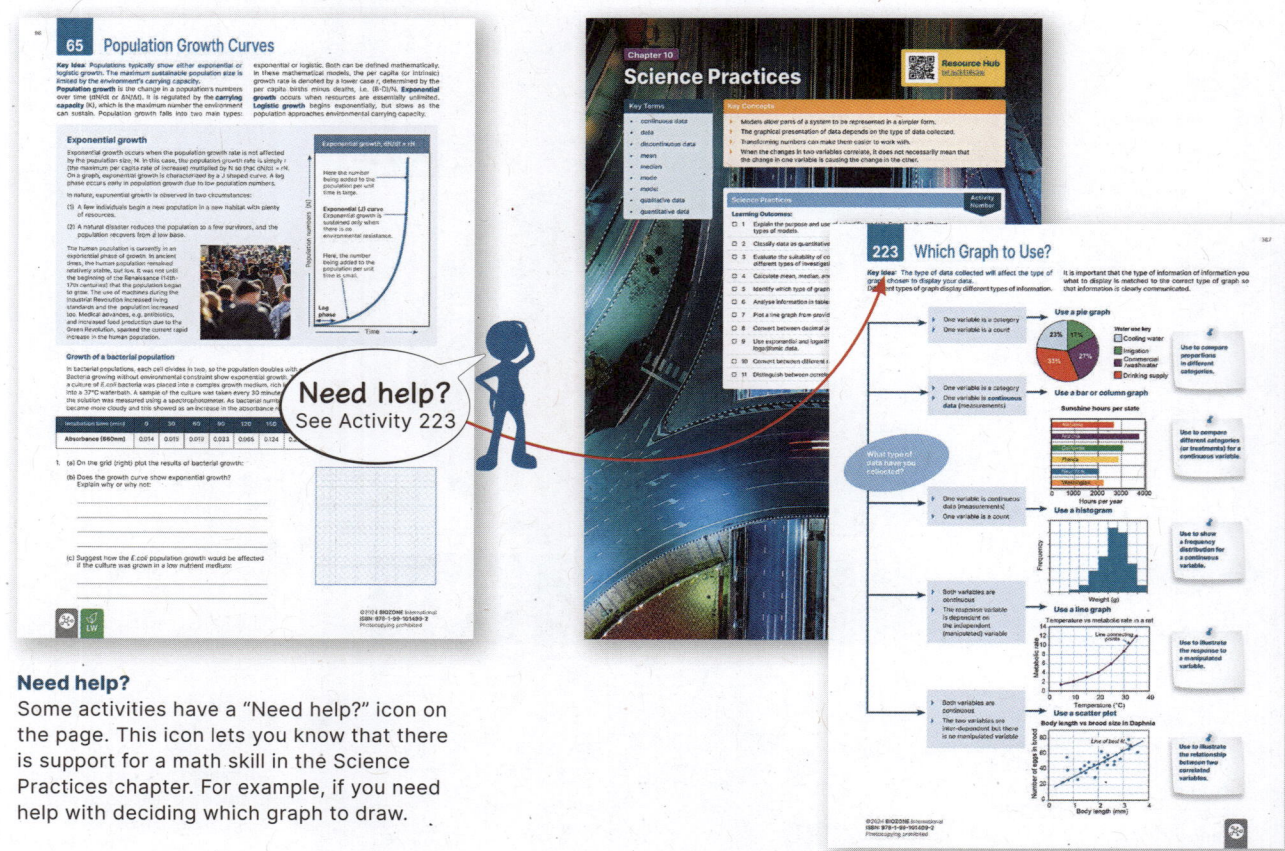

Need help?
Some activities have a "Need help?" icon on the page. This icon lets you know that there is support for a math skill in the Science Practices chapter. For example, if you need help with deciding which graph to draw.

Using BIOZONE's Resource Hub

▶ Most of the activities have interesting resources, such as videos and 3D models, to help you understand the content. A grey tab (right) on the activity page indicates there is support on the BIOZONE **Resource Hub** for the activity.

▶ Navigate to the **Resource Hub** either by bookmarking the link below, or by utilizing the bitly tag or QR code found on each chapter introduction (below, right).

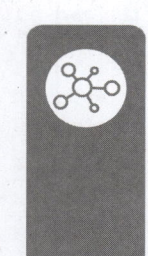

Step 1: Navigate to the BIOZONE **Resource Hub**

Step 2: Enter this code in the box displayed.

Step 3: Bookmark this page.

www.BIOZONEhub.com

ENS4-4092

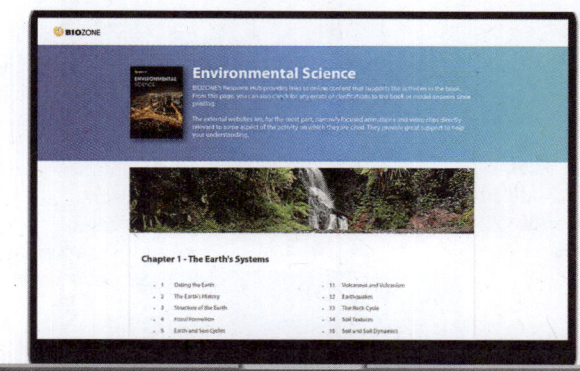

bit.ly/3LhnHRm

Use this bitly tag or QR code to directly access the BIOZONE Resource Hub.

Using the QR codes on activity pages

Some activities have QR codes on the pages (below). These link directly to informative and engaging 3D models or to live data sets. If your school does not let you use your phone in class, you can still access the models and data sets through the **Resource Hub**. Follow the steps above to access the resources.

Yellow QR codes on the pages (circled, below), link directly to informative and engaging 3D models. All models can be rotated and zoomed, and some contain informative notes.

Blue QR codes on the pages (circled, below), link directly to live data sets that are updated regularly, providing up-to-date data for some rapidly changing areas of environmental science.

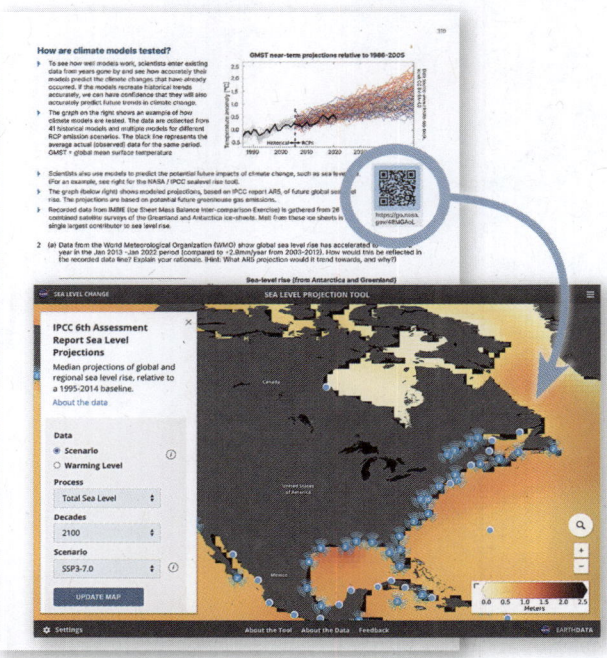

The Earth's Systems

Geological systems	• Age of the Earth • The Earth's surface • Plate tectonics
The atmosphere	• The atmosphere • Atmospheric circulation • Weather and climate
Global water resources	• Oceanic water • Fresh water
Rocks and soil	• Rocks and minerals • Soil dynamics

The Living World

Ecosystems	• Biomes and ecosystems • Energy flow • Species interactions
Natural ecosystem change	• Biogeochemical cycles • Ecosystem stability
Populations	• Features of populations • Population growth
Investigating ecosystems	• Sampling populations • Abiotic factors • Classification

The resources used by humans are a result of constantly changing global systems.

Complex systems arise as a result of the interactions between organisms and their environment.

Environmental Science

All of the Earth's systems are connected. Changes in one system may cause changes in other systems.

Environmental science is an interdisciplinary field of study involving both natural and social sciences.

Resources must be carefully managed to ensure they are available to future generations.

Understanding environmental systems is critical to understanding the environmental effects of human activities.

Global Resources

Food production	• Agriculture • Land management
Earth's Resources	• Fisheries • Irrigation
Energy sources	• Energy • Non-renewable energy • Renewable energy
Energy security	• Energy conservation • Energy threats • Energy storage

Global Change

Types of pollution	• Air pollution • Water pollution • Ozone depletion
Impacts and treatments	• Treating pollution • Impacts of pollution • Disasters
Conserving biodiversity	• Loss of biodiversity • Maintaining biodiversity
Climate change	• Climate change science • Tipping points

Geological Systems

The age of the Earth	• Measuring geological history • Using stratigraphy • Fossil formation and significance
The Earth's surface	• Location and cause of earthquakes • Volcanoes and volcanism • Fault formation and movements
Plate tectonics	• Plate boundaries and movement • Sea floor spreading and subduction zones • Movement of the mantle

The study of rocks provides information about past and present geological events.

The Atmosphere

The atmosphere	• Gaseous components of air • Layers of the atmosphere • Interactions of atmospheric layers
Atmospheric circulation	• Tricellular model • Effects on climate • The Coriolis effect • El Niño-Southern Oscillation
Weather	• Warm fronts and cold fronts • Cyclones and hurricanes

Movement of the atmosphere creates global and local weather patterns.

Earth Systems

The Earth's dynamic surface is a result of plate tectonics and weathering.

Changes and variations in the Earth's surface give rise to resources that can be exploited by humans for agricultural, industrial, and domestic uses.

Only a tiny proportion of the Earth's water is accessible for human use.

Soils are a complex mix of weathered rock and organic matter determined by geology and climate.

Oceanic water	• Surface movements • Thermohaline circulation • Coriolis effect
Freshwater	• Extent and location

Rocks and minerals	• The rock cycle • Rock types and significance • Weathering and erosion
Soil dynamics	• Features of a loam • Formation and features of horizons • Climatic influences on soil formation

Global Water Resources

Rocks and Soil

ON TOP OF THE WORLD

The Himalayas are a broad band of mountains forming a boundary between the Indian subcontinent to the South and the Tibetan plateau to the North. In geological terms, they are young mountains, having begun forming around 50 million years ago. The collision between two tectonic plates on which they sit continues to shape them today as one plate pushes against the other. Earthquakes are also relatively frequent in this seismically active area of Earth. There are more than 100 peaks exceeding 7,200 m in elevation, including Mount Everest (8,848.86 m). Mount Everest's height places its peak in the upper troposphere where it is exposed to the jet stream, with winds reaching 160 km/h.

The composition of a giant

Mt Everest itself is composed of limestone, marble and schist. Limestone rocks from near the top of the mountain were once marine sediments laid down around 500 million years ago. The rocks contain marine fossils, including trilobites, brachiopods, ostracods, and crinoids. Beneath the upper band of limestone, the pressure exerted by mountain building transformed limestone into marble, found in the 'yellow band', shown left.

Highest weather station in the world

In May 2022, a weather station was installed on Mt Everest at 8,810 m. It is the highest weather station in the world. At such heights, temperatures can drop to -40°C. Because the peak sits in the jet stream, wind speeds are commonly over 100 km/h, and wind gusts of over 250 km/h have been recorded there. The plume on this photo is snow and ice blasted off the summit by high winds.

Metamorphosis

The Himalayas illustrate various parts of the cycling of Earth's rocks. Pressure created during uplift of the mountains has metamorphosed limestones into marble, and sandstone, and mudstones into schist. The mountains undergo constant erosion via glaciers and weathering but, overall, are rising faster than they are being eroded.

🔍 Take a Deeper Look

- What geological processes build mountains?
- What reasons might there be for the Himalayas being, on average, so much higher than other mountains ranges around the world?
- What evidence is there for the age of the rocks that make up Mount Everest and the Himalayas?

Chapter 1 Earth Systems
The Earth's Systems

Resource Hub
bit.ly/3W9uDGx

Key Terms
- asthenosphere
- atmosphere
- convergent boundary
- core
- Coriolis effect
- crust
- divergent boundary
- ENSO
- fault
- horizon (soil)
- igneous
- intertropical convergence zone
- lithosphere
- loam
- mantle
- metamorphic
- plate boundary
- plate tectonics
- radiometric dating
- rock cycle
- sedimentary
- shield volcano
- soil
- strato-volcano
- thermohaline circulation
- transform boundary
- tricellular model

Key Concepts
- Earthquakes, volcanoes, and plate tectonics can all be linked to movement of the Earth's mantle.
- The components of soil depend upon the climate, rock type, and particle size.
- Interactions between the Earth's surface, rotation, and the Sun drive atmospheric circulation.
- Only a small percentage of the Earth's water is accessible for human use.

The Earth System

Learning Outcomes:

	#	Outcome	Activity Number
☐	1	Describe the difference between radiometric and relative dating.	1
☐	2	Describe how Earth's atmosphere has changed over time and interpret a simple timeline of Earth's history showing the development of life forms.	2
☐	3	Describe the internal structure of the Earth.	3
☐	4	Explain how fossils develop.	4
☐	5	Explain how the motion of Earth in its orbit influences the planet's climate.	5
☐	6	Describe the difference between Earth's oceanic and continental crust and define the terms tectonic plate, and transform, divergent and convergent boundary.	6-7
☐	7	Explain how movement in the asthenosphere drives movement of tectonic plates. Describe how earthquakes are generated. Define continental drift with reference to evidence for previous joining of current continents.	8-10
☐	8	Describe the differences between four types of volcano.	11
☐	9	Explain the differences between S and P waves resulting from earthquakes.	12

Soil Dynamics

	#	Outcome	Activity Number
☐	10	Describe the rock cycle with reference to igneous, metamorphic and sedimentary rocks. Explain the difference between rocks and minerals.	13
☐	11	Explain how the size of soil particles influences the type of loam formed.	14
☐	12	Describe how climate and local conditions affect soil development.	15

Atmosphere, Climate and Water

	#	Outcome	Activity Number
☐	13	Describe the composition of the Earth's atmosphere.	16
☐	14	Describe how the Sun affects the atmosphere's circulation with reference to the tricellular model. Describe the Coriolis effect and how it influences Earth's weather systems.	17
☐	15	Explain how the El Niño Southern Oscillation develops and its effect on weather patterns in named regions of the world.	18

Global Resources

	#	Outcome	Activity Number
☐	16	Describe the nature and extent of the Earth's freshwater and salt water resources.	19
☐	17	Describe the nature and extent of the Earth's oceans, including reference to deepwater and surface circulation.	20

1 Dating the Earth

Key Idea: The Earth can be dated accurately using radiometric dating. Relative dating uses the layering of rocks to age events relative to each other.

One of the most accurate ways of dating the Earth is **radiometric dating**. Many of the heavier elements, e.g. uranium and thorium, decay over time; their atoms break apart into smaller atoms. The rate of decay depends only on the original element's isotope, e.g. uranium-238 or uranium-235. For uranium-238, it takes 4.468 billion years for half the sample to decay into lead-206 (called the half-life). For uranium-235 (which decays into lead-207) the half-life is 700 million years. The ratios of uranium-238 to lead-206 and uranium-235 to lead-207 in a sample can therefore be used to determine the time since the sample formed. Relative dating uses the law of superposition, which states that the deeper layers or rock at a site will be older than the layers above it. Therefore objects, e.g. fossils, in deeper layers will be older than objects in shallower layers. Thus the order of events and relative age of rock layers (strata), even at sites that are distant from each other, can be determined by studying the content of the layers. Incursions of layers that can be radiometrically dated can help give ages to the layers.

Radiometric dating

The very oldest material dated on Earth is a **zircon** crystal found in a metamorphosed sandstone from Western Australia. It is dated to 4.4 billion years old, just a hundred million years after the Earth formed. Due to Earth's dynamic history, minerals older than this are unlikely to be found. However, other bodies in the solar system have remained essentially unchanged since its formation. Dating them can therefore tell us the age of the solar system, and how long ago Earth first formed.

SEM image of a zircon crystal

The Jack Hills formation (Australia), where the oldest minerals on Earth have been found

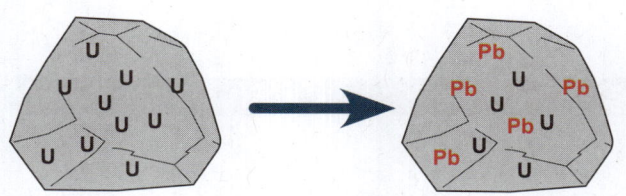
Unstable atoms, e.g. uranium, change into stable atoms, e.g. lead, following a long but predictable decay series involving many radioactive elements.

Zircons are crystals containing the elements zirconium, silicon, and oxygen, with the formula $ZrSiO_4$. They form when molten rock cools. Zircons also contain traces of uranium when they form and can therefore be dated radiometrically. Uranium has a similar electron structure to zirconium and so sometime gets incorporated into the zircon crystal structure, but lead does not. The uranium decays over time to lead. Therefore the ratio of U-238 to Pb-206 and U-235 to Pb-207 can tell us how long ago the crystal formed.

Relative dating

Relative dating establishes the sequential (relative) order of past events in a rock profile, but it cannot provide an absolute date for an event. Each rock layer (stratum) is unique in terms of the type of rock (**sedimentary** or volcanic) and the type of fossils it contains. By building up an index of fossils and the rocks they are found in, it is possible to determine the order of rock layers and events even at sites very distant from each other.

Limestone layers in Jurassic rocks

Fossils in rock layers, Japan

1. Outline the differences between radiometric and relative dating: _____
   _____
   _____
   _____
   _____
   _____

2. Potassium-argon dating is a commonly used radiometric dating technique. Potassium-40 decays into calcium 89% of the time and argon about 11% of the time and has a half-life of 1.26 billion years. Argon is inert and so is not contained in most minerals. Its mass in minerals is therefore related to the decay of potassium.

   A mineral was found with the following ratio of atoms: potassium-40: 50 atoms, calcium: 45 atoms, argon: 5 atoms

   (a) How many half-lives have passed since the mineral formed? _____

   (b) How old is the mineral? _____

2 The Earth's History

Key Idea: The Earth is 4.54 billion years old and has undergone continual change over that time.

The Earth, along with the rest of the solar system, formed about 4.5 billion years ago. During the very early stages of its formation, the Earth grew by accretion of rocks, ice, and dust from the solar nebula. Very soon after this initial formation, a collision with a smaller protoplanet formed the Moon. Cooling of the surface and formation of the oceans ended about 4 billion years ago (end of the Hadean eon). Life appeared very soon after this, with evidence of the earliest life appearing at around 3.5 billion years ago. The evolution of oxygen producing photosynthetic organisms modified the early **atmosphere**. Initially composed of mostly nitrogen, methane, ammonia, and hydrogen, it changed to one of mostly nitrogen and oxygen over the course of a billion years (the Great Oxidation Event). The supercontinent Rodinia broke up around 800 million years ago (mya), with the supercontinent Pangea then forming about 300 mya. This broke up about 180 mya, forming Laurasia and Gondwana, which themselves eventually broke up into the continents we know today.

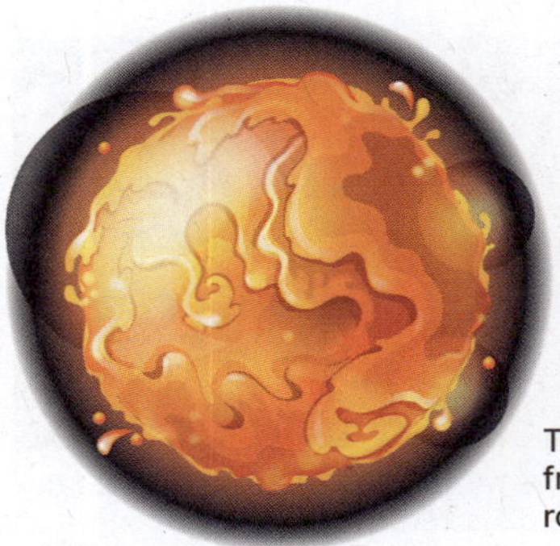

The Earth formed from the accretion of rocks, ice and dust.

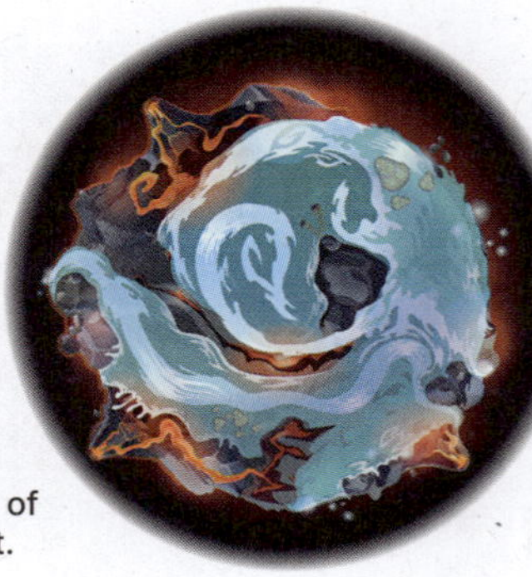

It is believed that much of Earth's water was delivered by comets and icy asteroid impacts. The latest evidence also indicates that the hydrogen within the Earth also played a large role in the formation of Earth's water.

Life appeared within a billion years of the formation of the Earth, and within 500 million years of the Earth cooling.

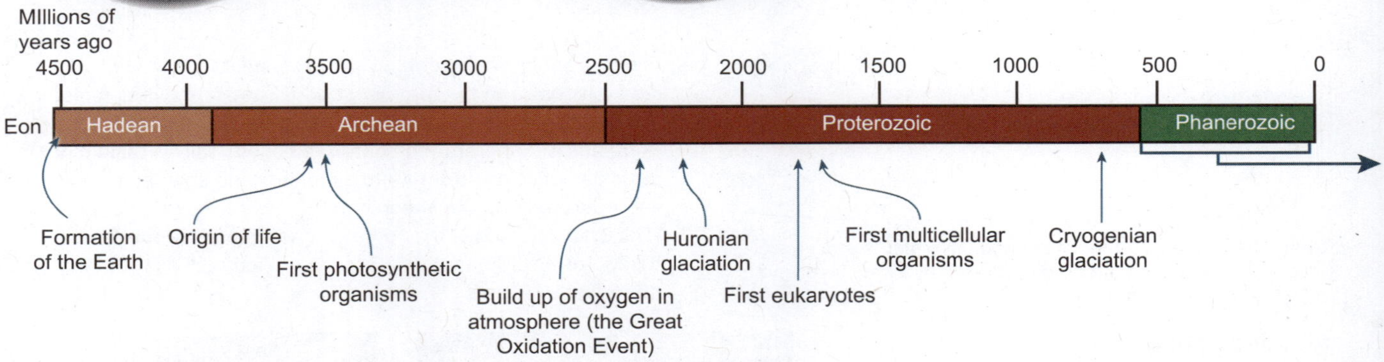

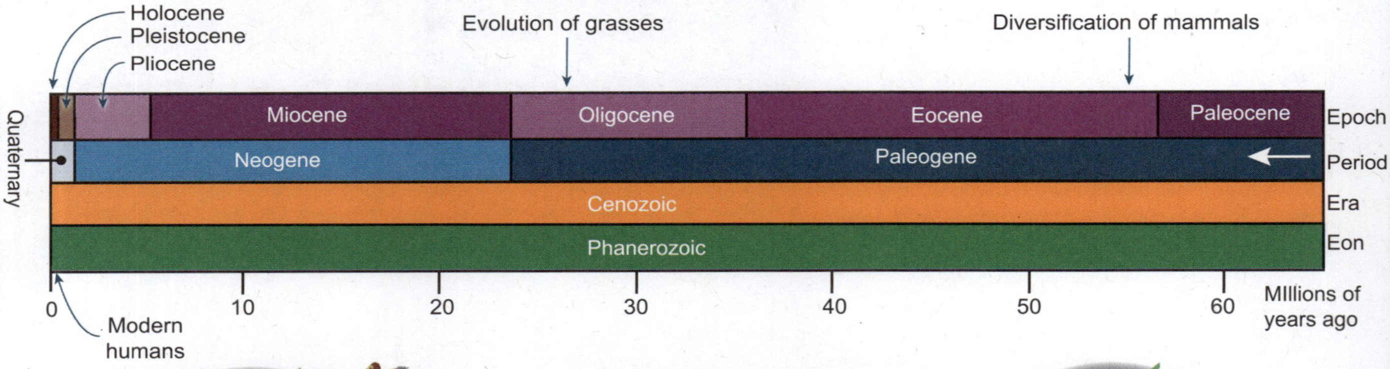

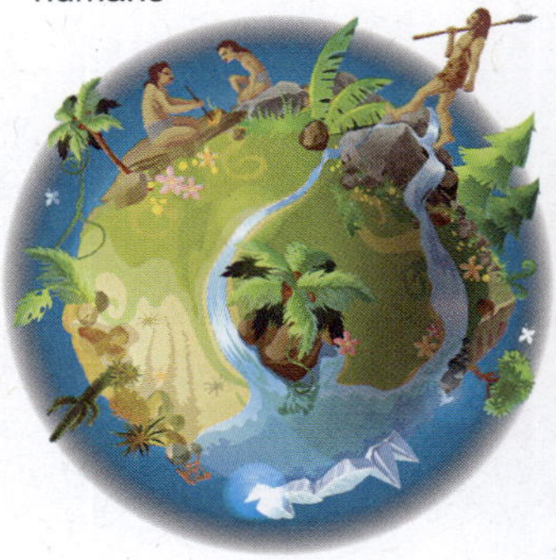

After the extinction of the dinosaurs, the mammals rapidly evolved to take their place and fill the many vacant niches.

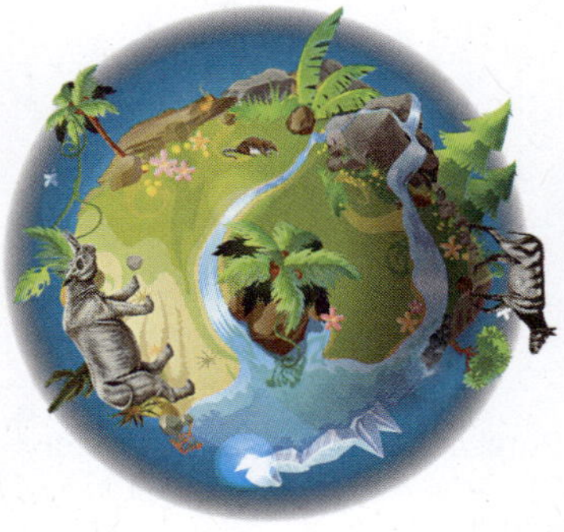

Human ancestors evolved about 3 million years ago, with anatomically modern humans appearing about 300,000 years ago. Humans now dominate the globe, with almost every aspect of our society affecting the environment in some way.

1. Modern humans evolved about 300,000 years ago. How much is this as a percentage of the age of the Earth?

_____

2. What caused the formation of the Moon? _____

Relationship of life and Earth

The Earth has been through numerous climatic episodes and changes in its dominant life forms. These leave traces in the rock layers, many of which are the basis for the division of Earth's history into the different eons, eras, and periods shown below. The evolution of Earth's climate and its life are interrelated. Evidence shows that the Earth has been through periods called 'Snowball Earths' in which most of the Earth's surface was covered by ice. It is thought that the first of these, the Huronian glaciation, may have been in part caused by the loss of atmospheric methane, a potent greenhouse gas, when it reacted with the oxygen being produced by the newly evolved oxygenic photosynthetic cells. This snowball Earth may have triggered the evolution of multicellular life. A second snowball event, the Cryogenian glaciation, may have occurred just prior to the Cambrian period.

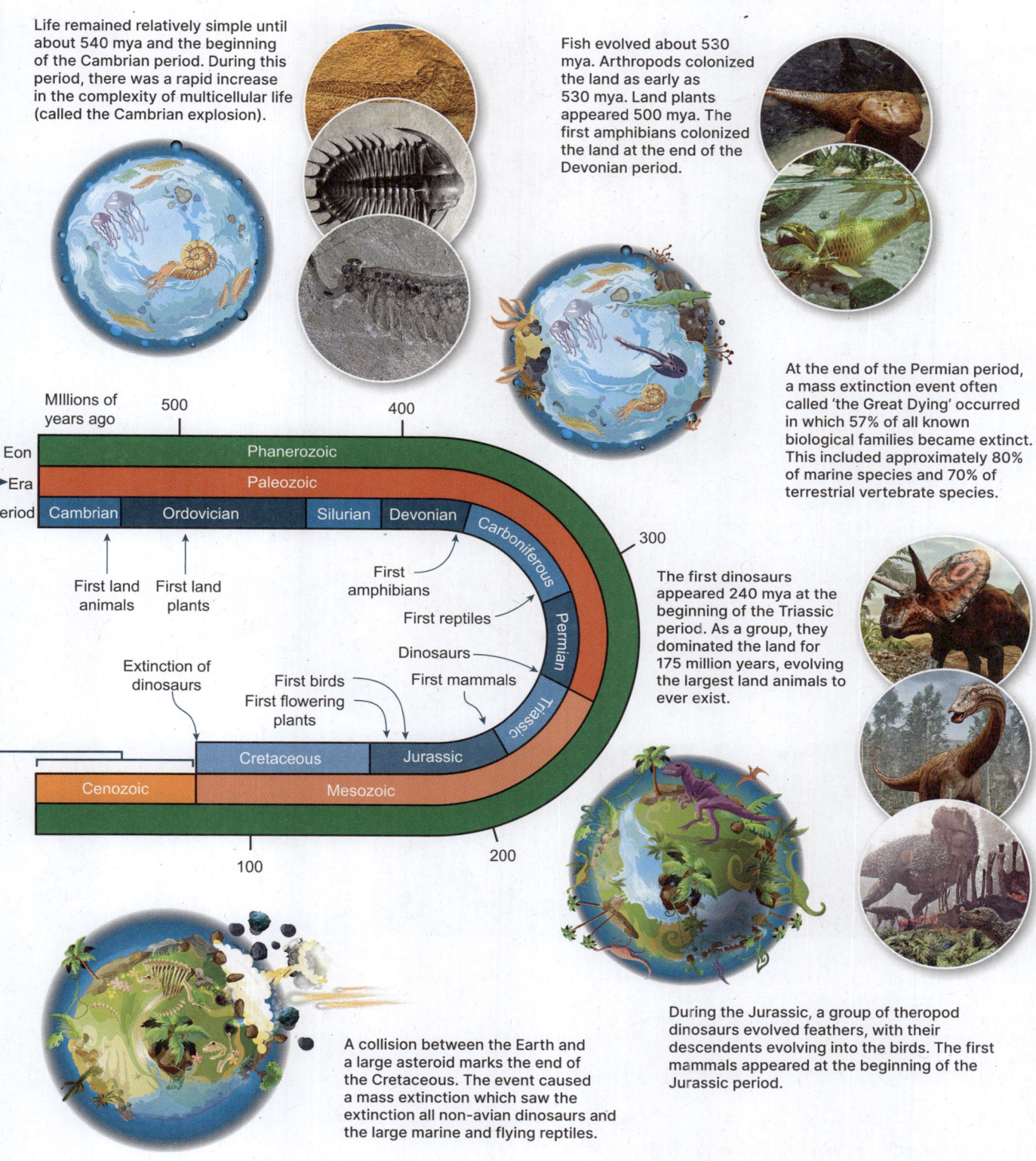

3. When did life first appear on Earth? _____

4. What caused the change from Earth's early, low oxygen atmosphere to its modern high oxygen atmosphere?

_____

3 Structure of the Earth

Key Idea: The Earth has several distinct layers, each with its own specific properties.

The Earth is layered due to the density of different materials in it. The Earth's **crust** has a density of about 3 g/cm³ and the **core** has a density of about 12 g/cm³. Movement of convection currents in the **mantle** shifts the plates of the Earth's crust, while movement of the outer core produces the Earth's magnetic field.

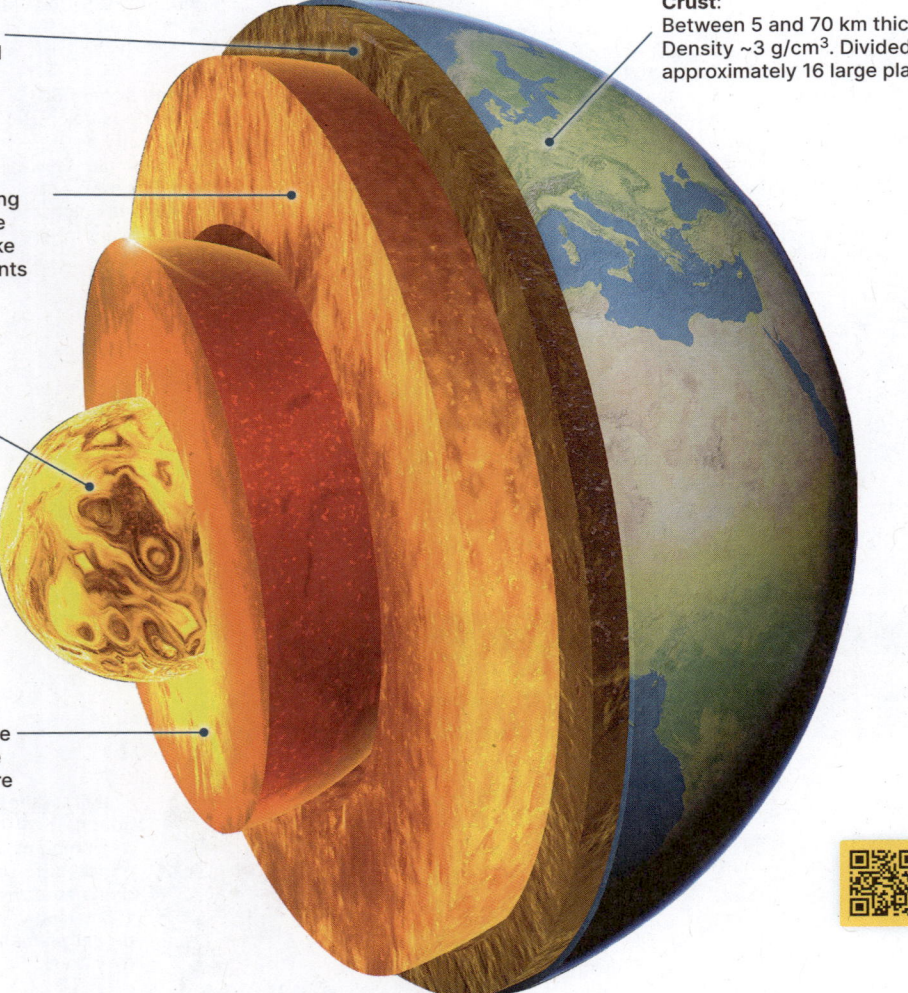

Upper mantle:
Solid layer about 400 km thick, with a transition layer between the upper and lower mantle of around 300 km. The temperature reaches about 1000°C.

Lower mantle:
Approximately 2000 km thick, extending to 2900 km below the surface. Like the upper mantle, it is solid but behaves like a viscous liquid, with convection currents slowly moving the mantle about.

Inner core:
Solid. The temperature at the core is about 4700°C. The core is made of mostly iron and nickel and is about 1200 km in diameter.

Outer core:
Liquid. The movement of the outer core produces Earth's magnetic field. Made mostly of iron and nickel, the outer core is 2200 km thick and reaches 4000°C.

Crust:
Between 5 and 70 km thick. Density ~3 g/cm³. Divided into approximately 16 large plates.

Evidence for the Earth's structure comes from many different sources. One is the timing of seismic waves produced by earthquakes. Evidence from these waves shows that the outer core must be liquid.

The Earth's magnetic field also indicates the outer core is liquid and surrounds a solid, iron/nickel inner core. Evidence from magnetic minerals shows the magnetic field reverses polarity every few million years.

The Earth is geologically active due to the residual heat from its formation. This heat causes mantle movements which move the crustal plates about, resulting in earthquakes and mountain building.

1. What causes the Earth to have different internal layers? _____

2. Identify whether each of the following is liquid or solid:

    (a) Mantle: _____ (b) Outer core: _____ (c) Inner core: _____

3. What produces the Earth's magnetic field? _____

4 Fossil Formation

Key Idea: Fossils are the remains of long-dead organisms that have escaped decay and have, after many years, become part of the Earth's crust.

A fossil may be the preserved remains of the organism itself, the impression of it in the sediment (a mold), or marks made by it during its lifetime (trace fossils). For fossilization to occur, rapid burial of the organism is required, usually in water-borne sediment. This is followed by chemical alteration, whereby minerals are added or removed. Fossilization requires the normal processes of decay to be permanently stopped. This can occur if the organism's remains are isolated from the air or water and decomposing microbes are prevented from breaking them down. Fossils provide a record of the appearance and extinction of organisms, from species to whole taxonomic groups. Once this record is calibrated against a time scale by use of a broad range of dating techniques, it is possible to build up a picture of the evolutionary changes that have taken place.

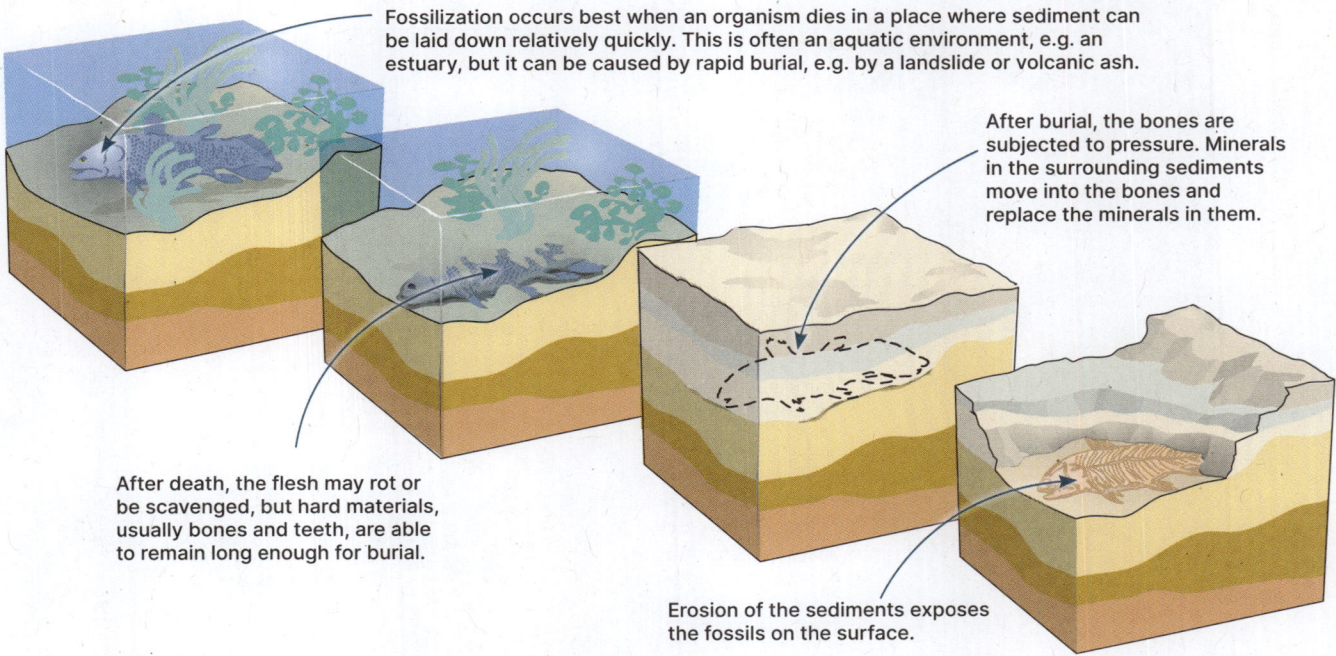

Fossilization occurs best when an organism dies in a place where sediment can be laid down relatively quickly. This is often an aquatic environment, e.g. an estuary, but it can be caused by rapid burial, e.g. by a landslide or volcanic ash.

After burial, the bones are subjected to pressure. Minerals in the surrounding sediments move into the bones and replace the minerals in them.

After death, the flesh may rot or be scavenged, but hard materials, usually bones and teeth, are able to remain long enough for burial.

Erosion of the sediments exposes the fossils on the surface.

A fossil may be the preserved remains of the organism itself (as in the shells below, left), the impression of it in the sediment (a mold and compression-impression fossils), or marks made by it during its lifetime (trace fossils).

Fossil shells: The fossil record is biased towards organisms with hard parts. Shells fossilize more easily than soft tissues.

Compression fossil: This fern frond shows traces of carbon and wax from the original plant, Carboniferous (USA).

Trace fossil: A dinosaur footprint in Lower Jurassic rock, SW Utah.

1. Describe how a fossil forms: _____
_____
_____
_____

2. Explain why the rapid burial of an organism is important in the formation of fossils: _____
_____
_____
_____

Sedimentary rock profile

This diagram represents a cutting through layers of **sedimentary** rock in which fossils are exposed. Fossils are the remains or impressions of plants or animals that become trapped in the sediments after death. Layers of sedimentary rock are arranged in the order by which they were deposited, with the most recent layers near the surface (unless they have been disturbed).

Recent fossils are found in more recent sediments
The more recent the layer of rock, the more resemblance there is between the fossils found in it and living forms.

Many extinct species
The number of extinct species is far greater than the number of species living today.

Fossil types differ in each stratum
Fossils in any given layer of sedimentary rock are generally quite different from fossils in other layers.

More primitive fossils are found in older sediments
Phyla are represented by more generalized forms in the older layers, and not by specialized forms (such as those alive today).

New fossil types mark changes in the environment. In the rocks marking the end of one geological period, it is common to find new fossil that become dominant in the next. Each geological time period had an environment very different from those before and after. The boundaries of these coincided with considerable environmental change and the creation of new niches. These produced new selection pressures and resulted in diversification of surviving genera.

A case study in the fossil record

The evolution of many present-day species can be very well reconstructed. African and Asian elephants have descended from a diverse group known as proboscideans (named for their long trunks). The first, pig-sized, trunkless members of this group lived in Africa 40 million years ago. From Africa, their descendants invaded all continents except Antarctica and Australia. As the group evolved, they became larger: an effective evolutionary response to deter predators. Examples of extinct members of this group are illustrated below.

Columbian mammoth
Pleistocene, Costa Rica to northern US. Range overlap with woolly mammoths in the North.

~4 m at the shoulder

Deinotherium
Miocene-Pleistocene, Asia, Africa

~4 m at the shoulder

Gomphotherium
Miocene, Europe, Africa

~ 3 m at the shoulder

Platybelodon
One of several genera of shovel-tuskers. Middle Miocene, Northern Asia, Europe, Africa

~3 m at the shoulder

3. Explain why the fossil record is biased towards marine organisms with hard parts: _____
_____
_____
_____

4. Fossils tell us much about the organisms that lived in the past. Suggest what other information they might provide:
_____
_____

5. Discuss the use of fossils as indicators of environmental change: _____
_____
_____
_____
_____

5 Earth and Sun Cycles

Key Idea: The movement of the Earth and Moon around the Sun influence many climate systems.

Of all celestial bodies, the Sun has the greatest influence on the Earth, affecting its movements, determining the day-night and seasonal cycles, driving climatic systems and longer term climate cycles, and providing the energy for most life on the planet. The Sun also plays a part in tidal movement on Earth, modifying the effect of the Moon to produce monthly variations in the tidal range. The Sun emits various types of radiation but most is absorbed high in the Earth's **atmosphere**. Only visible light, some infra-red radiation, and some ultraviolet light reach the surface in significant amounts. Visible light is pivotal to the producer base of Earth's biological systems but infra-red is also important because it heats the atmosphere, oceans, and land. The intensity of solar radiation is not uniform around the Earth and this uneven heating effect, together with the Earth's rotation, produce the global patterns of wind and ocean circulation that profoundly influence the Earth's climate.

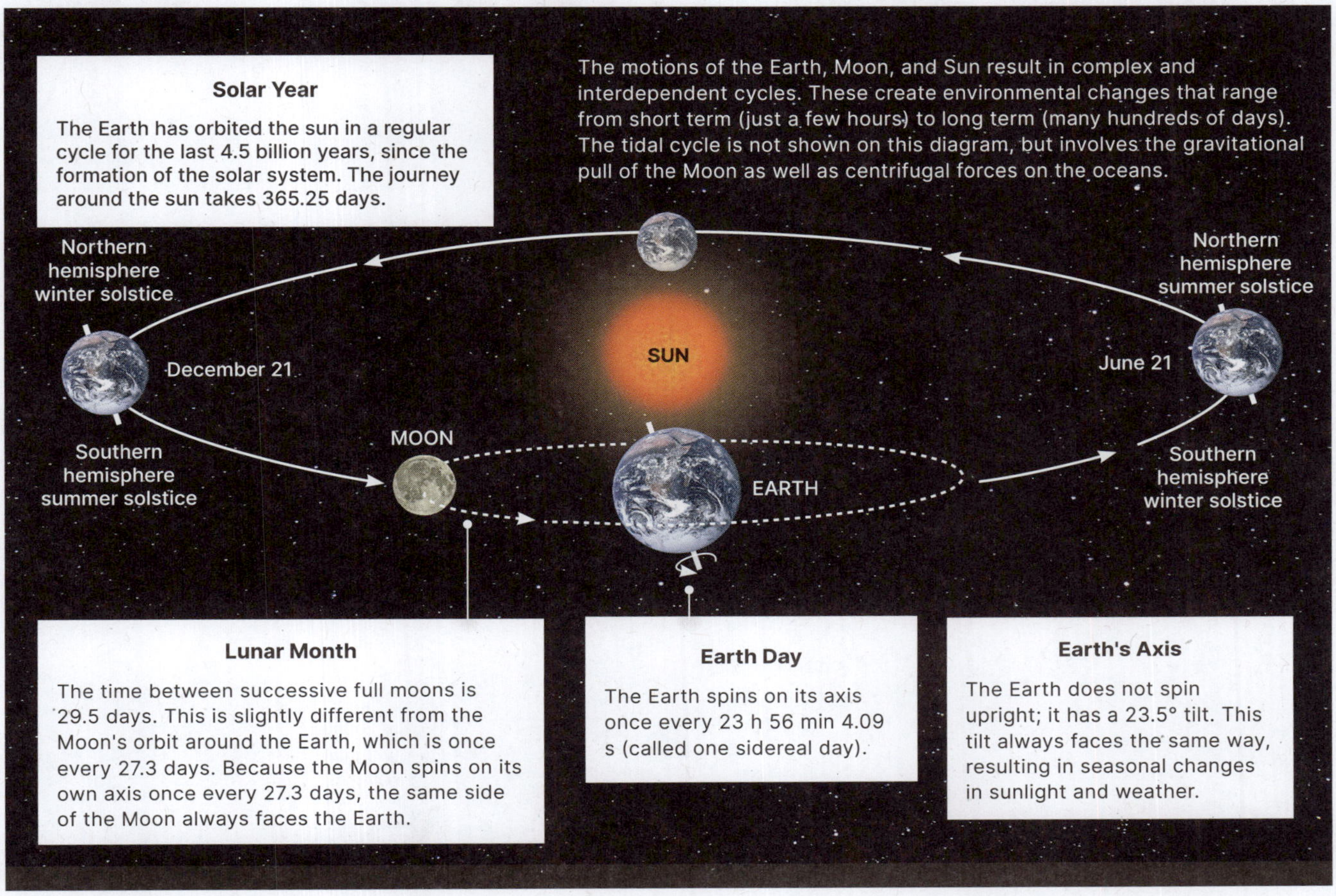

Solar Year

The Earth has orbited the sun in a regular cycle for the last 4.5 billion years, since the formation of the solar system. The journey around the sun takes 365.25 days.

The motions of the Earth, Moon, and Sun result in complex and interdependent cycles. These create environmental changes that range from short term (just a few hours) to long term (many hundreds of days). The tidal cycle is not shown on this diagram, but involves the gravitational pull of the Moon as well as centrifugal forces on the oceans.

Northern hemisphere winter solstice — December 21
Southern hemisphere summer solstice

Northern hemisphere summer solstice — June 21
Southern hemisphere winter solstice

Lunar Month

The time between successive full moons is 29.5 days. This is slightly different from the Moon's orbit around the Earth, which is once every 27.3 days. Because the Moon spins on its own axis once every 27.3 days, the same side of the Moon always faces the Earth.

Earth Day

The Earth spins on its axis once every 23 h 56 min 4.09 s (called one sidereal day).

Earth's Axis

The Earth does not spin upright; it has a 23.5° tilt. This tilt always faces the same way, resulting in seasonal changes in sunlight and weather.

1. (a) The solar year is based on: _____

   (b) The seasons experienced on Earth are caused by: _____

   (c) The time for the Earth to complete one rotation on its axis is: _____

   (d) The movement of the Earth's atmosphere is due to: _____

2. (a) Explain why tropical regions receive a greater input of solar radiation than the poles:

   _____
   _____
   _____

   (b) Describe the consequences of this to the Earth's climate: _____

   _____
   _____
   _____

©2025 BIOZONE International
ISBN: 978-1-99-101409-2
Photocopying prohibited

Long term orbital cycles

▸ Earth is not stable in space. It is affected by the gravity of the Sun, the planets, and the Moon. These influences affect the tilt of the Earth's axis and the shape of its orbit. The changes in orbit and tilt can combine to cause extreme changes in the Earth's climate, e.g. producing ice ages. The Earth experiences three main orbital and rotational cycles:

Axial tilt (obliquity)

Relative to its orbital axis, the Earth's rotational axis is at an angle of about 23.4°. This angle is responsible for seasonal changes. The angle is not constant. It varies between 22.1° and 24.5° over a period of ~41,000 years. The greater the degree of tilting, the greater the difference between the summer and winter seasons. There is great variation in the axial tilt of the planets in the solar system.

Planet	Axial tilt	Planet	Axial tilt
Mercury	0.1°	Jupiter	3°
Venus	177°	Saturn	27°
Earth	23.4°	Uranus	98°
Mars	25°	Neptune	30°

Precession

Like a spinning top, Earth wobbles about on its axis. This causes the Earth's axis to describe a cone in space. An entire cycle takes about 26,000 years. Precession alters the direction the Earth's axis is pointing. For example, during June, the Northern Hemisphere is pointed towards the Sun. However, 13,000 years ago the occurrence of the seasons was opposite to what they are today.

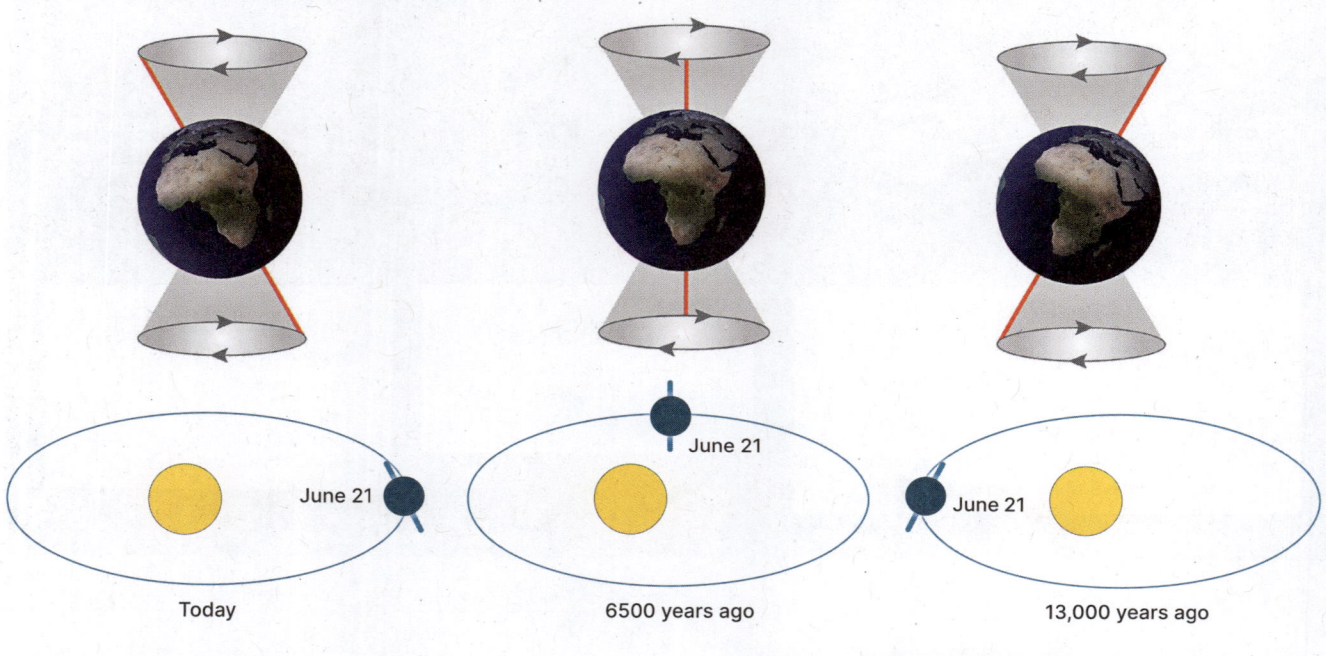

Orbital eccentricity

The Earth's elliptical orbit changes its eccentricity from very nearly circular (eccentricity of 0.000055) to slightly elliptical (eccentricity 0.0679). The changes in eccentricity have a cycle of about 100,000 years. Circular orbits tend to make the differences between the seasons rather mild, while more elliptical orbits exacerbate the difference as during one solstice the Earth will be closer to the Sun than the other. Currently, the Earth is about 3 million kilometers closer to the Sun during the northern hemisphere winter than in the northern hemisphere summer.

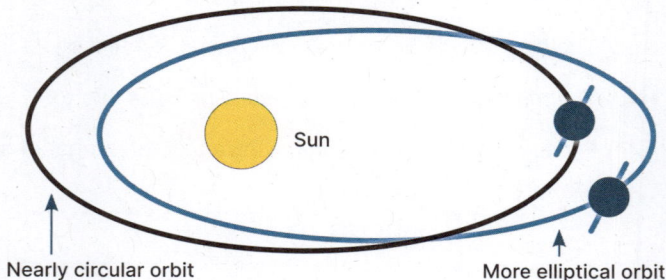

3. How might the long term orbital cycles of the Earth cause extremes in climate, such as ice ages?

_____
_____
_____
_____

6 The Earth's Crust

Key Idea: The different rocks of the Earth's crust have different properties based on their formation.

The Earth has a layered structure comprising a solid inner **core**, a liquid outer core, a highly viscous **mantle**, and an outer, silicate solid **crust**. The Earth's crust is thin compared to the bulk of the Earth, averaging just 25-70 km thick below the continents and about 10 km thick below the oceans. Overall, the crust is less dense than the mantle, being relatively rich in lighter minerals such as silicon, calcium, and aluminum. The crust is in constant change as a result of ocean formation, mountain building, and volcanism. It supports the biosphere, the hydrosphere, and the **atmosphere**.

The **continental crust** is made up of **igneous**, **metamorphic**, and **sedimentary** rocks. It is not recycled within the Earth to the same extent as oceanic crust, so some continental rocks are up to 4 billion years old. Parts of the continental crust go through repeated cycles of erosion, formation into sedimentary rock, and metamorphosis. The crust 'floats' on the mantle at a level determined by its thickness and density. The more mass there is above sea-level, the deeper the crust must extend down in support.

Water precipitated from the atmosphere forms rivers and lakes which flow back to the ocean, eroding the landscape in the process.

The Earth's persistent oceans of liquid water cycle moisture through the atmosphere to the land and back again.

Sedimentary rocks
Sediments eroded from continents and compressed into rock can be later lifted and exposed in mountains and other formations.

The **oceanic crust** makes up more than two thirds of the Earth's surface and is composed of relatively dense, basalt-rich rocks underlying a thin layer of sediment. The oceanic crust is being continually formed from mantle material within long rifts called spreading ridges. As a result it is relatively young; even the oldest parts of the ocean floor are no more than 200 million years old.

Igneous rocks, such as basalt (left), form a major component of the crust and are essentially unchanged since their formation.

1. Describe two differences between the oceanic and continental crust: _____

2. Explain the difference in thickness and relative positions of continental and oceanic crusts: _____

3. Explain why the Earth's crust is described as a dynamic structure: _____

7 Plate Boundaries

Key Idea: When tectonic plates meet, they form either convergent, divergent, or transform boundaries.

The outer rock layer of the Earth, comprising the **crust** and upper **mantle**, is called the **lithosphere** and it is broken up into seven large, continent-sized tectonic plates and about a dozen smaller plates. Throughout geological time, these plates have moved about the Earth's surface, shuffling continents, opening and closing oceans, and building mountains. The size of the lithospheric plates is constantly changing, with some expanding and some getting smaller. These changes occur along plate boundaries which are marked by well-defined zones of seismic and volcanic activity. Plate growth occurs at **divergent boundaries** along sea floor spreading ridges, e.g. the Mid-Atlantic Ridge and the Red Sea, whereas plate attrition occurs at **convergent boundaries** marked by deep ocean trenches and subduction zones. Divergent and convergent zones make up approximately 80% of plate boundaries. The remaining 20% are transform boundaries, where two plates slide past one another with little change in the size of either plate.

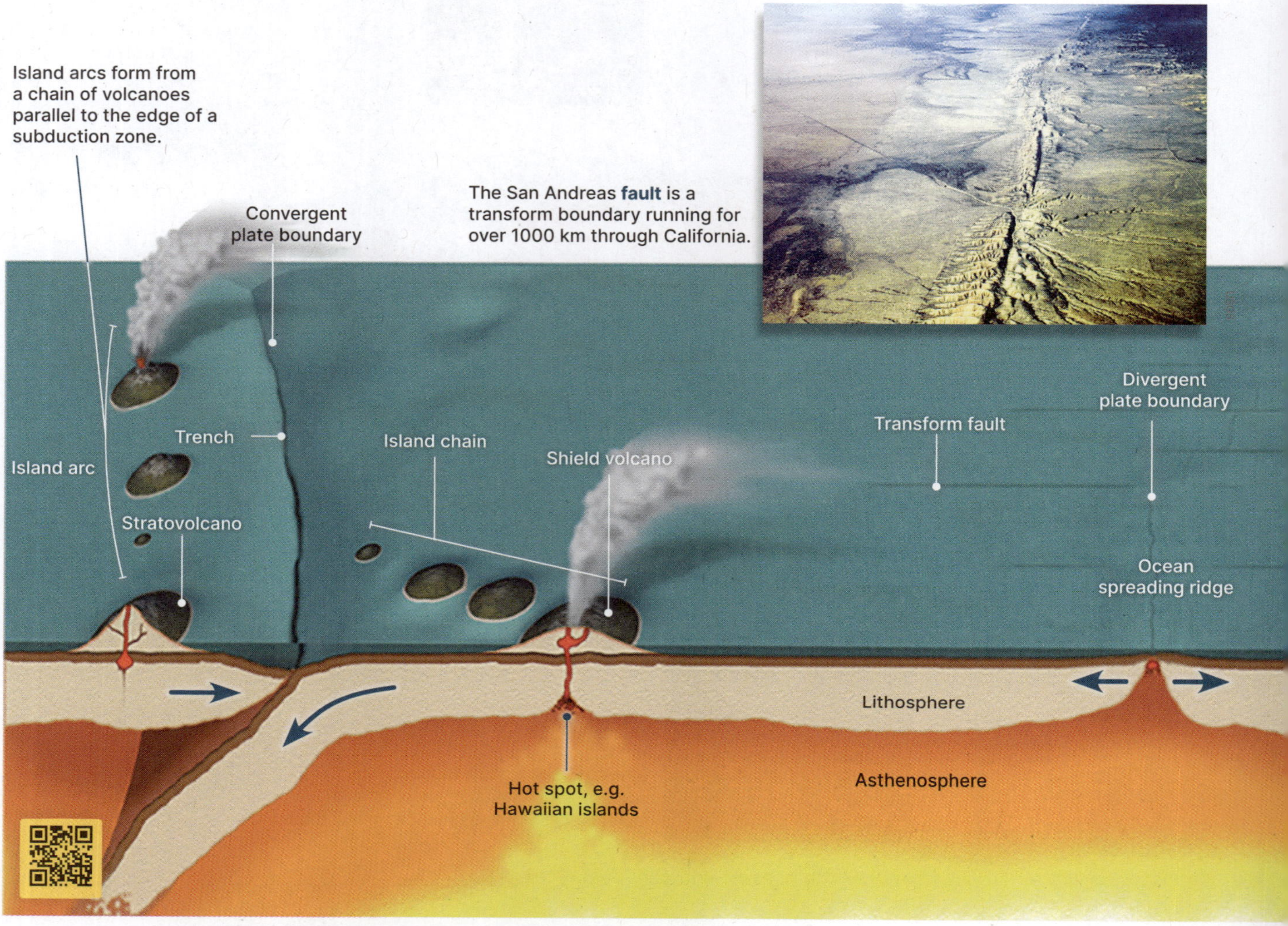

1. Describe what is happening at each of the following plate boundaries, and identify an example in each case:

   (a) Convergent plate boundary: _____
   _____
   _____

   (b) Divergent plate boundary: _____
   _____
   _____

   (c) Transform plate boundary: _____
   _____
   _____

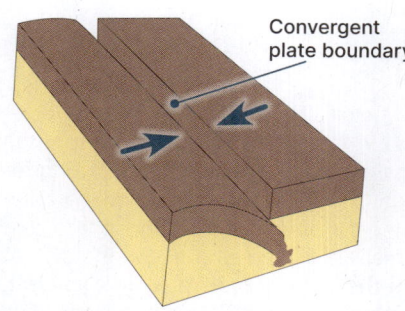 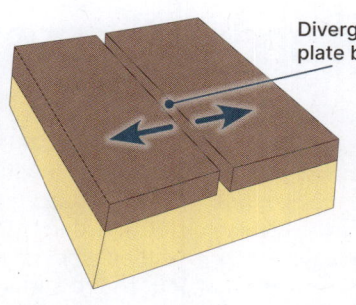 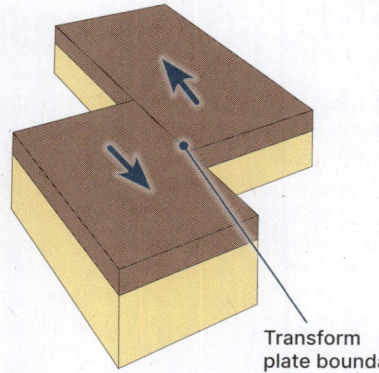

Plate boundaries moving towards each other are called convergent plate boundaries. Where oceanic **crust** and continental crust meet, the oceanic crust will subduct under the continental crust, creating a subduction zone. Volcanoes normally form along the continental border of a subduction zone. When continental crusts collide, huge mountain ranges such as the Himalayas can form.

Divergent plate boundaries form where the tectonic plates are moving away from each other. These are commonly found along the mid ocean ridges, but occasionally are seen on land, as in the Great Rift Valley and Iceland. Divergent boundaries are also known as constructive boundaries as they produce new crust from the upwelling of magma.

Transform boundaries are formed when the tectonic plates are moving past each other. They are, therefore, neither constructive nor destructive. Examples include the San Andreas fault in California and the Alpine Fault in New Zealand.

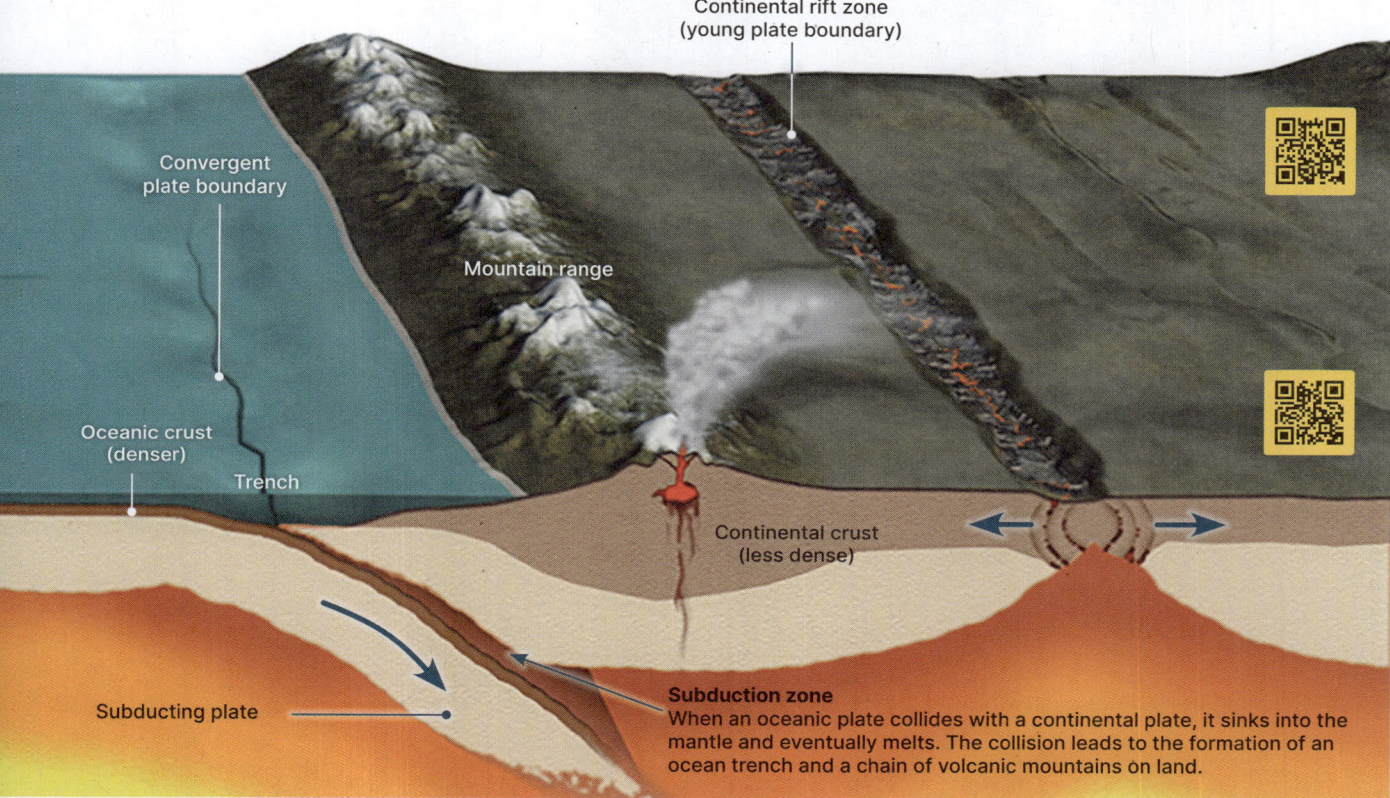

2. Identify the type of plate boundary at which each of the following occurs:

   (a) Mountain building: _____   (c) Creation of new ocean floor: _____

   (b) Subduction: _____   (d) Island arc: _____

3. (a) Explain why the oceanic crust subducts under the continental crust in a subduction zone:
   _____
   _____

   (b) What causes volcanoes to form along the continental plate boundary of a subduction zone?
   _____
   _____
   _____

8 The Lithosphere and Asthenosphere

Key Idea: The lithosphere and asthenosphere are two distinct regions of the mantle.

The **lithosphere** (lithos = 'stone') comprises the **crust** and the uppermost part of the **mantle**. It is both rigid and solid, and broken up into tectonic plates. The lithosphere can be divided into continental lithosphere, which contains relatively light minerals, and oceanic lithosphere, which contains much denser minerals. The lithosphere ranges from 400 km thick over the continents to 70 km thick in the oceans. The **asthenosphere** (asthenes = 'weak') lies below the lithosphere. This layer of rock is viscous and plastic (semi-fluid), it changes through plastic deformation, slowly moving about and so allowing for plate tectonic movement. The asthenosphere is only around 100 km thick. The boundary between the lithosphere and asthenosphere is thermal. The lithosphere conducts heat out to the surface while the asthenosphere retains its heat. The crust and the mantle are chemically distinct, forming a compositional boundary.

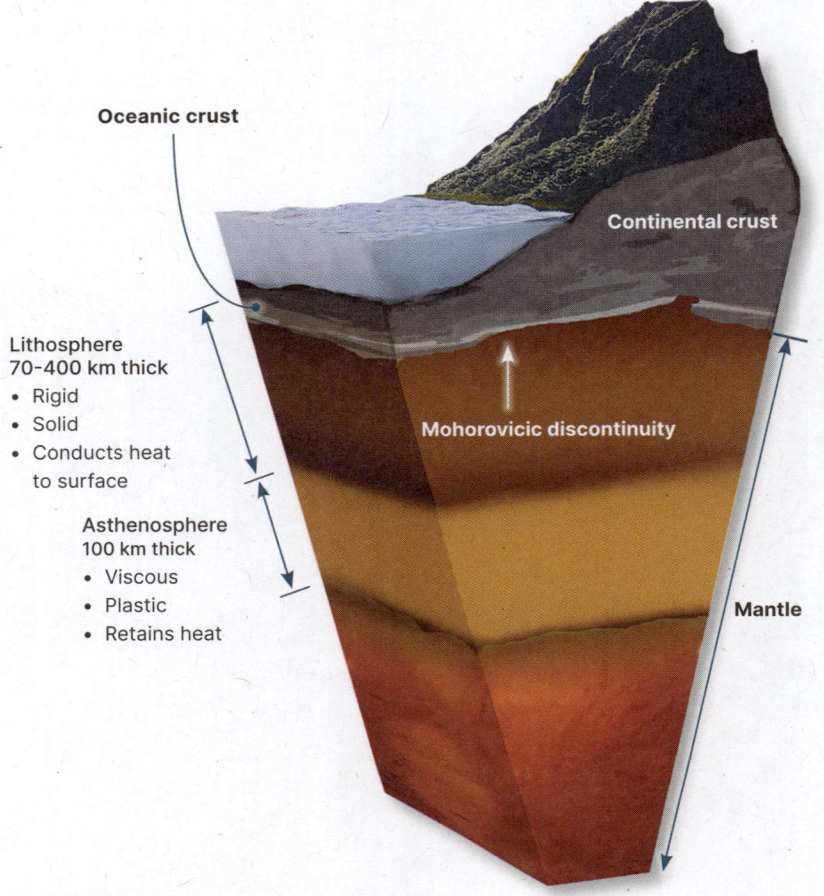

Locating the layers

The Mohorovicic discontinuity marks the boundary between the crust and the mantle. Seismic waves from earthquakes increase in velocity when travelling below this boundary.

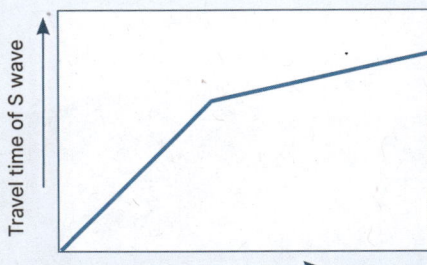

Experiments reveal more detail about the upper layers of the Earth

In an experiment in New Zealand, geologists exploded charges of TNT to produce seismic waves and recorded their echoes from layers of rock in the Earth. The echoes revealed a jelly-like layer of rock just 10 km thick between the rigid upper layer of the Earth (lithosphere) and the more plastic lower layer (asthenosphere). This research solved the problem of how the lithosphere could move about on the surface of the asthenosphere.

1. (a) Describe the structure of the lithosphere: _____

   _____

   (b) Describe the structure of the asthenosphere: _____

   _____

2. Describe how the lithosphere-asthenosphere boundary (LAB) can be detected: _____

   _____

3. The continental lithosphere is much older than the oceanic lithosphere. Explain why this is the case:

   _____

9 Mechanism of Plate Movement

Key Idea: Convection currents in the mantle cause the movement of the tectonic plates.

Evidence from earthquakes, volcanoes, and land formations has helped formulate the theory of plate tectonics, which describes the large scale movement of the Earth's crustal plates. The key principle of the theory is that the rigid plates are able to ride on the fluid-like underlying **asthenosphere**. The energy for this movement comes from dissipation of heat from the mantle. It is the movement of the plates that produces phenomena such as earthquakes and volcanism.

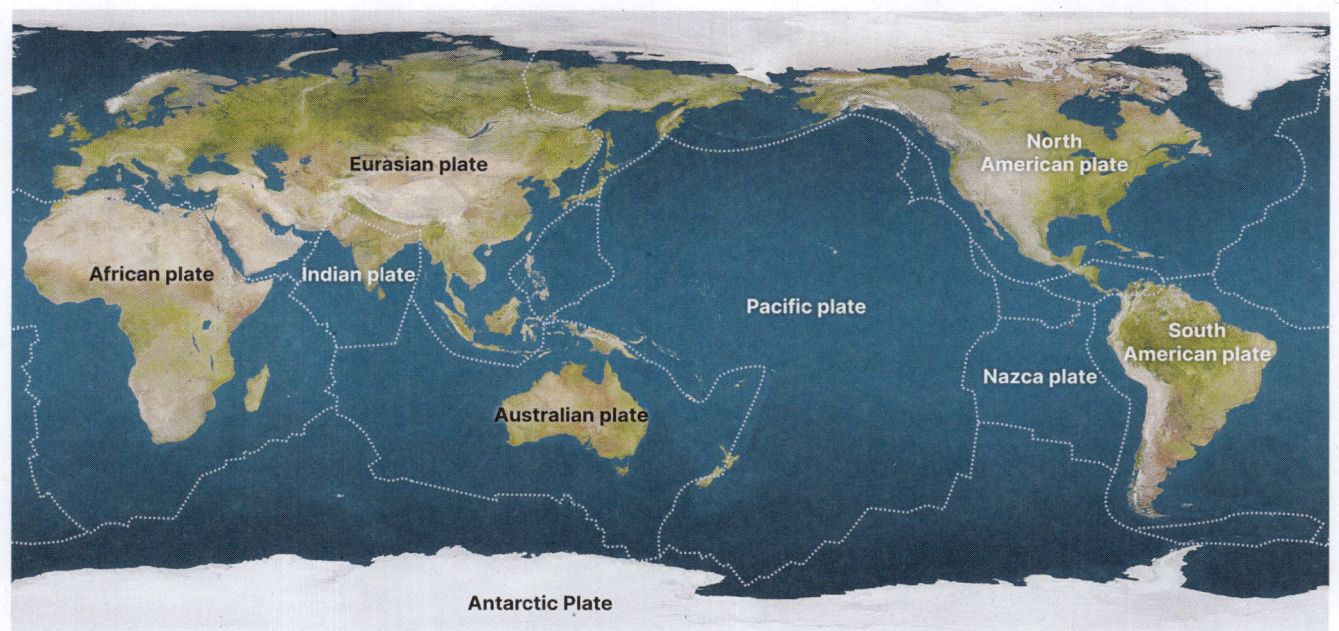

▸ The evidence for past plate movements has come from several sources: mapping of plate boundaries, the discovery of sea floor spreading, measurement of the direction and rate of plate movement, and geological evidence such as the distribution of ancient mountain chains, unusual deposits, and fossils. The size of the plates is constantly changing, with some expanding and some getting smaller. The extent of the tectonic plates is shown in the diagram above. The Pacific plate is by far the largest, measuring 103 million km^2.

The mechanism of plate movement

▸ The relatively cool lithosphere covers the hotter, plastic, and more fluid asthenosphere. Heat from the mantle drives two kinds of asthenospheric movement: convection and mantle plumes. Plate motion is partly driven by the weight of cold, dense plates sinking into the mantle at trenches. This heavier, cooler material, sinking under the influence of gravity, displaces heated material, which rises as mantle plumes.

▸ The movements of the tectonic plates puts the brittle rock of the crust under strain, creating faults where rocks fracture and slip past each other. Earthquakes are caused by energy release during rapid slippage along faults. Consequently, the Earth's major earthquake (and volcanic) zones occur along plate boundaries.

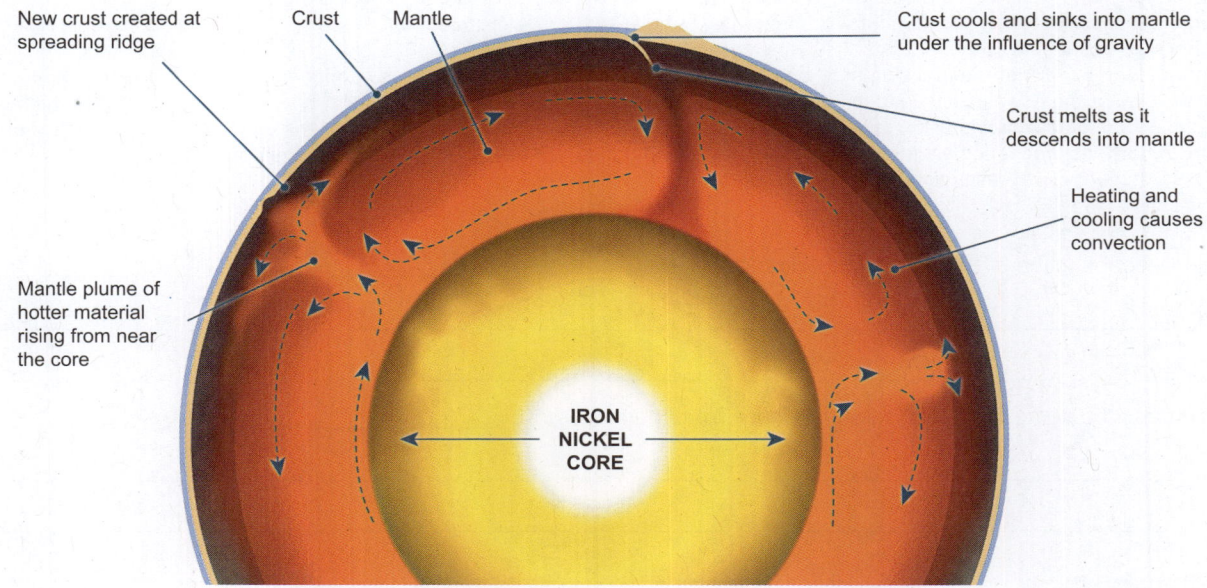

10 Continental Drift

Key Idea: Continental drift is the movement of the continents due to their being embedded in the tectonic plates.

Continental drift (the movement of the Earth's continents relative to each other) is a continuing and measurable phenomenon. Movements of up to 2-11 cm a year have been recorded between continents by use of laser technology. The movements of the Earth's 12 major crustal plates are driven by a geological process known as **plate tectonics**. Some continents are drifting apart while others are moving together. Many lines of evidence show that these continents were once joined as 'supercontintents' including matching continental shelves, and large areas where rocks of the same age span the former continental boundaries. One supercontinent, Gondwana, was made up of the continents Africa, South America, Antarctica, Australia, and India some 200 million years ago.

Current and former positions of Africa and South America

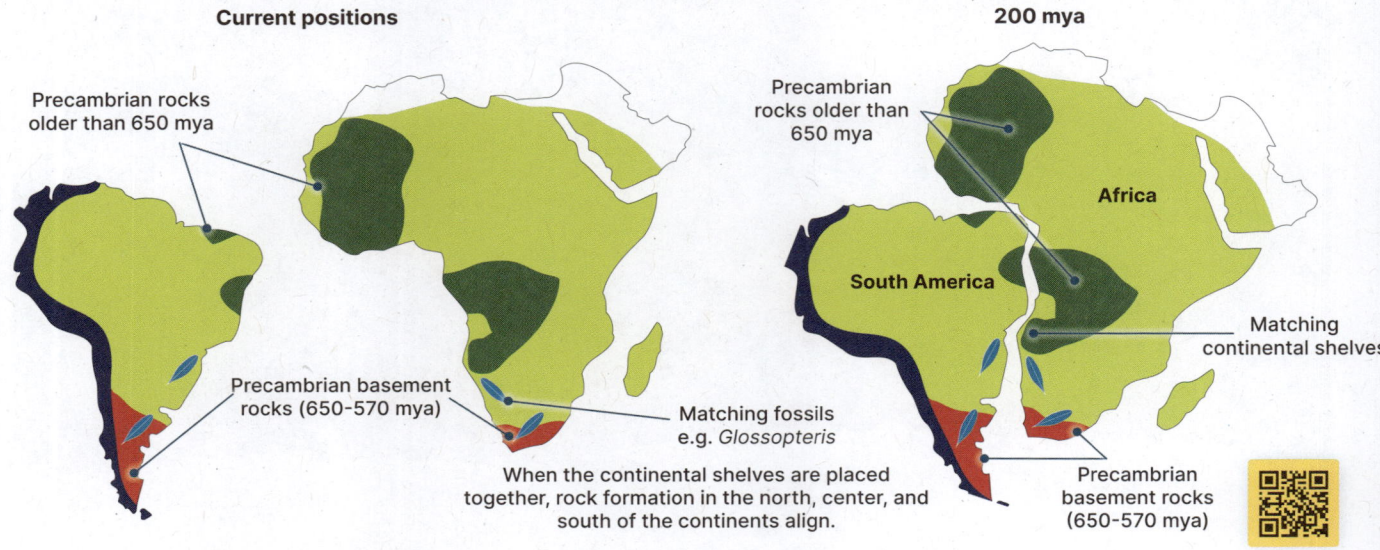

Evidence for continental drift

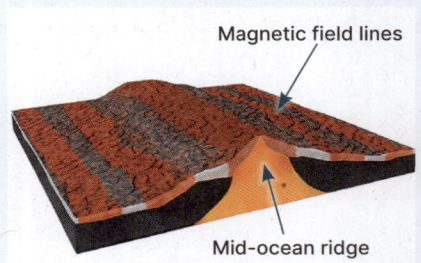

Africa began to separate from South America around 130 mya. The two continents are now separated by more than 2500 km of ocean. Measurements of rocks on either side of the mid-Atlantic ridge show that the rocks on the African side of the ridge exactly match those on the South American side of the ridge in both the magnetic field strength and orientation, as well as in age.

There is a large amount of biological evidence for the existence of Gondwana. Fossils from the plant *Glossopteris* and from the mammal-like reptile *Lystrosaurus* have been found in every continent associated with Gondwana. Ratites, the order of birds including the ostrich, rhea, emu, cassowary, kiwi, and the extinct elephant bird and moa, are also found associated with Gondwanan continents.

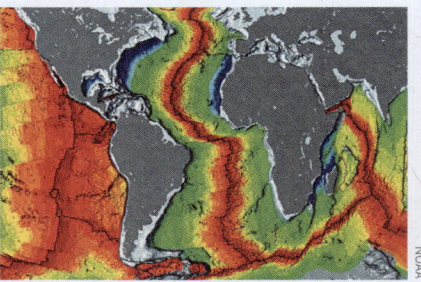

Geological evidence abounds for continental drift and the existence of Gondwana. Geological formations, including rock formations, impressions left by glaciations, the matching of continental shelves, and the position of mid-ocean ridges, all point to the fact that the continents were once joined as one supercontinent.

1. (a) Describe some of the geological evidence for continental drift: _____
   _____
   _____

   (b) Describe some of the biological evidence for continental drift: _____
   _____
   _____
   _____

©2025 BIOZONE International
ISBN: 978-1-99-101409-2
Photocopying prohibited

11 Volcanoes and Volcanism

Key Idea: Volcanoes form when magma reaches the surface, erupting as lava through vents, and forming cone-shaped mountains as the ejected material falls back to the ground. Volcanoes generally form along the edge of the tectonic plate overriding a subduction zone. Cracks in the **crust** caused by the buckling of the overriding plate allow magma to move towards the surface and form a magma chamber. Further cracks in the crust allow the magma to be ejected from the chamber and form a volcano. Magma reaching the surface is termed lava and its chemical properties depend upon the composition of both the original magma and the crust through which it passes. Lava can be placed into three general groups: basaltic, andesitic, and rhyolitc. Basaltic lava contains 48%-58% silica, is basic and is very fluid, with a temperature of around 1160°C. Rhyolitic lava contains 65% silica and above, and is acidic and viscous with a temperature of around 900°C. Andesite is intermediate between these lava types. Volcanoes formed from these lavas have very different and quite distinct properties, ranging from relatively flat and unexplosive to extremely steep and violently explosive.

Types of volcanoes

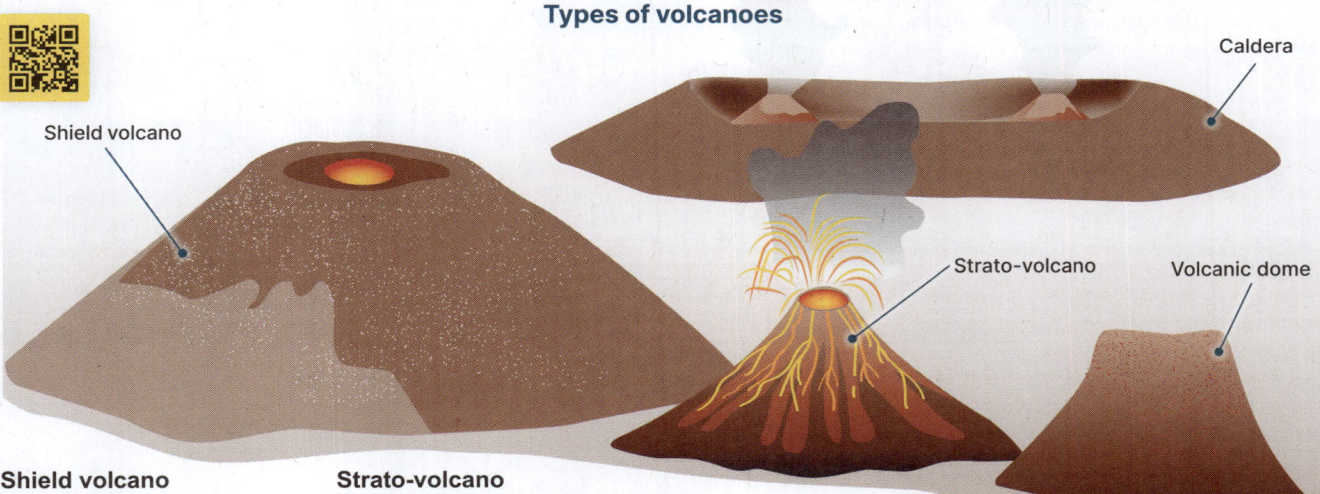

Shield volcano
Shield volcanoes are formed from fluid basalt lava. The lava contains low levels of silica and runs freely down the sides of the volcano, causing its base to spread out without increasing its height. These volcanoes tend to have an exceptionally large basal radius and can be extremely large. Examples include Mt Kilauea (Hawaii) and Skjaldbreiður (Iceland).

Strato-volcano
Strato-volcanoes tend to have steep upper slopes and a shallower gradient on the lower slopes. The lava ejected is commonly andesitic and relatively viscous, containing moderate levels of silica. This causes the lava to build up on the upper slopes before rolling down to the lower slopes and so forming the characteristic shape. Mt Fuji (Japan) is a classic example.

Volcanic domes
Dome volcanoes form from viscous, silica-rich, rhyolitic magma that slowly oozes from the vent or may build up slowly underground. This magma traps a large amount of gas which is not easily released and so eruptions are often extremely violent. Examples include Mt Tarawera (New Zealand) and Chaitén (Chile).

Caldera
Calderas are formed by the collapse of a magma chamber after a major eruption. The collapsed crater forms a characteristic ring fault. These can be very large and may remain active or become dormant after the eruption. Examples include Santorini caldera (Greece), and Yellowstone caldera (US).

Strato-volcano structure

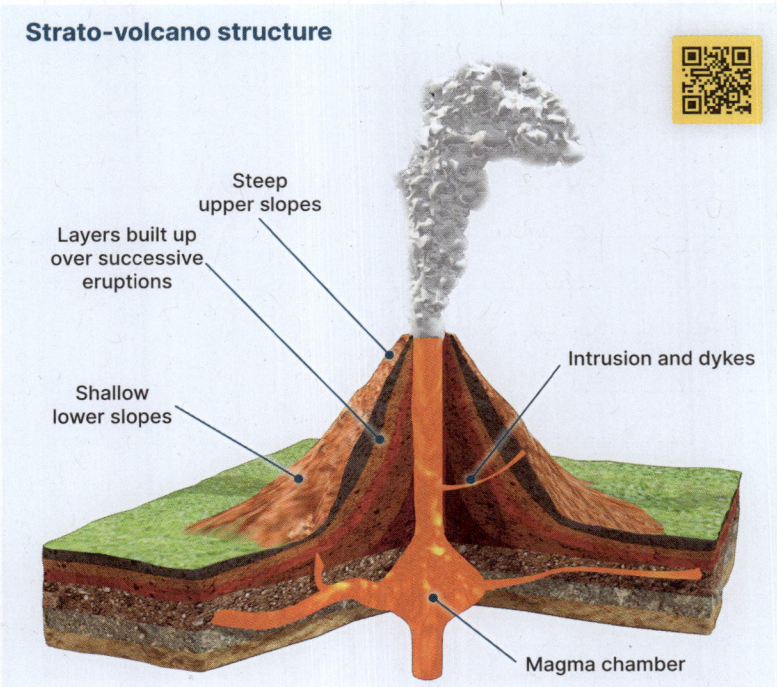

The largest active volcano on Earth is Mauna Loa in Hawaii. It is a shield volcano sitting above the Hawaiian hotspot and has been active for about 700,000 years. Mauna Loa makes up more than half the area of the island of Hawaii and, from its base, is over 9000 m tall. Its mass is close to the maximum the oceanic crust can support. Indeed, it depresses the crust by 8 km.

On Mars, the shield volcano Olympus Mons stands over 21 km high. Due to Mars' lacking tectonic activity, the volcano was able to build up over a stationary hotspot.

©2025 BIOZONE International
ISBN: 978-1-99-101409-2
Photocopying prohibited

Volcanoes in the Pacific

Tectonic plate boundaries

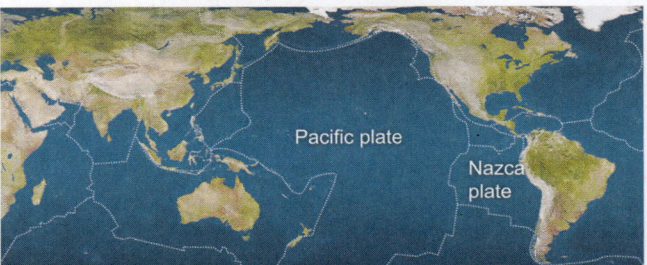

Pacific Ring of Fire

Subduction zones around the Pacific rim

Volcanoes around the Pacific rim

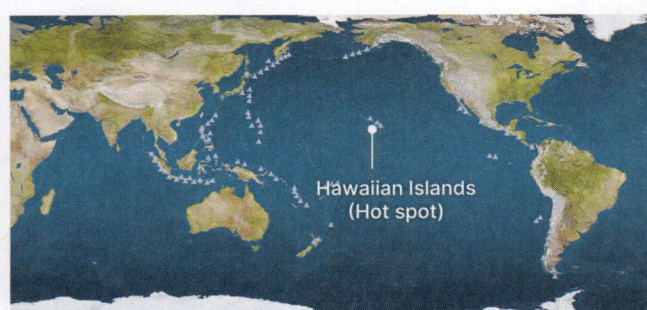

▶ Much of the northern and western edges of the Pacific plate form subduction zones. This produces an area that is seismically and volcanically extremely active. Along with parts of the Nazca plate, it forms the Pacific Ring of Fire. Around three quarters of the world's active and dormant volcanoes are found around the edge of the ring and nearly 90% of all earthquake activity is located there.

1. Match the following lava types to the appropriate volcano below: basaltic, andesitic, rhyolitic.

    (a) Strato-volcano: _____ (b) Dome: _____ (c) Shield: _____

2. Where are volcanoes commonly located? _____
   _____

3. Describe the formation of shield and strato-volcanoes, explaining how each develops its distinctive shape:
   _____
   _____
   _____
   _____

4. Explain how the Hawaiian islands were formed: _____
   _____
   _____
   _____

5. Explain the relationship between the Pacific Ring of Fire, volcanism, and subduction zones:
   _____
   _____
   _____

6. How does the type of magma affect the nature of a volcano? _____
   _____
   _____
   _____

12 Earthquakes

Key Idea: Earthquakes occur when the crust moves along fault lines as built-up strain is released.

As the tectonic plates grind against each other, friction causes them to become locked in place. The stresses produced are transferred to the rocks around the edges of the tectonic plates, causing the **crust** to buckle and deform. This produces cracks and faults in the crust. As the strain on the rocks increases, so too does the energy stored in them. Eventually, this strain will overcome friction and the rocks on either side of a fault or **plate boundary** will move past each other. The stored energy will be released as a series of waves, causing the ground to move both up and down and back and forth. The more energy stored, the larger these waves will be, and the more destructive the earthquake produced.

▸ The point along a fault line where an earthquake occurs is called the focus. The point on the surface above the focus where the center of the earthquake is felt is called the epicenter.

▸ The deeper the focus, the less violent the earthquake will feel at the epicenter. The more energy released at the focus, the more violent the earthquake will feel at the epicenter.

▸ The energy released by an earthquake is measured by the Richter scale (local magnitude scale (ML)) or the similar Moment Magnitude Scale (MMS). The scale (from 1-10) is logarithmic, with each step being approximately 10 times greater than the step before it. Earthquakes below magnitude 4 are extremely common and often not even noticed. Earthquakes above magnitude 6 are much rarer and can cause violent shaking and damage to buildings.

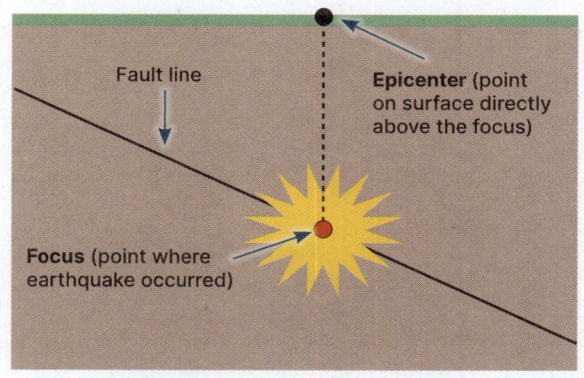

Seismic waves

Movement of the ground along fault lines in the Earth's crust causes earthquakes. During an earthquake, two types of ground wave are produced: compressional, or P-waves and shear, or S-waves.

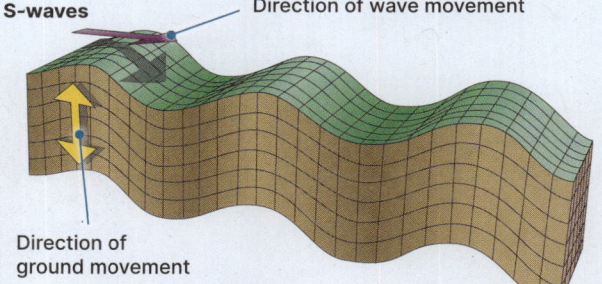

P-waves are compression waves (similar to sound waves in air). P-waves are the fastest moving wave from an earthquake and are therefore the first to arrive at a seismograph. P-waves can travel through all media, whether liquid or solid.

S-waves are transverse waves. They move the ground perpendicular to their direction of travel. They are unable to travel through liquids. They move more slowly than P waves, arriving at a seismograph some time after the P wave.

Earthquakes can produce extremely violent shaking. The damage above was produced by an earthquake with a magnitude of 6.5.

Fault lines can easily be seen where a line of sediments has been shifted by movements of the crust, as shown above.

Movement of the crust during an undersea earthquake can displace large volumes of water, causing massive tsunami waves.

1. How do earthquakes occur? _____

2. If an earthquake occurred 200 km from you, which would feel first, P waves or S waves? _____

3. How many times larger is an earthquake with a magnitude of 5 than one with a magnitude of 3? _____

4. Where do earthquakes most commonly occur? _____

13 The Rock Cycle

Key Idea: Rocks are formed, eroded, and reformed in a continuous process called the rock cycle.

The Earth's many rock types are grouped together according to the way they formed as **igneous**, **metamorphic**, and **sedimentary** rocks (as well as meteorites). These rocks form in a continuous cycle. Volcanism creates rocks at the Earth's surface. Erosion (the breakdown and deposition of rocks) of these and other surface rocks produces sediments, which burial transforms into sedimentary rocks. Heat and pressure within the Earth can then transform pre-existing rocks to form metamorphic rocks such as slate and schist. When rocks are exposed at the surface, they are then subjected to the physical, chemical, and biological processes of weathering. This cycle of rock formation, exposure, erosion, and deposition is known as the **rock cycle**.

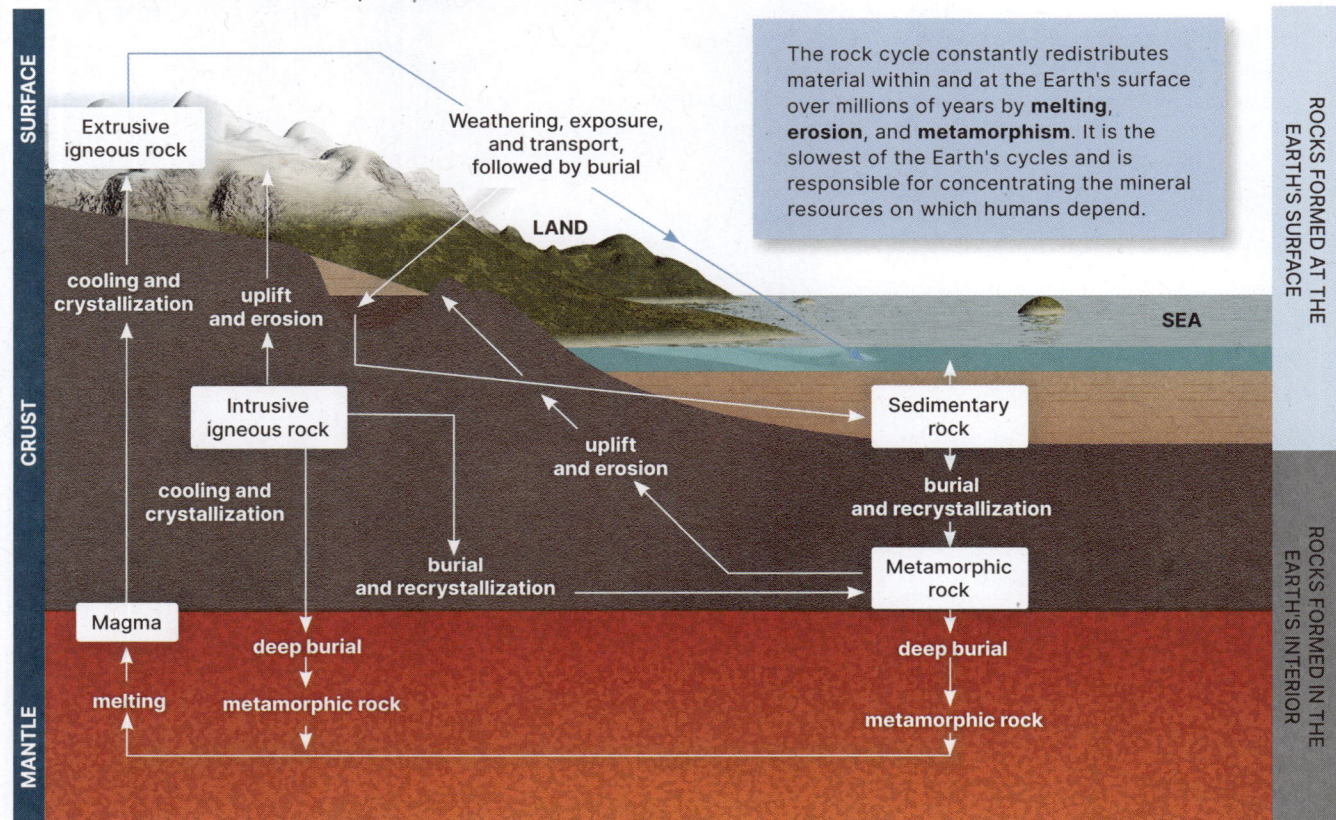

The rock cycle constantly redistributes material within and at the Earth's surface over millions of years by **melting**, **erosion**, and **metamorphism**. It is the slowest of the Earth's cycles and is responsible for concentrating the mineral resources on which humans depend.

These formations, known as hoodoos, are composed of soft sedimentary rock topped by a piece of harder, less easily-eroded rock that protects the column from the elements. Hoodoo shapes are affected by the erosional patterns of alternating hard and softer rock layers, while different minerals produce color.

Water and ice are powerful agents of erosion. Water lifts and transports rock fragments, and the freezing and thawing splits rocks apart. The flow of seawater millions of years ago, together with ice wedging and collapse along joints in the rock, has resulted in the formation of this spectacular arch in Utah, USA.

Salt weathering of rock produces a distinctive honeycombing effect. Seawater penetrates the rock and then evaporates to leave behind salt crystals which expand to produce holes in the rock surface. Softer parts of the rock are eroded at a greater rate than harder parts.

1. Using appropriate examples, distinguish between igneous, sedimentary, and metamorphic rocks:

___
___
___
___

Rocks and minerals of the Earth's crust

Rocks

The Earth's **crust** is made up of solid, naturally occurring assemblages of minerals called rocks. The huge diversity of the Earth's rocks has developed over thousands of millions of years through igneous activity, changes in form (metamorphism), and the formation of sediments and sedimentary rocks.

Igneous rocks solidify from volcanic magma and they vary in composition from basalt to granite and in texture from rapidly cooled glasses, such as obsidian, to slowly cooled coarse grains, such as granite. They may be intrusive (formed underground e.g. granite) or extrusive (formed above ground e.g. pumice).

Metamorphic rocks result when pre-existing rock is transformed by heat and pressure. Metamorphic rocks are classified by texture and composition. Examples include marble (metamorphosed limestone), slate (metamorphosed shale), schist, and gneiss. Gneisses include some of the oldest known rocks on Earth.

Sedimentary rocks form when sediments accumulate in different depositional environments and then become compressed into brittle, layered rocks, e.g. shale, sandstone, limestone, and conglomerate. Sedimentary rocks cover about 73% of the Earth's land surface, but make up only 8% of the crust by volume.

Minerals

▸ A mineral is a naturally occurring substance with a distinct chemical formula, crystal structure, and set of characteristics. Well known examples include diamonds, emeralds, and sapphires, but also less glamorous examples such as native copper, chalcanthite (hydrated copper sulphate), and native sulfur.

▸ The most common minerals in the Earth's crust are the feldspars (minerals containing aluminium tectosilicate ($AlSi_3O_8$)), and quartz, silicon dioxide (SiO_2).

Minerals are normally found as tiny crystals within rocks or alone in sands and gravel, eroded from larger structures many years earlier. Under the right temperatures, pressures, and chemical environment, minerals can form large crystal structures, many of which are sought after for their aesthetic value alone. The rarer the mineral the more valuable it is, e.g. diamonds only form under very specific conditions and can only be mined in a very few places around the world. There are more than 5000 types of mineral but only around 30 are relatively common. Most are compounds of two or more elements, e.g. halite, sodium chloride, but a few can be found as singular elements, e.g. gold, copper, and sulfur. Aside from larger examples being used in jewellery, many minerals are economically important and are used in everything from paint to ceramics, fertilizers, and abrasives.

2. Distinguish between weathering and erosion and describe the role of these processes in the rock cycle:

_____

_____

3. The Earth's mineral resources are produced by recycling processes, but are essentially non-renewable. Explain:

_____

_____

4. What is the difference between a mineral and a rock? _____

_____

_____

_____

14 Soil Textures

Key Idea: Soil texture depends on the amount of sand, silt, and clay present.

Soils are a mixture of sand, clay, silt, and organic matter. A **loam** contains a 40/40/20 mix of sand, silt, and clay and is considered the ideal soil for cultivating crops. Soils with too much clay hold water, become heavy, and are difficult to work, whereas soils with too much sand allow water to drain away too quickly. A loam contains enough clay to bind the water and hold it in place, but also enough sand to create spaces between the particles, allowing air to penetrate and water to drain. Because of these features, a loam is able to retain nutrients and humus better than other soil types.

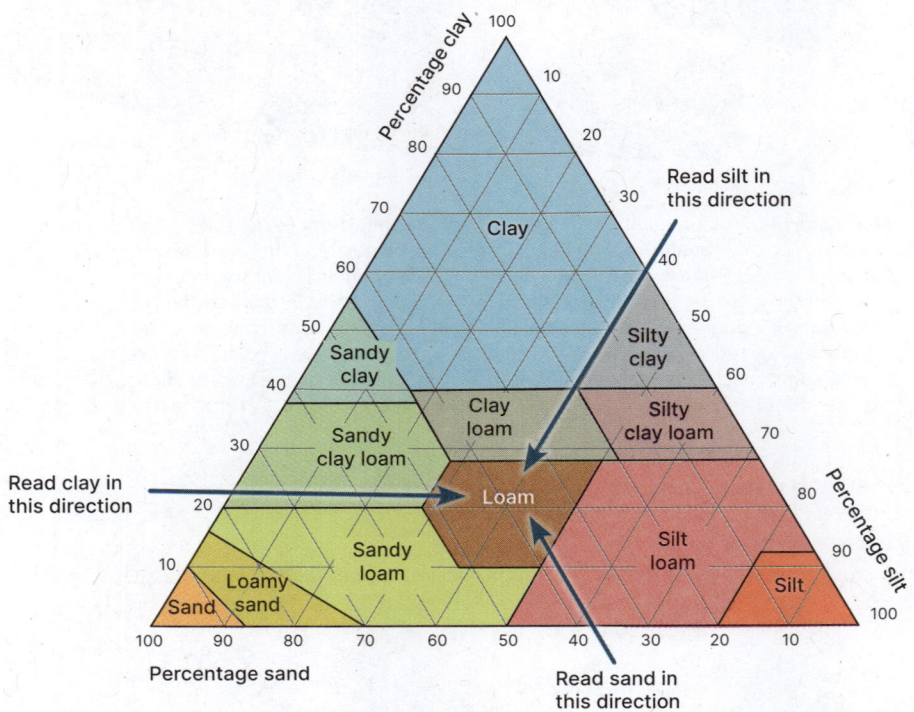

Soil sample 1

Soil sample 2

The percentage of sand, silt, and clay and therefore the type of soil can easily be measured by mixing a sample with water and letting it settle.

A loam consists of around 20% clay, 40% sand, and 40% silt. Around this point, various other loams exist which are named after their primary components. For example, a sandy loam consists of around 65% sand, 35% silt, and 10% clay.

	Clay	Silt	Sand	Loam
Nutrient holding capacity	++	+	0	+
Water infiltration capacity	0	+	++	+
Water holding capacity	++	+	0	+
Aeration	0	+	++	+
Workability	0	+	++	+

0 = low   + = medium   ++ = high

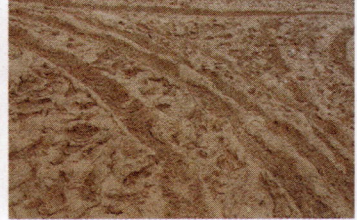

Loams are easily worked...   ...while silts and clays can be muddy.

The capacity of soil to be worked and produce viable crops depends on the mixture of particles within it. Silt provides moderate capacity in all areas, but by itself does not provide good soil as it too easily turns to mud when wet and is blown away by winds when dry. Loam consists of a variety of particle sizes and so remains more consistent in texture when both wet and dry.

1. Explain the term loam and how it applies to soil: _____
   _____
   _____

2. Using the scale on soil samples 1 and 2 above, calculate the percentage of sand, silt and clay in each sample and then use the soil triangle to identify the type of soil:

   Soil sample 1: % sand: _____ % silt: _____ % clay: _____ Soil type: _____

   Soil sample 2: % sand: _____ % silt: _____ % clay: _____ Soil type: _____

3. Explain why loamy soils are more easily worked and produce better crops than other soil types:
   _____
   _____

15 Soil and Soil Dynamics

Key Idea: Soils are a mixture of rock fragments and organic matter. The type of soil will depend on the environment.

Soils are a complex mixture of unconsolidated, weathered rock and organic material. Soils are essential to terrestrial life. Plants require soil and the microbial populations responsible for recycling organic wastes live in the soil and contribute to its fertility. Soils are named and classified on the basis of physical and chemical properties in their **horizons** (layers).

Soils have three basic horizons: A, B, and C. Soils and their horizons differ widely and are grouped according to their characteristics. These are determined by the underlying parent rock, the age of the soil, and the conditions under which the soil developed. A few soils weather directly from the underlying rocks and these residual soils have the same general chemistry as the original rocks. More commonly, soils form in materials that have moved in from elsewhere.

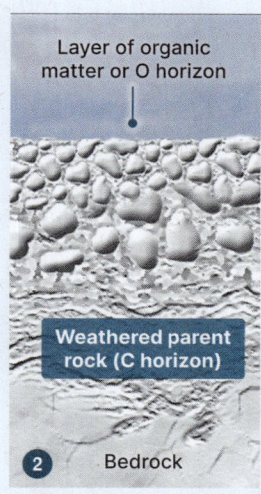
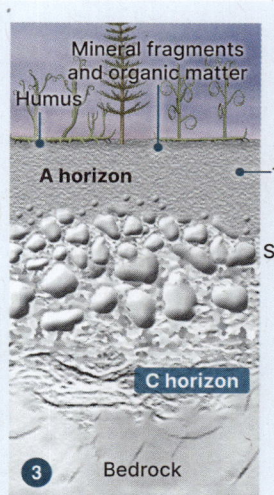

1. The parent rock is broken down by weathering to form a **regolith** which overlies the solid bedrock. The soil that forms is part of the regolith.

2. Plants establish and organic material builds up on the surface. This organic material aids the disintegration of the parent material.

3. As the mineral and organic content mix, horizons begin to form, with humus-rich layers at the surface and mineral-rich layers at the base.

4. Horizons are well developed in mature soils. The final characteristics of the soil are determined by the regional conditions and the rock type.

Influences on soil development

The character and composition of the parent material is important in determining soil properties. Parent materials include volcanic deposits, and sediments deposited by wind, water, or glaciers.

The occurrence of freeze-thaw and wet-dry cycles, as well as average temperature and moisture levels are important in soil development. Climate also affects vegetation, which in turn influences soil development.

Plants, animals, fungi, and bacteria help to create a soil both through their activities and by adding organic matter to the soil. Moist soils with a high organic content tend to be higher in biological activity.

The topography (hilliness) of the land influences soil development by affecting soil moisture and tendency towards erosion. Soils in steep regions are more prone to loss of the topsoil and erosion of the subsoil.

1. Explain the role of weathering in soil formation: _____
   _____

2. Discuss the influence of climate, rock type, and topography on the characteristics of a mature soil:
   _____
   _____
   _____

Soils in different climates

Soils are formed by the breakdown of rock and the mixing of inorganic and organic material. The soil profile is a series of horizontal layers that differ in composition and physical properties. Each recognizable layer is called a horizon. Soils have three basic **horizons** (A,B,C). The A horizon is the topsoil, which is rich in organic matter. If there is also a layer of litter (undecomposed or partly decomposed organic matter), this is called the O horizon, but it is often absent. The B horizon is a subsoil containing clay and soluble minerals. The C horizon is made up of weathered parent material and rock fragments. These horizons may be variously developed depending on whether or not the soil is mature. Mature soils have had enough time to develop distinct horizons. Immature soils have horizons that are lacking.

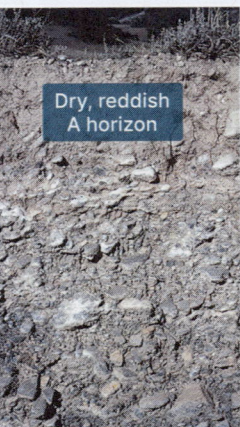

Arid regions

Desert soils are alkaline mineral soils with variable amounts of clay, low levels of organic matter, and poorly developed vertical profiles.

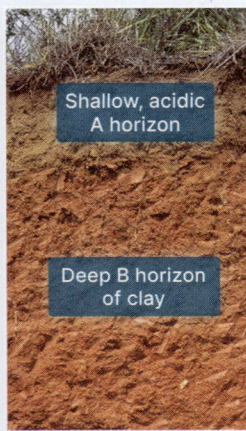

Humid tropics

Tropical soils: Leaching and chemical weathering make these soils acidic. Aluminum and iron oxides accumulate in the deep B horizon.

Mid latitudes

Grassland soils: Mature, alkaline, deep, well drained soils. They are typically nutrient-rich and productive with a high organic content.

Soil texture

Soil texture depends on the amount of each size of mineral particle in the soil (sand, silt, and clay sized particles). Coarse textured soils are dominated by sand, medium textures by silt, and fine textured soils by clay.

Sand feels gritty   **Silt** feels silky   **Clay** feels sticky

→ decreasing particle size

Polar regions

Very low temperatures slow the decomposition of organic matter and maintain the permafrost layer in these frozen soils.

Temperate

Weathered forest soils: Well developed soils with a deep organic layer and accumulated clay at lower levels.

Seasonally wet

Swelling soils: Marked seasonal rainfall results in deep cracks as the soil alternately swells and shrinks.

---

3. Describe the role of soil organisms in soil structure and development: _____

4. Identify which feature of a soil would most influence its:

   (a) Fertility: _____ (b) Water-holding capacity: _____

5. Explain how the characteristics described below arise in each of the following soil types:

   (a) Accumulation of organic matter in the frozen soils of the Arctic: _____

   (b) Shallow A horizon and poorly developed vertical profile of a desert soil: _____

6. Identify the different soil layers and their features: _____

16 The Atmosphere

Key Idea: The atmosphere can be divided into several layers. It becomes thinner and colder as altitude increases.

The Earth's **atmosphere** is a layer of gases surrounding the globe and retained by gravity. It contains roughly 78% nitrogen, 20.95% oxygen, 0.93% argon, 0.038% carbon dioxide, trace amounts of other gases, and a variable amount (average around 1%) of water vapor. This mixture of gases, known as air, protects life on Earth by absorbing ultraviolet radiation and reducing temperature extremes between day and night. The atmosphere consists of layers around the Earth, each one defined by the way temperature changes within its limits. The outermost troposphere thins slowly, fading into space with no boundary.

Cloud marks a westerly jet stream across the Red Sea

Jet streams are narrow, winding ribbons of strong wind in the upper troposphere. They mark the boundary between air masses at different temperatures. Cloud forms in the air that is lifted as it is driven into the core of the jet stream.

The Northern and Southern Lights, Aurora Borealis and Aurora Australis respectively, appear in the thermosphere. Typically, auroras appear either as a diffuse glow or as 'curtains' that extend in an approximately East to West direction.

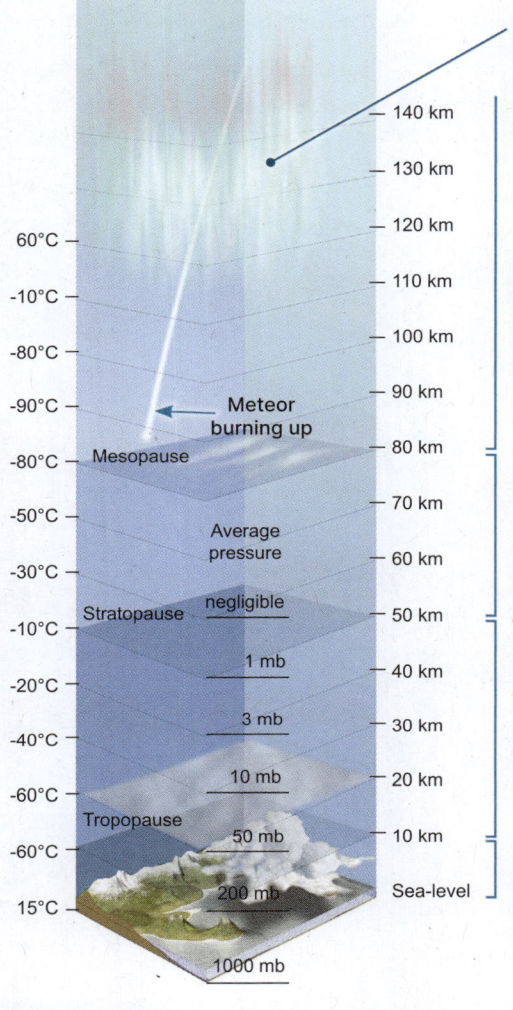

Aurora caused by collisions between protons and electrons from the Sun and the nitrogen and oxygen atoms in the atmosphere.

Thermosphere
This layer extends as high as 1000 km. Temperature increases rapidly after about 88 km.

Mesosphere
Temperature is constant in the lower mesosphere, but decreases steadily with height above 56 km.

Stratosphere
Temperature is stable to 20 km, then increases due to absorption of UV by the thin layer of ozone.

Troposphere
Air mixes vertically and horizontally. All weather occurs in this layer.

1. Describe two important roles of the atmosphere: _____

2. Explain what drives the atmospheric circulation: _____

3. Describe a characteristic feature and environmental issue for each of the following layers of the atmosphere:

   (a) Troposphere: _____

   (b) Stratosphere: _____

©2025 **BIOZONE** International
ISBN: 978-1-99-101409-2
Photocopying prohibited

17 Atmospheric Circulation

Key Idea: The circulation of the atmosphere is related to the heat from the sun and the rotation of the Earth.

The air of the **atmosphere** moves in response to heating from the Sun and, globally, the atmospheric circulation transports warmth from equatorial areas to high latitudes and returning, cooler air to the tropics. It is the interaction of the atmosphere and the oceans that creates the Earth's longe-term pattern of atmospheric conditions we call climate (as opposed to shorter-term weather). Heated air moving towards the poles from the equator does not flow in a single, uniform convection current. The rotation of the Earth causes air travelling across the globe to deflect to the left or right depending on the hemisphere and direction of travel. This deflection is called the **Coriolis effect** and is responsible for the direction of movement of large-scale weather systems in both hemispheres. The interactions of atmospheric systems are so complex that predicting weather for more than a few days in advance is extremely difficult and complex.

The tricellular model of atmospheric circulation

The temperature differential between the poles, combined with the rotation of the Earth, produces a series of atmospheric cells. This model of atmospheric circulation, with three cells in each hemisphere, is known as the **tricellular model**.

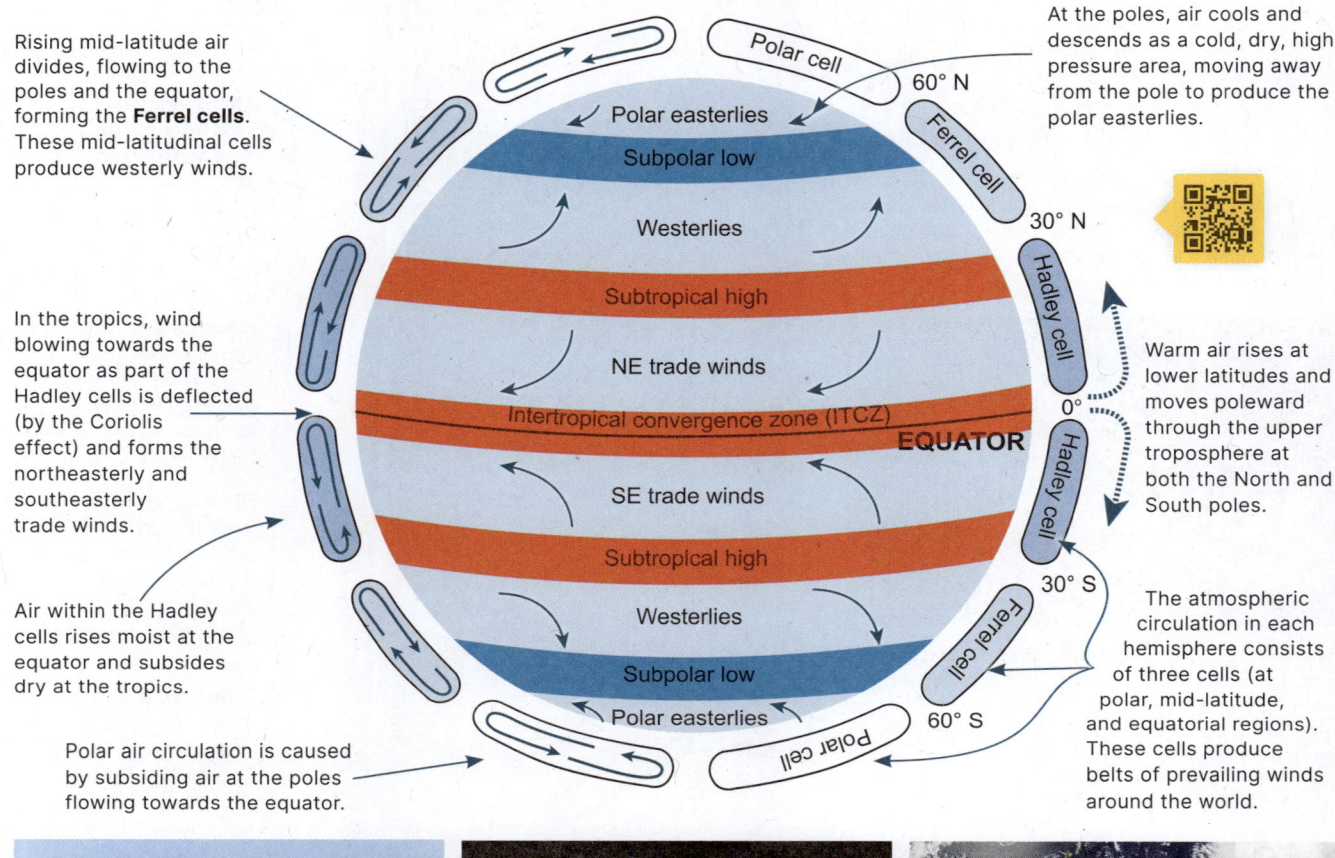

Rising mid-latitude air divides, flowing to the poles and the equator, forming the **Ferrel cells**. These mid-latitudinal cells produce westerly winds.

In the tropics, wind blowing towards the equator as part of the Hadley cells is deflected (by the Coriolis effect) and forms the northeasterly and southeasterly trade winds.

Air within the Hadley cells rises moist at the equator and subsides dry at the tropics.

Polar air circulation is caused by subsiding air at the poles flowing towards the equator.

At the poles, air cools and descends as a cold, dry, high pressure area, moving away from the pole to produce the polar easterlies.

Warm air rises at lower latitudes and moves poleward through the upper troposphere at both the North and South poles.

The atmospheric circulation in each hemisphere consists of three cells (at polar, mid-latitude, and equatorial regions). These cells produce belts of prevailing winds around the world.

Dunes form in sandy deserts and polar regions, where prevailing winds create drifts with characteristic shapes. Sand or snow grains are blown up the slope and fall down the far side to create sinuous crests extending for great distances.

The **intertropical convergence zone** marks the meeting of trade winds at the equator. It is characterized by varying, often calm winds, as well as violent thunderstorms. The position of the ITCZ can drastically affect the rainfall in equatorial nations.

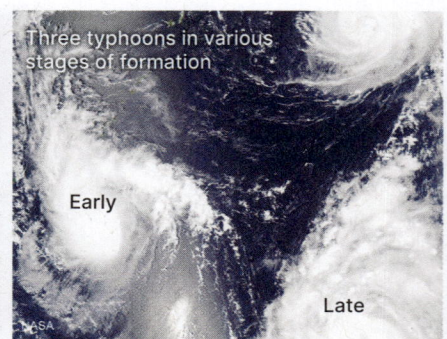

Tropical cyclones (also called typhoons or hurricanes), are low pressure systems that develop mainly over warm seas where winds start the air spiralling, producing low surface pressure into which air accelerates.

1. Explain what is meant by a prevailing wind: _____
   _____
   _____

The Coriolis effect

Air flowing towards, or away from, the equator follows a curved path that swings it to the right in the northern hemisphere and to the left in the southern hemisphere (right).

This phenomenon, known as the Coriolis effect, is caused by the anticlockwise rotation of the Earth about its axis, so as air moves across the Earth's surface, the surface itself is moving but at a different speed. The magnitude of the Coriolis effect depends on the latitude and the speed of the moving air. It is greatest at the poles and is responsible for the direction of the rotation of large hurricanes.

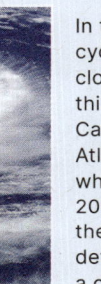

Air flows from high pressure to low pressure (see inset). In the northern hemisphere, the Coriolis effect deflects this moving air to the right, causing cyclonic (low pressure) systems to rotate counter-clockwise, seen here in a low pressure system over Iceland. Cyclonic weather is usually dull, with grey cloud and persistent rain.

In the southern hemisphere, cyclonic systems spiral in a clockwise direction, seen in this photograph of cyclone Catarina, a rare, South Atlantic tropical cyclone which hit Brazil in March 2004. As air rushes into the low pressure area, it is deflected to the left, causing a clockwise spiral.

Frontal weather

A weather front marks the boundary between two air-masses at different densities. A front is about 100-200 km wide and slopes where warm and cool air masses collide. A front appears on a weather map as a line with triangles (cold front) or semicircles (warm front) attached.

In a cold front, cold air undercuts warm air, forcing it steeply upwards along the line of the front and triggering the formation of towering cumulus clouds and rain. Cold fronts are often associated with low pressure systems and unsettled weather conditions. As the front passes, the cold air mass behind it can cause a rapid drop in temperature.

Gradual slope of a warm front

In a warm front, warm air rises over cold air more gradually, producing flattened, stratus-type clouds. Warm fronts produce low intensity rainfall that may last for some time and preceed warm weather. Because it moves more quickly, a cold front will eventually overtake a warm front, creating an occlusion.

2. Describe some of the physical and biological effects of prevailing winds: _____
_____
_____
_____
_____

3. Explain the role of the Coriolis effect in creating the prevailing winds in different regions of the globe:
_____
_____
_____
_____

4. In the spaces provided below, draw schematic diagrams to show:

   (a) The movement of a cold front into an area of warm air:

   (b) The movement of air in a Southern Hemisphere cyclone and a Northern Hemisphere hurricane:

18 El Niño and La Niña

Key Idea: El-Niño is a major climate pattern that affects regions around the Pacific ocean.

The El-Niño-Southern Oscillation, which has a periodicity of three to seven years, is a climate cycle that occurs in the Pacific region. El-Niño years cause a reversal of the normal climate regime and are connected to events such as the collapse of fisheries stocks, flooding in the Mississippi Valley, drought-induced crop failures and forest fires in Australia and Indonesia. An intensification of the normal situation, called La Niña, often occurs soon after an El Niño event.

In non-El Niño conditions, a low pressure system over Australia draws the southeast trade winds across the eastern Pacific from a high pressure system over South America. This system produces rain in the area of Australasia and dry conditions on the coast of South America.

In an El Niño event (right), the pressure systems over Australia and South America are weakened or reversed, beginning with a rise in air pressure over the Indian Ocean, Indonesia, and Australia. Warm waters block the nutrient upwelling along the west coast of the Americas. El Niño brings drought to Indonesia and northeastern South America, while heavy rain over Peru and Chile causes the deserts to bloom.

El Niño weather pattern

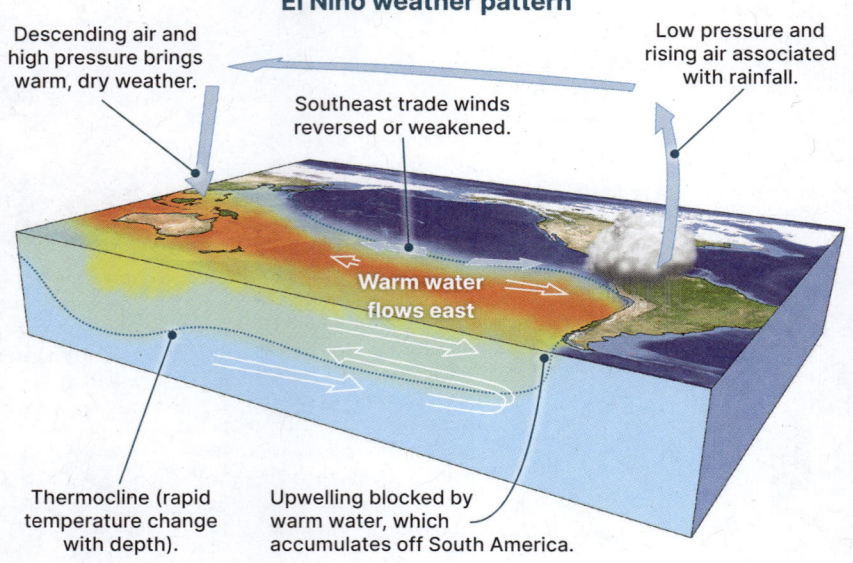

- Descending air and high pressure brings warm, dry weather.
- Southeast trade winds reversed or weakened.
- Low pressure and rising air associated with rainfall.
- Warm water flows east
- Thermocline (rapid temperature change with depth).
- Upwelling blocked by warm water, which accumulates off South America.

During non-El Niño years, cool, nutrient-rich waters along the South American coast sustain huge populations of fish such as anchovy. During El Niño events, warm waters reduce nutrient supply, and fish populations either crash or move to feeding grounds elsewhere.

El Niño events bring more rain to deserts in parts of South America and Baja California. On the islands of the Gulf of California, plant cover increases from 0-4% during non-El Niño years to 54-89% during El Niño years. In Northern Chile, plant cover increases over five times during El Niño.

During La Niña, the southeastern trade winds intensify, blowing warm water closer to Asia than normal. Cold, nutrient rich waters well up along the coast of the Americas. Winter temperatures in the southern states are warmer than usual and the hurricane season is more severe.

La Niña conditions bring cold waters to the surface near the Americas. This tends to push the jet stream over North America further North. This results in droughts in the southern US and more rain and cooler temperatures in the Pacific Northwest.

1. Describe the events that cause El Niño conditions and its effects on ocean circulation:

   _____

   _____

   _____

2. Describe the effect of an El Niño year on:

   (a) The climate of the western coast of South America: _____

   _____

   _____

   (b) The climate of Indonesia and Australia: _____

   _____

   _____

19 Water

Key Idea: Most of the Earth's water is in the oceans. Only a small percentage is easily accessible to humans.

The total volume of water on Earth is estimated at 1.386 billion km^3. 97% of the Earth's water is found in the oceans and is saline. Only 3% of water on Earth is freshwater, and only about 1% of that water is accessible, liquid freshwater (being available for use in rivers and lakes etc). The Great Lakes in North America and Lake Baikal in Russia hold about 45% of the world's accessible freshwater. It's estimated the Earth's **mantle** holds 5 times as much water as the Earth's oceans.

The Earth's water

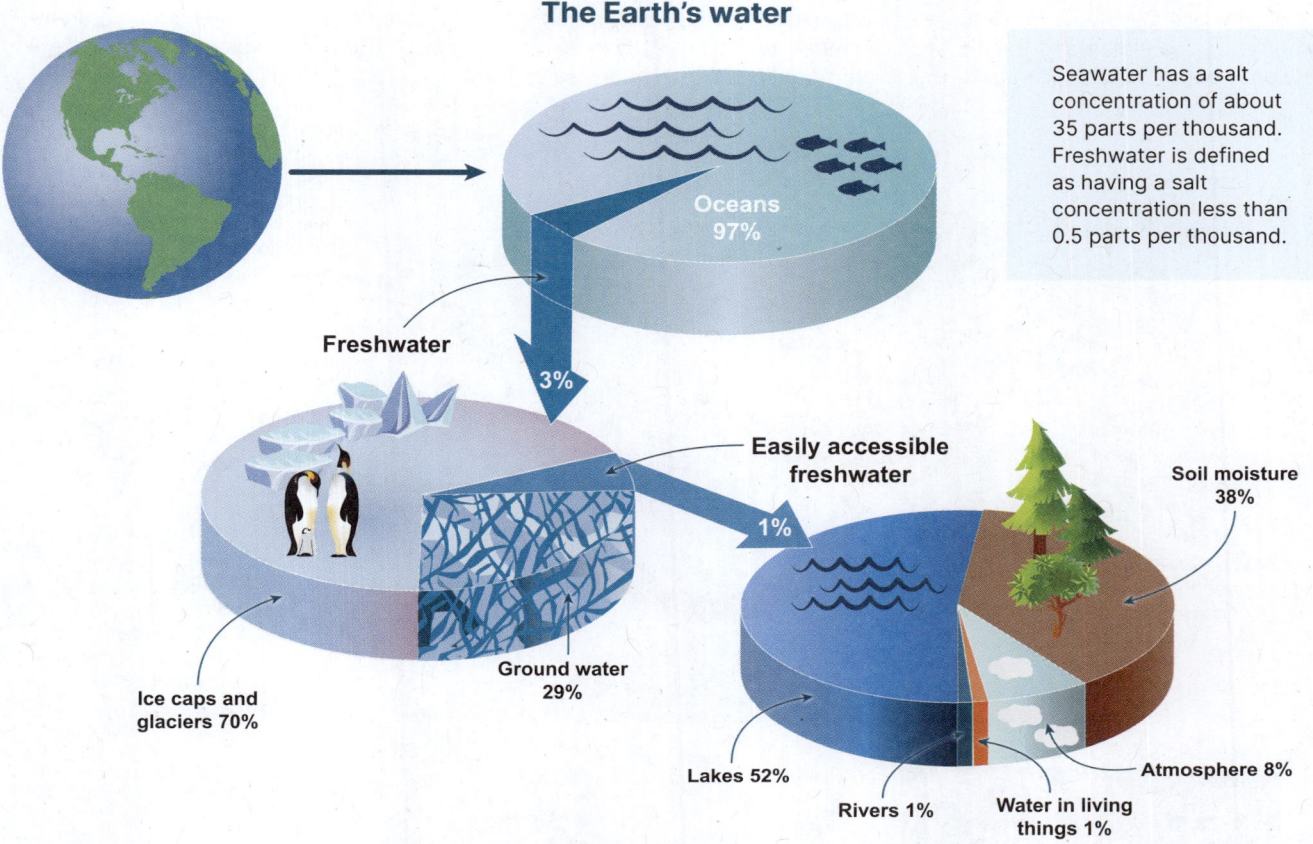

Seawater has a salt concentration of about 35 parts per thousand. Freshwater is defined as having a salt concentration less than 0.5 parts per thousand.

The Great Lakes of North America cover 244,106 km^2 and contain 21% of the world's accessible surface water. They formed about 14,000 years ago at the end of the Last Glacial Period.

The Amazon River is the largest river on Earth by discharge. It discharges more water than the next seven largest rivers combined. The Amazon is about 6400 km long.

The majority of Earth's freshwater is held in the polar icecaps. The Antarctic ice cap contains about 30 million km^3 of ice, enough to raise the level of the oceans by nearly 60m if it were to melt.

1. Calculate the following:

    (a) The volume of freshwater on Earth: _____

    (b) The volume of accessible freshwater: _____

2. What is accessible freshwater? _____

3. Water has a mass of 1 kg per liter. There is 1000 L per cubic meter. What is the mass of the Antarctic ice cap?

20 Ocean Circulation and Currents

Key Idea: Throughout the oceans, there is a constant circulation of water, both across the surface and at depth. Surface circulation, much of which is in the form of circular gyres, is driven by winds. In contrast, the deep-water ocean currents (the **thermohaline circulation**) is driven by the cooling and sinking of water masses in polar and subpolar regions. Cold water circulates through the Atlantic, penetrating the Indian and Pacific oceans, before returning as warm, upper ocean currents to the South Atlantic. Deep water currents move slowly and once a body of water sinks, it may spend hundreds of years away from the surface. The polar oceans comprise the Arctic Ocean in the northern hemisphere and the Southern Ocean in the South. They differ from other oceans in having vast amounts of ice, in various forms, floating in them. This ice coverage has an important stabilizing effect on global climate, insulating large areas of the oceans from solar radiation in summer and preventing heat loss in winter.

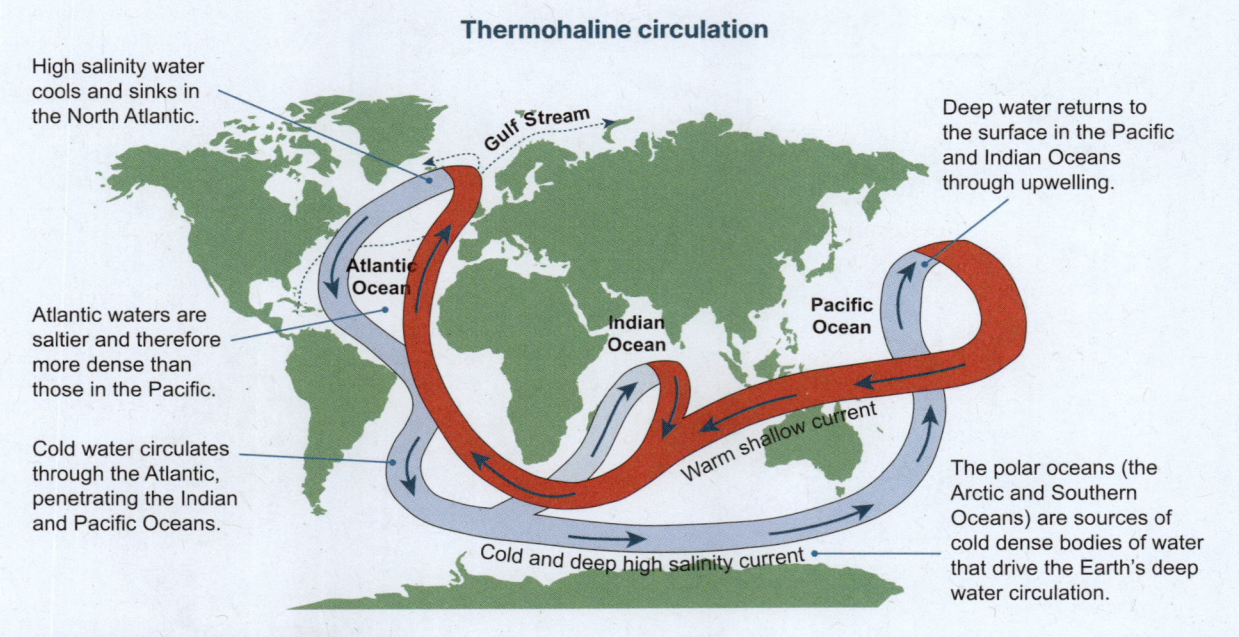

Thermohaline circulation

- High salinity water cools and sinks in the North Atlantic.
- Atlantic waters are saltier and therefore more dense than those in the Pacific.
- Cold water circulates through the Atlantic, penetrating the Indian and Pacific Oceans.
- Deep water returns to the surface in the Pacific and Indian Oceans through upwelling.
- The polar oceans (the Arctic and Southern Oceans) are sources of cold dense bodies of water that drive the Earth's deep water circulation.

Southern ocean

The Southern Ocean encircles Antarctica and is covered in ice for much of the year. Complex currents in the Southern Ocean produce rich upwelling zones that support abundant plankton and complex food webs.

Arctic Ocean

The vast amounts of ice associated with the polar oceans has an important, stabilizing effect on the global climate, insulating large areas of oceans from solar radiation in the summer and preventing heat loss in winter.

Icebreaker, Arctic sea ice

Satellite observations show that the Arctic sea ice is melting earlier and more rapidly than previously reported. The loss of ice cover will dramatically reduce the surface albedo (reflectivity) in the Arctic region.

1. Explain the basis of the Earth's thermohaline circulation: _____
   _____
   _____
   _____

2. Explain how thermohaline circulation could influence global climate: _____
   _____
   _____
   _____
   _____

Surface circulation in the oceans

The surface circulation of the oceans is driven by winds, but modified by the Coriolis effect. In the northern hemisphere, the Coriolis effect deflects the wind-driven water movements slightly to the right and in the southern hemisphere to the left. Drag accentuates the Coriolis effect so that the average water motion in the top few hundred metres of the ocean surface is almost at right angle to the wnd direction. The overall effect is a pattern of large scale, circular movements of water, or gyres, which rotate clockwise in the northern hemisphere and anticlockwise in the southern hemisphere (below and right). These currents carry warm water away from the equator and colder water towards it.

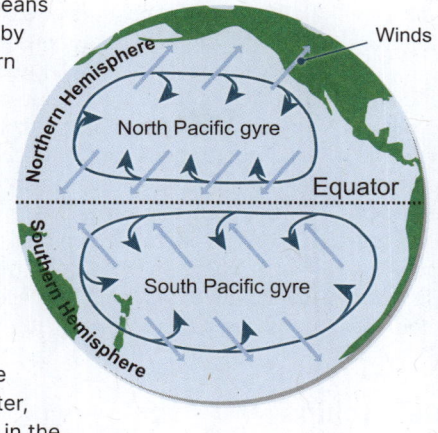

Overall pattern of surface currents

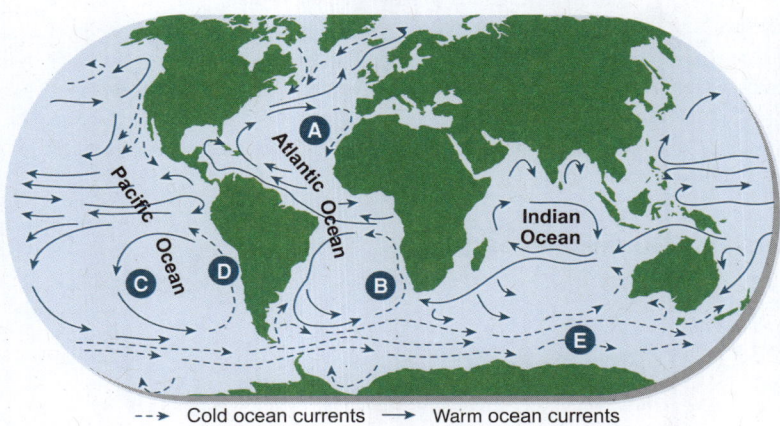

--→ Cold ocean currents   → Warm ocean currents

Local currents and upwelling

Local currents and vertical transport of water (upwelling and downwelling) are important phenomena around coastal regions. Upwelling in particular has important biological effects because it returns nutrients to surface waters, which promotes the growth of plankton.

Local currents are the result of interactions between tidal forces and coastlines. The whirlpools (vortices), seen above at Saltstraumen in Norway, are created by exceptionally strong tidal movements as water forces its way through a long, narrow strait.

Plankton blooms, seen here as bright spots around the coast of England and Ireland, often occur in upwelling zones as nutrients are brought to the surface. Upwelling occurs to replace the seawater that is moved offshore by surface circulation.

3. Contrast the mechanisms operating to drive deep water and surface water circulation:
   _____
   _____

4. Match each description below with its appropriate letter on the above diagram, 'Overall Pattern of Surface Currents':

   (a) Antarctic circumpolar current: _____   (d) North Atlantic gyre: _____

   (b) Peru current: _____   (e) South Atlantic gyre: _____

   (c) South Pacific gyre: _____

5. Describe a similarity between atmospheric circulation and surface ocean circulation patterns:
   _____
   _____

6. (a) Describe the biological importance of upwelling in coastal regions: _____
   _____

   (b) Explain how normal patterns of upwelling are affected during an El Niño year: _____
   _____
   _____
   _____

21 Earth's Past Climate

Key Idea: The Earth's past climate has undergone significant changes due to natural climate forcings: factors altering Earth's climate system.

Over the past 500 million years, the Earth has experienced multiple fluctuations in climate patterns. Around 500 million years ago, the Earth was in the midst of the Cambrian period, characterized by relatively warm temperatures and a diverse array of marine life. As the Earth progressed into the Permian period, it witnessed a shift towards cooler temperatures and the formation of vast ice sheets, leading to the first major ice age. This period was followed by the Triassic and Jurassic periods, which saw a return to warmer conditions and the flourishing of dinosaurs and other reptiles. However, towards the end of the Cretaceous period, the Earth experienced another significant cooling event, resulting in the formation of glaciers and the onset of the second major ice age. Since then, the Earth's climate has gradually warmed, with fluctuations along the way, leading to the emergence of mammals and eventually the rise of modern humans. Throughout this time, precipitation levels have varied, with periods of increased rainfall and others marked by aridity. These changes in temperature, ice ages, and precipitation levels have played a crucial role in shaping the Earth's ecosystems and the evolution of life on our planet.

Past climate changes and natural climate forcings

▸ The Earth's climate has undergone continuous changes throughout its history, although these changes have occurred over long geological timescales.

▸ Over the course of Earth's existence, there have been periods characterized by the presence of ice caps as well as periods without them. Additionally, many climates of the past were significantly different from the current climate we experience today.

▸ These climate changes are influenced by natural climate forcings such as variations in solar cycles, radiation levels from the sun, volcanic activity releasing aerosols, and fluctuations in greenhouse gases from both organic and inorganic sources.

▸ The movement of land masses through continental drift has played a significant role in shaping Earth's climate. For example, the formation of large land masses like Pangea around 300 million years ago and swampy, low-lying islands in the Carboniferous period 50 million years prior had a direct impact on climate patterns.

▸ These changes in climate, in turn, have had a close relationship with the emergence or extinction of species and groups, as they are highly influenced by the prevailing climatic conditions of the time.

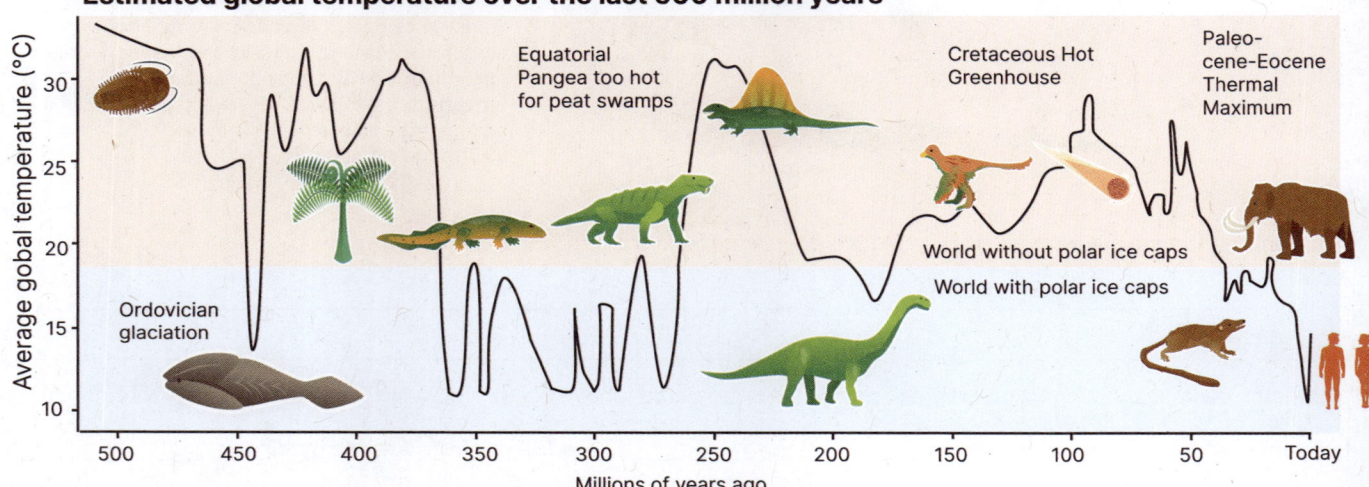

1. Discuss how the Earth's climate has changed over the past 500 million years:

    _____
    _____
    _____
    _____

2. Suggest what factors / climate forcings could have influenced Earth's temperature over the past 500 years:

    _____
    _____

Ordovician glaciation - 460 million years ago

▸ Scientific research suggests that the Ordovician glaciation began as a result of the tectonic uplift of the Appalachian mountain range, which is currently located in the United States.

▸ Before the onset of the glaciation event, there were significant volcanic eruptions that released a substantial quantity of CO_2 into the atmosphere. This CO_2 subsequently precipitated as acid rain onto the weathering basalt rock of the uplifted Appalachians. The carbon from the CO_2 was then absorbed into rock formations, leading to the formation of limestone (calcium carbonate). As a result of these processes, glaciation occurred across various supercontinents, including Gondwana.

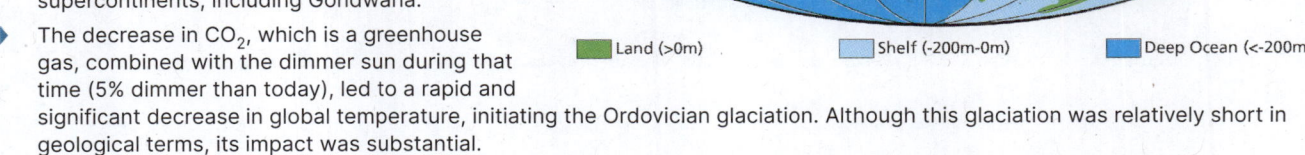

Paleogeography of the Middle Ordovician (458 MYA)

▸ The decrease in CO_2, which is a greenhouse gas, combined with the dimmer sun during that time (5% dimmer than today), led to a rapid and significant decrease in global temperature, initiating the Ordovician glaciation. Although this glaciation was relatively short in geological terms, its impact was substantial.

▸ Intense volcanic activity released significant quantities of greenhouse gases into the atmosphere, bringing an end to the relatively short 20 million year ice age. This volcanic activity resulted in the melting of the glaciation and a rapid increase in global temperature.

The Cretaceous Hot Greenhouse - 90 million years ago

▸ During the era commonly referred to as the 'age of dinosaurs', the Earth underwent a period of exceptionally high global temperatures known as the Hot Greenhouse, or scientifically, the Cretaceous Thermal Maximum (CTM).

▸ The climate during this period was strongly influenced by atmospheric CO_2 levels, which exceeded 1000 ppm, nearly double the current levels today.

▸ Scientists rely on the presence of fossilized glassy foraminifera, a type of plankton found in the ocean, as a crucial indicator of the elevated temperatures that prevailed on Earth during that time.

▸ After the supercontinent of Pangaea fragmented, the Atlantic Ocean expanded, resulting in the release of significant amounts of underwater methane clathrates, which are crystalline structures. When this methane reacted with atmospheric oxygen, it led to the formation of CO_2.

A reconstruction of 'Hothouse' Antarctica with a hypsilophodont-like dinosaur species, *Diluvicursor*, often found in temperate ecosystems in the late Cretaceous.

▸ Antarctic polar surface waters were estimated to have temperatures of at least 27 °C, which is believed to be the primary factor preventing the formation of ice sheets on the continent. Fossils of dinosaurs, as well as other animals and plants adapted to warm climates, have been discovered beneath the ice in Antarctica and the Arctic circle.

▸ The end of the Cretaceous Hot Greenhouse period was marked by a decrease in volcanic activity and the occurrence of phytoplankton blooms. These blooms absorbed CO_2 from the atmosphere, leading to a reduction in its levels. However, this process also resulted in low oxygen (anoxic) conditions in the oceans.

3. What led to the Ordovician glaciation, 460 million years ago? _____

4. What was the Cretaceous Hot Greenhouse? _____

22 Did You Get It?

1. (a) Use the following list to label the diagram below: *continental crust, mid ocean ridge, subduction zone, convergent boundary, divergent boundary, hot spot volcano, island chain, oceanic crust.*

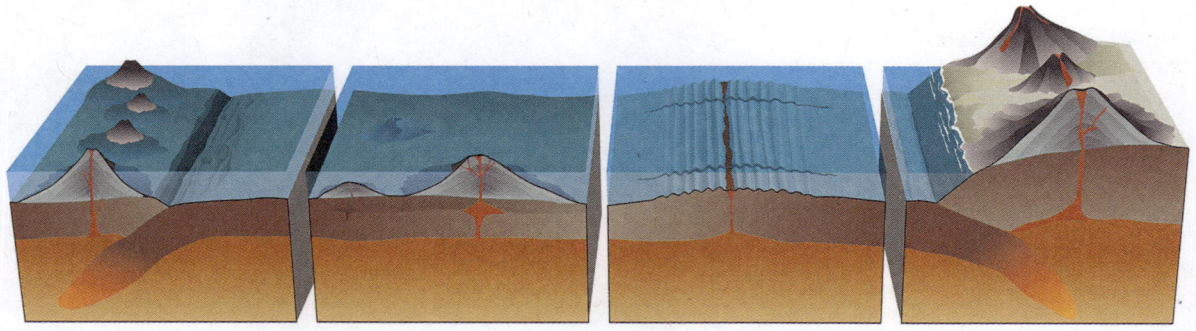

   (b) Draw arrows on the diagram to show the direction of the crust and convection currents:

2. Use the graph on the right to answer the following questions:

   (a) What type of soils are shown in the diagrams?

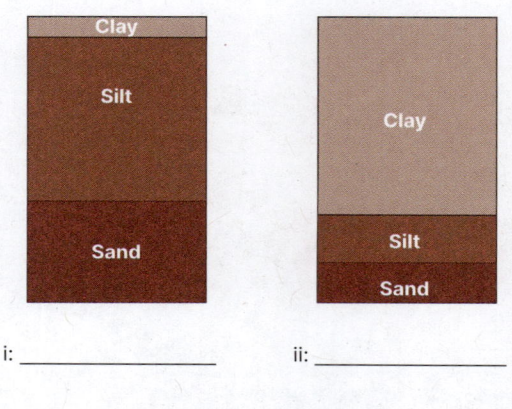

   i: _____   ii: _____

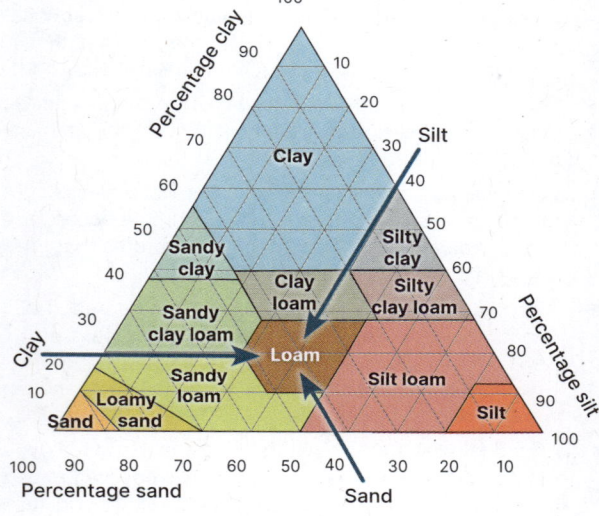

   (b) A soil was analysed and found to have 30% clay, 60% silt, and 10% sand. Identify the type of soil:

   _____

   (c) Another soil was analysed and found to have 20% clay, 70% silt, and 10% sand. Identify the type of soil:

   _____

3. Describe the origins on Earth of the following rock types:

   (a) Igneous: _____

   (b) Sedimentary: _____

   (c) Metamorphic: _____

   _____

4. Describe how the heating of the planet by the Sun and the planet's rotation affects the movement of the atmosphere:

   _____
   _____
   _____
   _____

Ecosystems

Biomes & ecosystems
- Distribution and features of biomes
- Habitat and niche
- Environmental gradients
- Photosynthesis and respiration

Energy flow
- Food webs and food chains
- Trophic structure
- Ecological pyramids
- Measuring productivity

Species interactions
- Interspecific competition
- Intraspecific competition

Both biotic and abiotic factors interact to form the environment in which we live.

Natural Ecosystem Change

Bio-geochemical cycles
- Importance of biogeochemical cycles
- Carbon, nitrogen, phosphorus and oxygen cycling
- Water cycle
- Bacteria and biogeochemical cycles

Ecosystem stability
- Time scales of change
- Primary and secondary succession
- Key species and their importance
- Environmental change

Large and small natural changes occur on both long and short time scales.

The Living World

Ecological principles can be applied at any scale and to any environment.

Studying the various aspects of ecology develops an overall understanding of the interactions between and within the biotic and abiotic worlds, and our place in them

Population size and growth is regulated by both biotic and abiotic factors.

Features of populations
- Population density
- Population distribution
- Age structure and its effects

Population growth
- Population regulation
- Natality, mortality, and migration
- Limiting factors
- Survivorship curves
- Humans and demography

Population

Sampling populations helps to analyze current trends and predict future ones.

Sampling populations
- Sampling and collecting equipment
- Quantitative sampling methods
- Qualitative sampling methods
- Recording data
- Diversity indices

Abiotic factors
- Monitoring equipment
- Measuring environmental change

Classification
- Plant and animal keys

Investigating Ecosystems

THE GANNET COLONY

The Australasian gannet population has been increasing since the 1950s. Breeding pairs in Australia are estimated to be over 20,000 while in New Zealand, breeding pairs are estimated at over 55,000. The Cape Kidnappers colony in New Zealand was reported to have 100 pairs in 1855 but now boasts over 6500 pairs. In Australia, the Black Pyramid colony grew from 500 pairs in 1960 to over 12,000 pairs in 1998. Conversely, other colonies have shrunk, with the Cat Island population near Tasmania falling from as many as 10,000 pairs in 1908, to just 450 pairs by 1947, and to virtually none by 1980.

Breeding season

The adult Australasian gannet (*Morus serrator*) spends half the year at sea but during the breeding season between July and February they form large colonies. Most of these are on offshore islands but there are also several mainland colonies.

Nutrient redistribution

Seabirds redistribute nutrients from the sea to the land via droppings and nesting material, e.g. seaweed. Like many other seabirds that form large colonies, gannet colonies build up large deposits nitrogen and phosphorus-rich guano over time. Excretion into the sea aids nutrient cycling.

Tertiary consumers

Gannets feed on pilchards and other small fish and squid. Pilchards feed on zooplankton which, in turn, feed on phytoplankton. Gannet chicks and eggs are preyed upon by the Pacific gull (*Larus pacificus*) and adult birds are occasionally preyed upon by the southern giant petrel (*Macronectes giganteus*).

🔍 Take a Deeper Look

- The growth and reduction of Australasian gannet colonies is well documented. Use data from websites of the provided spreadsheets to graph the growth or reduction of various colonies.
- Are there any patterns to the shape of the graphs?
- What factors might influence the growth and reduction of seabird populations?

Australasian gannet colony, Muriwai, near Auckland, New Zealand

Chapter 2 Ecosystems
Ecosystems

Resource Hub
bit.ly/4cyoAkw

Key Terms

- abiotic factors
- biome
- biosphere
- biotic factors
- carbon cycle
- cellular respiration
- climax community
- commensalism
- community
- consumer
- decomposer
- detritivore
- ecological niche
- ecosystem
- eutrophication
- food chain
- food web
- habitat
- interspecific
- intraspecific
- keystone species
- mutualism
- nitrogen cycle
- nutrient cycling
- parasitism
- phosphorus cycle
- photosynthesis
- pioneer species
- primary succession
- producer
- resilience
- secondary succession
- stability
- trophic level
- water cycle

Key Concepts

- Both biotic and abiotic factors contribute to the characteristics of an ecosystem.
- Interactions between organisms and their environment affect an organism's ability to exploit resources in their environment.
- The number of trophic levels in an ecosystem is limited by the amount of energy available to each trophic level.
- Nutrients and elements constantly cycle through the ecosystem in a complex series of reactions and interactions.
- Ecological succession is a natural process by which an ecosystem changes over time.

Ecosystems and habitats

Learning Outcomes:

		Activity Number
☐ 1	Identify the components and scales of ecosystems.	22, 23
☐ 2	Describe the Earth's major biomes and explain how they are classified according to vegetation type. Describe the effect of latitude and local climate in determining the distribution of biomes.	24 - 27
☐ 3	Describe different types of gradient in the abiotic environment. Explain the role of these environmental gradients on species distribution.	28 - 32
☐ 4	Describe components of an organism's ecological niche, distinguishing between the fundamental and the realized niche.	33 - 34

Energy in ecosystems

☐ 5	Explain how photosynthesis provides the main route by which energy enters an ecosystem and how respiration uses this energy.	35 - 38
☐ 6	Explain how trophic levels are used to describe energy flow in an ecosystem. Explain how energy is transferred between these levels.	39 - 41
☐ 7	Construct diagrams, including food chains and food webs, to explain the flow of energy through an ecosystem.	39 - 43
☐ 8	Distinguish between producers, primary and secondary consumers, detritivores, and saprotrophs.	39 - 43

Species interactions

☐ 9	Explain the nature of interspecific and intraspecific interactions within communities. Include examples of competition, mutualism, commensalism, and exploitation.	44
☐ 10	Investigate how species interactions in an ecosystem affect stability. Describe the scale of ecosystem change over time.	45, 46

Ecosystem stability and change

☐ 11	Describe processes involved in the carbon cycle. Describe how human activity may intervene in various aspects of the carbon cycle.	47, 48
☐ 12	Describe processes involved in the nitrogen cycle. Describe the role of microorgansims including nitrifying bacteria, nitrogen-fixing bacteria, and denitrifying bacteria.	49
☐ 13	Describe the effects on human activity on the nitrogen cycle, including eutrophication.	49
☐ 14	Describe the processes involved in the oxygen cycle and explain the link between the oxygen cycle and the carbon cycle.	50
☐ 15	Describe the processes involved in the hydrological (water) cycle, including water cycling between various reservoirs.	51
☐ 16	Describe the processes involved in the phosphorus cycle, using arrows to show the direction of nutrient flow. Contrast the phosphorus cycle with other biogeochemical cycles.	52
☐ 17	Use examples to help describe the process of ecological succession.	53 - 56

23 Components of an Ecosystem

Key Idea: Ecosystems are composed of biotic (living) and abitoic (non-living) factors.

The concept of the **ecosystem** was developed to describe the way groups of organisms are predictably found together in their physical environment. A **community** comprises all the organisms within an ecosystem. Both abiotic and **biotic factors** affect the organisms in a community, influencing their distribution and their survival, growth, and reproduction.

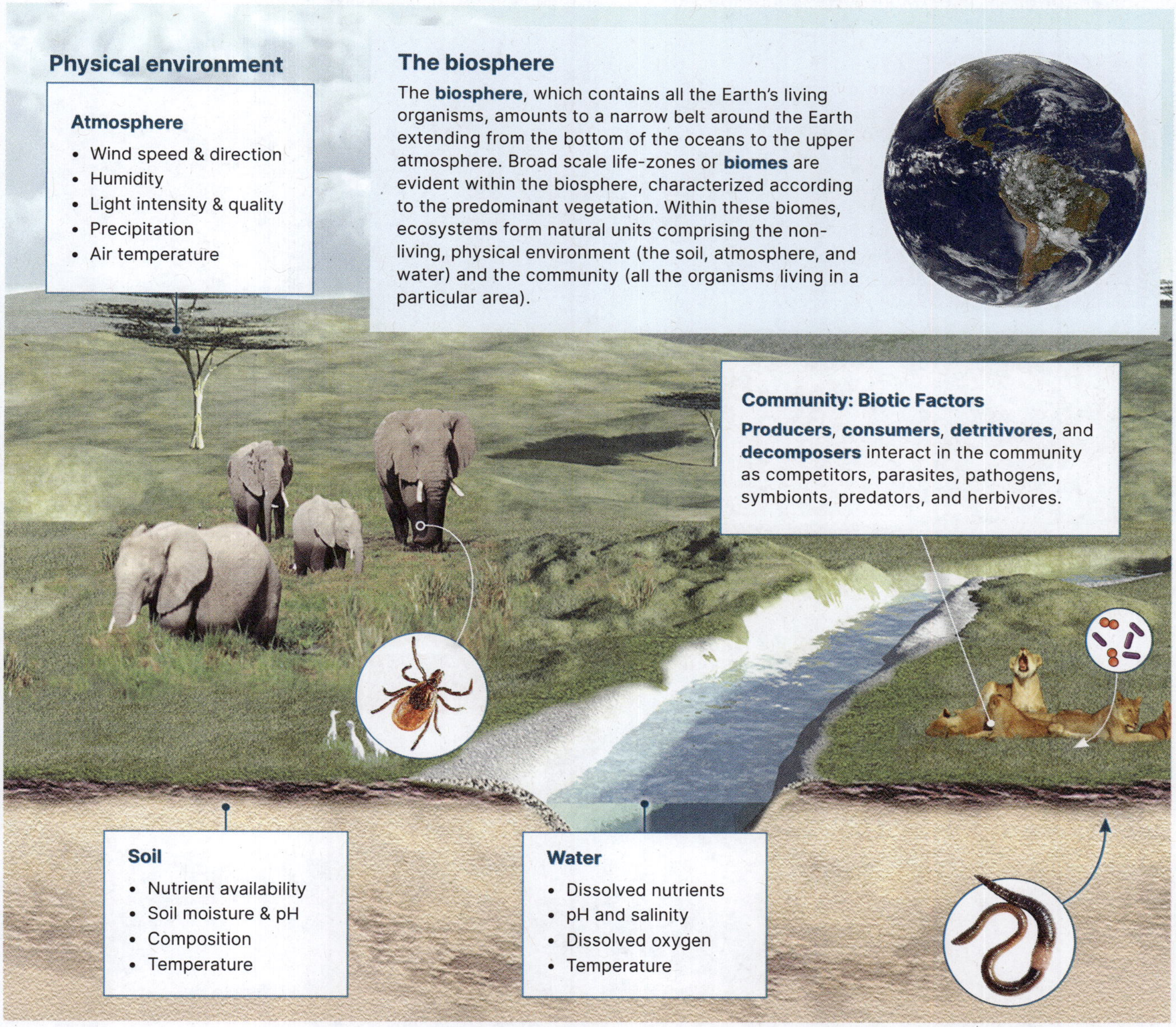

Physical environment

Atmosphere
- Wind speed & direction
- Humidity
- Light intensity & quality
- Precipitation
- Air temperature

The biosphere

The **biosphere**, which contains all the Earth's living organisms, amounts to a narrow belt around the Earth extending from the bottom of the oceans to the upper atmosphere. Broad scale life-zones or **biomes** are evident within the biosphere, characterized according to the predominant vegetation. Within these biomes, ecosystems form natural units comprising the non-living, physical environment (the soil, atmosphere, and water) and the community (all the organisms living in a particular area).

Community: Biotic Factors

Producers, **consumers**, **detritivores**, and **decomposers** interact in the community as competitors, parasites, pathogens, symbionts, predators, and herbivores.

Soil
- Nutrient availability
- Soil moisture & pH
- Composition
- Temperature

Water
- Dissolved nutrients
- pH and salinity
- Dissolved oxygen
- Temperature

1. Choose the letter of the term that corresponds to each of the statements below:

   A Community   **B** Population   **C** Ecosystem   **D** Physical factor

   (a) All the green tree frogs present in a rainforest: _____

   (b) An entire forest: _____

   (c) The humidity in a rainforest: _____

   (d) A community of organisms and their environment: _____

   (e) An association of different species interacting together: _____

2. Distinguish between biotic and abiotic factors: _____

24 Scales of Ecosystems

Key Idea: Ecosystems have no fixed boundaries and so can vary in size.

Ecosystems can be any size. The only limit is the size determined by the human observer. For example, a tree can be thought of as an ecosystem if we ignore the individual comings and goings of animals and look at the system as a whole. However, the tree may be part of a larger ecosystem, a forest, which again is part of a larger **biome**, and so on until we encompass the entire **biosphere**: the narrow belt around the Earth containing all living organisms.

Ecosystems can be on vastly different scales. Yosemite National Park in northern California covers 3000 km². Large parts of it are covered in mixed coniferous forests. The forest ecosystem comprises various tree species, e.g. Douglas fir, giant sequoia, and black oak. There are over 250 species of vertebrates including deer, bears, mountain lions, and a variety of bird life.

The ecosystem of a tree can be quite varied. The tree provides energy and materials for insects and other invertebrates that live on or in it. Bacteria and fungi decompose leaves and dead material on the tree or in the soil. The tree provides roosts for birds and fruit or seeds as a food source.

Within the forested areas, there are clearings that consist of grasses and scrub with the occasional isolated tree. These areas provide good grazing for deer and open hunting areas for owls.

Tidal rock pools are micro-ecosystems. Each one is slightly different from the next, with different species assemblages and **abiotic factors**. The ocean in the background is an ecosystem on a vastly larger scale.

The border of a garden or back yard can be used to define an ecosystem. Gardens can provide quite different ecosystems, ranging from tropical to dry depending on the type of plants and watering system.

Animals can be ecosystems in the same way as trees. All animals carry populations of microbes in their gut or on their bodies. Invertebrates, such as lice, may live in the fur and spend their entire lifecycle there.

1. Describe the borders that would define each of the three Yosemite ecosystems described above:

    (a) _____

    _____

    (b) _____

    _____

    (c) _____

    _____

25 Factors Affecting Biome Distribution

Key Idea: The circulation of the Earth's atmosphere produces large and specific climatic areas on either side of the equator. **Biomes** represent large areas with the same or similar climate and vegetation characteristics. These biomes exist in part because of the arrangement of weather conditions around the planet. The Earth is circled in the northern and southern hemispheres by three air cells. The interaction of these cells plays a major role in the formation of biomes. The cells form areas of rising or descending air, affecting the amount of rainfall. Surface features, such as oceans and mountain ranges, affect the final positions and size of these biomes but four general areas in each hemisphere can be identified.

Earth's climate and biomes

- Biomes are closely related to the major air cells that circle the Earth and are reflected in the northern and southern hemispheres.
- The Earth's biomes are the largest, geographically-based, biotic communities that can be conveniently recognized.
- Biomes are large areas where the vegetation type shares a particular suite of physical requirements.
- Terrestrial biomes are recognized for all the major climatic regions of the world. They are classified by their predominant vegetation type. Biomes are closely related to the major air cells that circle the Earth and are reflected in the northern and southern hemispheres.

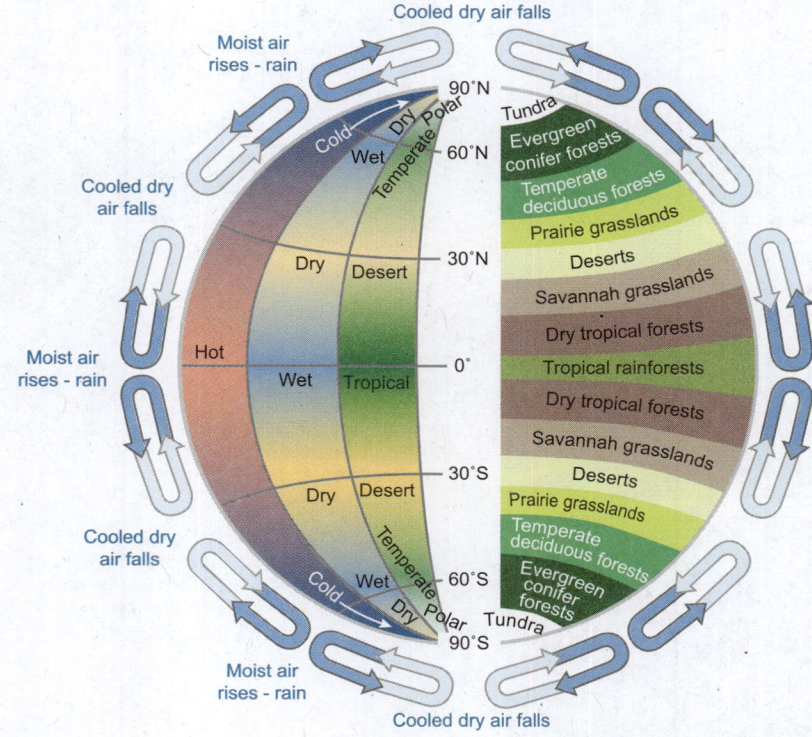

Biomes and landscapes

Climate is heavily modified by the landscape. Where there are large mountain ranges, wind is deflected upwards causing rain on the windward side and a rain shadow on the leeward side. The biome that results from this is considerably different from the one that may have appeared with no wind deflection. Large expanses of ocean and flat land also change the climate by modifying air temperatures and the amount of rainfall.

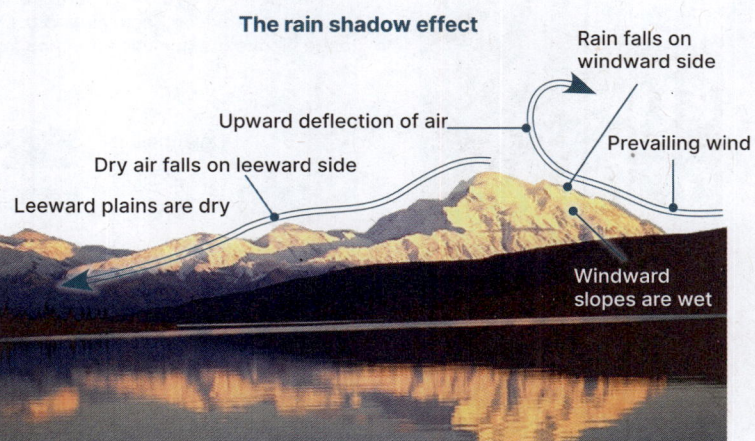

The rain shadow effect

Leeward plains

Windward slopes

1. Explain why the pattern of biomes is reflected in the northern and southern hemispheres:

___

2. What kind of features might prevent these patterns from matching exactly?

___

26 The World's Terrestrial Biomes

Key Idea: The climate plays an important role in determining the location of Earth's biomes.

Global patterns of vegetation distribution are closely related to climate. Although they are complex, major vegetation **biomes** can be recognized. These are large areas where a distinctive vegetation type has formed in response to a particular physical environment. Biomes have characteristic features, but the boundaries between them are not distinct. The same biome may occur in widely separated regions wherever the climatic and soil conditions are similar.

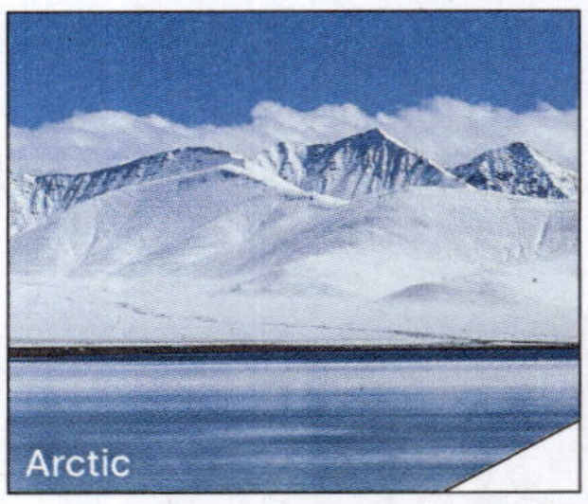
Arctic

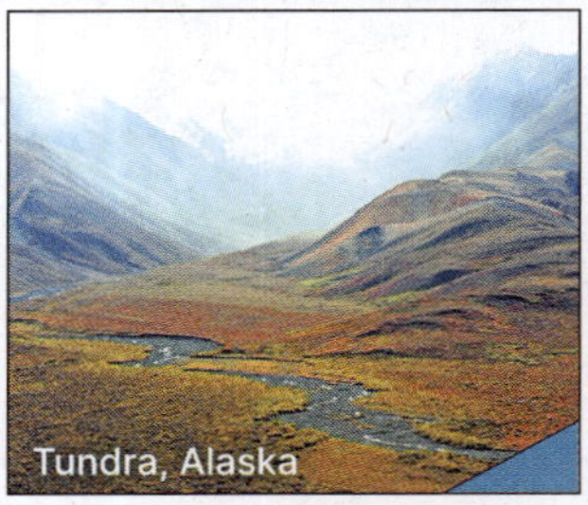
Tundra, Alaska

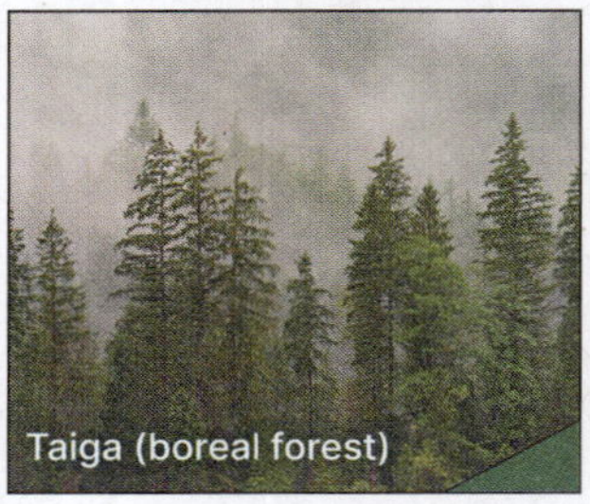
Taiga (boreal forest)

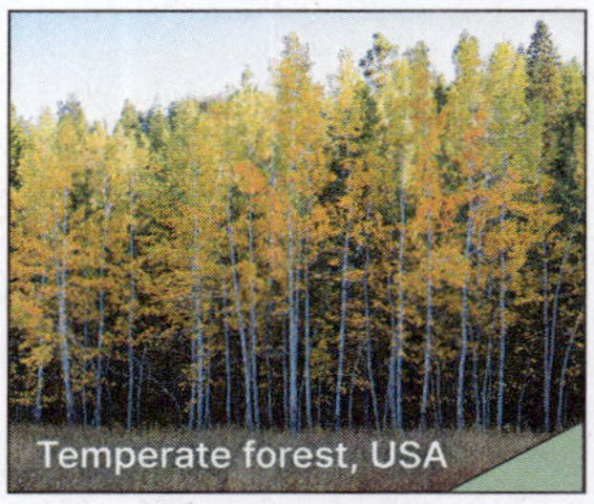
Temperate forest, USA

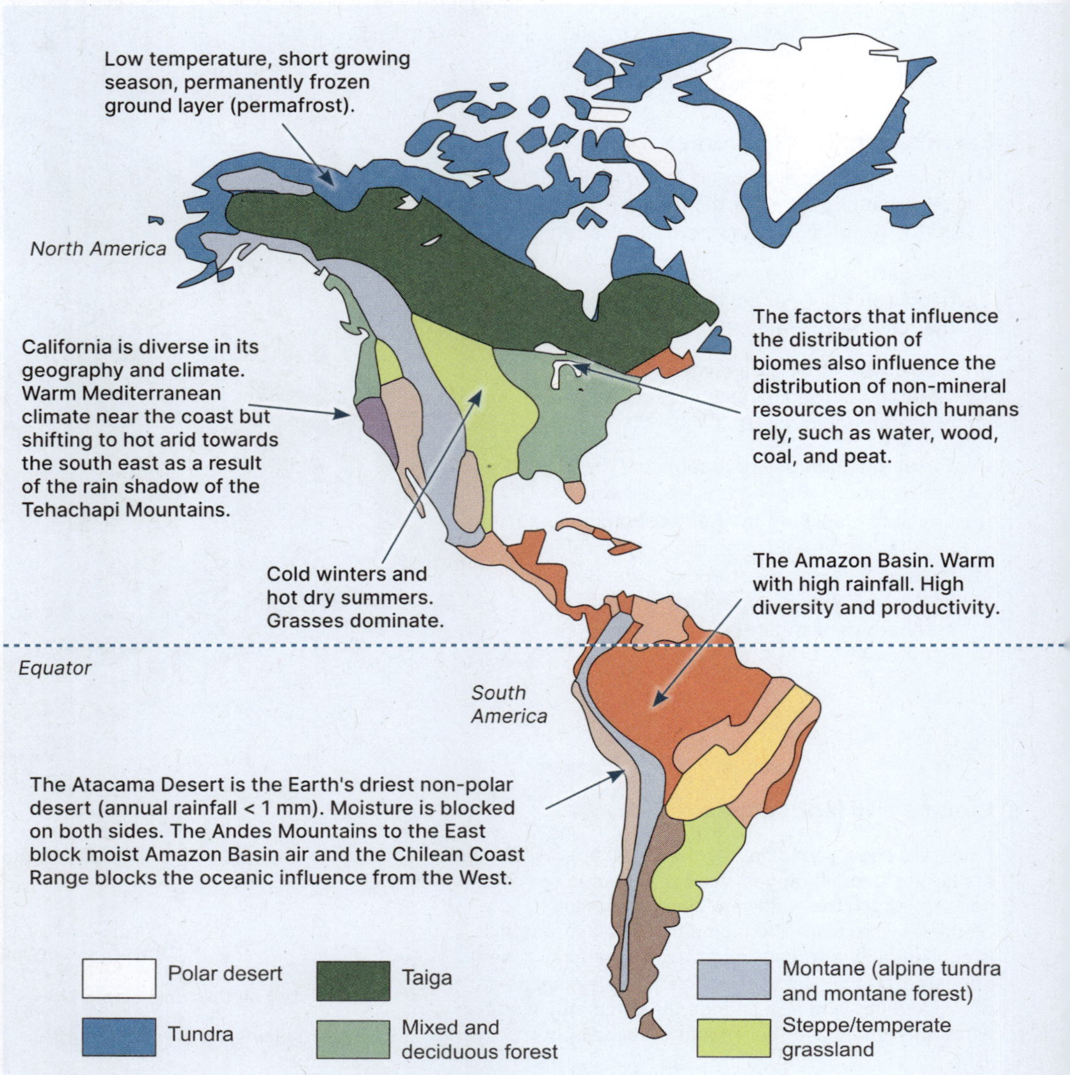

Low temperature, short growing season, permanently frozen ground layer (permafrost).
North America
California is diverse in its geography and climate. Warm Mediterranean climate near the coast but shifting to hot arid towards the south east as a result of the rain shadow of the Tehachapi Mountains.
The factors that influence the distribution of biomes also influence the distribution of non-mineral resources on which humans rely, such as water, wood, coal, and peat.
Cold winters and hot dry summers. Grasses dominate.
The Amazon Basin. Warm with high rainfall. High diversity and productivity.
Equator
South America
The Atacama Desert is the Earth's driest non-polar desert (annual rainfall < 1 mm). Moisture is blocked on both sides. The Andes Mountains to the East block moist Amazon Basin air and the Chilean Coast Range blocks the oceanic influence from the West.
Polar desert
Tundra
Taiga
Mixed and deciduous forest
Montane (alpine tundra and montane forest)
Steppe/temperate grassland

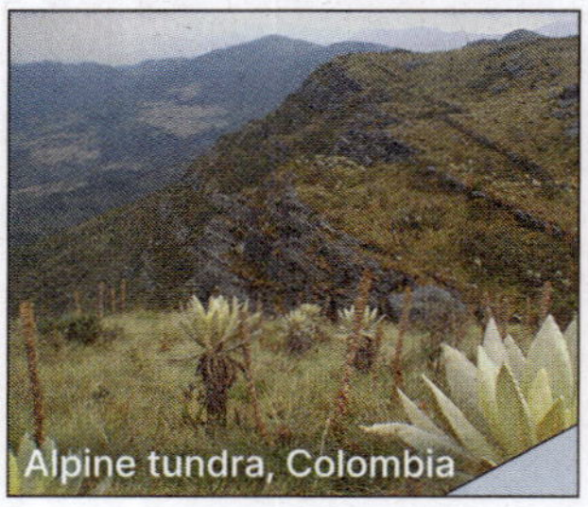
Alpine tundra, Colombia

Friedrich Kircher cc 3.0

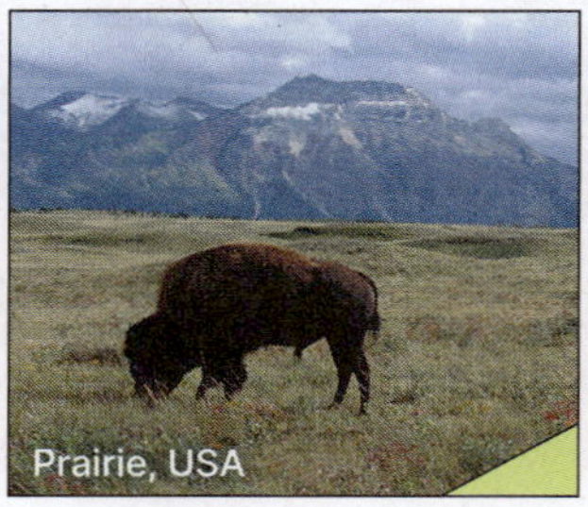
Prairie, USA

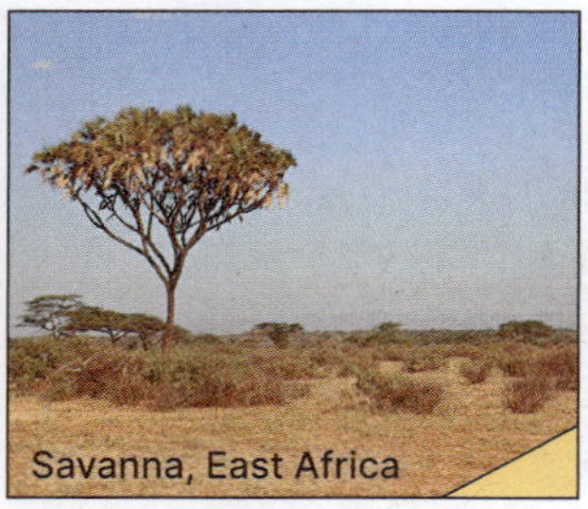
Savanna, East Africa

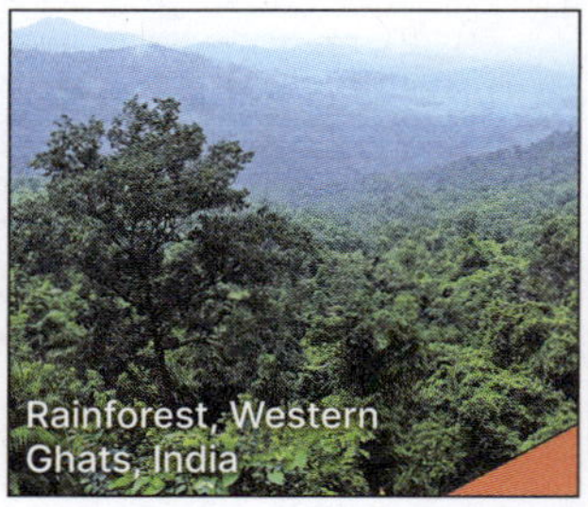
Rainforest, Western Ghats, India

1. Explain the distribution of deserts and semi-desert areas in northern parts of Asia and in the west of North and South America (away from equatorial regions):

____

____

2. Suggest what abiotic factor(s) limit the northern extent of boreal forest: _______________

____

____

LW

ISBN: 978-1-99-101409-2

Vegetation patterns are determined largely by climate, which in turn is heavily influenced by topography and proximity to the ocean. Large mountain ranges cause wind to deflect upwards. This causes rain to fall on the windward side and creates a drier 'rain shadow' on the leeward side. Rain shadowing governs the occurrence of many deserts globally, and some of the world's driest regions, including the Atacama Desert in Chile and Death Valley in California, are in rain shadows. Wherever they occur, montane regions are associated with their own altitude-adapted vegetation. Biome classification may vary considerably and is not necessarily static as environments shift under patterns of changing climate and human influence. However, most classifications recognize desert, tundra, grassland, and forest types and distinguish them on the basis of latitude.

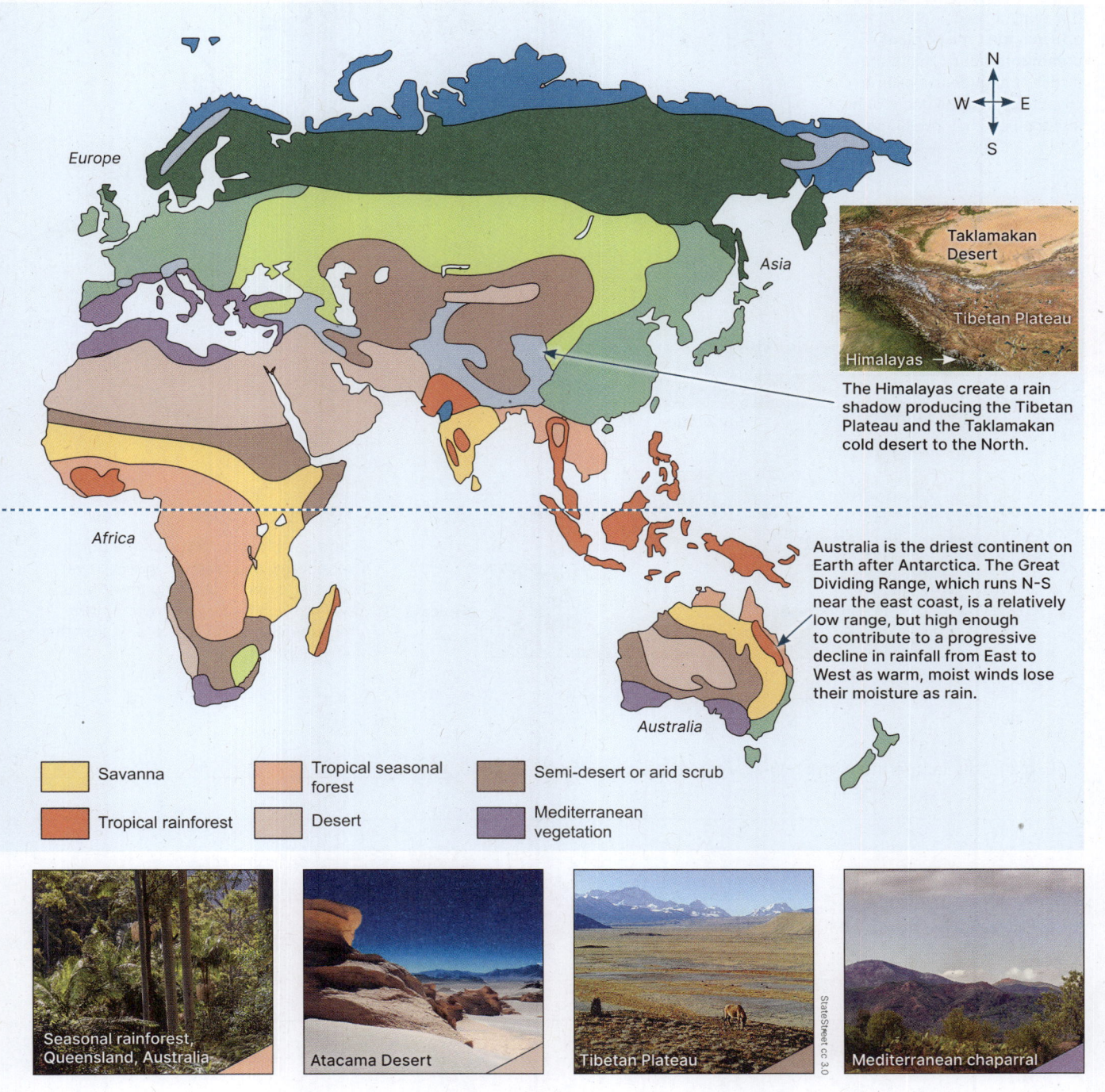

3. Explain how variations in climate, geography, latitude, and altitude influence the global distribution of non-mineral ecosystem resources, such as water and wood:

_____
_____
_____
_____
_____

27 The Effect of Temperature on Biomes

Key Idea: Temperature and rainfall play an important role in determining the geographical location of terrestrial biomes. Temperature and precipitation are excellent predictors of **biome** distribution. Temperature decreases from the equator to the poles. Temperature and precipitation act together as limiting factors to determine the type of desert, grassland, or forest biome in a region. Latitude directly affects solar input and temperature.

Within a single latitudinal region, the level of precipitation (rainfall) governs the type of plant **community** found. Note that the effect of altitude is similar to that of latitude (ice will occur at high altitudes even at low-latitudes).

As the Earth curves towards the poles, solar energy is spread out over an ever increasing area. This energy must also travel through a greater amount of the atmosphere, expending more energy than at low latitudes.

1. Explain how temperature and rainfall affect the distribution of biomes: _____
   _____
   _____
   _____

2. Explain why biomes are not evenly distributed about the globe: _____
   _____
   _____

3. Explain how the landscape can modify climate: _____
   _____
   _____
   _____

4. Explain why higher latitudes receive less solar energy than lower latitudes: _____
   _____
   _____

28 Aquatic Biomes

Key Idea: The major aquatic biomes can be either marine or freshwater and are influenced by local conditions.

Water covers ~70% of Earth's surface, so aquatic **biomes** are a major component of the global environment. Aquatic biomes include all those environments that are dominated by water. These environments include deep oceans, shallow seas and reefs, swamps and estuaries, and rivers and lakes. Aquatic biomes include some of the most productive and the least productive **ecosystems** in the world (production being the amount of accumulated biomass). The characteristics of different aquatic ecosystems depend on where they occur on the globe and the local conditions that shape them.

Marine biomes
oceans, estuaries, and reefs

Freshwater biomes
lakes, ponds, rivers, and freshwater wetlands

The open oceans are characterized by saline (salty) waters, waves, and currents. Five oceanic divisions are recognized but they are all interconnected as one global ocean.

The prairie potholes of the Midwest (USA) are shallow wetlands resulting from glaciation. Glaciation, tectonic events, and volcanic activity have formed the world's largest lakes.

Lake Tahoe, California

Lakes are inland depressions containing standing water. Their distribution is not uniform but dependent on geology and geography. They are an important source of drinking water.

Estuaries are regions where fresh water of rivers meets tidal flows from the ocean.

Streams and rivers are characterized by continuously flowing water.

Wetlands include marshes, bogs, fens, and swamps. Bogs and fens have peaty soils.

Coral reefs occur in tropical and subtropical regions and are of biological (not geological) origin. They are made from living organisms.

- Aquatic biomes are broadly categorized as marine and freshwater. The marine biomes include coral reefs and estuaries, whereas the freshwater biomes include rivers, lakes, and wetlands. Oceans cover more than 70% of the Earth's surface and contain 97% of the Earth's water. Less than 1% of the Earth's water is freshwater, an essential resource for plant and animal life.

- Temperature, salinity, tides, waves, and currents determine the characteristics and functioning of marine ecosystems. The most productive marine biomes are coral reefs, coastal waters, and estuaries and coastal marshes, because these are regions with high light and large nutrient fluxes.

- Freshwater biomes may be contained within a basin or flowing and are heavily influenced by geological formation, topography, and land use. Wetland systems are varied and often difficult to categorize. They are best described according to soil type, vegetation, and hydrology.

1. Describe the defining features of each of the following marine biomes:

    (a) Open oceans: _____

    (b) Estuaries: _____

    (c) Coral reefs: _____

2. Identify the main difference between lakes and ponds, and rivers and streams: _____
   _____

29 Physical Factors and Gradients

Key Idea: Gradients in the physical environment influence the range of physical conditions and may create microhabitats. Gradients in **abiotic factors** are found in all environments. They create microhabitats and microclimates (see definitions below) within a larger area and influence patterns of species distribution. Organisms can exploit the microclimates produced by physical gradients and so occupy apparently inhospitable environments, e.g. frogs living in deserts.

A desert environment

Deserts experience extremes in temperature and humidity but they are not uniform with respect to these factors. The diagram below gives hypothetical values for temperature and humidity for typical microclimates in a desert environment at midday.

> **Microclimate**: The climate of a very small or restricted area.
>
> **Microhabitat**: A habitat of limited extent, which differs in its characteristics from the surrounding, more extensive, habitat.

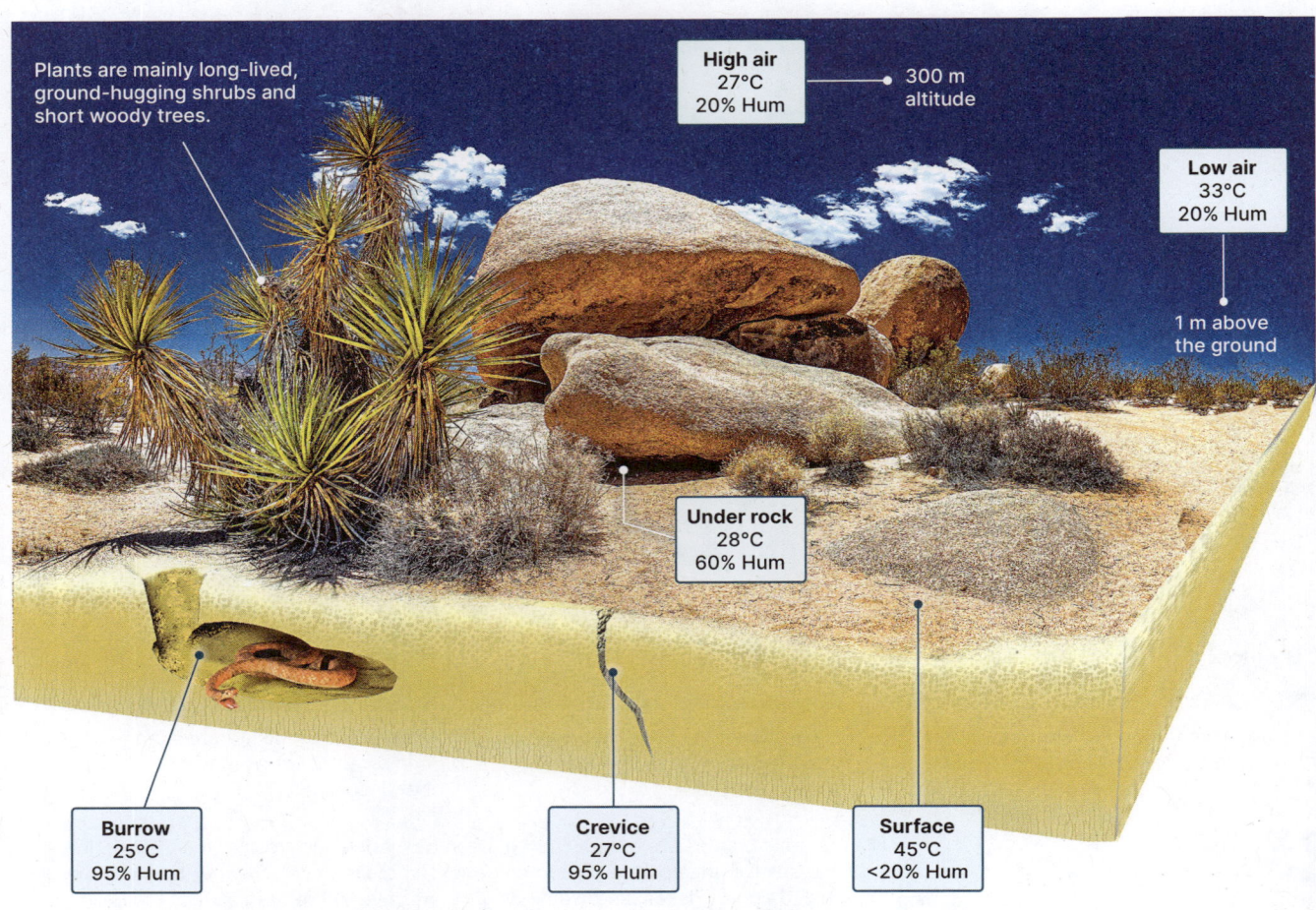

High air: 27°C, 20% Hum — 300 m altitude
Low air: 33°C, 20% Hum — 1 m above the ground
Under rock: 28°C, 60% Hum
Burrow: 25°C, 95% Hum
Crevice: 27°C, 95% Hum
Surface: 45°C, <20% Hum

Plants are mainly long-lived, ground-hugging shrubs and short woody trees.

1. (a) Study the diagram above and describe the general microhabitats where humidity is highest:
   _____

   (b) How do these microhabitats enable land animals to survive the extreme high temperatures of midday:
   _____
   _____

2. Desert surfaces not only receive more solar radiation than humid regions, they lose much more heat at night, so night-time temperatures can be very low. How does this advantage both the plants and animals living there?
   _____
   _____

3. Suggest why there are relatively few large mammals in deserts: _____
   _____
   _____

30 Physical Factors in a Forest

Key Idea: Environmental gradients can occur with altitude and with horizontal distance, e.g. along a shore, but they can also arise as a result of vertical distance from the ground. A forest often has a layered structure because vegetation grows to different heights. The light quantity and quality, humidity, wind speed, and temperature change gradually from the canopy to the forest floor. These changes are a consequence of the modification of the physical environment by the species present at each layer.

Light: 70%
Wind: 15 km/h
Humid: 67%

Light: 50%
Wind: 12 km/h
Humid: 75%

Light: 12%
Wind: 9 km/h
Humid: 80%

Light: 6%
Wind: 5 km/h
Humid: 85%

Light: 1%
Wind: 3 km/h
Humid: 90%

Light: 0%
Wind: 0 km/h
Humid: 98%

Canopy

Leaf litter

A datalogger fitted with suitable probes was used to gather data on wind speed (Wind), humidity (Humid), and light intensity (Light) for each layer (left). Light intensity is given as a percentage of full sunlight.

Tropical rainforests are complex communities with a vertical structure that divides the vegetation into layers. This pattern of vertical layering is called stratification.

1. Describe the general trend from the canopy to the leaf litter for each of the following:

    (a) Light intensity: _____

    (b) Wind speed: _____

    (c) Humidity: _____

2. Explain why each of these factors changes as the distance from the canopy increases:

    (a) Light intensity: _____

    (b) Wind speed: _____

    (c) Humidity: _____

3. What other feature of light, other than intensity, will also change with distance from the canopy and why?

31 Stratification in a Forest

Key Idea: Forest communities throughout the world show a pattern of vertical layering called stratification which arises as a result of tree species growing to different heights.
The enormous diversity of species found in rainforests can be supported because the vertical structure itself creates variation in physical factors and microclimates. Fallen logs and cavities in trees also add to vertical structure and enhance biodiversity. While a general stratification pattern can be described, species composition varies from region to region according to the altitude, rainfall, light levels, soil type etc.

Canopy
The canopy intercepts most of the direct sunlight. Canopy trees that grow taller than the canopy layer are called emergents.

Subcanopy
Sometimes called the understory. This lower level of smaller trees is not always present.

Epiphytes and lianes
Epiphytes are plants that have no contact with the soil but grow in crevices in the branches and trunks of larger trees. Lianes are rooted in the ground, but clamber into the canopy. This layer includes ferns and orchids.

Shrub layer
A layer of plants 1-3 m tall. Includes seedlings less than 1 m tall and shade adapted, low growing plants such as ferns.

Ground layer
Includes mosses, fungi, lichens, dead leaves, and debris. This layer may also incorporate some of the plants from the shrub layer (ferns and shrubs).

1. Using examples, explain why a forest with a strong pattern of stratification might provide a greater diversity of habitats than a forest without such a vertical structure:

   _____
   _____
   _____
   _____

2. Predict the impact of deliberate removal (logging) of emergents and large canopy trees on the community composition. Consider how logging alters the physical environment and how existing and colonizing species might respond to this:

   _____
   _____
   _____
   _____

32 Physical Factors on a Rocky Shore

Key Idea: A rocky shore can have pronounced gradients. Gradients in **abiotic factors** are found in almost every environment. They influence **habitats** and create microclimates, and are important in determining **community** patterns. This activity examines the physical gradients and microclimates that might be found on a rocky shore, where the community often shows clear zones of species distribution.

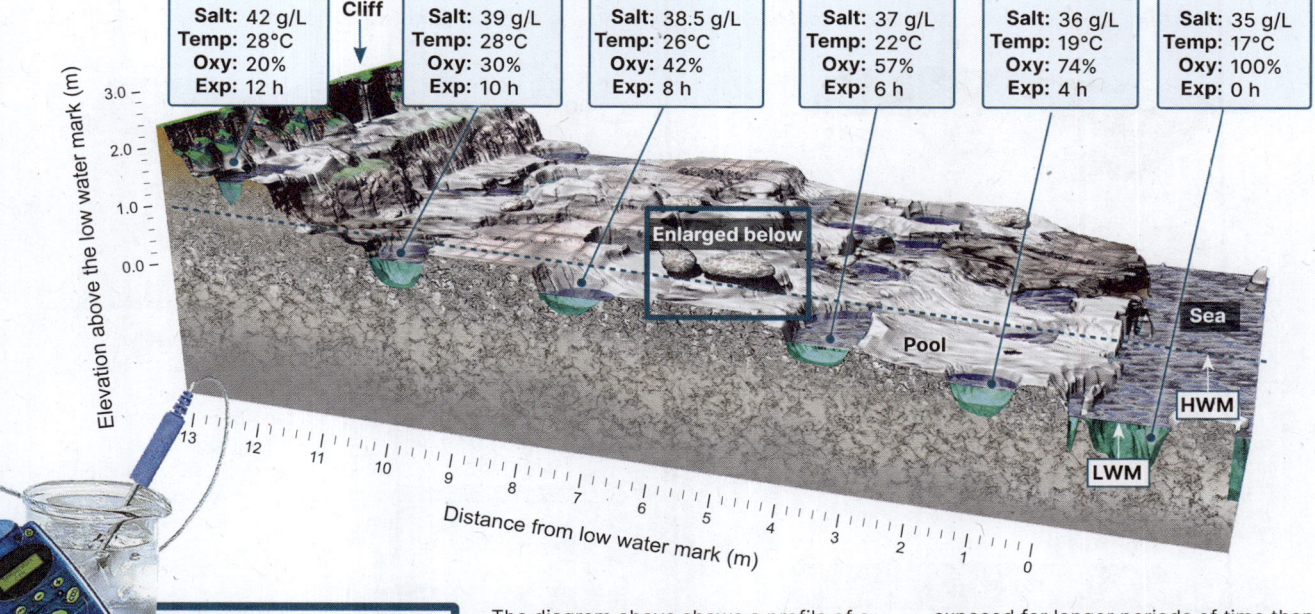

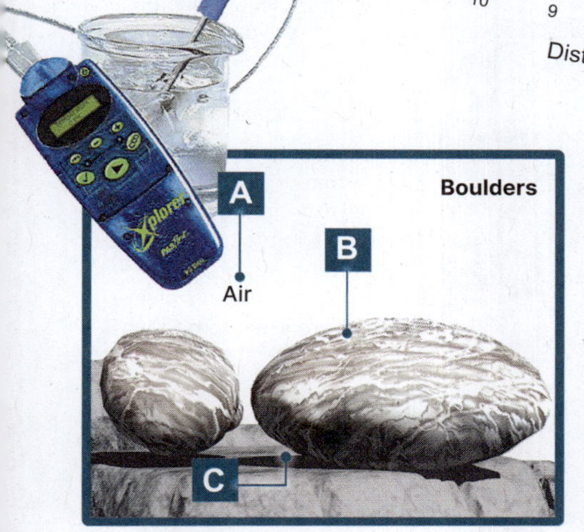

The diagram above shows a profile of a rock platform at low tide. The high water mark (**HWM**) shown is the average height of the spring tide. In reality, the high tide level varies with the phases of the moon. The low water mark (**LWM**) is an average level subject to the same variations. The rock pools vary in size, depth, and position on the platform. They are isolated at different elevations, trapping water from the ocean for time periods that may be relatively brief or up to 10-12 hours duration. Pools near the **HWM** are exposed for longer periods of time than those near the **LWM**. The difference in exposure times results in some of the physical factors exhibiting a gradient and the factor's value gradually changes over a horizontal and/or vertical distance. Physical factors sampled in the pools include salinity, or the amount of dissolved salts (g) per liter (**Salt**), temperature (**Temp**), dissolved oxygen compared to that of open ocean water (**Oxy**), and exposure, or the amount of time isolated from the ocean water (**Exp**).

1. Describe the environmental gradient (general trend) from the low water mark (**LWM**) to the high water mark (**HWM**) for:

    (a) Salinity: _____

    (b) Temperature: _____

    (c) Dissolved oxygen: _____

    (d) Exposure: _____

2. Rock pools above the normal high water mark (HWM), such as the uppermost pool in the diagram above, can have wide extremes of salinity. What abiotic conditions might cause these pools to have very high salinity?

    _____

    _____

3. (a) The inset diagram (above, left) is an enlarged view of two boulders on the rock platform. How might the abiotic factors listed below differ at each of the labelled points A, B, and C?

    Mechanical force of wave action: _____

    _____

    Surface temperature when exposed: _____

    _____

    (b) State the term given to these localized variations in physical conditions: _____

33 Physical Factors in a Small Lake

Key Idea: In lakes, temperature and light change with depth. Oxbow lakes are formed from old river meanders that have become isolated from the main channel following a change in the river's course. They are shallow (~2-9 m deep) but may be deep enough to develop temporary temperature gradients from top to bottom (below). Oxbows are commonly very productive and this can influence values for **abiotic factors** such as dissolved oxygen and light penetration, which can vary widely with depth and distance to the shore. Typical values for water temperature (**Temp**), dissolved oxygen (**Oxygen**), and light penetration as a percentage of the light striking the surface (**Light**) are indicated below.

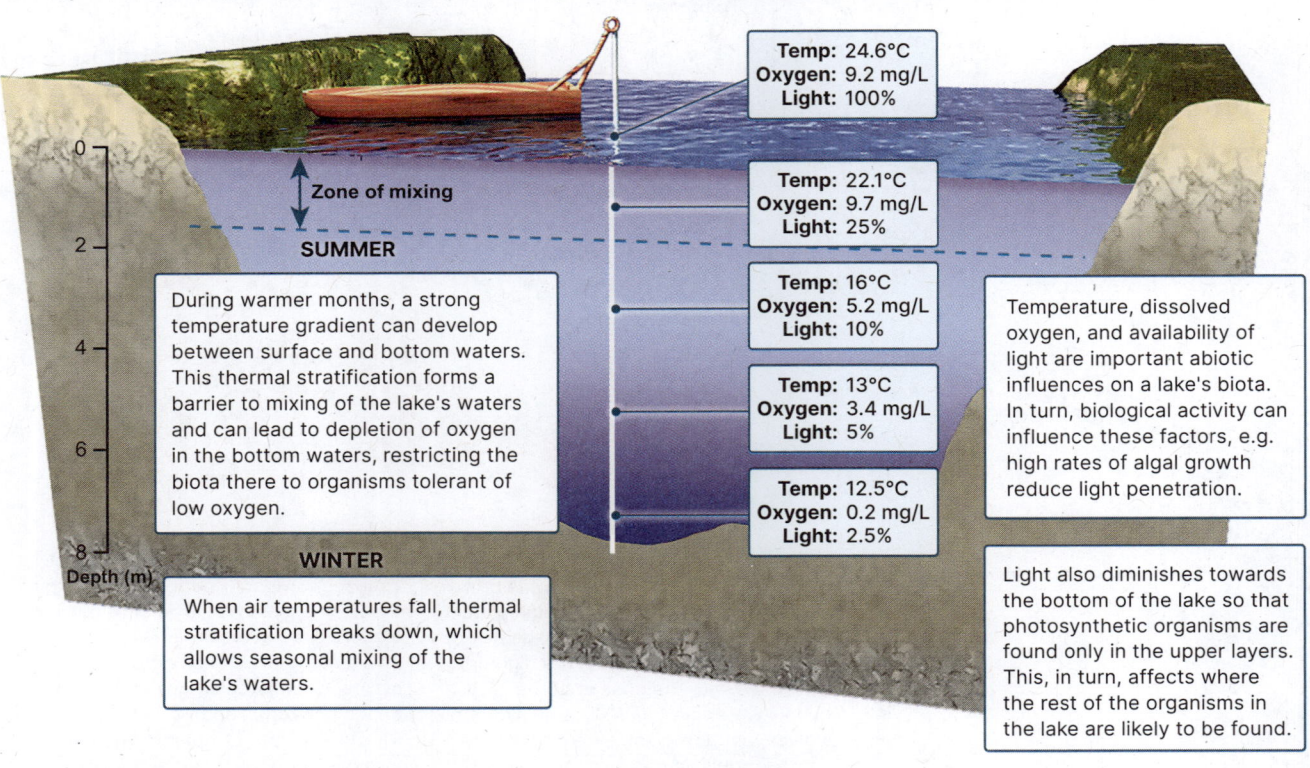

During warmer months, a strong temperature gradient can develop between surface and bottom waters. This thermal stratification forms a barrier to mixing of the lake's waters and can lead to depletion of oxygen in the bottom waters, restricting the biota there to organisms tolerant of low oxygen.

When air temperatures fall, thermal stratification breaks down, which allows seasonal mixing of the lake's waters.

Temperature, dissolved oxygen, and availability of light are important abiotic influences on a lake's biota. In turn, biological activity can influence these factors, e.g. high rates of algal growth reduce light penetration.

Light also diminishes towards the bottom of the lake so that photosynthetic organisms are found only in the upper layers. This, in turn, affects where the rest of the organisms in the lake are likely to be found.

1. With respect to the diagram above, describe the environmental gradient (general trend) from surface to lake bottom for:

    (a) Water temperature: _____

    (b) Dissolved oxygen: _____

    (c) Light penetration: _____

2. During the summer months, the warm surface waters are mixed by wind action. Deeper, cool waters are isolated from this surface water. This sudden change in the temperature profile is called a thermocline, which itself is a further barrier to the mixing of shallow and deeper water.

    (a) What is the effect of the thermocline on the dissolved oxygen at the bottom of the lake?

    _____

    (b) What causes the oxygen level to drop to very low levels near the lake bottom?

    _____

3. Many of these shallow lakes can undergo great changes in their salinity (sodium, magnesium, and calcium chlorides).

    (a) What event could reduce the salinity of a small lake relatively quickly? _____

    (b) What process would gradually increase the salinity of a small lake? _____

4. What is the general effect of physical gradients on the distribution of organisms in habitats?

    _____

34 Habitat

Key Idea: A population lives within a habitat. The tolerance to changes within the habitat, e.g. temperature, varies between individuals.

The environment in which a species population (or a individual organism) lives (including all the physical and **biotic factors**) is termed its **habitat**. Within a prescribed habitat, each species population has a range of tolerance to variations in its physical and chemical environment. Within the population, individuals will have slightly different tolerance ranges based on small differences in genetic make-up, age, and health. The wider an organism's tolerance range for a given abiotic factor, e.g. temperature or salinity, the more likely it is that the organism will be able to survive variations in that factor. Species dispersal is also strongly influenced by tolerance range. The wider the tolerance range of a species, the more widely dispersed the organism is likely to be. As well as a tolerance range, organisms have a narrower optimum range within which they function best. This may vary from one stage of an organism's development to another or from one season to another. Every species has its own optimum range. Organisms will usually be most abundant where the **abiotic factors** are closest to the optimum range.

Habitat occupation and tolerance range

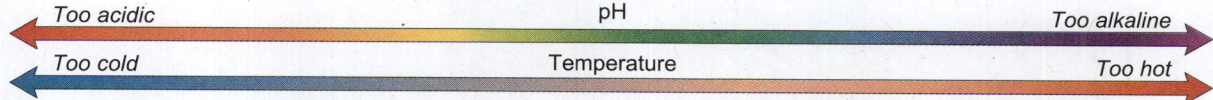

The law of tolerances states that *"for each abiotic factor, a species population (or organism) has a tolerance range within which it can survive. Toward the extremes of this range, that abiotic factor tends to limit the organism's ability to survive"*.

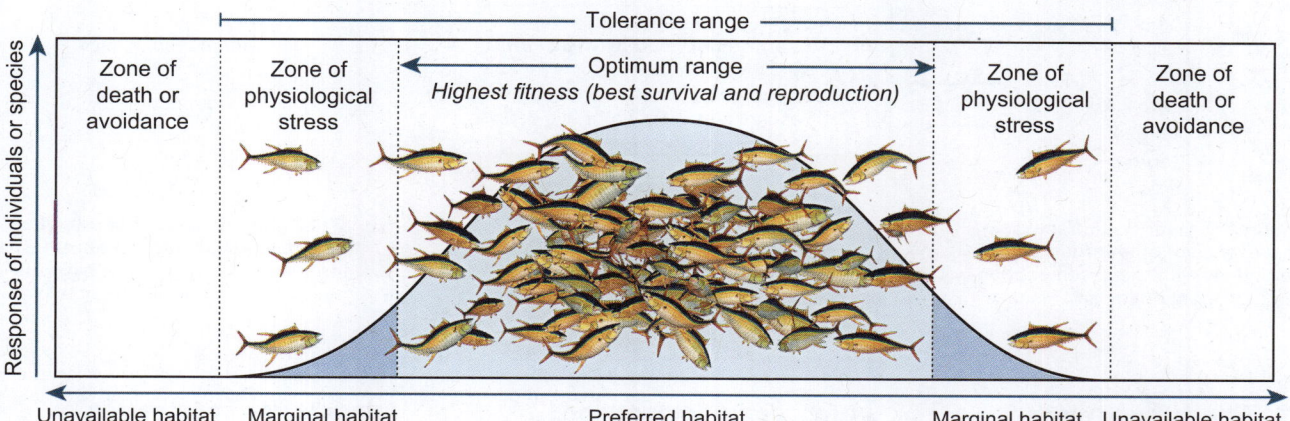

Some species, e.g. the zander, tolerate a wider range of temperatures than others, e.g. salmonids. Even within a species, tolerance can change seasonally as individuals acclimatize to slowly changing conditions, e.g. from winter to summer.

Tolerance for a particular environmental factor can be broad for some life stages and very narrow for others. Adult Atlantic blue crabs, for example, survive in a wide range of salinities, but eggs and larvae survive only in salinities above 23 ppt.

Temperature can affect species range. A mature tree may grow outside its natural range but its seedlings may not survive. The NZ kauri tree is commonly found in parks throughout NZ but below 37° S it is too cold for its seeds to naturally germinate.

1. Identify the range in the diagram above in which most of the species population is found. Explain why this is the case:

   _____

   _____

2. Describe some probable stresses on an organism forced into a marginal habitat:

   _____

   _____

   _____

   _____

35 Ecological Niche

Key Idea: An organism's niche describes its functional role within its environment.

The **ecological niche** describes the functional position of a species in its **ecosystem**, how it responds to the distribution of resources and how it, in turn, alters those resources for other species. The full range of environmental conditions (biological and physical) under which an organism can exist describes its fundamental niche. As a result of direct and indirect interactions with other organisms, species are usually forced to occupy a niche that is narrower than this. This is termed the realized niche. From the concept of the niche arose the idea that two species with the same niche requirements could not coexist because they would compete for the same resources and one would exclude the other. This is known as Gause's Competitive Exclusion Principle. If two species compete for some of the same resources, e.g. food, their resource use curves will overlap. Within the zone of overlap, competition will be intense.

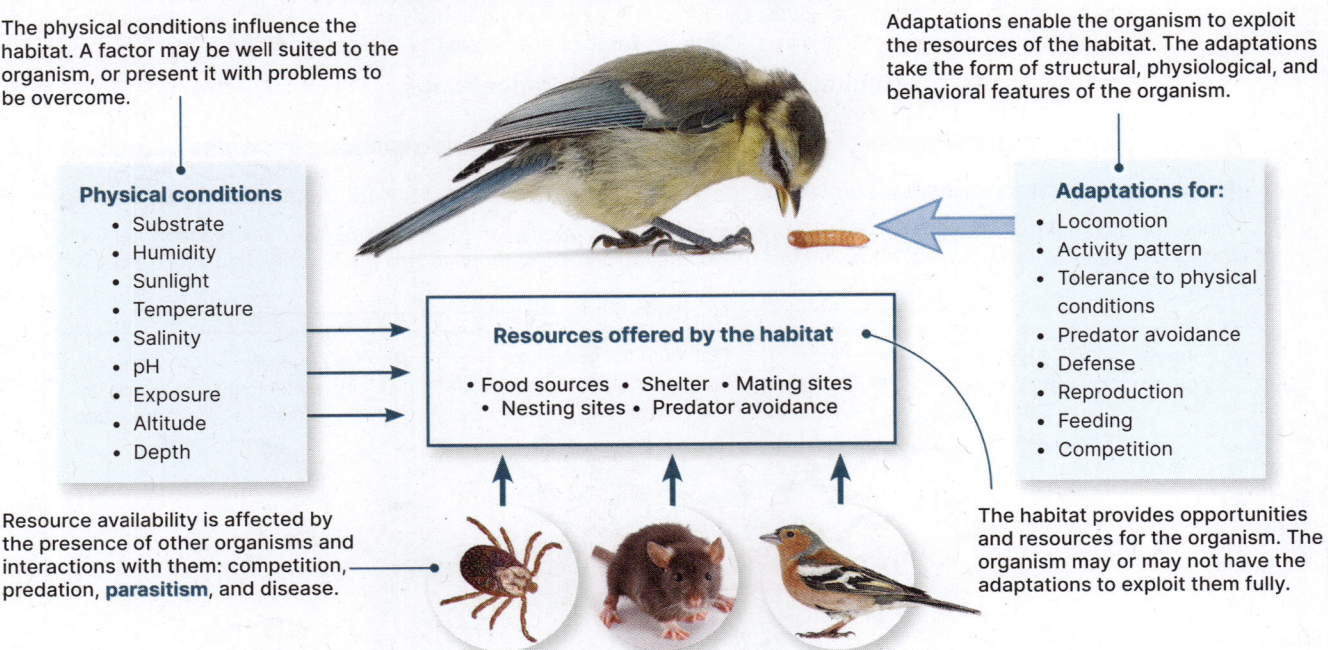

The physical conditions influence the habitat. A factor may be well suited to the organism, or present it with problems to be overcome.

Physical conditions
- Substrate
- Humidity
- Sunlight
- Temperature
- Salinity
- pH
- Exposure
- Altitude
- Depth

Resources offered by the habitat
- Food sources • Shelter • Mating sites
- Nesting sites • Predator avoidance

Adaptations enable the organism to exploit the resources of the habitat. The adaptations take the form of structural, physiological, and behavioral features of the organism.

Adaptations for:
- Locomotion
- Activity pattern
- Tolerance to physical conditions
- Predator avoidance
- Defense
- Reproduction
- Feeding
- Competition

Resource availability is affected by the presence of other organisms and interactions with them: competition, predation, **parasitism**, and disease.

The habitat provides opportunities and resources for the organism. The organism may or may not have the adaptations to exploit them fully.

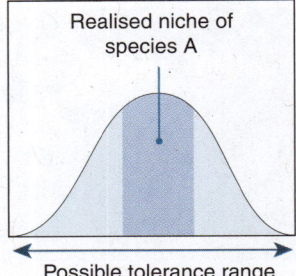

The realized niche

Realised niche of species A

Possible tolerance range

The tolerance range represents the fundamental niche a species could exploit. The actual or realized niche of a species is narrower than this because of competition with other species.

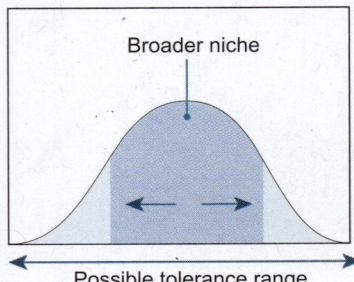

Intraspecific competition

Broader niche

Possible tolerance range

Competition is intense between individuals of the same species because they exploit the same resources. Individuals must exploit resources at the extremes of their tolerance range and the realized niche expands.

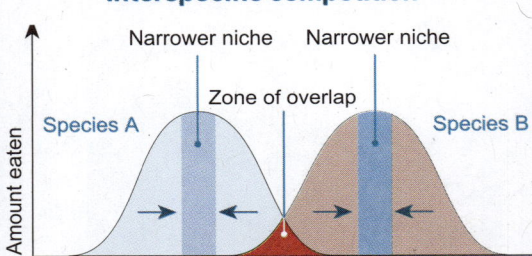

Interspecific competition

Narrower niche   Narrower niche
Zone of overlap
Species A   Species B
Amount eaten
Resource use as measured by food item size

If two (or more) species compete for some of the same resources, their resource use curves will overlap. Within the zone of overlap, resource competition will be intense and selection will favor niche specialization so that one or both species occupy a narrower niche.

1. Explain in what way the realized niche could be regarded as flexible: _____

2. Explain the contrasting effects of interspecific competition and intraspecific competition on niche breadth: _____

Competitive exclusion in barnacles

Seashores provide a wide range of **habitats** and opportunities. This results in a high diversity and abundance of organisms competing for the benefits of living there. Competition in species of barnacle, a common crustacean on rocky shores, was studied by J.H. Connell in Scotland. By removing one barnacle species and observing the effect on another, it was possible to determine the extent of the fundamental niches and compare them to the realized niches.

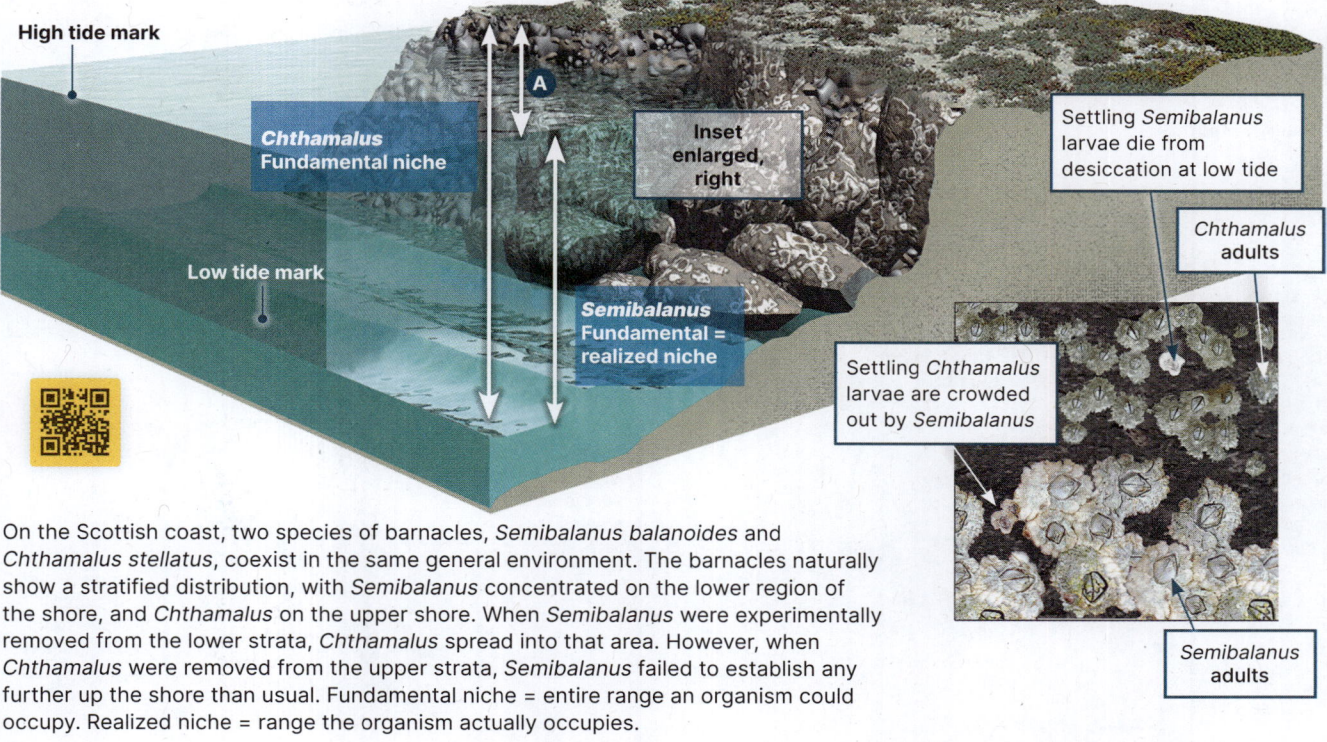

On the Scottish coast, two species of barnacles, *Semibalanus balanoides* and *Chthamalus stellatus*, coexist in the same general environment. The barnacles naturally show a stratified distribution, with *Semibalanus* concentrated on the lower region of the shore, and *Chthamalus* on the upper shore. When *Semibalanus* were experimentally removed from the lower strata, *Chthamalus* spread into that area. However, when *Chthamalus* were removed from the upper strata, *Semibalanus* failed to establish any further up the shore than usual. Fundamental niche = entire range an organism could occupy. Realized niche = range the organism actually occupies.

3. (a) In the example of the barnacles above, describe what is represented by the zone labelled with the arrow A:

   _____
   _____
   _____

   (b) Outline the evidence for the barnacle distribution being the result of competitive exclusion:

   _____
   _____
   _____

4. (a) What keeps *Semibalanus* larvae from establishing at higher shore levels? _____
   _____

   (b) What is the consequence of this to the realized niche compared to the fundamental niche of *Semibalanus*?

   _____
   _____
   _____

5. There are many studies underway about the effect global warming and rising sea levels might have on marine communities. What effect might a rise in sea level have on the *Chthamalus*/*Semibalanus* community?

   _____
   _____
   _____
   _____
   _____

36 Energy Inputs and Outputs

Key Idea: Energy is lost at each step of a food chain.
Within **ecosystems**, organisms are assigned to **trophic levels** based on the way in which they obtain their energy. **Producers** or autotrophs manufacture their own food from simple inorganic substances. Most producers utilize sunlight as their energy source for this, but some use simple chemicals. The **consumers** or heterotrophs (herbivores, carnivores, omnivores, **decomposers**, and **detritivores**), obtain their energy from other organisms. Energy flows through trophic levels rather inefficiently, with only 5-20% of usable energy being transferred to the subsequent level. Energy not used for metabolic processes is lost as heat.

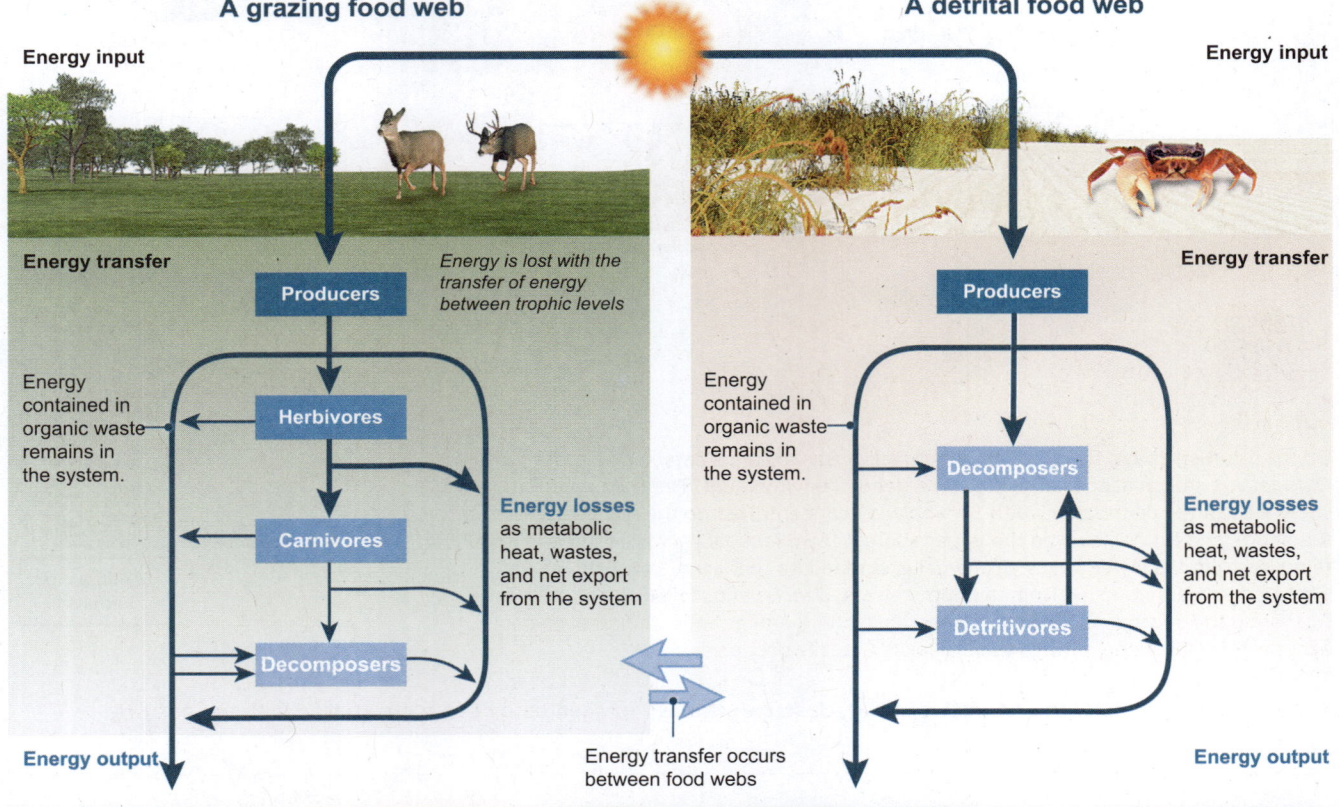

Producers (algae, green plants, and some bacteria) make their own food using simple inorganic carbon sources, e.g. CO_2. Sunlight is the most common energy source for this process.

Consumers (animals, non-photosynthetic protists, and some bacteria) rely on other living organisms or organic particulate matter for their energy and their source of carbon. First order consumers, such as aphids (left), feed on producers. Second (and higher) order consumers, such as ladybugs (center) eat other consumers. **Detritivores** consume (ingest and digest) detritus (decomposing organic material) from every trophic level. In doing so, they contribute to decomposition and the recycling of nutrients. Common detritivores include woodlice, millipedes (right), and many terrestrial worms.

Decomposers, e.g fungi, obtain their energy and carbon from the extracellular breakdown of dead organic matter (DOM). Decomposers play a central role in **nutrient cycling**.

1. Describe how energy is transferred through ecosystems: _____

2. With respect to energy flow, describe a major difference between a detrital and a grazing food web: _____

37 Plants as Producers

Key Idea: Photosynthesis by producers (photoautotrophs) directly and indirectly sustains almost all heterotrophic life. Plants, algae, and some bacteria are **producers** (they make their own food). They capture light energy and convert it into sugars through a process called **photosynthesis**. The chemical energy stored in this food fuels the reactions that sustain life. Heterotrophs rely on producers directly or indirectly for their energy. The photosynthesis that occurs in the oceans is vital to the Earth's functioning, providing oxygen and absorbing carbon dioxide.

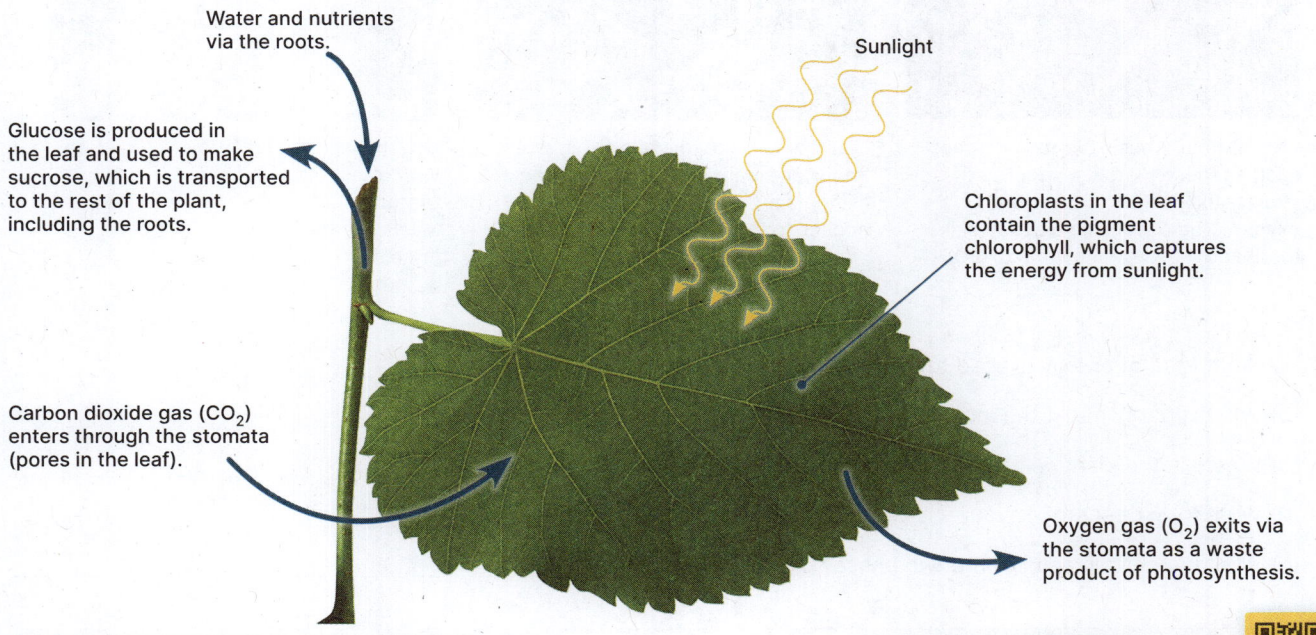

- Water and nutrients via the roots.
- Glucose is produced in the leaf and used to make sucrose, which is transported to the rest of the plant, including the roots.
- Carbon dioxide gas (CO$_2$) enters through the stomata (pores in the leaf).
- Sunlight
- Chloroplasts in the leaf contain the pigment chlorophyll, which captures the energy from sunlight.
- Oxygen gas (O$_2$) exits via the stomata as a waste product of photosynthesis.

Requirements for photosynthesis

Plants need only a few raw materials to make their own food:

- Light energy from the sun
- Chlorophyll absorbs light energy
- CO$_2$ gas is reduced to carbohydrate
- Water is split to provide the electrons for the fixation of carbon as carbohydrate

Photosynthesis is not a single process but two complex processes (the light dependent and light independent reactions), each with multiple steps.

Production of carbohydrate (light independent reactions) occurs in the fluid stroma of chloroplast. This is commonly called carbon 'fixation'.

Energy capture occurs in the inner membranes of the chloroplast and requires sunlight (light dependent phase).

Photosynthesis equation	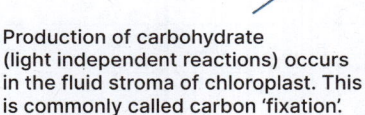

$$6CO_2 + 12H_2O \xrightarrow[\text{Chlorophyll}]{\text{Light}} C_6H_{12}O_6 + 6O_2 + 6H_2O$$

1. Explain why photosynthesis is so important for life on Earth: _____

2. Explain why plant leaves appear green: _____

3. (a) Identify the raw materials (inputs) for photosynthesis: _____

   (b) Identify the products (outputs) of photosynthesis: _____

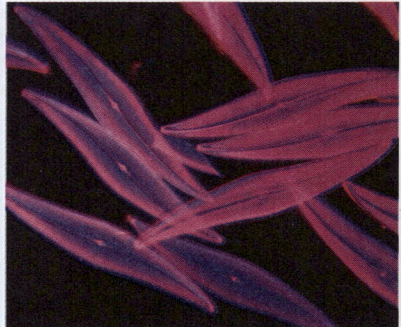

The photosynthesis of marine algae, such as these diatoms, supplies a substantial portion of the world's oxygen. The oceans also act as sinks for absorbing large amounts of CO_2.

Macroalgae, like this giant kelp, are important marine producers. Algae living near the ocean surface get access to light for photosynthesis.

On land, vascular plants, such as trees with transport vessels, are the main producers of food. Plants at different levels in a forest receive different intensities and quality of light.

4. (a) What form of energy is used to drive photosynthesis? _____

   (b) What is the name of the molecule that captures this energy? _____

   (c) Where in the plant cell does photosynthesis take place? _____

   (d) What form of energy is your answer to (a) converted into? _____

   (e) What happens in each of the two phases of photosynthesis? _____
   _____

5. (a) Primary production is the production of carbon compounds from carbon dioxide, generally by photosynthesis. Study the graph (right) showing primary production in the oceans. Describe what the graph is showing:

   _____
   _____
   _____

   (b) Explain the shape of the curves described in (a): _____
   _____
   _____
   _____

   (c) About 90% of all marine life lives in the photic zone (the depth to which light penetrates). Suggest why this is so:

   _____
   _____
   _____
   _____

   Ocean primary production

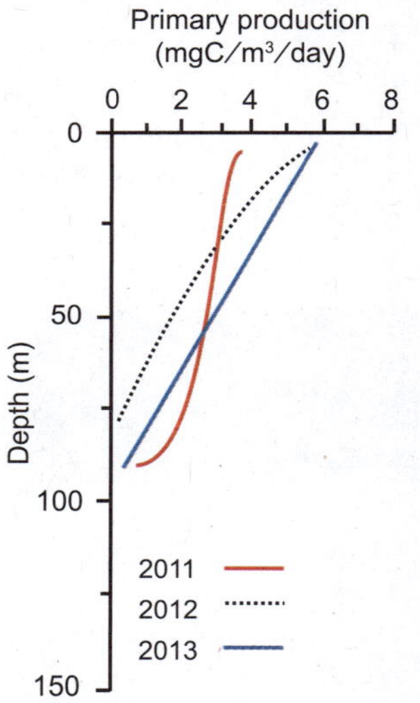

   Primary production (mgC/m³/day)

   2011 —
   2012 ·····
   2013 —

6. Where do marine phytoplankton get their carbon dioxide from? _____
   _____

7. (a) How does the light in a forest change from the canopy to the forest floor?
   _____

   (b) How might this affect how plants can photosynthesize and how might they overcome this?
   _____
   _____

38 Measuring Primary Productivity

Key Idea: We can use different techniques to estimate the gross and net primary productivity of an ecosystem.

Estimating **ecosystem** productivity is very problematic because it is technically difficult to measure the rates of **photosynthesis** and **cellular respiration** directly. In terrestrial systems, plant biomass (dry weight) gives a good indication of the difference between Gross Primary Productivity and respiration. In aquatic systems, productivity is often estimated indirectly from the quantities of oxygen released or carbon dioxide used in production (the light and dark bottle method). This method is useful but cannot account for the respiration by bacteria and so may underestimate productivity.

Leaf area index (LAI)
Leaf area measures the upper leaf surface area of a canopy to the ground area. It can be used to measure primary productivity.

Measuring productivity
Measuring gross primary productivity (GPP) can be difficult due to the effect of ongoing respiration, which uses up some of the organic material produced (glucose). One method for measuring GPP is to measure the difference in production between plants kept in the dark and those in the light. A simple method for measuring GPP in phytoplankton is illustrated below.

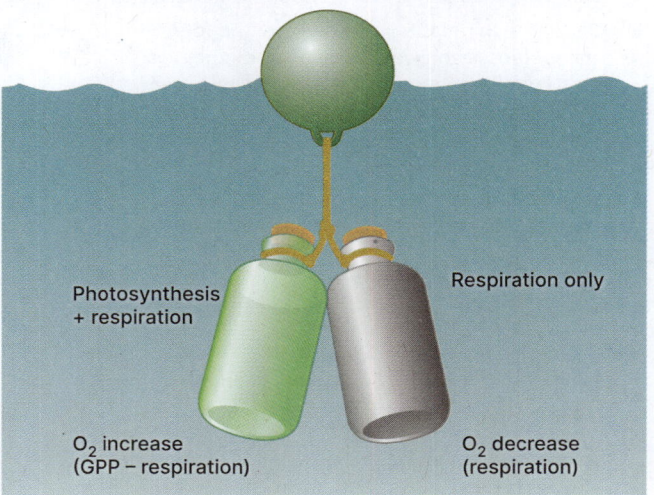

Relative growth rate (R)
Relative growth rate is the gain in mass of plant tissue per unit time.

$$R = \frac{\text{Increase in dry mass in unit time}}{\text{Original dry mass of the plant}}$$

Harvestable dry biomass
Used for commercial purposes, it is the dry mass of crop available for sale or use.

Net assimilation rate (NAR)
NAR is the increase in plant weight per unit of leaf area per unit time. Essentially, it is the balance between carbon gain from photosynthesis and carbon loss from respiration.

$$NAR = \frac{\text{Increase in dry mass in unit time}}{\text{Leaf area}}$$

Two bottles are lowered into the ocean or lake to a specified depth, filled with water, and stoppered. One bottle is transparent, the other is opaque. The O_2 concentration of the water around the bottles is measured and the bottles are left for a specified amount of time. The phytoplankton in the transparent bottle will photosynthesize, increasing the O_2 concentration, and respire, using some of that O_2. The phytoplankton in the opaque bottle will only respire. The final, measured difference in O_2 between the bottles gives the amount of O_2 produced by the phytoplankton in the specified time (including that used for respiration). The amount of O_2 used allows the amount of glucose produced to be determined, and therefore the GPP of the phytoplankton.

1. Why can measuring GPP be difficult? _____
_____
_____

2. An experiment was carried out to measure the gross primary production of a lake. The lake was initially measured to have 8 mg O_2/L. A clear flask and an opaque flask were lowered into the lake, filled and stoppered. When the flasks were retrieved it was found the clear flask contained 10 mg O_2/L while the opaque contained 5 mg O_2/L.

    (a) How much O_2 was used (respired) in the opaque flask? _____

    (b) What is the net O_2 production in the clear flask? _____

    (c) What is the gross O_2 production in the system? _____

3. Why do productivity measurements use dry biomass? _____
_____
_____

39 Cellular Respiration

Key Idea: Cellular respiration is the process by which the energy in glucose is transferred to ATP.

Cellular respiration is the process by which organisms break down energy rich molecules, e.g. glucose, and generate the energy transfer molecule adenosine triphosphate (ATP). All living cells respire in order to exist, although the substrates they use may vary. Aerobic respiration requires oxygen. Forms of cellular respiration that do not require oxygen are said to be anaerobic. Some plants and animals can generate ATP anaerobically for short periods of time. Other organisms use only anaerobic respiration and live in oxygen-free environments. For these organisms, there is some other final electron acceptor other than oxygen, e.g. nitrate or Fe^{2+}. Respiration plays a part in the cycling of carbon by adding it to oxygen and releasing it to the atmosphere as carbon dioxide.

▸ Cellular respiration is the process of extracting the energy stored in the chemical bonds in glucose and storing it in ATP molecules. The process includes many chemical reactions, some of which produce ATP molecules and some that prepare molecules for further chemical reactions.

▸ Cellular respiration can be divided into four major steps, each with its own set of chemical reactions. The four steps are: **glycolysis**, the **link reaction**, the **Krebs cycle**, and the **electron transport chain** (ETC). Every step, except the link reaction, produces ATP.

▸ Glycolysis occurs in the cytoplasm; the other steps take place within the mitochondrion.

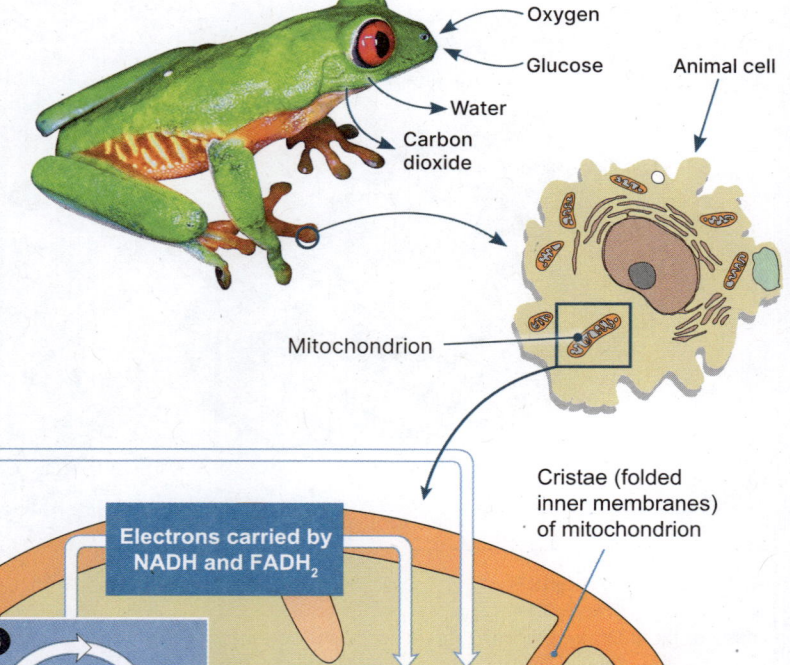

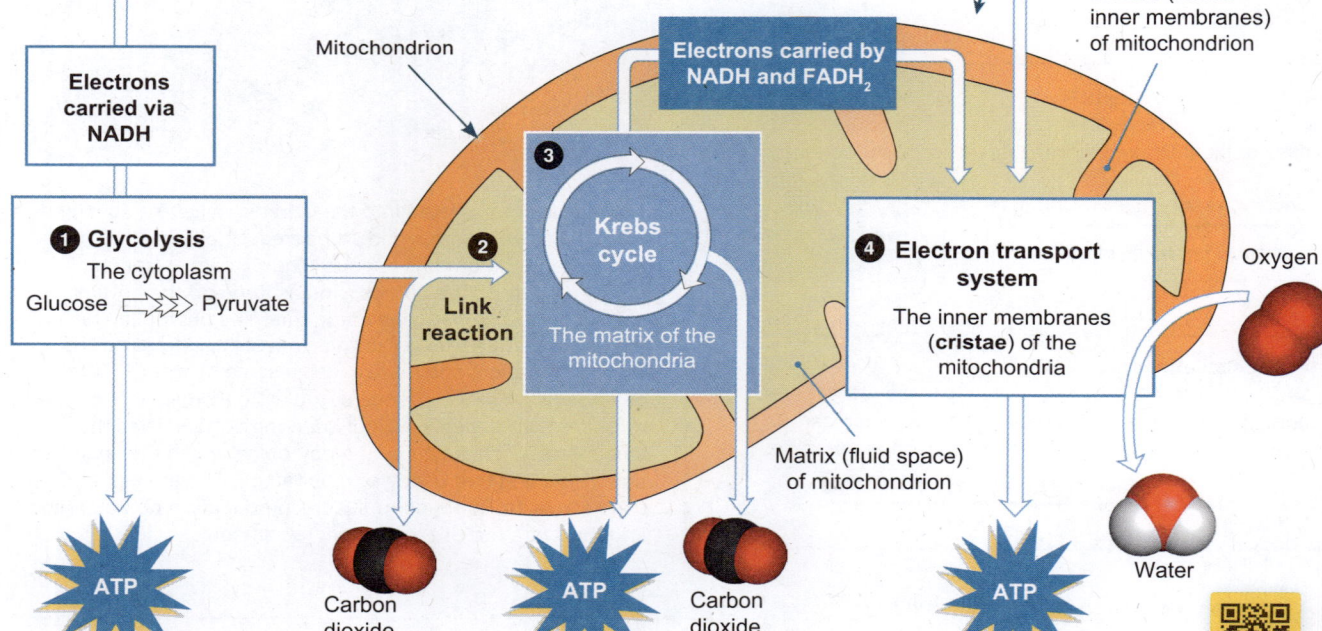

1. Which of the four steps in cellular respiration yield ATP? _____

2. (a) What are the main reactants (inputs) for cellular respiration? _____

   (b) What are the main products (outputs) of cellular respiration? _____

3. Which of the four steps of cellular respiration occur in the mitochondria? _____

4. What is the importance of ATP? _____

40 Food Chains

Key Idea: Every ecosystem has a trophic structure: a hierarchy of feeding relationships that determines the pathways for energy flow and nutrient cycling.

In an **ecosystem**, organisms are assigned to **trophic levels** based on the way in which they obtain their energy. **Producers** (autotrophs) manufacture their own food from simple, inorganic substances. This producer level ultimately supports all other levels. The **consumers** or heterotrophs, obtain their energy from other organisms. Consumers are ranked according to the trophic level they occupy, i.e. first order (primary), second order (secondary), and third order (tertiary). The sequence of organisms, each of which is a source of food for the next, is called a **food chain**. Energy flows through trophic levels rather inefficiently, with only 5-20% of usable energy being transferred to the subsequent level. For this reason, food chains seldom have more than six links. Those organisms whose food is obtained through the same number of links belong to the same trophic level. Note that some consumers may feed at several different trophic levels, and many primary consumers eat many plant species. The different food chains in an ecosystem therefore tend to form complex webs of interactions (**food webs**).

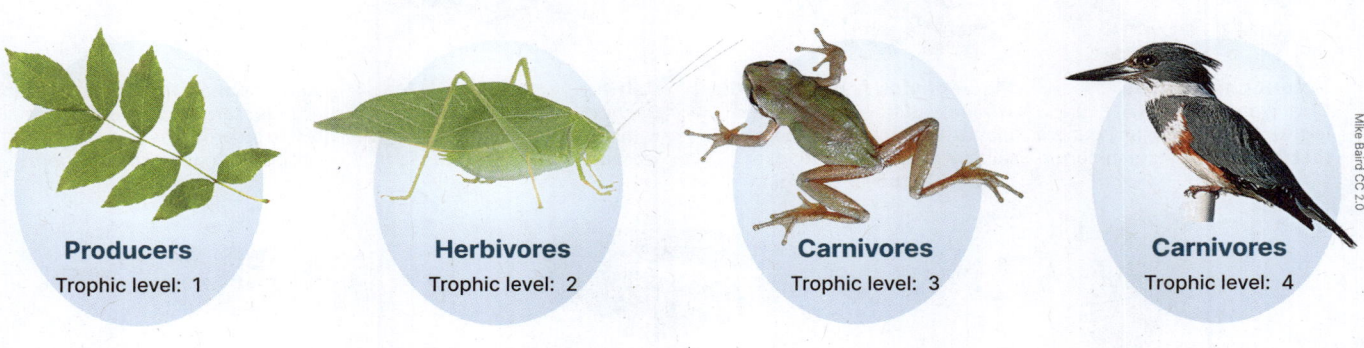

The diagram above represents the basic elements of a food chain. In the questions below, you are asked to add to the diagram the features that indicate the flow of energy through the community of organisms.

1. (a) What is the original energy source for this food chain? _____

   (b) Draw arrows on the diagram above to show how the energy flows through the organisms in the food chain. Label each arrow with the process involved in the energy transfer. Draw arrows to show how energy is lost by respiration.

2. Explain why the amount of energy available reduces at each successive trophic level in a food chain:
   _____

3. Discuss the trophic structure of ecosystems, including reference to food chains and trophic levels:
   _____
   _____
   _____
   _____
   _____
   _____

4. Explain what you could infer about the trophic level(s) of the kingfisher if it was found to eat both katydids and frogs:
   _____
   _____

41 Food Webs

Key Idea: Food chains intersect to form food webs. Food webs can be complex, with many interwoven parts.

The organisms that inhabit lakes will vary from one location to the next. The organisms illustrated below are typical of those found in lakes. For the sake of simplicity, only fourteen organisms are represented here. Real lake communities may have hundreds of different species interacting together. Your task is to assemble the organisms below into a **food web** in a way that illustrates their position in **trophic levels** and their relative positions in **food chains**. Detritus is the accumulated debris of dead organisms from within the lake and washed in from the surrounding lake margins and streams. It contains the remains of land plants (such as leaves), algae, zooplankton, and larger animals in various stages of decay. Detritus forms a layer at the bottom of the lake that provides a rich food source for any animal that can exploit it.

Feeding requirements of lake organisms

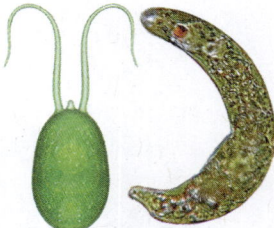

Autotrophic protists
Chlamydomonas (left), *Euglena* (right) Two of the many genera that form the phytoplankton.

Macrophytes (various species)
A variety of flowering aquatic plants are adapted for being submerged, free-floating, or growing at the lake margin.

Detritus
Decaying organic matter from within the lake itself or it may be washed in from the lake margins.

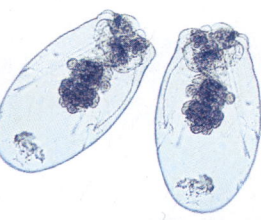

Asplanchna (planktonic rotifer)
A large, carnivorous rotifer that feeds on protozoa and young zooplankton (e.g. small *Daphnia*).

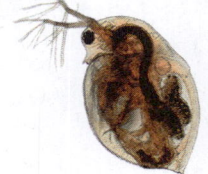

Daphnia
Small freshwater crustacean that forms part of the zooplankton. It feeds on planktonic algae by filtering them from the water with its limbs.

Leech (*Glossiphonia*)
Leeches are fluid feeding predators of smaller invertebrates, including rotifers, small pond snails and worms.

Three-spined stickleback (*Gasterosteus*)
A common fish of freshwater ponds and lakes. It feeds mainly on small invertebrates such as *Daphnia* and insect larvae.

Diving beetle (*Dytiscus*)
Diving beetles feed on aquatic insect larvae and adult insects blown into the lake community. The will also eat organic detritus collected from the bottom mud.

Carp (*Cyprinus*)
A heavy bodied freshwater fish that feeds mainly on bottom living insect larvae and snails, but will also take some plant material (not algae).

Dragonfly larva
Large aquatic insect larvae that are voracious predators of small invertebrates including *Hydra*, *Daphnia*, other insect larvae, and leeches.

Great pond snail (*Limnaea*)
Omnivorous pond snail, eating both plant and animal material, living or dead, although the main diet is aquatic macrophytes.

Herbivorous water beetles (e.g. *Hydrophilus*)
Feed on water plants, although the young beetle larvae are carnivorous, feeding primarily on small pond snails.

Protozan (e.g. *Paramecium*)
Ciliated protozoa such as *Paramecium* feed primarily on bacteria and microscopic green algae such as *Chlamydomonas*.

Pike (*Esox lucius*)
A top ambush predator of all smaller fish and amphibians. They are also opportunistic predators of rodents and small birds that end up in the water.

Mosquito larva (*Culex* spp.)
The larvae of most mosquito species feed on planktonic algae and small protozoans before passing through a pupal stage and undergoing metamorphosis into adult mosquitoes.

Hydra
A small carnivorous cnidarian that captures small prey items, e.g. small *Daphnia* and insect larvae, using its stinging cells on the tentacles.

©2025 **BIOZONE** International
ISBN: 978-1-99-101409-2
Photocopying prohibited

1. From the information provided for the lake food web components on the previous page, construct ten different food chains (using their names only) to show the feeding relationships between the organisms. Some food chains may be shorter than others and most species will appear in more than one food chain. An example has been completed for you.

Example 1: Macrophyte → Herbivorous water beetle → Three-spined stickleback → Pike

(a) _____
(b) _____
(c) _____
(d) _____
(e) _____
(f) _____
(g) _____
(h) _____
(i) _____
(j) _____

2. Use the food chains that you have created above to help you to draw up a complete food web for this community. Use only the supplied information to draw arrows showing the flow of energy between species. (NOTE: Only energy from (not to) the detritus is required)

   Label each species with the following codes to indicate its trophic level and status: Indicate:
   • Diet type: P = **Producer**, H = **Herbivore**, C = **Carnivore**, O = **Omnivore** (Note: based on the information given).
   • Position in the food chain as a consumer (1st, 2nd, 3rd, 4th order consumer): 1–4 (does not include producers).

   Example: Mosquito larva is C2

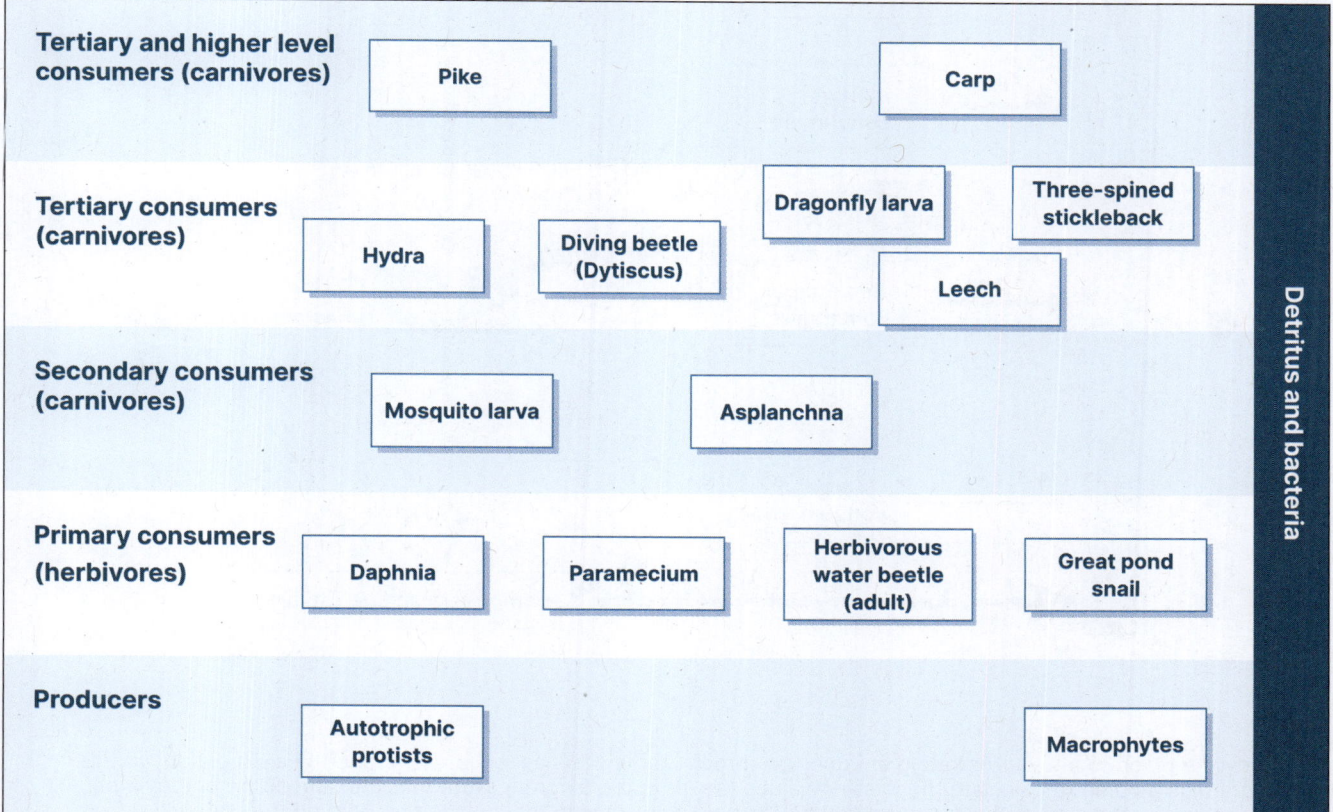

42 Energy Flow in an Ecosystem

Key Idea: Most of the energy entering an ecosystem is lost in energy transfers between trophic levels.

Energy cannot be created or destroyed. It can only be transformed from one form, e.g. light energy, to another, e.g. chemical energy in the bonds of molecules. This means that the flow of energy through an **ecosystem** can be measured. Each time energy is transferred from one **trophic level** to the next, e.g. by eating, some energy is given out as heat to the environment as a result of **cellular respiration**. Living organisms cannot convert heat to other forms of energy, so this heat is effectively lost from the system and the amount of energy available to one trophic level is always less than the amount at the previous level. Potentially, we account for the transfer of energy from its input (as solar radiation) to its release as heat from organisms because energy is conserved. The percentage of energy transferred from one trophic level to the next is the trophic efficiency. It varies between 5% and 20% and measures the efficiency of energy transfer. An average figure of 10% trophic efficiency is often used. This is called the ten percent rule.

Energy flow through an ecosystem

NOTE: Numbers represent **kilojoules** of energy per square meter per year ($kJ/m^2/yr$)

The energy available to each trophic level will always equal the amount entering that trophic level, minus total losses to that level (due to metabolic activity, death, excretion etc). Energy lost as heat will be lost from the ecosystem. Other losses become part of the detritus and may be used by other organisms in the ecosystem.

- Sunlight falling on plant surfaces: 7,000,000
- Light absorbed by plants: 1,700,000
- Producers: 87,400
- Primary consumers
- Secondary consumers
- Tertiary consumers
- Detritus
- Decomposers and detritivores

Arrows and values:
- 50,450 (heat loss from producers)
- 7,800 (heat loss from primary consumers)
- 1,600 (primary → secondary consumers)
- 1,330 (heat loss from secondary consumers)
- 90 (secondary → tertiary consumers)
- 55 (heat loss from tertiary consumers)
- 4,600 (primary consumers → detritus)
- 22,950 (producers → detritus)
- 2,000
- 10,465
- 19,300 (detritus → decomposers)
- 19,200 (decomposers heat loss)

1. Study the diagram above illustrating energy flow through a hypothetical ecosystem. Use the example above this diagram as a guide to calculate the missing values (a)–(d) in the diagram. Note that the sum of the energy inputs always equals the sum of the energy outputs. Place your answers in the spaces provided on the diagram.

©2025 BIOZONE International
ISBN: 978-1-99-101409-2
Photocopying prohibited

2. What is the original source of energy for this ecosystem? _____

3. Identify the processes occurring at the points labelled A – G on the diagram:

   A. _____ E. _____

   B. _____ F. _____

   C. _____ G. _____

   D. _____

4. (a) Calculate the percentage of light energy falling on the plants that is absorbed at point **A**:

   Light absorbed by plants ÷ sunlight falling on plant surfaces x 100 = _____

   (b) What happens to the light energy that is not absorbed? _____
   _____

5. (a) Calculate the percentage of light energy absorbed that is actually converted (fixed) into producer energy:

   Producers ÷ light absorbed by plants x 100 = _____

   (b) How much light energy is absorbed but not fixed: _____

   (c) Account for the difference between the amount of energy absorbed and the amount actually fixed by producers:
   _____
   _____
   _____

6. Of the total amount of energy fixed by producers in this ecosystem (at point **A**), calculate:

   (a) The total amount that ended up as metabolic waste heat (in kJ): _____

   (b) The percentage of the energy fixed that ended up as waste heat: _____

7. (a) State the groups for which detritus is an energy source: _____

   (b) How could detritus be removed or added to an ecosystem? _____
   _____

8. Under certain conditions, decomposition rates can be very low or even zero, allowing detritus to accumulate:

   (a) From your knowledge of biological processes, what conditions might slow decomposition rates?
   _____

   (b) What are the consequences of this lack of decomposer activity to the energy flow? _____
   _____

   (c) Add an additional arrow to the diagram on the previous page to illustrate your answer.

   (d) Describe three examples of materials that have resulted from a lack of decomposer activity on detrital material:
   _____
   _____
   _____

9. The ten percent rule states that the total energy content of a trophic level in an ecosystem is only about one-tenth (or 10%) that of the preceding level. For each of the trophic levels in the diagram on the previous page, determine the amount of energy passed on to the next trophic level as a percentage:

   (a) Producer to primary consumer: _____

   (b) Primary consumer to secondary consumer: _____

   (c) Secondary consumer to tertiary consumer: _____

   (d) Which of these transfers is the most efficient? _____

43 Production and Trophic Efficiency

Key Idea: Primary production measures the amount of material produced by plants. Productivity is expressed as a rate.

What do we mean when we say that an **ecosystem** has a high (or low) productivity? The energy accumulated by plants or other **producers** in an ecosystem (or measured area) is called primary production. It is the first energy storage step in an ecosystem. All of the sunlight energy that is fixed as chemical energy is the gross primary production (GPP). However, some of this energy is required by the producers themselves for **respiration**. Subtracting respiration from GPP gives the net primary production (NPP). This represents the energy or biomass available to the primary **consumers** in the ecosystem. Note that 'production' refers to a quantity of material. Productivity, which is more meaningful in biological systems is a rate, usually expressed as grams or kJ per m^2 (or per m^3) per year. Having made that distinction, you will often see the terms used interchangeably because production values are usually given for a set time period.

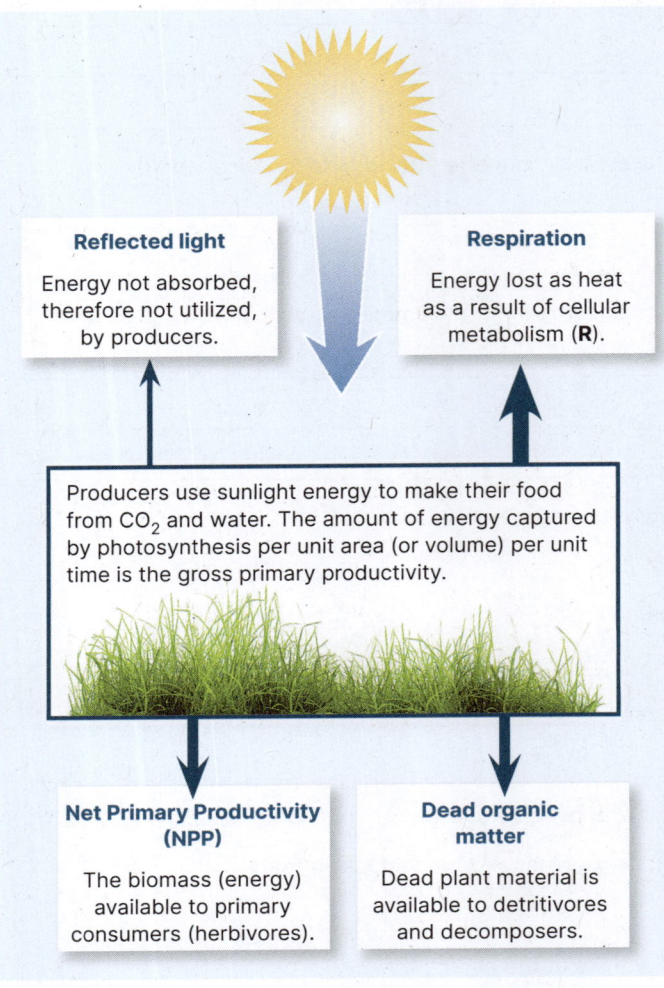

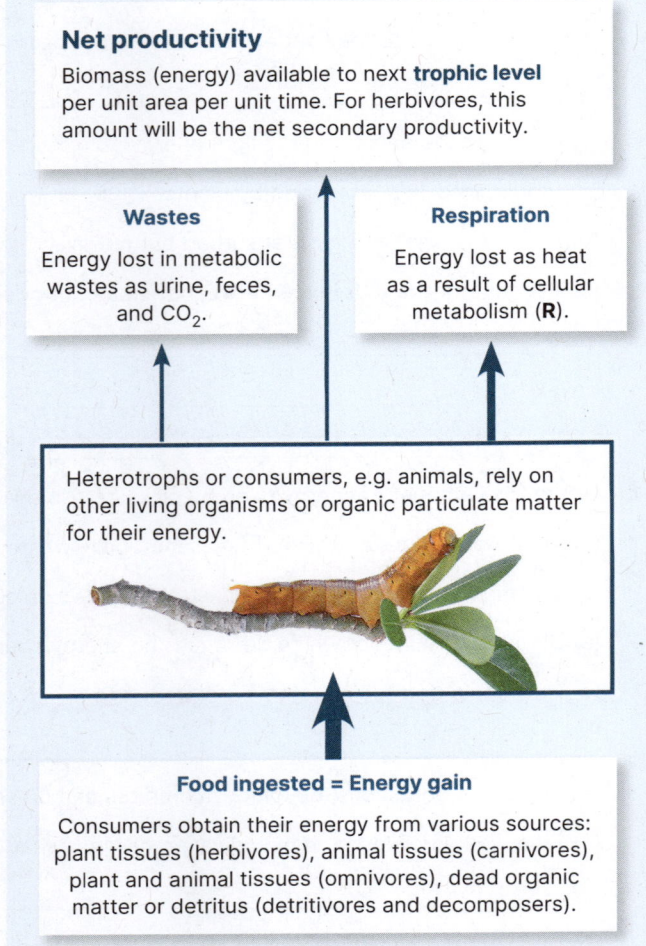

1. (a) Explain the difference between gross primary productivity (GPP) and net primary productivity (NPP):
   _____
   _____

   (b) Write a simple word equation to show how NPP is derived from GPP: _____
   _____

   (c) What factors do you think could influence the GPP of ecosystems? _____
   _____
   _____

   (d) Why do you think it is important to distinguish between GPP and NPP when studying the productivity of ecosystems?
   _____
   _____

Measuring productivity

The gross primary productivity of an ecosystem will depend on the capacity of the producers to capture and fix carbon in organic compounds. In most ecosystems, this is limited by constraints on photosynthesis (availability of light, nutrients, or water for example). The net primary productivity is then determined by how much of this goes into plant biomass per unit time, after respiratory needs are met. This will be the amount available to the next trophic level. It is difficult to measure productivity, but it is often estimated from the harvestable dry biomass or standing crop (the net primary production).

Agriculture and productivity

Increasing net productivity in agriculture (increasing yield) is a matter of manipulating and maximizing energy flow through a reduced number of trophic levels. On a farm, the simplest way to increase the net primary productivity is to produce a monoculture. Monocultures reduce competition between the desirable crop and weed species, allowing crops to put more energy into biomass. Other agricultural practices designed to increase productivity in crops include pest (herbivore) control and spraying to reduce disease. Higher productivity in feed-crops also allows greater secondary productivity, e.g. in livestock. Here, similar agricultural practices make sure the energy from feed-crops is efficiently assimilated by livestock.

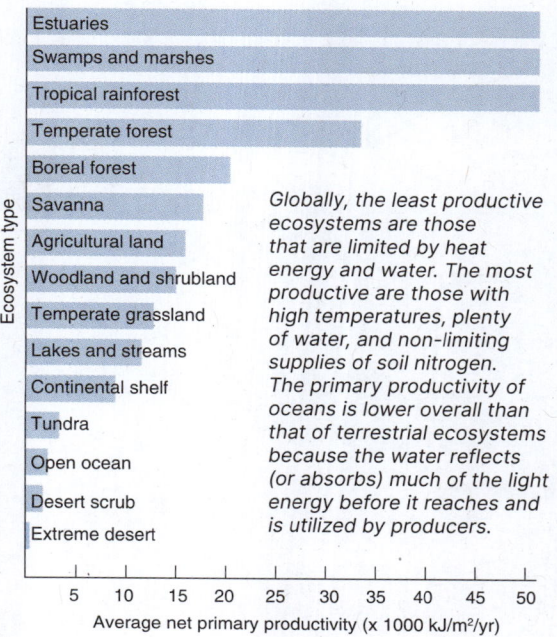

Globally, the least productive ecosystems are those that are limited by heat energy and water. The most productive are those with high temperatures, plenty of water, and non-limiting supplies of soil nitrogen. The primary productivity of oceans is lower overall than that of terrestrial ecosystems because the water reflects (or absorbs) much of the light energy before it reaches and is utilized by producers.

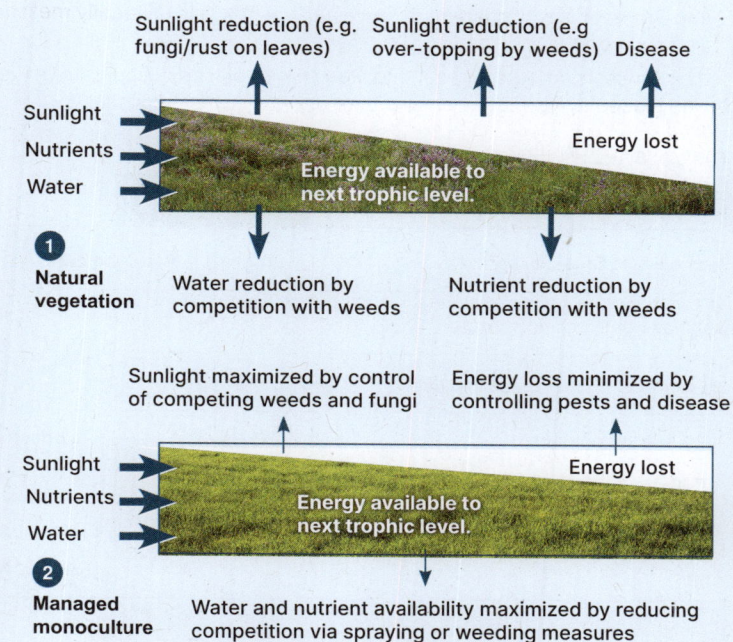

2. With reference to the bar graph above:

   (a) Suggest why tropical rainforests are among the most productive terrestrial ecosystems, while tundra and desert ecosystems are among the least productive:

   _____
   _____

   (b) Suggest why, amongst aquatic ecosystems, the NPP of the open ocean is low relative to that of coastal systems:

   _____
   _____
   _____

3. How could a farmer maximize the net primary productivity of a particular crop?

   _____
   _____
   _____

4. Explain the contrasting net primary productivities of intensive agricultural land and extreme desert:

   _____
   _____
   _____
   _____

Investigating secondary productivity

Consumers eat producers and use the energy and material to generate their own biomass. The efficiency of this process can be investigated by measuring the biomass of the producer and consumer. In the experiment below, students determined the net secondary production and respiratory losses using 12 day old cabbage white butterfly larvae feeding on Brussels sprouts. Of the NPP from the Brussels sprouts that is consumed by the larvae, some will be used in cellular respiration, some will be available to secondary consumers (the net secondary production), and some will be lost as egested waste products called frass.

The method

- The wet mass of ten, 12 day old larvae, and approximately 30 g Brussels sprouts was accurately measured and recorded.
- The larvae and Brussels sprouts were placed into an aerated container. After three days, the container was disassembled and the wet mass of the Brussels sprouts, larvae, and waste (frass) was individually measured and recorded.
- The Brussels sprouts, larvae, and waste were placed in separate containers and placed in a drying oven and their dry mass was recorded.

Cabbage white caterpillar (larva)

Note: We assume the proportion of biomass of Brussels sprouts and caterpillars on day 1 is the same as the calculated value from day 3.

Table 1: Brussels sprouts

	Day 1	Day 3	
Wet mass of Brussels sprouts	30 g	11 g	g consumed =
Dry mass of Brussels sprouts	–	2.2 g	
Plant proportion biomass (dry/wet)			
Plant energy consumed (wet mass x proportion biomass x 18.2 kJ)			kJ consumed per 10 larvae =
Plant energy consumed ÷ no. of larvae			kJ consumed per larva (E) =

Table 3: Frass

	Day 3
Dry mass frass from 10 larvae	0.5 g
Frass energy (waste) = frass dry mass x 19.87 kJ	
Energy from frass from 1 larva (W)	

Table 2: Caterpillars (larvae)

	Day 1	Day 3	
Wet mass of 10 larvae	0.3 g	1.8 g	g gained =
Wet mass per larva			g gained per larva =
Dry mass of 10 larvae	–	0.27 g	
Larva proportion biomass (dry/wet)			
Energy production per larva (wet mass x proportion biomass x 23.0 kJ)			kJ gained per larva (S) =

Table 4: Respiration

kJ consumed per larva (E) =	
kJ gained per larva (S) =	
Energy (in kJ) from frass from 1 larva (W) =	
Respiratory losses (in kJ) per larva =	

5. Complete the calculations in tables 1-4 above.

6. (a) Write the net secondary production per larva value here: _____

(b) Write the equation to calculate the percentage efficiency of energy transfer from producers to consumers (use the notation provided) and calculate the value here:
_____

(c) Is this value roughly what you would expect? Explain: _____

7. (a) Calculate the approximate gain in dry biomass of all 10 larvae over 3 days (in grams): _____

(b) Calculate the approximate rate of secondary production for the larvae in grams per day:
_____

44 Ecological Pyramids

Key Idea: Ecological pyramids show the biomass, numbers, or energy in each trophic level.

The **trophic levels** of any **ecosystem** can be arranged in pyramids of increasing trophic level. The first trophic level is placed at the bottom and subsequent trophic levels are stacked on top in their 'feeding sequence'. Ecological pyramids can illustrate changes in the numbers, biomass (weight), or energy content of organisms at each level. Each of these three kinds of pyramid tells us something different about the flow of energy and movement of materials between one trophic level and the next. The type of pyramid you choose in order to express information about an ecosystem will depend on what particular features of the ecosystem you are interested in and the type of data you have collected.

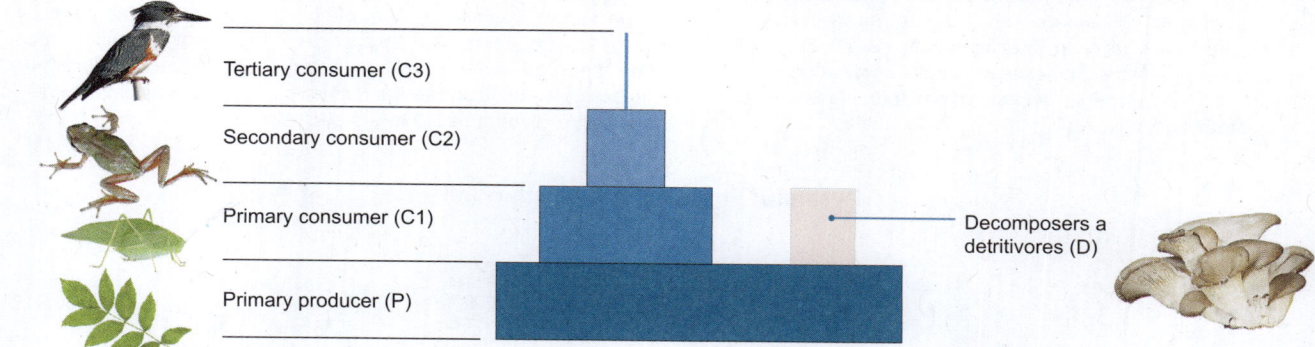

Tertiary consumer (C3)
Secondary consumer (C2)
Primary consumer (C1)
Primary producer (P)
Decomposers a detritivores (D)

The generalized ecological pyramid pictured above shows a conventional pyramid shape, with a large number (or biomass) of **producers** forming the base for an increasingly small number (or biomass) of **consumers**. **Decomposers** are placed at the level of the primary consumers and off to the side. They may obtain energy from many different trophic levels and so do not fit into the conventional pyramid structure. For any particular ecosystem at any one time, the shape of this typical pyramid can vary greatly, depending on whether the trophic relationships are expressed as numbers, biomass, or energy (below).

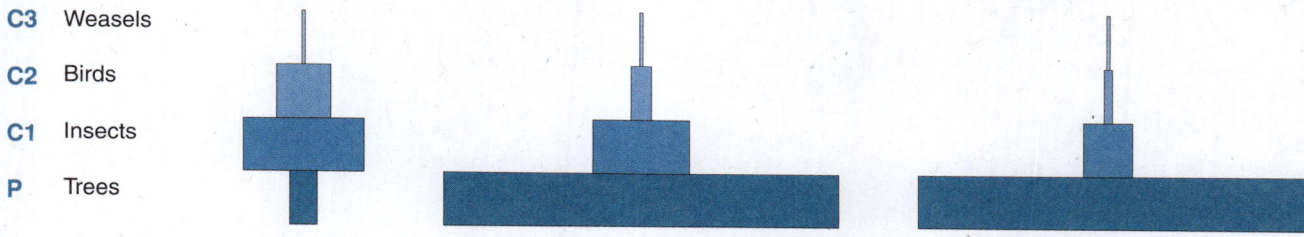

C3 Weasels
C2 Birds
C1 Insects
P Trees

Numbers in a forest community

Pyramids of numbers display the number of individual organisms at each trophic level. The pyramid above has few producers, but they may be of a very large size, e.g. trees. This gives an 'inverted pyramid' although not all pyramids of numbers are like this.

Biomass in a forest community

Biomass pyramids measure the 'weight' of biological material at each trophic level. Water content of organisms varies, so 'dry weight' is often used. Organism size is taken into account, so meaningful comparisons of different trophic levels are possible.

Energy in a forest community

Pyramids of energy are often very similar to biomass pyramids. The energy content at each trophic level is generally comparable to the biomass, i.e. similar amounts of dry biomass tend to have about the same energy content.

1. Describe what the three types of ecological pyramids measure:

    (a) Number pyramid: _____

    (b) Biomass pyramid: _____

    (c) Energy pyramid: _____

2. Explain the advantage of using a biomass or energy pyramid rather than a pyramid of numbers to express the relationship between different trophic levels:

    _____
    _____
    _____

3. Explain why it is possible for the forest community (on the next page) to have very few producers supporting a large number of consumers:

    _____
    _____

©2025 BIOZONE International
ISBN: 978-1-99-101409-2
Photocopying prohibited

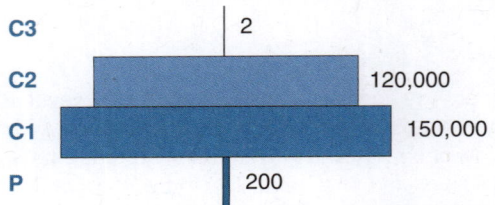
Pyramid of numbers: forest community
In a forest community, a few producers may support a large number of consumers. This is due to the large size of the producers: large trees can support many individual consumer organisms. The example above shows the numbers at each trophic level for an oak forest in England in an area of 10 m^2.

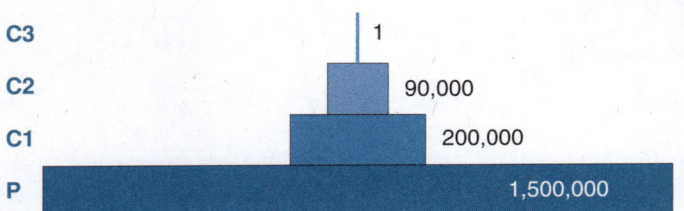
Pyramid of numbers: grassland community
In a grassland community, a large number of producers are required to support a much smaller number of consumers. This is due to the small size of the producers. Grass plants can support only a few individual consumer organisms and take time to recover from grazing pressure. The example above shows the numbers at each trophic level for a derelict grassland area (10 m^2) in Michigan, United States.

Pyramids for a plankton community

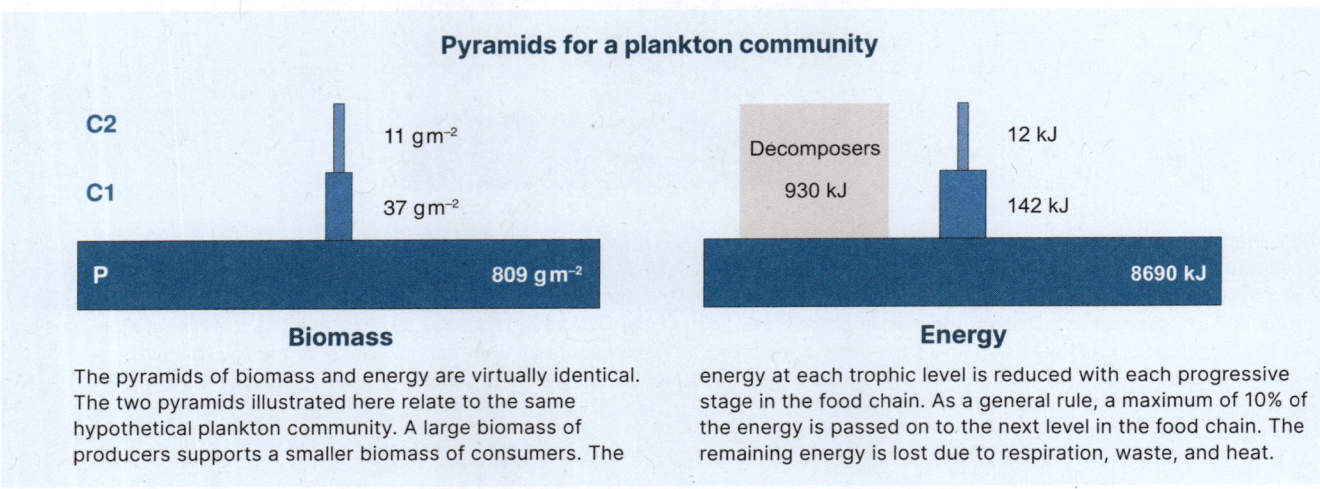

Biomass
The pyramids of biomass and energy are virtually identical. The two pyramids illustrated here relate to the same hypothetical plankton community. A large biomass of producers supports a smaller biomass of consumers. The

Energy
energy at each trophic level is reduced with each progressive stage in the food chain. As a general rule, a maximum of 10% of the energy is passed on to the next level in the food chain. The remaining energy is lost due to respiration, waste, and heat.

4. Determine the energy transfer between trophic levels in the plankton community example in the above diagram:

   (a) Between producers and the primary consumers: _____

   (b) Between the primary consumers and the secondary consumers: _____

   (c) Explain why the energy passed on from the producer to primary consumers is considerably less than the normally expected 10% occurring in most other communities (describe where the rest of the energy was lost to):
   _____
   _____

   (d) After the producers, which trophic group has the greatest energy content: _____

   (e) Give a likely explanation why this is the case: _____
   _____
   _____

An unusual biomass pyramid

The biomass pyramids of some ecosystems appear rather unusual, with an inverted shape. The first trophic level has a lower biomass than the second level. What this pyramid does not show, is the rate at which the producers (algae) are reproducing in order to support the larger biomass of consumers.

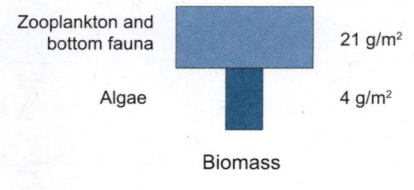

5. Give a possible explanation of how a small biomass of producers (algae) can support a larger biomass of consumers (zooplankton):
   _____
   _____
   _____

45 Species Interactions in Communities

Key Idea: Species in a community interact in specific ways, producing ecosystems with specific characteristics.

The particular characteristics of **ecosystems** and their communities arise as a result of the physical environment, the species that live there, and the complex interactions occurring between those species. Species interactions may involve only occasional or indirect contact (predation or competition) or they may involve close association or symbiosis. Symbiosis is a term that encompasses a variety of interactions involving close species contact. There are three types of symbiosis: **parasitism** (a form of exploitation), **mutualism**, and **commensalism**.

Type of interaction between species				
Mutualism	**Commensalism**	**Parasitism**	**Predation**	**Competition**
A symbiosis in which both species benefit. If both species depend on the symbiosis for survival, the mutualism is obligate. Mutualism can involve more than two species. **Examples**: Flowering plants and their insect pollinators. The flowers are pollinated and the insect gains food. Ruminants and their rumen protozoa and bacteria. The microbes digest the cellulose in plant material and produce short-chain fatty acids, which the ruminant uses as an energy source.	A symbiosis in which one species benefits and the other is unaffected. It is likely that most commensal relationships involve some small benefit to the apparently neutral party. **Example**: The squat anemone shrimp (or sexy shrimp) lives among the tentacles of sea anemones where it gains protection and scavenges scraps of food from the anemone. The anemone appears to be neither harmed, nor gain any benefit.	A symbiotic relationship in which the parasite lives in or on the host, taking all its nutrition from it. The host is harmed but not usually killed, at least not directly. Parasites may have multiple hosts and their transmission is often linked to food webs. **Example**: Parasitic tongue-replacing isopods cut the blood supply to the tongue of the host fish, causing it to fall off. The parasite attaches to what is left of the tongue, feeding on blood or mucus.	A predator kills the prey and eats it. Predators may take a range of species as prey or they may prey exclusively on one other species. Predation is a consumer-resource interaction and a type of exploitation. **Examples**: Praying mantis consuming insect prey. Canada lynx eating snowshoe hare. The ochre sea star feeding on its primary prey, mussels. They also eat chitons, barnacles, and limpets.	Individuals of the same or different species compete for the same, limited resources. Both parties are detrimentally affected. **Examples**: Neighboring plants of the same and different species compete for light and soil nutrients. Vultures compete for the remains of a carcass. Insectivorous birds compete for suitable food in a forest. Tree-nesting birds with similar requirements compete for nest sites.
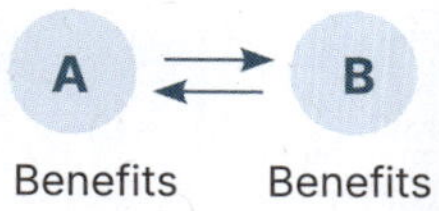 A ⇌ B Benefits Benefits				

1. For the purposes of this exercise, assume that species A in the diagram represents humans. Briefly describe an example of our interaction with another species (B in the diagram above) that matches each of the following interaction types:

    (a) Mutualism: _____

    (b) Exploitation: _____

    (c) Competition: _____

©2025 **BIOZONE** International
ISBN: 978-1-99-101409-2
Photocopying prohibited

Examples of interactions between different species are illustrated below. For each example, identify the type of interaction and explain how each species in the relationship is affected.

2. The honeyeaters are a diverse family of small to medium-sized nectar-feeding birds common in Australia. Many Australian plant species, including proteas and myrtles, are pollinated by honeyeaters.

    (a) Identify this type of interaction: _____

    (b) Describe how each species is affected (benefits/harmed/no effect):
    _____
    _____

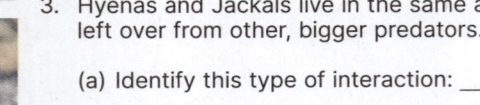

3. Hyenas and Jackals live in the same area of Africa. They both scavenge food left over from other, bigger predators.

    (a) Identify this type of interaction: _____

    (b) Describe how each species is affected (benefits/harmed/no effect):
    _____
    _____
    _____
    _____

4. Hyenas will kill and scavenge a range of species. They form large groups and attack and kill large animals, such as wildebeest, but will also scavenge carrion or drive other animals off their kills.

    (a) Identify this type of interaction: _____

    (b) Describe how each species is affected (benefits/harmed/no effect):
    _____
    _____
    _____
    _____

5. Ticks are obligate hematophages and must obtain blood to pass from one life stage to the next. Ticks attach to the outside of hosts where they suck blood and fluids and cause irritation.

    (a) Identify this type of interaction: _____

    (b) Describe how each species is affected (benefits/harmed/no effect):
    _____
    _____
    _____
    _____

6. Large herbivores expose insects in the vegetation as they graze. The cattle egret, which is widespread in tropical and subtropical regions, follows the herbivores as they graze, feeding on the insects disturbed by the herbivore.

    (a) Identify this type of interaction: _____

    (b) Describe how each species is affected (benefits/harmed/no effect):
    _____
    _____
    _____

7. Explain the similarities and differences between a predator and a parasite: _____
    _____
    _____

46 Ecosystem Stability

Key Idea: The greater the species diversity in an ecosystem, the greater the stability of that ecosystem will be.

Ecological theory suggests that all species in an **ecosystem** contribute in some way to ecosystem function. Therefore, species loss past a certain point is likely to have a detrimental effect on the functioning of the ecosystem and on its ability to resist change (its **stability**). Although many species still await discovery, we do know that the rate of species extinction is increasing. This loss of biodiversity has serious implications for the long term stability of many ecosystems.

Rainforest

Deforestation

Rainforests (above left) represent the highest diversity systems on Earth. Whilst they are generally resistant to disturbance, once degraded, (above right) they have little ability to recover. The diversity of ecosystems at low latitudes is generally higher than that at high latitudes, where climates are harsher, niches are broader, and systems may be dependent on a small number of key species.

The concept of ecosystem stability

The stability of an ecosystem refers to its apparently unchanging nature over time. Ecosystem stability has various components, including inertia (the ability to resist disturbance) and **resilience** (the ability to recover from external disturbances).

Evidence from both experimental and natural systems indicates that the most diverse ecosystems are also the most stable. This correlation is presumed to be a consequence of the large number of biotic interactions operating to buffer diverse systems against change. However, there is uncertainty over what level of diversity provides insurance against catastrophe. Ecosystems are very complex and stability probably relies more on the differential responses of all its species to variable conditions. Current thinking emphasizes the role of multiple factors, including diversity, in dictating stability.

Monoculture of soy beans

Natural grassland

Single species crops (far left), represent low diversity systems that can be vulnerable to disease, pests, and disturbance. In contrast, natural grasslands (left) may appear on the surface to be homogeneous, but contain many species which vary in their predominance seasonally. Although they may be easily disturbed, e.g. by burning, they are very resilient and usually recover quickly.

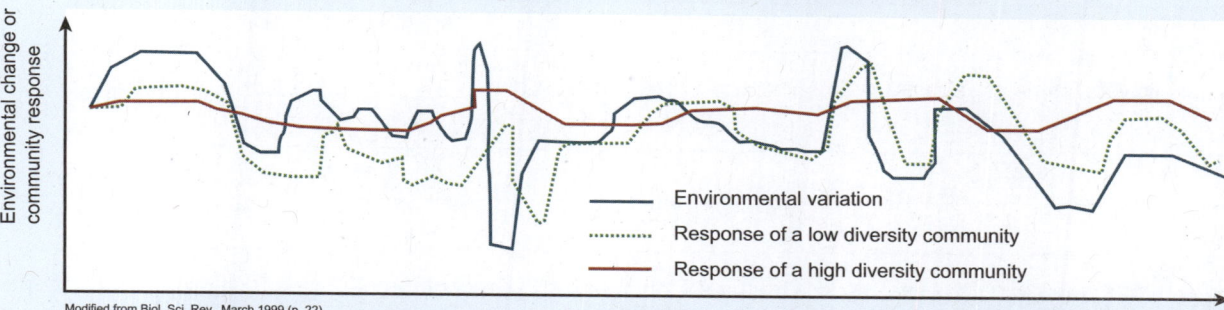
Community response to environmental change

In models of ecosystem function, higher species diversity increases the stability of ecosystem functions such as productivity and **nutrient cycling**. In the graph above, note how the low diversity system varies more consistently with the environmental variation, whereas the high diversity system is buffered against major fluctuations. In any one ecosystem, some species may be more influential than others in the stability of the system. Such keystone (key) species have a disproportionate effect on ecosystem function due to their pivotal role in some ecosystem functions such as nutrient recycling or production of plant biomass.

Elephants can change the entire vegetation structure of areas into which they migrate. Their pattern of grazing on taller plant species promotes a predominance of lower growing grasses with small leaves.

Termites are among the few larger soil organisms able (through a symbiosis with microbes) to break down plant cellulose. They have a profound effect on the rate of nutrient processing in tropical environments.

The starfish *Pisaster* occurs along the coasts of North America where it feeds on mussels. If it is removed, the mussels dominate, crowding out most algae and leading to a decrease in the number of herbivore species.

Keystone species in North America

Gray wolf

Beaver, *Castor canadensis*

Sea otter, *Enhydra lutris*

Quaking aspen

Gray or timber wolves (*Canis lupus*) are a keystone predator and were once widespread in North American ecosystems. Historically, wolves were eliminated from Yellowstone National Park because of their perceived threat to humans and livestock. As a result, elk populations increased to the point that they adversely affected other flora and fauna. Wolves have since been reintroduced to the park and balance is returning to the ecosystem.

Two smaller mammals are also important **keystone species** in North America. Beavers (top) play a crucial role in biodiversity and many species, including 43% of North America's endangered species, depend partly or entirely on beaver ponds. Sea otters are also critical to ecosystem function. When their numbers were decimated by the fur trade, sea urchin populations exploded and the kelp forests, on which many species depend, were destroyed.

Quaking aspen (*Populus tremuloides*) is one of the most widely distributed tree species in North America and aspen communities are among the most biologically diverse in the region, with a rich understory flora supporting an abundance of wildlife. Moose, elk, deer, black bear, and snowshoe hare browse its bark, and aspen groves support up to 34 species of birds, including ruffed grouse, which depends heavily on aspen for its winter survival.

1. How is ecosystem stability linked to species diversity? _____

2. Explain why keystone species are so important to ecosystem function: _____

3. For each of the following species, discuss features of their biology that contribute to their position as keystone species:

   (a) Sea otter: _____

   (b) Beaver: _____

   (c) Gray wolf: _____

   (d) Quaking aspen: _____

47 The Scale of Environmental Change

Key Idea: Environmental changes occur on many scales. Environmental changes come from three sources: the **biosphere** itself, geological forces (crustal movements and plate tectonics), and cosmic forces (the movement of the Moon around the Earth, and the Earth and planets around the Sun). All three forces can cause cycles, steady states, and trends (directional changes) in the environment. Environmental trends such as climate cooling cause long term changes in communities. Some short-term cycles may also influence patterns of behavior and growth in many species, regulating internal, cyclical behavior patterns, called biological rhythms.

Climatic change during the last 2-3 million years has involved cycles of glacial and interglacial conditions. These cycles are largely the result of an interplay between astronomical cycles and atmospheric CO_2 concentrations.

Volcanic eruptions may have a large effect on local biological communities. They may also cause prolonged changes to regional and global weather, e.g. Mount Pinatubo eruption, 1991.

Some weather patterns are responsible for subtle changes to ecosystems, such as the gradual onset of a drought. They may also provide large scale and forceful changes, such as those caused by hurricanes or cyclones.

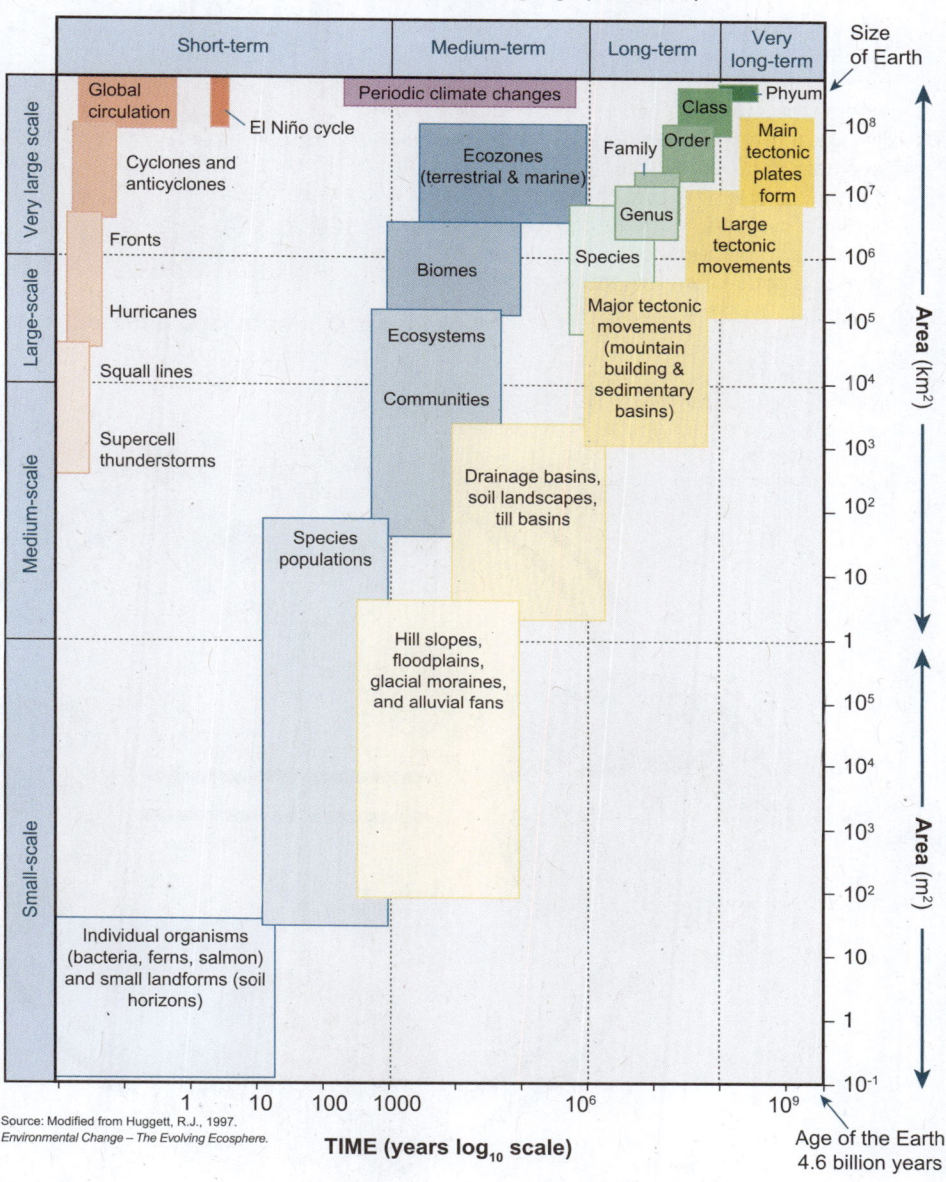

Time scale and geographic extent of environmental change
(X axis = time scale, Y axis = geographic extent)

Source: Modified from Huggett, R.J., 1997. *Environmental Change – The Evolving Ecosphere.*

1. Periodic, long term changes in the Earth's orbit, a change in the Sun's heat output, and continental drift may have been the cause of cycles of climate change in the distant past. These climate changes involved a cooling of the Earth.

    (a) Identify the term referring to these periods of global cooling: _____

    (b) Describe two changes to the landscape that occurred during this period: _____

2. Identify the main causes of environmental change: _____

3. Identify the scale of the following ecological or geographical changes:

    (a) Supercell thunderstorm: _____

    (b) Biome: _____

©2025 BIOZONE International
ISBN: 978-1-99-101409-2
Photocopying prohibited

48 Cycles of Matter

Key Idea: Nutrients constantly cycle from the soil to plant and animal life and back.

Nutrient cycles move and transfer chemical elements, e.g. carbon, within and between **ecosystems**. The cycling of these elements is called a nutrient cycle, or a biogeochemical cycle. The term biogeochemical means that biological, geological, and chemical processes are involved in the chemical transformations. Each biogeochemical cycle has one or more reservoirs. These are large, usually abiotic, stores of the chemical element and smaller, more active pools where the nutrient cycles between the biotic (living) and abiotic components of an ecosystem (see diagram below). Energy, ultimately from the Sun, drives the cycling of matter within and between systems. Matter is conserved throughout all these transformations, although it may pass from one ecosystem to another.

Tropical rainforest (Amazon)

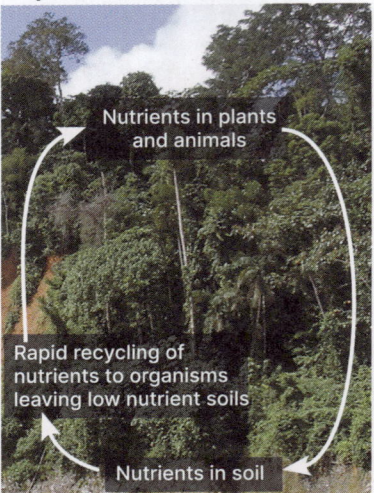

Nutrients in plants and animals
Rapid recycling of nutrients to organisms leaving low nutrient soils
Nutrients in soil

Temperate woodland (Illinois)

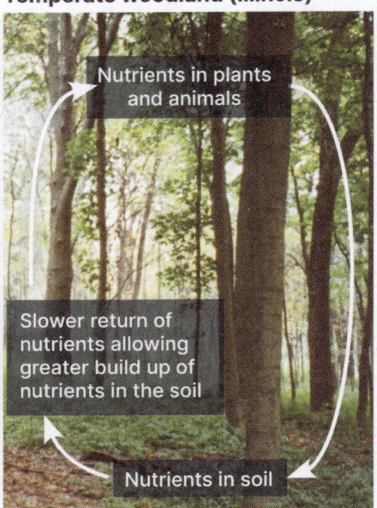

Nutrients in plants and animals
Slower return of nutrients allowing greater build up of nutrients in the soil
Nutrients in soil
Dustin M. Ramsey

Processes in a generalized biogeochemical cycle

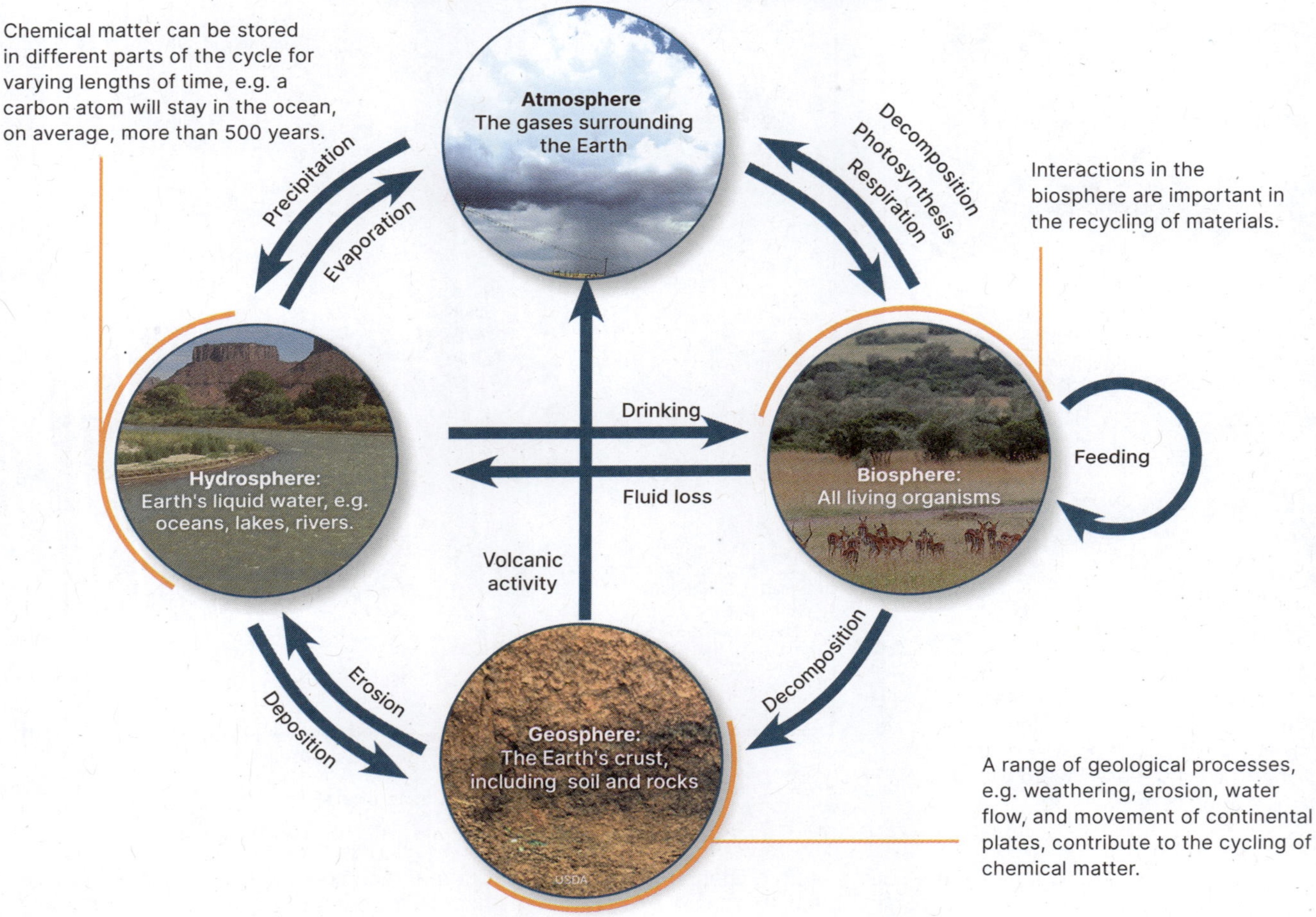
Chemical matter can be stored in different parts of the cycle for varying lengths of time, e.g. a carbon atom will stay in the ocean, on average, more than 500 years.
Atmosphere
The gases surrounding the Earth
Precipitation
Evaporation
Decomposition
Photosynthesis
Respiration
Interactions in the biosphere are important in the recycling of materials.
Hydrosphere:
Earth's liquid water, e.g. oceans, lakes, rivers.
Drinking
Fluid loss
Biosphere:
All living organisms
Feeding
Volcanic activity
Erosion
Deposition
Decomposition
Geosphere:
The Earth's crust, including soil and rocks
USDA
A range of geological processes, e.g. weathering, erosion, water flow, and movement of continental plates, contribute to the cycling of chemical matter.

1. What is a nutrient cycle? ______________________________

____

____

2. Suggest why it is important that matter is cycled through an ecosystem: ______________________________

____

____

____

LW

ISBN: 978-1-99-101409-2

49 The Carbon Cycle

Key Idea: Carbon cycles through the abiotic and biotic components of ecosystems.

Carbon is an essential element of life and is incorporated into the organic molecules that make up living organisms. Large quantities of carbon are stored in sinks, which include the atmosphere as carbon dioxide gas (CO_2), the ocean as carbonate and bicarbonate, and rocks such as coal and limestone. Carbon cycles between the biotic and abiotic environment. Carbon dioxide is converted by autotrophs into carbohydrates via **photosynthesis** and returned to the atmosphere as CO_2 through **respiration** (fluxes). These fluxes can be measured. Carbon may remain locked up in biotic or abiotic systems for long periods of time as, e.g. in wood or coal. Human activity has disturbed the balance of the **carbon cycle** (the global carbon budget) through activities such as combustion and deforestation.

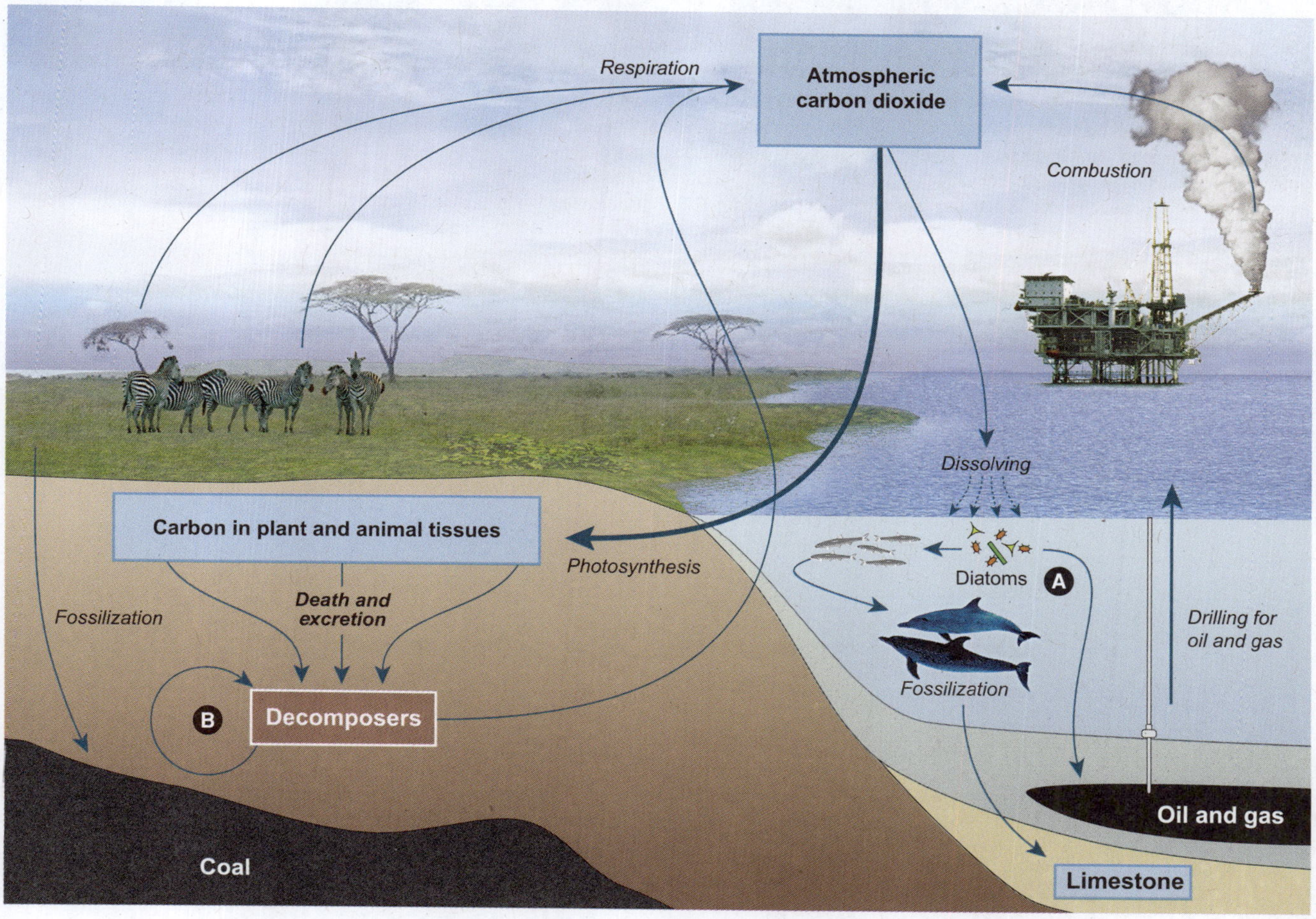

1. In the diagram above, add arrows and labels to show the following activities:

   (a) Dissolving of limestone by acid rain.

   (b) Release of carbon from the marine food chain.

   (c) Mining and burning of coal.

   (d) Burning of plant material.

2. (a) Name the processes that release carbon into the atmosphere: _____

   (b) In what form is the carbon released? _____

3. What would be the effect on carbon in the atmosphere if photosynthesis increased without any increase in respiration, combustion, or decomposition (dead trees are buried with little or no decomposition)?
   _____
   _____

4. Name the four geological reservoirs (sinks) in the diagram above that can act as a source of carbon:

   (a) _____   (c) _____

   (b) _____   (d) _____

5. (a) Identify the process carried out by diatoms at point [A]: _____

   (b) Identify the process carried out by decomposers at [B]: _____

Termite mound in rainforest

Dung beetle on animal dung

Bracket fungus on tree trunk

Termites: These insects play an important role in nutrient recycling. With the aid of symbiotic protozoans and bacteria in their guts, they can digest the tough cellulose of woody tissues in trees. Termites fulfil a vital function in breaking down the endless rain of debris in tropical rainforests.

Dung beetles: Beetles play a major role in the decomposition of animal dung. Some beetles merely eat the dung, but true dung beetles, such as the scarabs and geotrupes, bury the dung and lay their eggs in it to provide food for the beetle grubs during their development.

Fungi: Together with decomposing bacteria, fungi perform an important role in breaking down dead plant matter in the leaf litter of forests. Some mycorrhizal fungi have been found to link up to the root systems of trees where an exchange of nutrients occurs in a mutualistic relationship.

6. Predict the consequences to carbon cycling if there were no decomposers present in an ecosystem: _____
   _____
   _____

7. Explain how each of the three organisms listed below has a role to play in the carbon cycle:

   (a) Dung beetles: _____
   _____
   _____

   (b) Termites: _____
   _____
   _____

   (c) Fungi: _____
   _____
   _____

8. Using specific examples, explain the role of insects in carbon cycling: _____
   _____
   _____

9. In natural circumstances, accumulated reserves of carbon such as peat, coal, and oil represent a sink or natural diversion from the cycle. Eventually, the carbon in these sinks returns to the cycle through the action of geological processes which return deposits to the surface for oxidation.

   (a) Describe the effects of human activity on the amount of carbon stored in sinks: _____
   _____

   (b) Describe two global effects of this activity: _____
   _____
   _____

   (c) Suggest what could be done to prevent or alleviate these effects: _____
   _____
   _____

50 The Nitrogen Cycle

Key Idea: Nitrogen constantly cycles between the atmosphere and biosphere.

Nitrogen is an essential element in living things. It is a component of photosynthetic pigments in plants, and part of proteins and nucleic acids in all organisms. Plants obtain their nitrogen from the soil or via symbioses, whereas consumers obtain their nitrogen from other organisms, e.g. by eating or via symbioses. The Earth's atmosphere is about 80% nitrogen gas (N_2), but molecular nitrogen is so stable that it is directly available only to those few organisms (all of them bacteria) that can fix it (capture and combine it into another molecule).

Bacteria play a crucial role in nitrogen cycling, transferring nitrogen between the biotic and abiotic environments through nitrogen fixation and other nitrogen transformations. Humans intervene in the **nitrogen cycle** by producing, and applying to the land, large amounts of nitrogen fertilizer. Some applied fertilizer is from organic sources, e.g. green crops and manures, but much of it is inorganic, produced from atmospheric nitrogen using an energy-expensive industrial process. Overuse of nitrogen fertilizers may lead to excessive nutrient enrichment of water which may cause excess microbial growth called **eutrophication**.

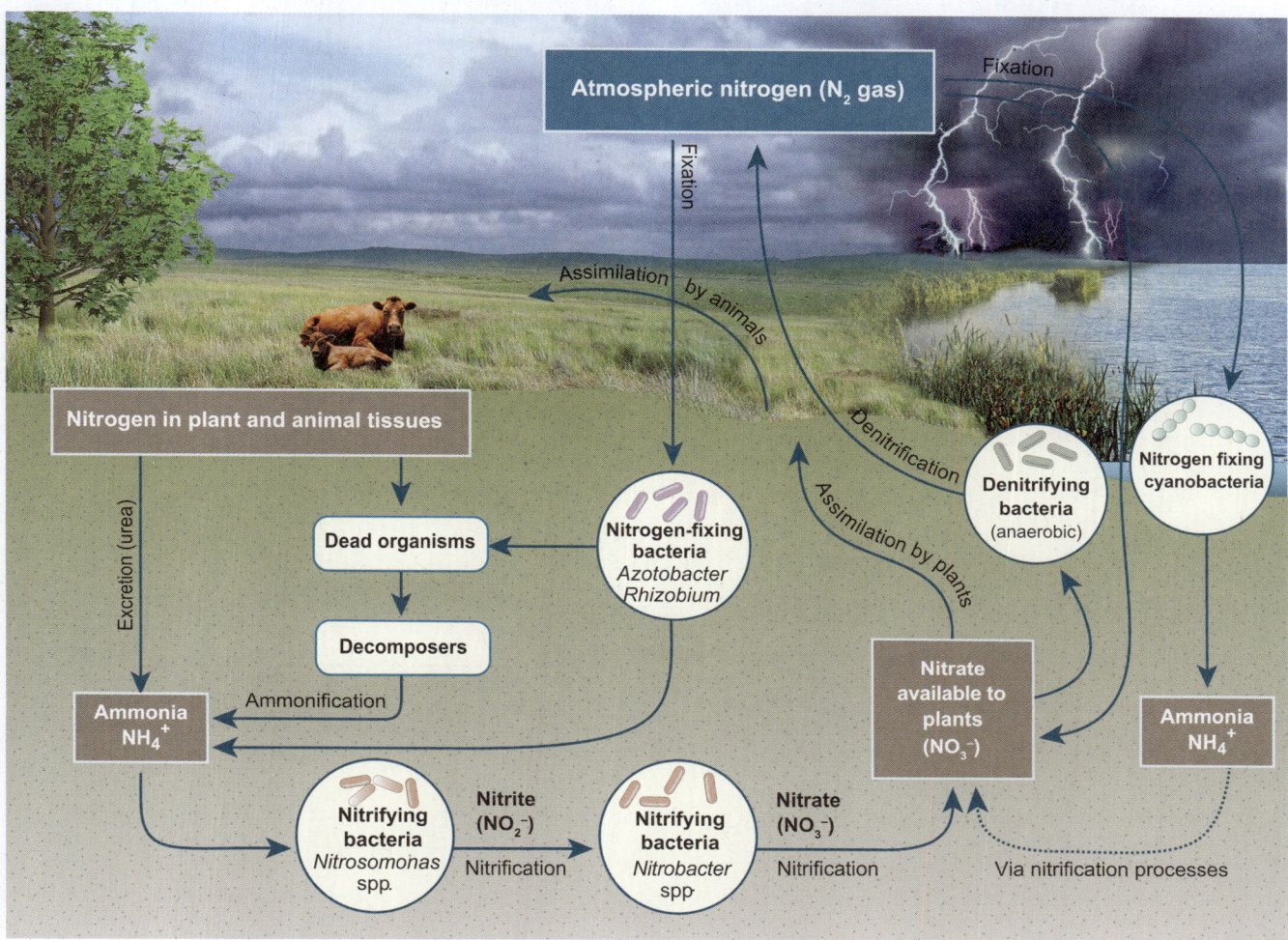

1. Describe five instances in the nitrogen cycle where bacterial action is important. Include the name of each of the processes and the changes to the form of nitrogen involved:

   (a) _____

   (b) _____

   (c) _____

   (d) _____

   (e) _____

Nitrogen fixation in root nodules

Root nodules are a root symbiosis between a higher plant and a bacterium. The bacteria fix atmospheric nitrogen and are extremely important to the nutrition of many plants, including the economically important legume family. Root nodules are extensions of the root tissue caused by entry of a bacterium. In legumes, this bacterium is *Rhizobium*. Other bacterial genera are involved in the root nodule symbioses in non-legume species.

The bacteria in these symbioses live in the nodule where they fix atmospheric nitrogen and provide the plant with most, or all, of its nitrogen requirements. In return, they have access to a rich supply of carbohydrate. The fixation of atmospheric nitrogen to ammonia occurs within the nodule using the enzyme nitrogenase. Nitrogenase is inhibited by oxygen and the nodule provides a low O_2 environment in which fixation can occur.

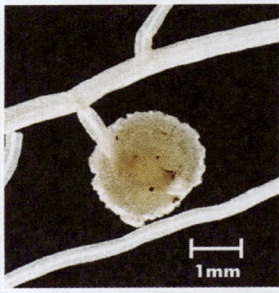

Two examples of legume nodules caused by *Rhizobium*. The photographs above show the size of a single nodule (left), and the nodules forming clusters around the roots of Acacia (right).

Human intervention in the nitrogen cycle

Until about sixty years ago, microbial nitrogen fixation (left) was the only mechanism by which nitrogen could be made available to plants. However, during WW II, Fritz Haber developed the Haber process, whereby nitrogen and hydrogen gas are combined to form gaseous ammonia. The ammonia is converted into ammonium salts and sold as inorganic fertilizer. Its application has revolutionized agriculture by increasing crop yields.

As well as adding nitrogen fertilizers to the land, humans use anaerobic bacteria to break down livestock wastes and release NH_3 into the soil. They also intervene in the nitrogen cycle by discharging effluent into waterways. Nitrogen is removed from the land through burning, which releases nitrogen oxides into the atmosphere. It is also lost by mining, harvesting crops, and irrigation, which leaches nitrate ions from the soil.

Two examples of human intervention in the nitrogen cycle. The photographs above show the aerial application of a commercial fertilizer (left), and the harvesting of an agricultural crop (right).

2. Identify the process that releases nitrogen gas into the atmosphere: _____

3. Identify the main geological reservoir that provides a source of nitrogen: _____

4. Identify the form in which nitrogen is available to most plants: _____

5. Describe how animals acquire the nitrogen they need: _____

6. Explain why farmers may plow a crop of legumes into the ground rather than harvest it:
_____
_____

7. How do humans intervene in the nitrogen cycle, and what are some of the effects of this intervention?
_____
_____
_____
_____
_____
_____
_____
_____
_____

51 The Oxygen Cycle

Key Idea: The oxygen cycle describes the movement of oxygen (O_2) through an ecosystem.

The oxygen cycle is closely linked to the **carbon cycle**. The oxygen cycle describes the movement of oxygen between the biotic and abiotic components of **ecosystems**. Photosynthesis is the main source of oxygen in ecosystems.

Oxygen is involved to some degree in all the other biogeochemical cycles, but is closely linked to the carbon cycle in particular. This is because most **producers** utilize carbon dioxide in **photosynthesis**, and produce oxygen as a waste product. The oxygen is used in **cellular respiration** and carbon dioxide is produced as a waste product.

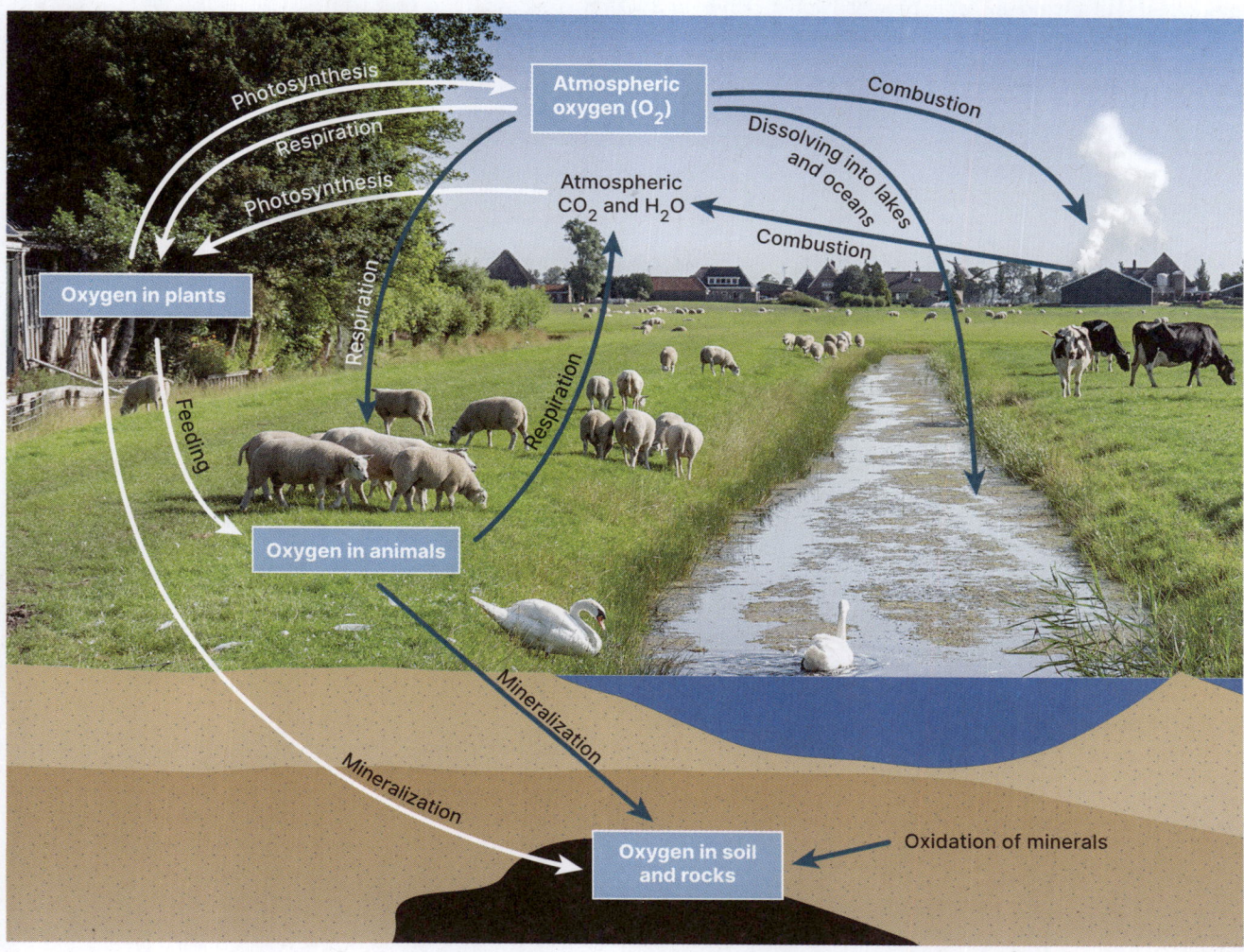

1. What is the main source of oxygen for the oxygen cycle? _____

2. (a) Why are the oxygen cycle and carbon cycle so closely linked? _____

   (b) In the space below, draw a diagram to show how the oxygen and carbon cycles are linked. Use appropriate chemical equations where needed.

52 The Water Cycle

Key Idea: Water moves through the environment via the water cycle. Surface water exists as ice, water, and vapor. Powered by energy from the Sun, the hydrologic cycle, collects, purifies, and distributes the Earth's fixed supply of water. The oceans are the Earth's largest reservoir of water, with much smaller reservoirs in the ice caps and groundwater. Water constantly changes states between liquid, vapor, and ice as it moves through the biotic and abiotic components of **ecosystems**. Besides replenishing inland water supplies, rainwater causes erosion and transports dissolved nutrients within and among ecosystems. On a global scale, evaporation (conversion of liquid water to gaseous vapor) exceeds precipitation over the oceans. This results in a net movement of water vapor (carried by winds) over the land. On land, precipitation exceeds evaporation. Some of this precipitation becomes locked up in snow and ice but most forms surface and groundwater systems that flow back to the sea, completing the major part of the cycle. Over the sea, most of the water vapor is due to evaporation alone. However on land, about 90% of the vapor results from transpiration.

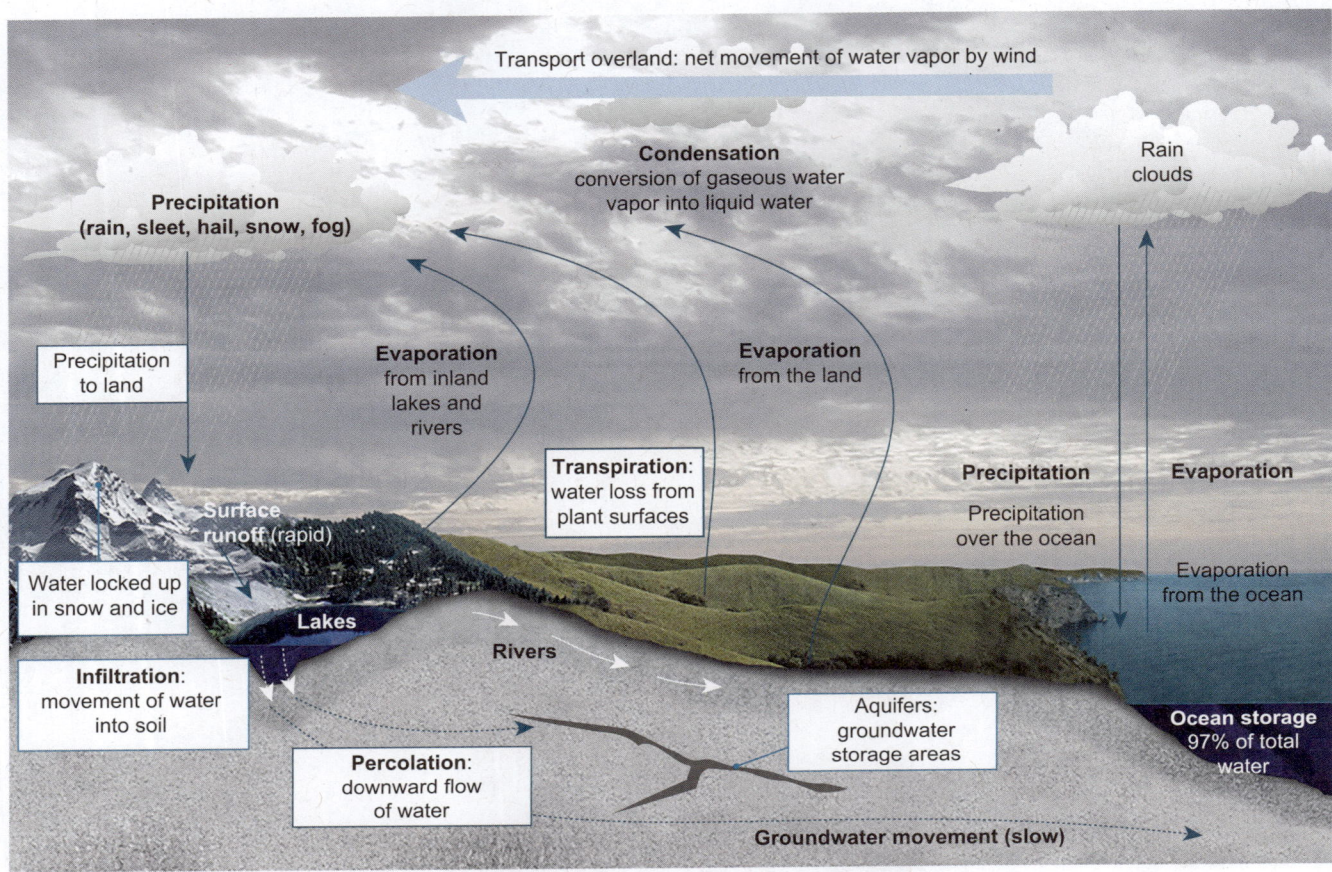

1. Identify the main reservoir for water on Earth: _____
2. Identify the main reservoirs for fresh water: _____
3. What is the ultimate source of energy for the processes involved in the hydrologic cycle? _____
4. Describe what is involved in each of the following processes and its role in the hydrologic cycle:

    (a) Evaporation: _____

    (b) Precipitation: _____

    (c) Condensation: _____

    (d) Transpiration: _____

5. Identify two ways in which water returns to the oceans from the land:

    (a) _____ (b) _____

53 The Phosphorus Cycle

Key Idea: Phosphorus is mostly stored in rocks and mineral deposits. A small amount is found in organic matter.

Phosphorus is an essential component of genetic and energy systems of living organisms and an important limiting factor to productivity in **ecosystems**. Phosphorus cycling has no atmospheric component and return from the ocean to the land is very slow. Its main reservoirs are rock and sediments, and small losses from terrestrial systems by leaching are generally balanced by gains from weathering of rock. In both aquatic and terrestrial ecosystems, phosphorus is cycled through **food webs**. Sedimentation may lock phosphorus away although it can become available again through long-term geological processes. Some phosphorus returns to the land as guano (phosphate-rich manure) but this return is small relative to the phosphate transferred to the oceans each year by natural processes and human activity.

Bacteria can immobilize inorganic phosphorus but can also convert organic to inorganic phosphorus by mineralization, returning phosphates to the soil. Phosphorus is lost from ecosystems through run-off, precipitation, and sedimentation.

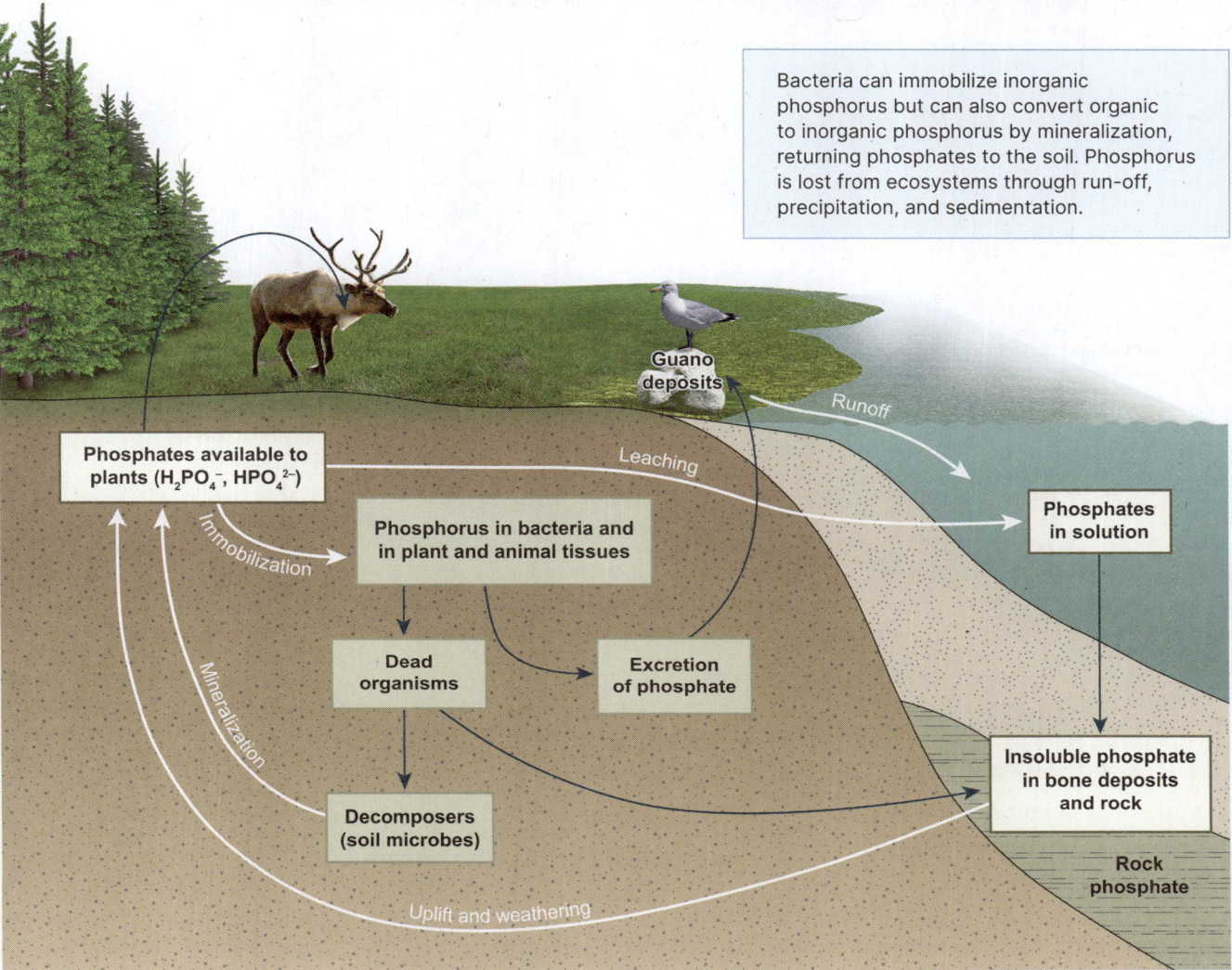

1. Identify the main reservoirs of phosphorus: _____

2. Explain why phosphorus is often a limiting factor in the productivity of ecosystems: _____

3. Identify two instances in the phosphorus cycle where bacteria are important and describe the transformations involved:

   (a) _____

   (b) _____

54 Primary Succession

Key Idea: Succession describes community changes that occur in an ecosystem as diversity returns to a natural state. Succession takes place as a result of the complex interactions between biotic and **abiotic factors**. Early communities modify the physical environment causing it to change. This in turn alters the biotic **community**, which further alters the physical environment, and so on. Each successive community makes the environment more favorable for the establishment of new species. An 'idealized' succession (or sere) proceeds in seral stages, until the formation of a climax (old growth) community, which is generally stable until further disturbance. Early successional communities are characterized by a low species diversity, a simple structure, and broad niches. In contrast, climax communities are complex, with a large number of species interactions, narrow niches, and high species diversity.

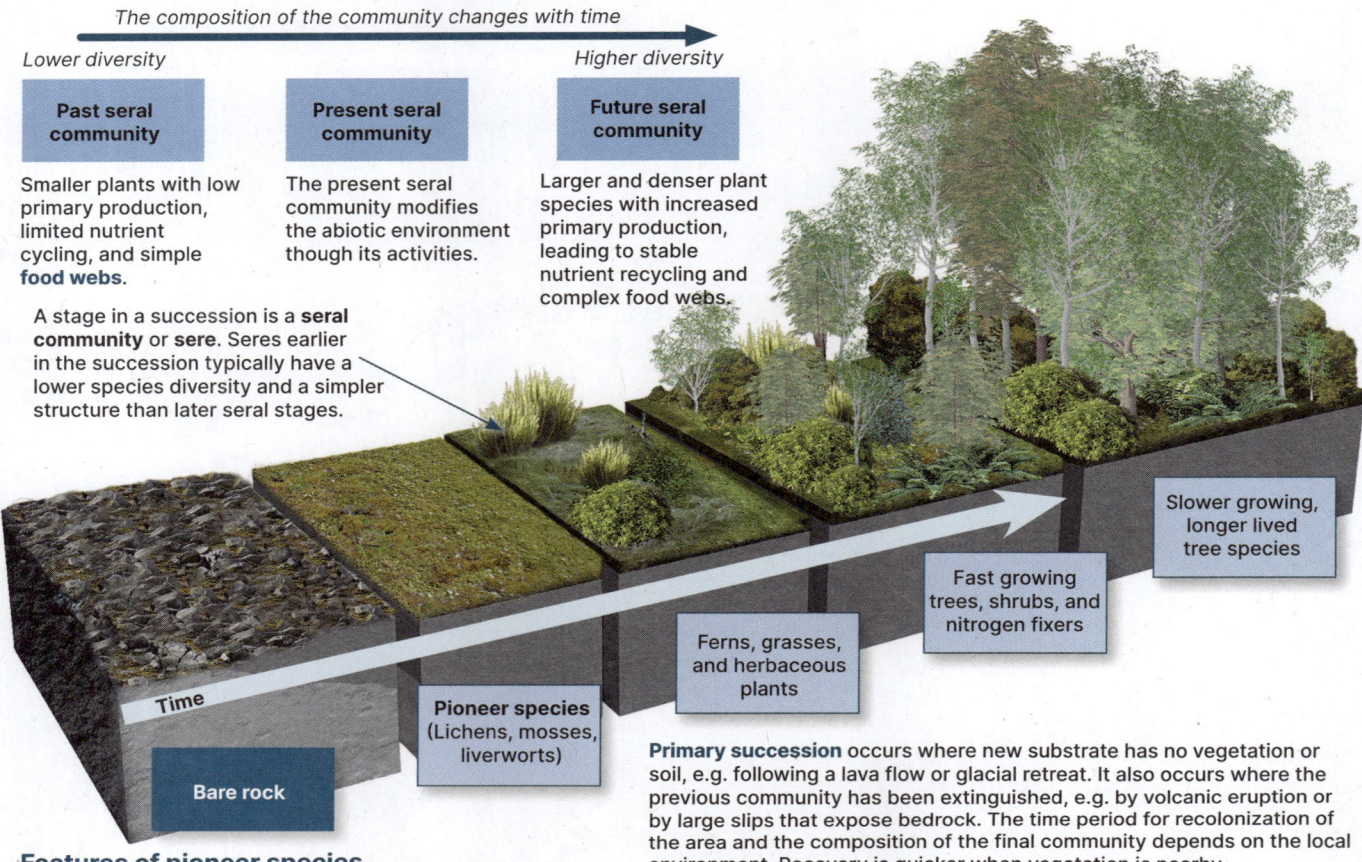

A stage in a succession is a **seral community** or **sere**. Seres earlier in the succession typically have a lower species diversity and a simpler structure than later seral stages.

Primary succession occurs where new substrate has no vegetation or soil, e.g. following a lava flow or glacial retreat. It also occurs where the previous community has been extinguished, e.g. by volcanic eruption or by large slips that expose bedrock. The time period for recolonization of the area and the composition of the final community depends on the local environment. Recovery is quicker when vegetation is nearby.

Features of pioneer species

The earliest **pioneer species** are microorganisms, e.g. cyanobacteria, and simple photosynthetic plants and algae. They are able to survive on exposed substrates lacking in nutrients and make their own food using sunlight energy. Even at this level, ecological associations are important. Lichens, which are important pioneers, are a symbiosis between fungi and algae. Associations between mosses and cyanobacteria (which can fix nitrogen) are also important. Pioneers begin the process of soil formation by breaking down the substrate and adding organic matter through their own death and decay. Their growth therefore creates a more favorable environment for vascular plant growth.

Lichen on bare rock

Note the vascular plants establishing in the crevices where soil is forming.

Moss on bare rock

Associations between mosses and cyanobacteria provides mosses with nitrogen.

1. Describe situations in which a primary succession is likely to occur: _____

_____

_____

2. (a) Identify pioneers during the colonization of bare rock: _____

   (b) Describe two important roles of species that are early colonizers of bare slopes: _____

   _____

   _____

55 Primary Succession: Surtsey Island

Key Idea: As a new island, Surtsey provided researchers with an opportunity to study primary succession in detail.

Surtsey Island is a volcanic island lying 33 km off the southern coast of Iceland. The island was formed over four years from 1963 to 1967 when a submarine volcano 130 m below the ocean surface built up an island that initially reached 174 m above sea level and covered 2.7 km². Erosion has since reduced the island to around 150 m above sea level and 1.4 km². The colonization of the island by plants and animals has been recorded since the island's formation. The first vascular plant there (sea rocket) was discovered in 1965, two years before the eruptions on the island ended. Since then, 69 plant species have colonized the island and there are a number of established seabird colonies.

Sea rocket

H. peploides

Plant colonization of Surtsey can be divided into four stages. The first stage (1965-1974) was dominated by shore plants colonizing the northern shores. The most successful of these was *Honckenya peploides*, which established on tephra sand and gravel flats. It set seed in 1971 and subsequently spread across the island. Carbon and nitrogen levels in the soil were recorded as being very low during this time. This initial colonization by shore plants was followed by a lag phase (1975-1984). Shore plants continued to establish but there were few new colonizers. This slowed the rate of succession.

P. annua

S. phylicifolia

After a gull colony established on the southern end of the island, a number of new plant species arrived (1985-1994). Vegetation within or near the colony expanded rapidly to about 3 ha, while populations outside the colony remained low but stable. Grasses such as *Poa annua* formed extensive patches of vegetation. After this rapid increase in plant species, the arrival of new colonizers again slowed (1995-2008). A second wave of colonizers began to establish following this slower phase and soil organic matter increased. The first bushy plant, the willow *Salix phylicifolia*, established in 1998. The area of vegetation near the gull colony expanded to about 10 ha.

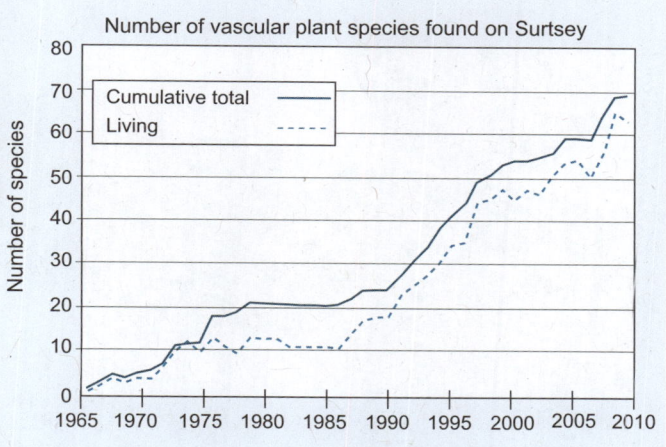

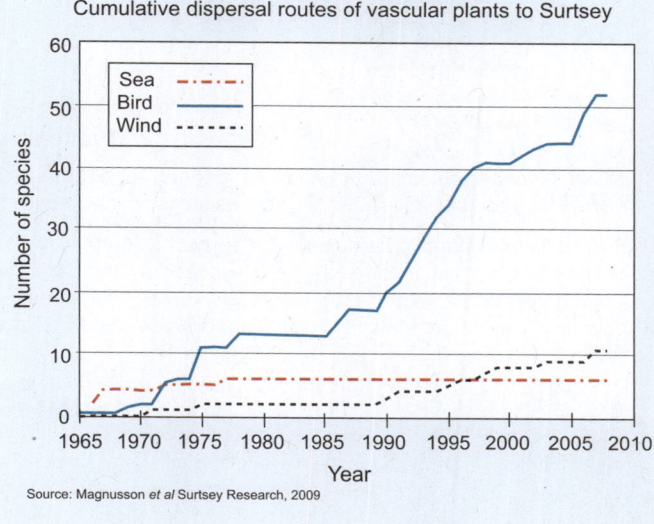

Source: Magnusson *et al* Surtsey Research, 2009

1. Explain why Surtsey provided ideal conditions for studying primary succession:

2. Why did the first colonizing plants established in the north of the island, while later colonizers established in the south?

3. (a) Identify the year the gull colony established: _____

   (b) What was the most common method for new plant species to arrive on the island: _____

   (c) What was the year of the arrival of the second wave of plant colonizers. Suggest a reason for this second wave:

56 Secondary Succession

Key Idea: A secondary succession takes place after a land clearance, e.g. following a fire or a landslide.

Land clearance resulting in **secondary succession** does not involve the loss of the soil, and the seeds and root stocks in it. As a result, secondary succession tends to be more rapid than **primary succession**, although the time scale depends on the species involved and the climate and edaphic (soil) factors. Secondary succession events may occur over a wide area (such as after a forest fire), or in smaller areas where single trees have fallen.

Secondary succession in cleared land

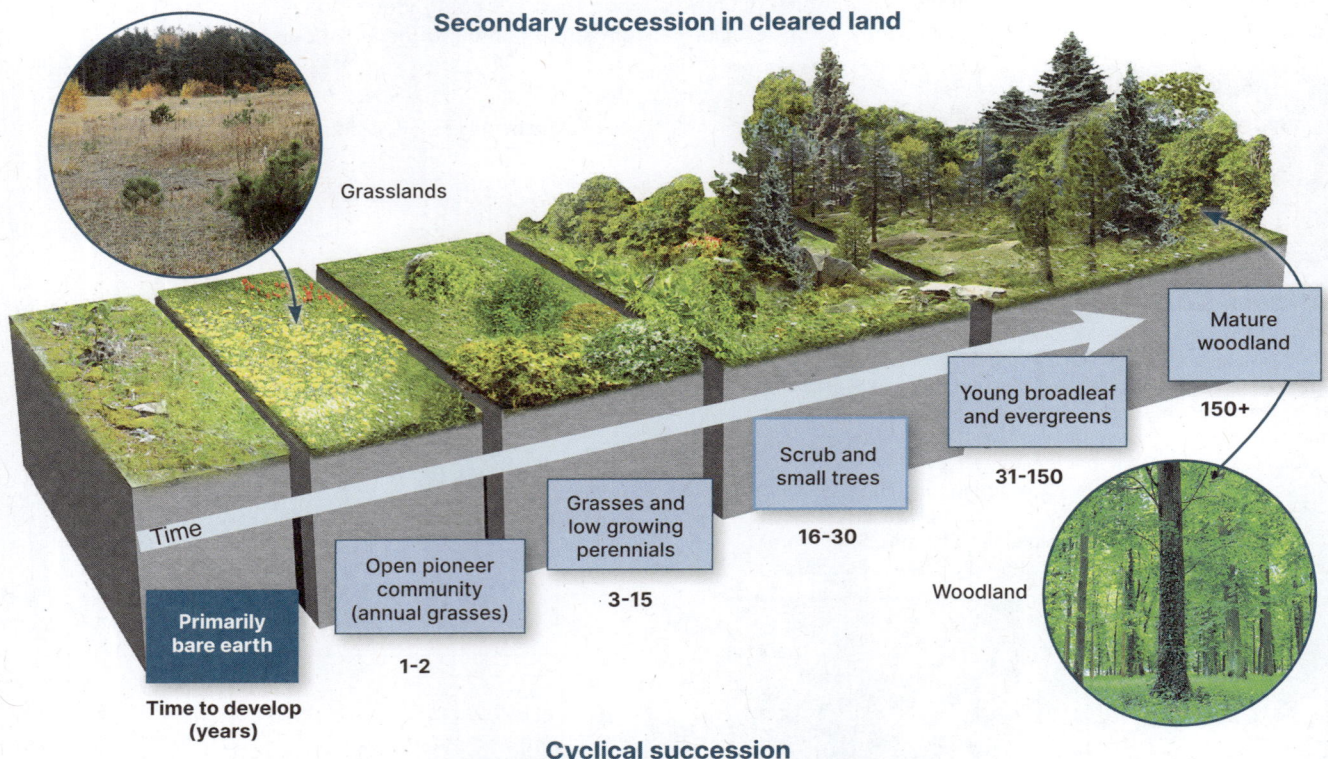

Cyclical succession

- Some events, such as natural wildfires, or death of old trees in forests, can change the composition of a **community** and may occur in repeated cycles. This cyclical succession can be in varying lengths depending on the community.

- Examples of cyclic succession include regular wildfires in the chaparral ecosystem in coastal California which occur naturally every 30-150 years. After each fire, a succession of different vegetation species replace each other in the same order. In the Sonoran desert in Southwestern US and Mexico, two plants, cholla and creosote, grow together, but cyclically replace each other with assistance from animals dispersing seeds. Regular wildfires are also common in parts of Australia (below).

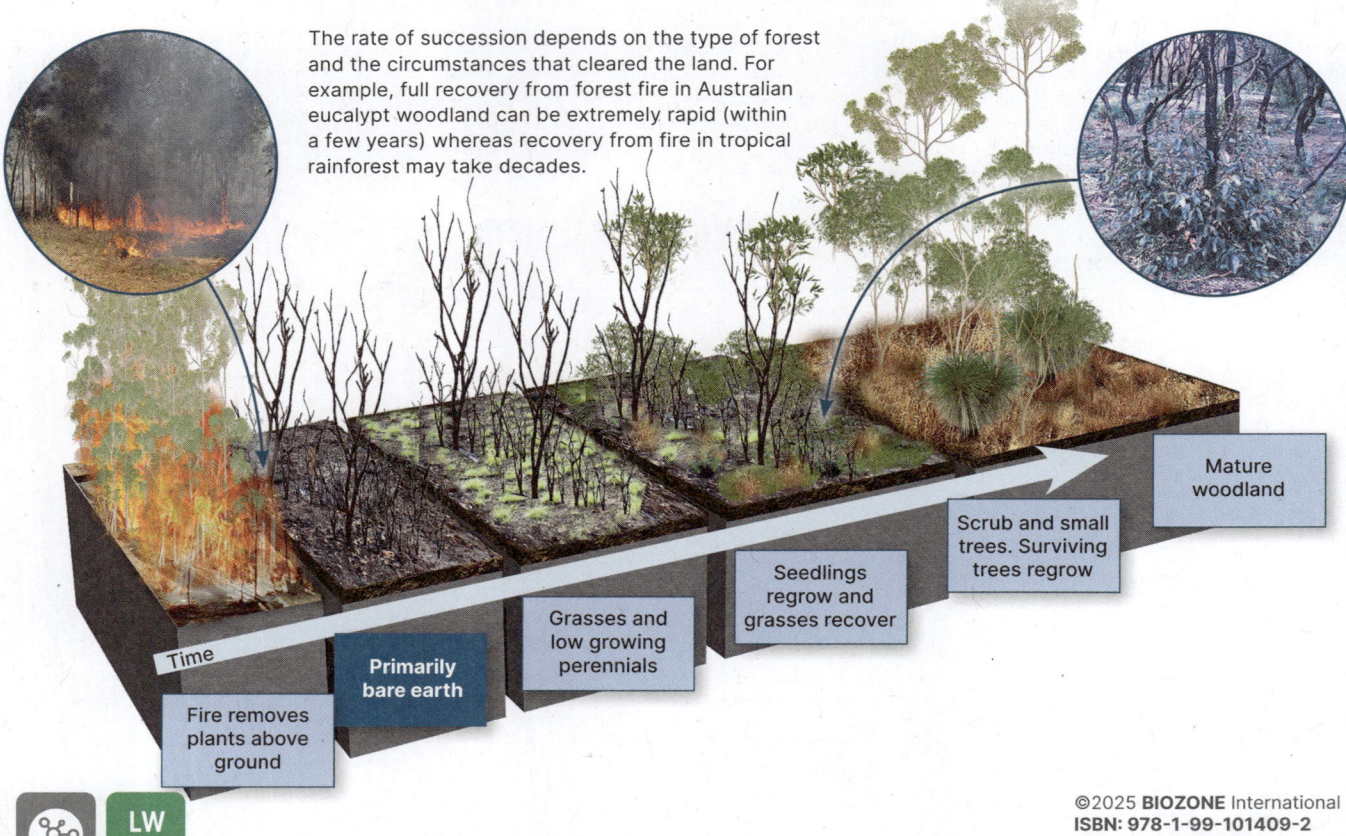

The rate of succession depends on the type of forest and the circumstances that cleared the land. For example, full recovery from forest fire in Australian eucalypt woodland can be extremely rapid (within a few years) whereas recovery from fire in tropical rainforest may take decades.

©2025 **BIOZONE** International
ISBN: 978-1-99-101409-2
Photocopying prohibited

Deflected succession

Humans (and sometimes nature) may deflect the natural course of succession, e.g. by mowing or fire, and the **climax community** that results will differ from the community that would occur if there had been no disturbance.

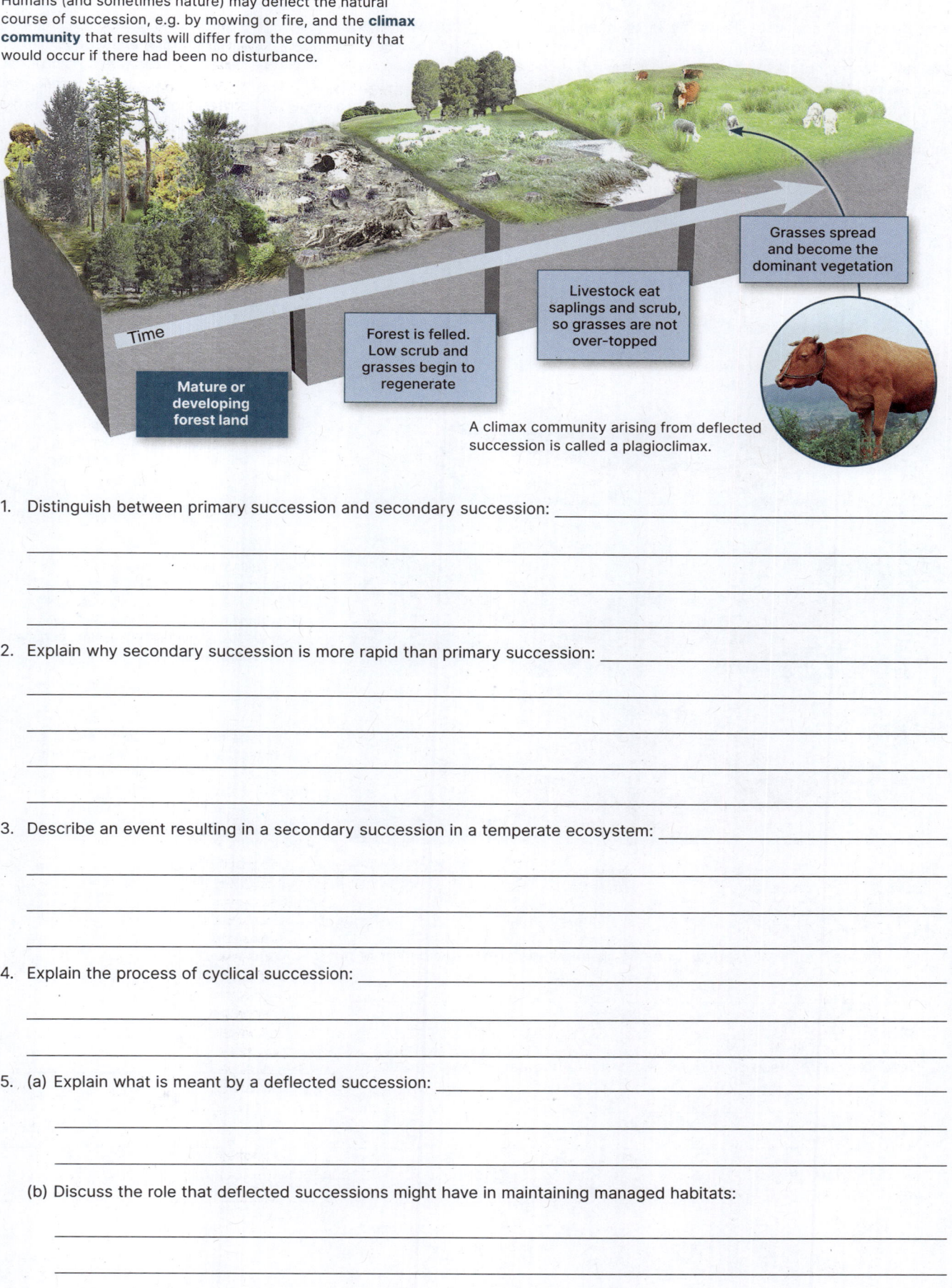

A climax community arising from deflected succession is called a plagioclimax.

1. Distinguish between primary succession and secondary succession: _____

2. Explain why secondary succession is more rapid than primary succession: _____

3. Describe an event resulting in a secondary succession in a temperate ecosystem: _____

4. Explain the process of cyclical succession: _____

5. (a) Explain what is meant by a deflected succession: _____

   (b) Discuss the role that deflected successions might have in maintaining managed habitats: _____

57 Wetland Succession

Key Idea: Wetlands can slowly fill in with sediment and transition to dry land over hundreds or thousands of years. Wetland areas present a special case of ecological succession. They are constantly changing as plant invasion of open water leads to siltation and infilling. This process is accelerated by **eutrophication**. In well drained areas, pasture or heath may develop as a result of succession from freshwater to dry land. When the soil conditions remain non-acid and poorly drained, a swamp will eventually develop into a seasonally dry fen. In special circumstances, an acid peat bog may develop. The domes of peat produce a hummocky landscape with a unique biota. Wetland peat ecosystems may take more than 5000 years to form but are easily destroyed by excavation and lowering of the water table.

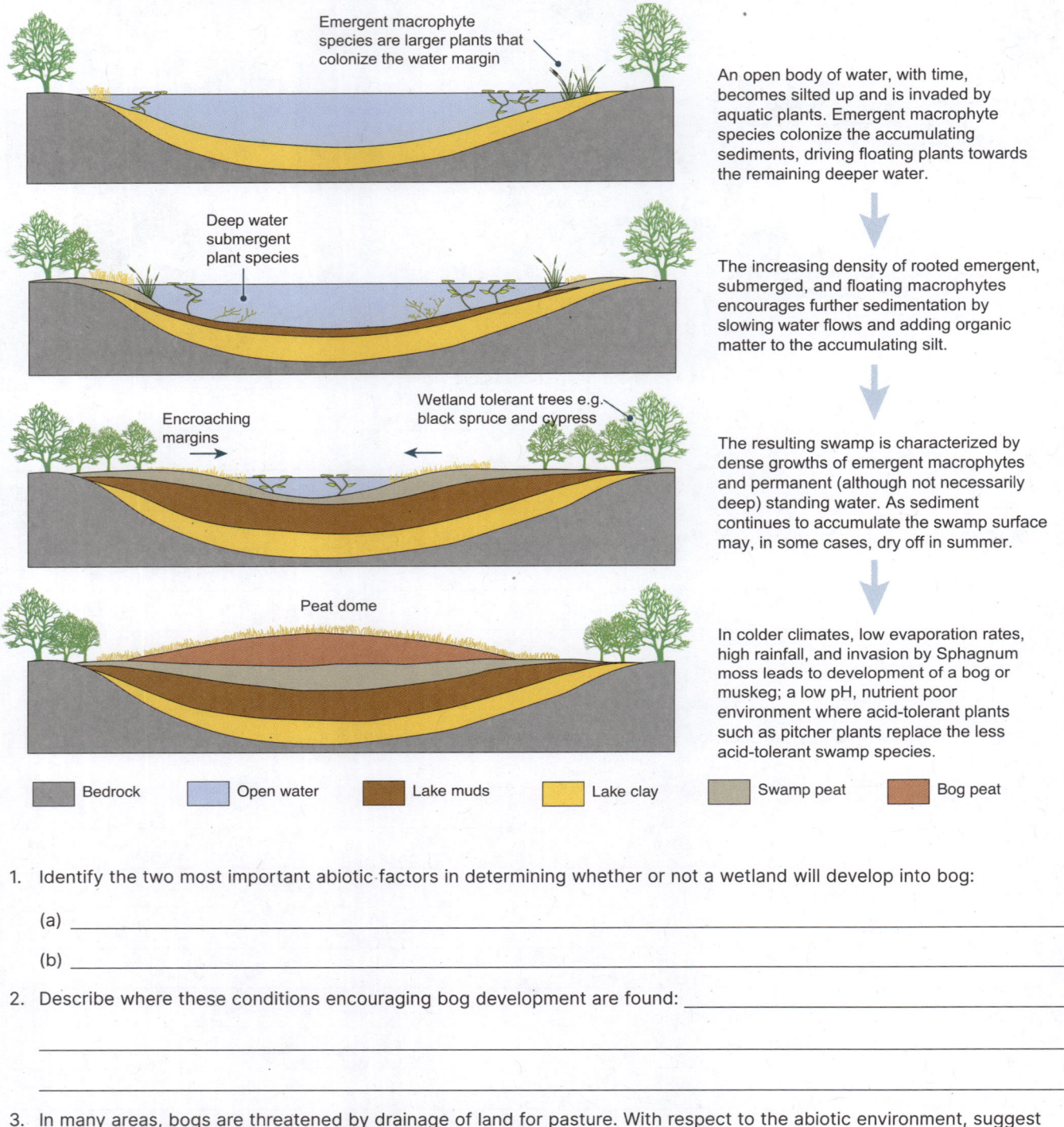

An open body of water, with time, becomes silted up and is invaded by aquatic plants. Emergent macrophyte species colonize the accumulating sediments, driving floating plants towards the remaining deeper water.

The increasing density of rooted emergent, submerged, and floating macrophytes encourages further sedimentation by slowing water flows and adding organic matter to the accumulating silt.

The resulting swamp is characterized by dense growths of emergent macrophytes and permanent (although not necessarily deep) standing water. As sediment continues to accumulate the swamp surface may, in some cases, dry off in summer.

In colder climates, low evaporation rates, high rainfall, and invasion by Sphagnum moss leads to development of a bog or muskeg; a low pH, nutrient poor environment where acid-tolerant plants such as pitcher plants replace the less acid-tolerant swamp species.

Legend: Bedrock | Open water | Lake muds | Lake clay | Swamp peat | Bog peat

1. Identify the two most important abiotic factors in determining whether or not a wetland will develop into bog:

    (a) _____

    (b) _____

2. Describe where these conditions encouraging bog development are found: _____

3. In many areas, bogs are threatened by drainage of land for pasture. With respect to the abiotic environment, suggest why land drainage threatens the viability of bog ecosystems:

58 Did You Get It?

1. The schematic below shows the movement of energy and minerals from producers to consumers.

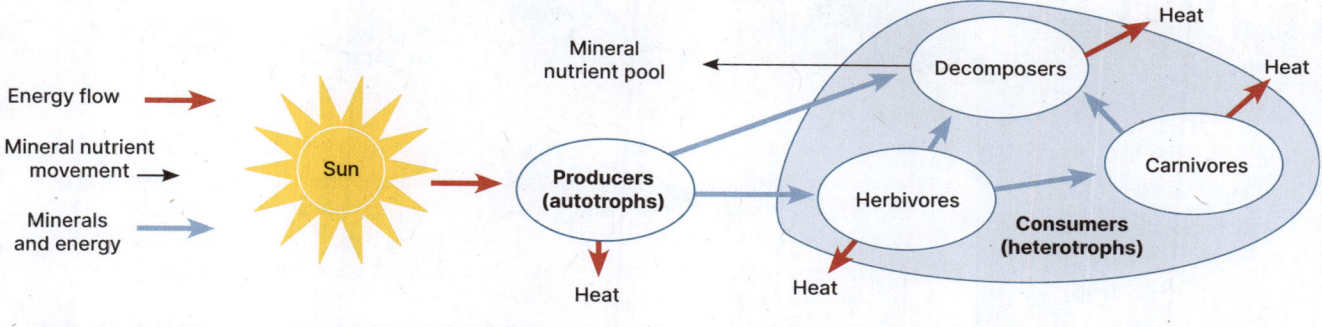

   (a) How are the movements of minerals and energy different? _____
   _____
   _____

   (b) What process is responsible for losses of energy from the system? _____

2. Draw a labeled diagram to illustrate the process of primary succession:

3. (a) Use the table below to draw the ecological pyramid of the grassland community:

Numbers in a grassland community	
Trophic level	Number of organisms
Producer	1,500,000
Primary consumer	200,000
Secondary consumer	90,000
Tertiary consumer	1

   (b) Which trophic level is the herbivore? _____

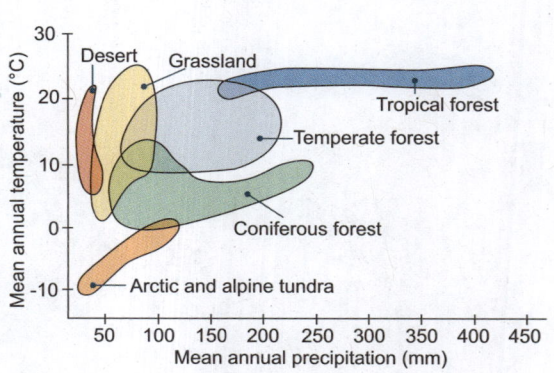

4. The graph on the left shows the annual temperature and rainfall of various biomes. What is a biome?
   _____
   _____
   _____
   _____
   _____

5. Which biome has the greatest rainfall per year?
   _____

Chapter 3 The Living World
Populations

Resource Hub
bit.ly/3LIF6Z7

Key Terms

- biotic potential
- carrying capacity (K)
- demographic transition model (DTM)
- density dependent factor
- density independent factor
- developed country
- developing country
- emigration
- exponential growth
- fertility rate
- immigration
- K-selected species
- life expectancy
- logistic growth
- mortality
- natality
- population
- population density
- population distribution
- population growth
- r-selected species
- survivorship curve
- urbanization

Key Concepts

- Population growth is regulated by density dependent and density independent factors.
- Population characteristics can be described and measured in a variety of ways including changes to natality, mortality, and net migration; and graphically using age structure pyramids and survivorship curves.
- Human population dynamics are shifting: growth rates are slowing, and distribution and density patterns are changing.

Features of Populations

Learning Outcomes: | **Activity Number**

☐ 1 Describe how populations can be measured and the factors that can affect the density and distributions of individuals within a population. — 58-59

☐ 2 Explain the difference between density dependent and density independent factors that affect populations and give examples. — 60

☐ 3 List the factors that affect population growth and calculate population changes from data. — 61

☐ 4 Interpret survivorship curves and describe population patterns from these. List the factors that affect life expectancy in humans and calculate human survivorship from data. — 62-63

☐ 5 Compare and contrast the shapes of exponential and logistic growth curves. Explain the meaning of the term 'carrying capacity'. Use a spreadsheet to model population growth from data. — 64-65

☐ 6 Describe the features of 'r' and 'K' selected species. — 66

Population Structure and Distribution

☐ 7 Compare and contrast population pyramids for human populations. Describe how population pyramids can be used to manage animal populations. — 67

☐ 8 Describe the distribution of humans around the world and explain how the revised Valeriepieris circle can be used to describe distribution. Explain why urbanization has been increasing in many countries. — 68

☐ 9 Explain how human populations have changed over time and describe the factors that affect birth rates in different parts of the world. — 69

☐ 10 Describe the five stages of the demographic transition model and analyse its strengths and weaknesses. — 70

☐ 11 Describe the effect of the human population on the Earth's resources. — 71

59 Features of Populations

Key Idea: A number of attributes can be measured and used to define and describe a population.

A **population** is a group of individuals of the same species living in a given area at a given time. Accurately describing key characteristics provides information about a population and its interactions within its environment. Population size, i.e. the total number of organisms in the population, is a basic measurement, but it is also useful to know the **population density**, i.e. the number of organisms per unit area. The density of a population is often a reflection of the carrying capacity of the environment. The **carrying capacity** is how many organisms an environment can support. Populations also have structure: particular ratios of different ages and sexes. These data enable us to determine whether the population is declining or increasing in size. We can also look at the distribution of organisms within their environment and determine what particular aspects of the habitat are favored over others. Some of the population attributes that we can measure or calculate are illustrated on the diagram below.

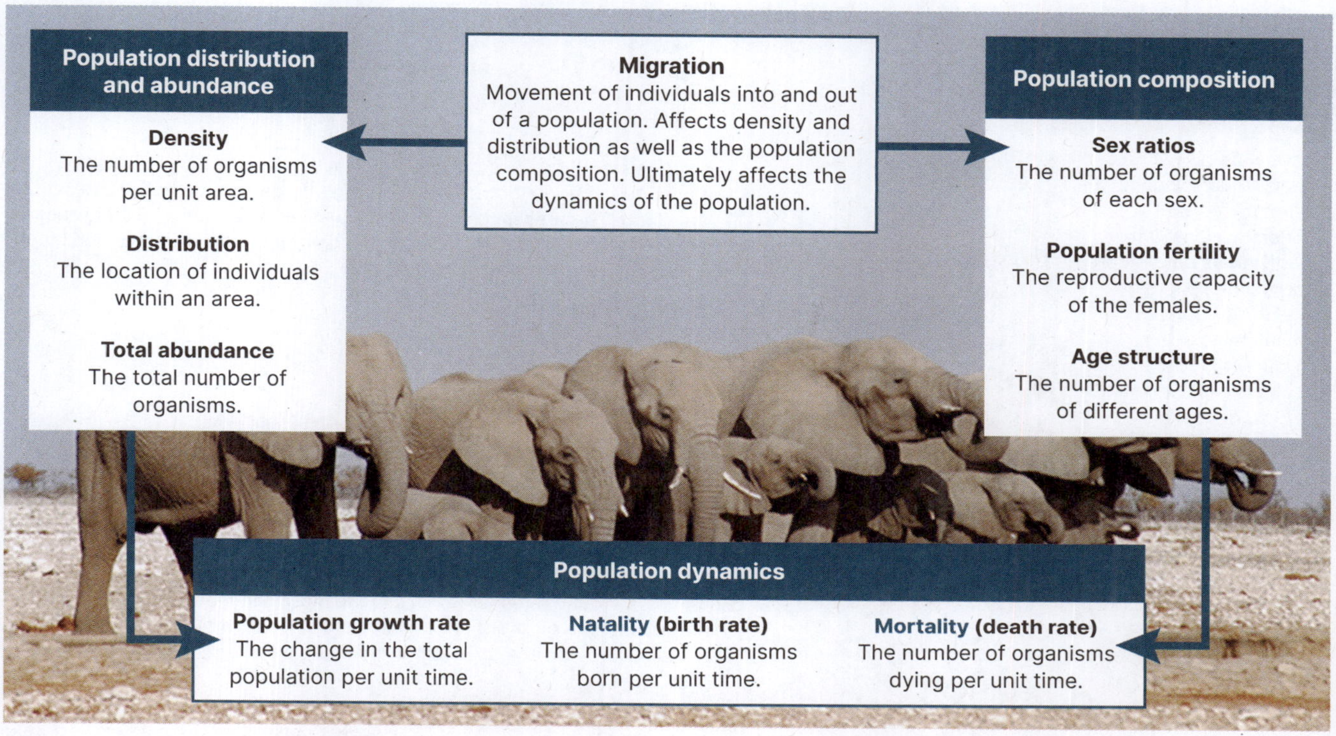

Population distribution and abundance

Density
The number of organisms per unit area.

Distribution
The location of individuals within an area.

Total abundance
The total number of organisms.

Migration
Movement of individuals into and out of a population. Affects density and distribution as well as the population composition. Ultimately affects the dynamics of the population.

Population composition

Sex ratios
The number of organisms of each sex.

Population fertility
The reproductive capacity of the females.

Age structure
The number of organisms of different ages.

Population dynamics

Population growth rate
The change in the total population per unit time.

Natality (birth rate)
The number of organisms born per unit time.

Mortality (death rate)
The number of organisms dying per unit time.

1. Define the term population: _____

2. List the population attributes that would be good indicators of whether the population is increasing or decreasing: _____

3. Describe how population growth rate, death rate (mortality), and birth rate (natality) would most likely be trending in a growing population: _____

4. (a) Identify the population attributes that can be measured directly from the population: _____

   (b) Identify the population attributes that must be calculated from the data collected: _____

60 Density and Distribution

Key Idea: Density measures the number of individuals within a given space. Distribution describes how those individuals are dispersed within the space.

Distribution and density are two interrelated properties of populations. **Population density** is the number of individuals per unit area (for land organisms) or volume (for aquatic organisms). **Population distribution** describes the spatial arrangement of organisms within a defined area. The three basic distribution patterns are random, clumped, and uniform. In the diagram below, the circles represent individuals of the same species. The diagrams can also represent populations of different species.

Low density

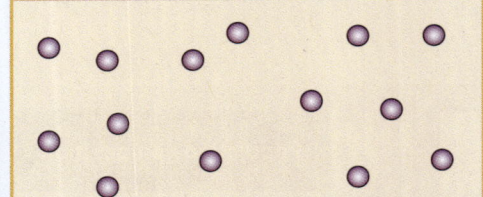

In low density populations, individuals are spaced well apart. There are only a few individuals per unit area or volume, e.g. highly territorial, solitary mammal species. Tigers (right) are solitary animals found at low densities.

High density

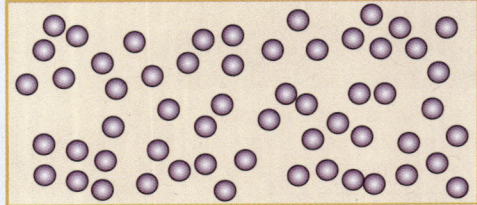

In high density populations, individuals are crowded together. There are many individuals per unit area or volume, e.g. colonial organisms, such as many corals. Termites (right) form highly organized, high density colonies.

Random distribution

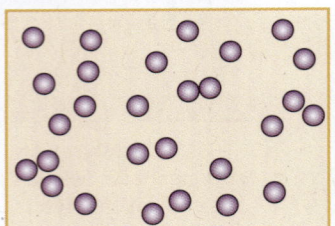 

Random distributions occur when the spacing between individuals is irregular due to unpredictable factors, e.g. the germination of seeds carried by the wind. The presence of one individual does not directly affect the location of any other individual. Random distributions are uncommon in animals but are often seen in plants.

Clumped distribution

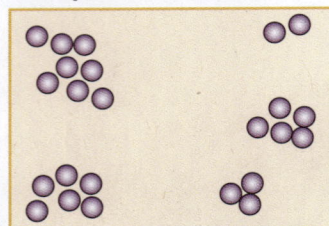 

Clumped distributions occur when individuals are grouped in patches, sometimes around a resource. This is the most common distribution pattern in nature and is typical of herd and social species. The presence of one individual increases the probability of finding another close by.

Uniform distribution

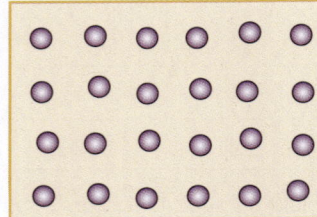 

In regular distribution, individuals are evenly spaced and the distance between them is maximized. Uniform distributions occur in territorial spaces, e.g. breeding colonies of penguins, above, but may also occur in plants that produce chemicals to inhibit the growth of other plants nearby.

1. Explain how the distribution and availability of resources might influence population density? _____

2. What type of distribution pattern would you expect to see when:

    (a) Resources are not evenly spaced out: _____
    (b) Resources are evenly spaced out: _____
    (c) Animals are social: _____
    (d) Animals are territorial: _____

3. Why do you think random distributions are uncommon in nature? _____

61 Population Regulation

Key Idea: Population size is regulated by density independent and density dependent factors.

Very few species show continued **exponential growth** where the population continues to grow rapidly at a constant rate. Population size is regulated by factors that limit population growth. The diagram below illustrates how population size can be regulated by environmental factors. **Density independent factors** are those that act independently of population density and affect all individuals more or less equally. **Density dependent factors** have a greater effect when the population density is higher. They become less important when the population density is low.

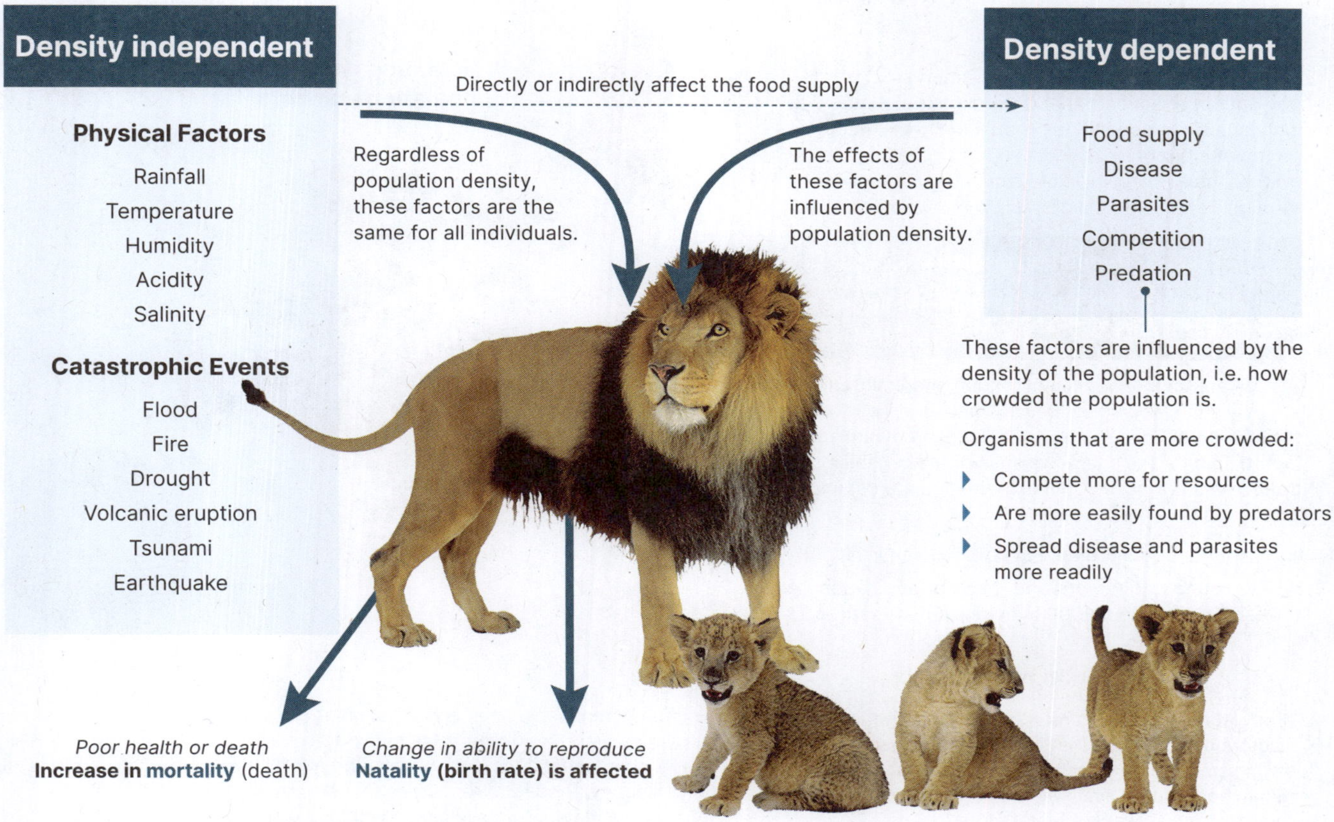

Density independent

Physical Factors
- Rainfall
- Temperature
- Humidity
- Acidity
- Salinity

Catastrophic Events
- Flood
- Fire
- Drought
- Volcanic eruption
- Tsunami
- Earthquake

Regardless of population density, these factors are the same for all individuals.

Directly or indirectly affect the food supply

The effects of these factors are influenced by population density.

Poor health or death
Increase in mortality (death)

Change in ability to reproduce
Natality (birth rate) is affected

Density dependent

- Food supply
- Disease
- Parasites
- Competition
- Predation

These factors are influenced by the density of the population, i.e. how crowded the population is.

Organisms that are more crowded:
▸ Compete more for resources
▸ Are more easily found by predators
▸ Spread disease and parasites more readily

1. Discuss the role of density dependent factors and density independent factors in population regulation. In your discussion, make it clear that you understand the meaning of each of these terms:

   _____
   _____
   _____
   _____
   _____
   _____

2. Explain how an increase in population density allows disease to have a greater influence in regulating population size:

   _____
   _____
   _____
   _____

3. In cooler climates, aphids go through a population increase during the summer months. In autumn/fall, population numbers decline steeply. Suggest how competition and cooler temperature regulate the population:

   _____
   _____
   _____

62 Population Growth

Key Idea: Changes in population size can be measured.

A **population** is a group of organisms of the same species living together in one geographical area. This area may be difficult to define as populations may be composed of widely dispersed individuals that come together infrequently, e.g. for mating. The number of individuals in a population may also fluctuate considerably over time. These changes make populations dynamic: populations gain individuals through births or **immigration**, and lose individuals through deaths and **emigration**. For a population in equilibrium, these factors balance out; there is no net change in the population. When losses exceed gains, the population declines.

Births, deaths, immigrations (movements into the population), and emigrations (movements out of the population) are events that determine the numbers of individuals in a population. Population growth depends on the number of individuals added to the population from births and immigration, minus the number lost through deaths and emigration. This is expressed as:

$$\text{Population growth} = \text{Births (B)} - \text{Deaths (D)} + \text{Immigration (I)} - \text{Emigration (E)}$$

The difference between immigration and emigration gives net migration. Ecologists usually measure the rate of these events. These rates are influenced by environmental factors and by the characteristics of the organisms themselves. Rates in population studies are commonly expressed in one of two ways:

- Numbers per unit time, e.g. 20,150 live births per year.
- Per capita rate (number per head of population), e.g. 122 live births per 1000 individuals per year (12.2%).

Limiting factors

Population size is also affected by limiting factors: factors or resources that control a process such as organism growth, or population growth or distribution. Examples include availability of food, predation pressure, or available habitat.

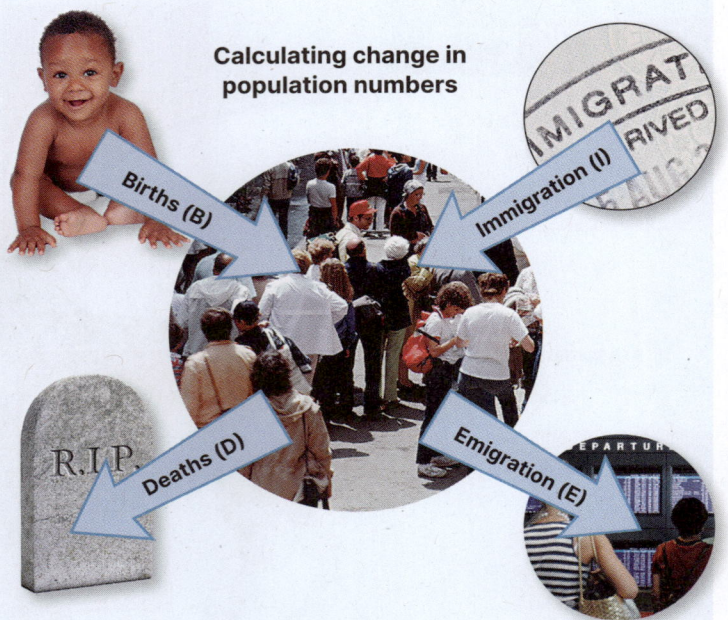

Calculating change in population numbers

Human populations often appear exempt from limiting factors because technology and medicine solve many problems. However, as the last arable (cultivated) land is used and agriculture reaches its limits of efficiency, food, and resources to maintain the growing human population may potentially act as limiting factors.

1. Define the following terms used to describe changes in population numbers:

   (a) Death rate (mortality): _____

   (b) Birth rate (natality): _____

   (c) Net migration rate: _____

2. Using the terms, B, D, I, and E (above), construct equations to express the following (the first is completed for you):

   (a) A population in equilibrium: $B + I = D + E$

   (b) A declining population: _____

   (c) An increasing population: _____

3. The rate of population change can be expressed as the interaction of these factors:

   Rate of population change = Birth rate − Death rate + Net migration rate (positive or negative)

   Using the formula above, determine the annual rate of population change for the United States and Japan in 2023:

	USA	Japan
Birth rate:	1.20%	0.70%
Death rate:	0.92%	1.15%
Net migration rate:	0.27%	0.05%

   (a) Rate of population change for USA = _____

   (b) Rate of population change for Japan = _____

   (c) Explain the biggest contributor to the result you obtained for Japan? _____
   _____

   (d) Calculate the rate of population change for your own country or state: _____

63 Survivorship Curves

Key Idea: A survivorship curve shows the number of individuals in a population that can be expected to survive to a specified age.

The **survivorship curve** depicts age-specific **mortality**. It is obtained by plotting the number of individuals in a particular cohort against time. Survivorship curves are standardized to start at 1000 and, as the **population** ages, the number of survivors progressively declines. The shape of a survivorship curve shows the life stages at which highest mortality occurs. Wherever the curve becomes steep, there is an increase in mortality. Survivorship curves in many populations fall into one of three patterns. Type I curves show the heaviest mortality late in life. Type II curves show a relatively constant mortality at all life stages. Type III curves show the highest mortality in early life stages. Many species exhibit a mix of two curve types, e.g., some birds have high chick mortality (Type III) but adult mortality is fairly constant (Type II).

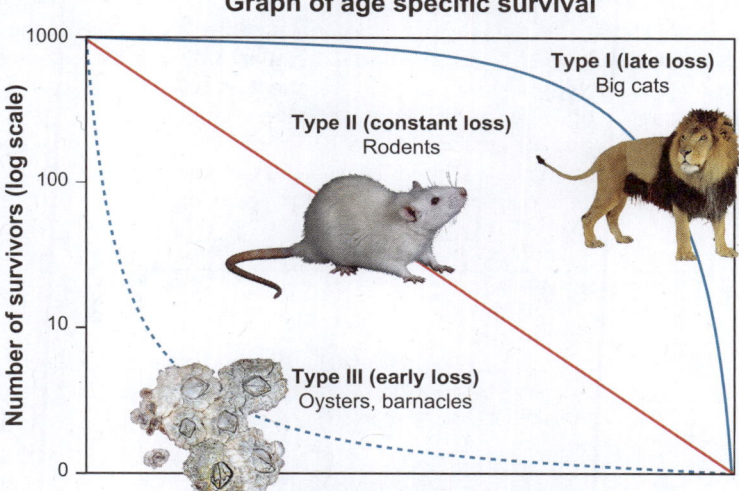

Above: Three basic types of survivorship curves and representative organisms for each. The vertical axis may be scaled arithmetically or logarithmically.

Hypothetical survivorship curves

Type I: Late loss survivorship curve

Mortality (death rate) is very low in the infant and juvenile years, and throughout most of adult life. Mortality increases rapidly in old age.
Examples: Humans (in developed countries) and many other large mammals, e.g. big cats, elephants.
Elephants (above) have a close matriarchal society and a long period of parental care. Elephants are long-lived and females usually produce just one calf.

Type II: Constant loss survivorship curve

Mortality is relatively constant through all life stages (no one age is more susceptible than another).
Examples: Some invertebrates such as *Hydra*, some birds, some annual plants, some lizards, and many rodents.
Rodents (above) are well known for their large litters and prolific breeding capacity. Individuals are lost from the population at a more or less constant rate.

Type III: Early loss survivorship curve

Mortality is very high during early life stages, followed by a very low death rate for the few individuals reaching adulthood.
Examples: Many fish (not mouth brooders) and most marine invertebrates, e.g. oysters, barnacles.
Despite vigilant parental care, many birds (above) suffer high juvenile losses (Type III). For those surviving to adulthood, deaths occur at a constant rate (Type II).

1. Match the following terms to the statements below: **A** Type I **B** Type II **C** Type III

    (a) Curve followed by organisms with high mortality rates early in life: _____

    (b) Human populations in developed countries follow this type of curve: _____

    (c) Organisms displaying constant loss over time show this curve: _____

2. How do species with type III survivorship ensure some juvenile survive the steep early loss rates?
   _____
   _____

3. Using birds as an example, discuss the following statement: 'Species do not have a standard survivorship curve. The curve depicts the nature of a population at a particular time and place and under certain environmental conditions.'
   _____
   _____
   _____

64 Life Expectancy and Survivorship in Humans

Key Idea: Life expectancy is the average number of years of life remaining at a given age. Many factors influence it.

Life expectancy is a measure of the number of years a person is expected to live for at any give age. It is based on the probability of living to the next year of life. It changes as an individual ages. For example, in the US, at birth, a human has a life expectancy of around 76 years. However, a 65 year old has a life expectancy of around 18.4 years, meaning they should live to the age of 83. In human societies, life expectancy is heavily dependent on aspects of socio-economic structure such as public health facilities, presence and treatment of endemic disease, and level of poverty. These factors are not static. Countries where war, famine, or disease are common, invariably have low life expectancies. Life expectancy is also affected by gender: international life expectancy for males is 71 compared with 76 for females.

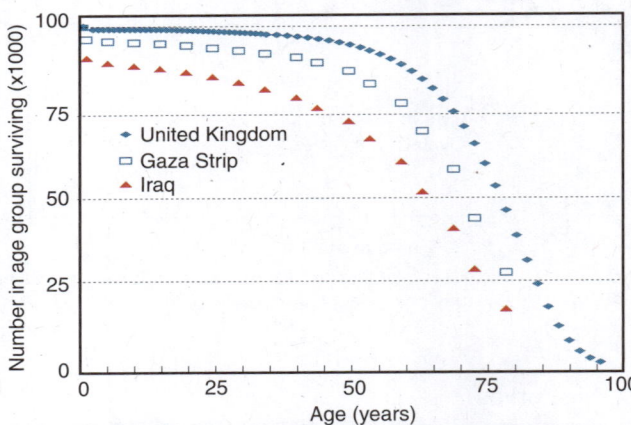

Human populations typically follow a type I **survivorship curve**. However, the average life expectancy can vary greatly between developed and developing nations. The average life expectancy can be estimated from suvivorship curves (above) and is the age at which 50% of the people in the sample are still alive.

War, civil instability, poor infrastructure, reduced social services, and poverty can greatly reduce life expectancy. Life expectancy in stable, industrialized nations is usually higher than in war-torn, poor, or developing nations. For example, life expectancy in Japan is around 84 at birth, but as low as 57 at birth in Somalia.

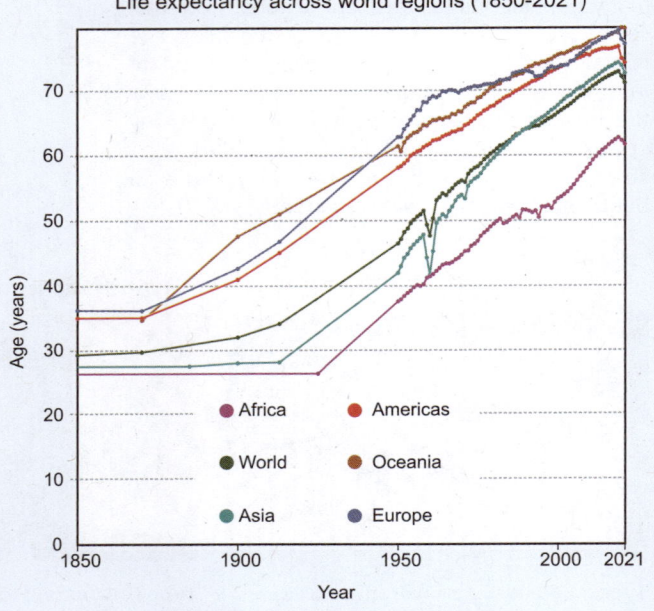

Life expectancy has increased worldwide. In 1900, the average life expectancy for a newborn was 32 years. This has more than doubled to 71 years in 2021. The rise is due to a number of factors including advances in medicine and technology, and improved standards of living in many regions. However, there is still large variability in life expectancy between the world's geographical regions. This can be attributed to variations in a number of factors including standard of living, healthcare, socio-economic factors, e.g. poverty, famine, and war. The graph above provides life expectancy data for the world as a whole and for each of the major geographical regions.

1. Estimate average life expectancy for: (a) United Kingdom: _____ (b) Gaza Strip: _____ (c) Iraq: _____

2. Identify some factors that lower life expectancy and survivorship of a human population: _____

3. (a) Explain the general world trend in life expectancy between 1900 and 2021: _____

   (b) Identify the geographical region with the lowest life expectancy. Suggest a reason for this:

Cemeteries are an excellent place to study changes in human demographics. Data collected from headstones can be used to calculate death rates and produce survivorship curves. It is also possible to compare survivorship curves over different periods and see how certain factors, e.g. war, medical advances, have altered survivorship.

The data (right) represents age of death data for males and females collected over two different time periods: pre-1950 and post-1950. The pre-1950s was characterized by two world wars, and the prevalence of diseases such as polio and tuberculosis (TB). The years post-1950 have also seen global conflict, but to a lesser degree than the pre-1950 period. Many advances in medicine, e.g. vaccines and widespread use of antibiotics, as well as technology have been made during this time.

The data used in this exercise has been collected from the online records of several cemeteries across five different states in the United States to provide representative data.

Pre-1950				Post 1950			
Males age at death		Females age at death		Males age at death		Females age at death	
81	5	9	4	80	31	92	76
40	0	76	18	81	78	46	92
54	24	0	71	79	56	44	96
70	70	78	2	81	86	70	65
75	39	69	63	8	80	80	54
64	71	6	1	30	64	71	87
45	27	46	84	88	41	88	82
22	64	60	68	90	76	65	80
71	0	84	58	84	17	51	90
62	41	75	19	64	40	80	85
89	77	43	24	60	79	87	63
31	21	64	62	71	74	76	58
10	1	67	52	62	46	63	89
42	75	42	29	63	71	33	56
1	50	39	8	83	90	99	86

Data source: http://www.interment.net/us/us/index.htm

4. Complete the table below using the cemetery data provided. The males pre-1950 data have been entered for you.

   (a) In the number of deaths column, record the deaths for each age category. You may wish to do a tally chart first.

   (b) Calculate the survivorship for each age category. For each column, enter the total number of individuals in the study (30) in the 0-9 age survivorship cell. This is the survivorship for the 0-9 age group. Subtract the number of deaths at age 0-9 from the survivorship value at age 0-9. This is the survivorship at the 10-19 age category. To calculate the survivorship for age 20-29, subtract the number of deaths at the age 10-19 age category from the survivorship value for age 10-19. Continue until you have completed the column.

Age	Males pre-1950		Females pre-1950		Males post 1950		Females post 1950	
	No. of deaths	Survivorship	No. of deaths	Survivorship	No. of deaths	Survivorship	No. of deaths	Survivorship
0-9	5	30	7	30	1	30	0	30
10-19	1	25	2	23	1	29	0	30
20-29	4	24	2	21	0	28	0	30
30-39	2	20	1	19	2	28	1	30
40-49	4	18	3	18	3	26	2	29
50-59	2	14	2	15	1	23	4	27
60-69	3	12	7	13	5	22	4	23
70-79	7	9	4	6	7	17	4	19
80-89	2	2	2	2	8	10	10	15
90-99	0	0	0	0	2	2	5	5
Total	30		30		30		30	

5. (a) On a separate piece of graph paper, construct a graph to compare the survivorship curves for each category. Attach the graph into this worktext once you have completed the activity.

   (b) What conclusions can you make about survivorship before 1950 and after 1950? _____

   (c) What factors might cause these differences? _____

65 Population Growth Curves

Key Idea: Populations typically show either exponential or logistic growth. The maximum sustainable population size is limited by the environment's carrying capacity.

Population growth is the change in a population's numbers over time (dN/dt or $\Delta N/\Delta t$). It is regulated by the **carrying capacity** (K), which is the maximum number the environment can sustain. Population growth falls into two main types: exponential or logistic. Both can be defined mathematically. In these mathematical models, the per capita (or intrinsic) growth rate is denoted by a lower case r, determined by the per capita births minus deaths, i.e. $(B-D)/N$. **Exponential growth** occurs when resources are essentially unlimited. **Logistic growth** begins exponentially, but slows as the population approaches environmental carrying capacity.

Exponential growth

Exponential growth occurs when the population growth rate is not affected by the population size, N. In this case, the population growth rate is simply r (the maximum per capita rate of increase) multiplied by N so that $dN/dt = rN$. On a graph, exponential growth is characterized by a J shaped curve. A lag phase occurs early in population growth due to low population numbers.

In nature, exponential growth is observed in two circumstances:

(1) A few individuals begin a new population in a new habitat with plenty of resources.

(2) A natural disaster reduces the population to a few survivors, and the population recovers from a low base.

The human population is currently in an exponential phase of growth. In ancient times, the human population remained relatively stable, but low. It was not until the beginning of the Renaissance (14th-17th centuries) that the population began to grow. The use of machines during the Industrial Revolution increased living standards and the population increased too. Medical advances, e.g. antibiotics, and increased food production due to the Green Revolution, sparked the current rapid increase in the human population.

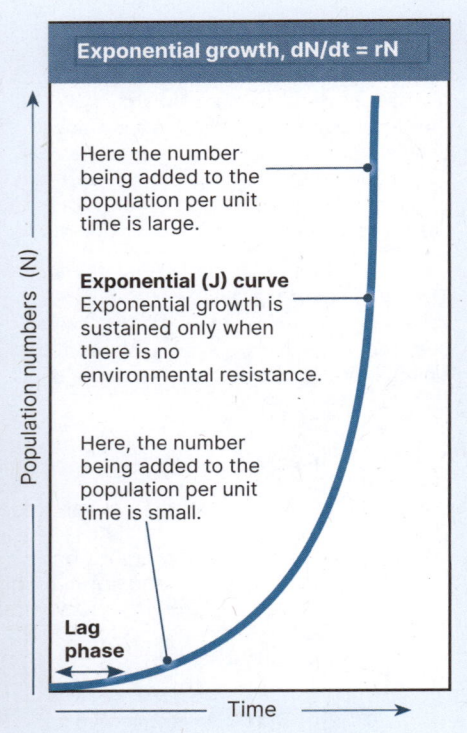

Exponential growth, $dN/dt = rN$

Here the number being added to the population per unit time is large.

Exponential (J) curve
Exponential growth is sustained only when there is no environmental resistance.

Here, the number being added to the population per unit time is small.

Lag phase

Population numbers (N) / Time

Growth of a bacterial population

In bacterial populations, each cell divides in two, so the population doubles with every generation. Bacteria growing without environmental constraint show exponential growth. To study bacterial growth, a culture of *E.coli* bacteria was placed into a complex growth medium, rich in nutrients, and placed into a 37°C waterbath. A sample of the culture was taken every 30 minutes and the absorbance of the solution was measured using a spectrophotometer. As bacterial numbers increased, the solution became more cloudy and this showed as an increase in the absorbance reading. The results are below.

Need help? See Activity 223

Incubation time (min)	0	30	60	90	120	150	180	210	240	270	300
Absorbance (660nm)	0.014	0.015	0.019	0.033	0.065	0.124	0.238	0.460	0.698	0.910	1.070

1. (a) On the grid (right) plot the results of bacterial growth:

   (b) Does the growth curve show exponential growth? Explain why or why not:

   _____

   _____

   _____

   (c) Suggest how the *E.coli* population growth would be affected if the culture was grown in a low nutrient medium:

   _____

   _____

Logistic growth

In nature, the population growth of most organisms follows a logistic growth curve. When entering a new environment, a founding population will enter a period of exponential growth. The maximum population size the environment can support is called the carrying capacity (K). As the population nears K and the resources become limiting, population growth slows.

Under the logistic growth model, dN/dt = rN is multiplied by the proportion of K that is left unfilled or unused. As the population increases, the proportion of K available decreases and individuals find it difficult to find or utilize space and resources. The rate of population increase therefore slows as population size approaches carrying capacity.

Occasionally, a population's growth rate may not slow as it approaches K. This usually occurs in rapidly breeding organisms when there is a time lag between the depression in resources and the population response. In this case, the population overshoots K and then declines again as it responds to low resource availability. In time, populations usually stabilize around K.

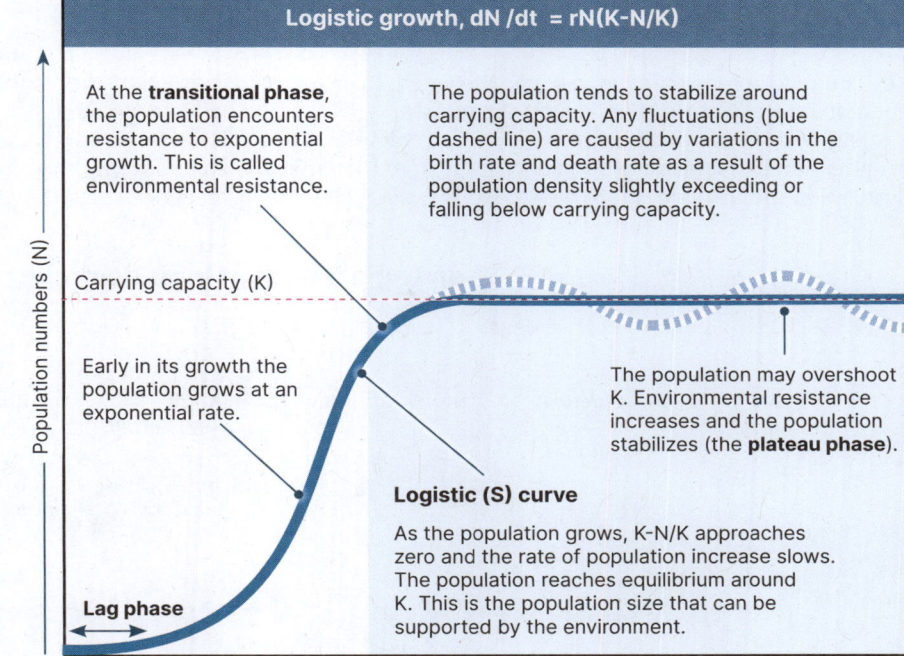

Logistic growth, dN/dt = rN(K-N/K)

At the **transitional phase**, the population encounters resistance to exponential growth. This is called environmental resistance.

The population tends to stabilize around carrying capacity. Any fluctuations (blue dashed line) are caused by variations in the birth rate and death rate as a result of the population density slightly exceeding or falling below carrying capacity.

Carrying capacity (K)

Early in its growth the population grows at an exponential rate.

The population may overshoot K. Environmental resistance increases and the population stabilizes (the **plateau phase**).

Logistic (S) curve

As the population grows, K-N/K approaches zero and the rate of population increase slows. The population reaches equilibrium around K. This is the population size that can be supported by the environment.

Lag phase

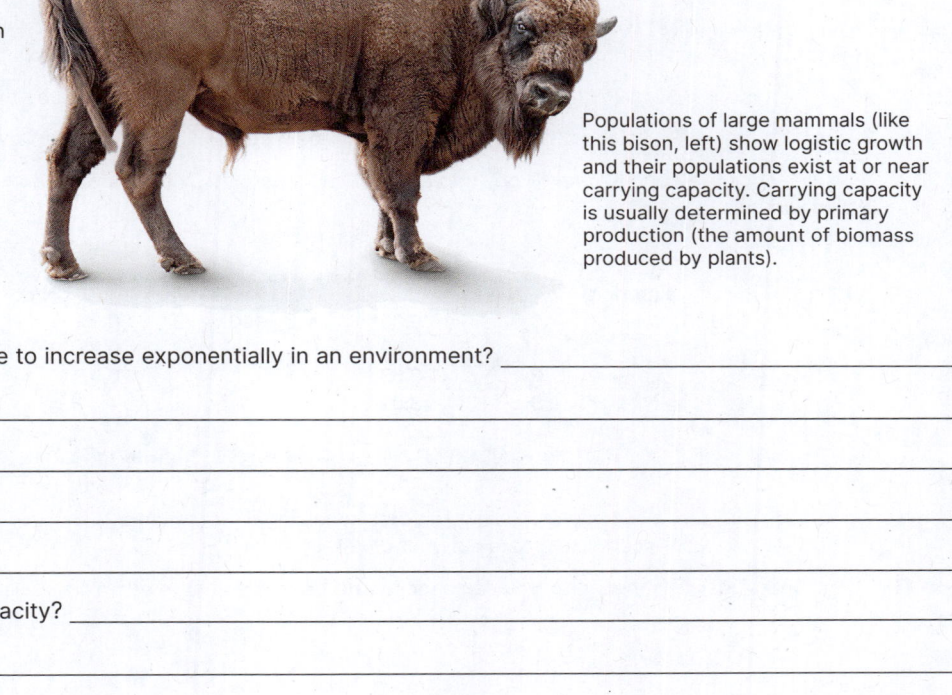

Populations of large mammals (like this bison, left) show logistic growth and their populations exist at or near carrying capacity. Carrying capacity is usually determined by primary production (the amount of biomass produced by plants).

2. Why don't populations continue to increase exponentially in an environment? _____
_____
_____
_____
_____

3. What is meant by carrying capacity? _____
_____
_____
_____

4. Describe and explain the phases of the logistic growth curve: _____
_____
_____
_____
_____
_____
_____

66 Modeling Population Growth

Key Idea: A spreadsheet model can be used to help us understand the factors affecting logistic growth.

Plotting a logistic growth curve on a spreadsheet can help in understanding **logistic growth** and the factors that affect **population growth** rate. In this exercise, you will follow the instructions below to create your own spreadsheet model of logistic growth for a hypothetical population of 2, where r is 0.15 and K 100. You can use Microsoft Excel or an equivalent spreadsheet program. Refer back to activity 65 if you need a reminder about the notations used in this activity.

Creating a model of logistic growth

1. In cells A1 to F1, add the headings r, t(period), N, K, K-N/K, and dN/dt, as shown in the image below.

	A	B	C	D	E	F	G
1	r	t (period)	N	K	K-N/K	dN/dt	
2	0.15	0	2	100	=($D$2-C2)/$D$2	=$A$2*C2*E2	
3		=B2+1	=C2+F2				
4							
5							
6							
7							
8							

Population at t_1 = population at t_0 + dN/dt (the amount of population change over 1 time period).

2. In cell A2, type 0.15, the value for r.
3. In cell B2, type 0.
4. In cell C2, type 2 (the initial population number).
5. In cell D2, type 100 (the carrying capacity).
6. In cell E2, type =($D$2-C2)/$D$2.
   This term, K-N/K, is the fraction of the carrying capacity that has not yet been "used up."
7. In cell F2, type =$A$2*C2*E2. This is the change in population number described by the logistic equation rN(K-N/K).
8. In cell B3, type =B2+1. In cell C3, type =C2+F2. Shift-select cells B3 and C3 and fill down.
9. Shift-select cells E2 and F2 and fill down to about 60 time periods. The cells will automatically calculate (the first few are shown below).

dN = change in numbers   dt = change over time
t_0 = time zero t_1 = one time period

	A	B	C	D	E	F	G
1	r	t (period)	N	K	K-N/K	dN/dt	
2		0.15	0	2.00	100	0.98	0.29
3			1	2.29		0.98	0.34
4			2	2.63		0.97	0.38
5			3	3.01			
6							

10. When your time series is complete, select the data in columns B (time) and C (Numbers) and choose < Insert < Chart < XY scatter to create a plot of dN/dt.
11. Under Chart Design in the menu, you can choose Add Chart Element to add axis labels and a title.

1. Describe the shape of the curve you have plotted: _____
2. Around which time period does the curve on the spreadsheet above begin to flatten out? _____
3. Use the logistic equation and mathematical reasoning to explain the changes in population growth rate (dN/dt):

67 r and K Selection

Key Idea: *r*-selected species have high biotic potentials (*r*) and typically show exponential growth. K-selected species have lower biotic potentials and exist near carrying capacity. The maximum rate at which a **population** can grow (its intrinsic rate of increase or *r*) is also called its **biotic potential**. It is a measure of reproductive capacity and is assigned a set value that is specific to the organism involved. Species with a high biotic potential are called **r-selected species**. They include algae, bacteria, rodents, many insects, and most annual plants. These species show life history features associated with rapid growth in disturbed environments and they usually exist well below the **carrying capacity** (K) of the environment. Species with lower biotic potential tend to exist at or near K. These **K-selected species**, which include most large mammals, birds of prey, and large, long-lived plants are forced, through their interactions with other species, to use resources more efficiently. These species have fewer offspring and longer lives, and put their energy into nurturing their young to reproductive age. Most organisms have reproductive patterns between these two extremes.

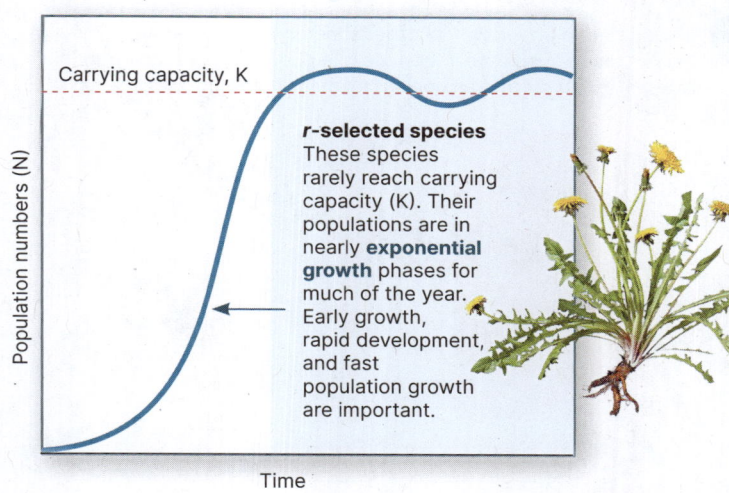

r-selected species
These species rarely reach carrying capacity (K). Their populations are in nearly **exponential growth** phases for much of the year. Early growth, rapid development, and fast population growth are important.

Features of r-selected species	
Climate	Variable and/or unpredictable
Mortality	Density independent
Survivorship	Often Type III (early loss)
Population size	Fluctuates wildly. Often below K
Competition	Variable, often lax. Generalist niche
Selection favors	Rapid development, high biotic potential, early reproduction, small body size, single (annual) reproduction pattern
Length of life	Short (usually less than one year)
Leads to:	Productivity

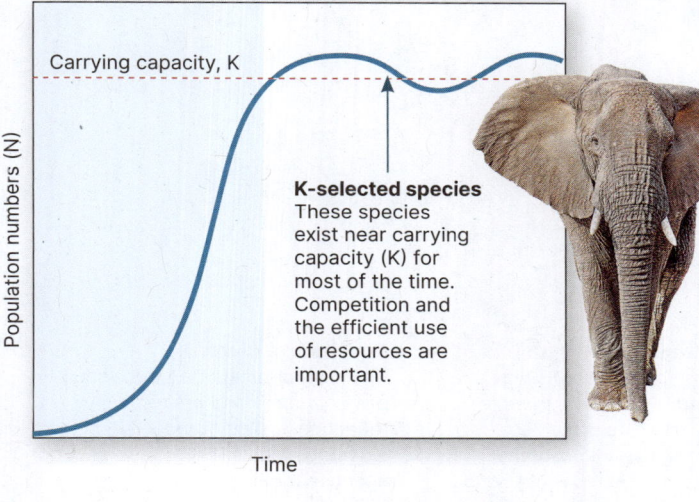

K-selected species
These species exist near carrying capacity (K) for most of the time. Competition and the efficient use of resources are important.

Features of K-selected species	
Climate	Fairly constant and/or predictable
Mortality	Density dependent
Survivorship	Types I or II (late or constant loss)
Population size	Fairly constant in time. Near equilibrium with the environment
Competition	Usually high. Specialist niche
Selection favors	Slower development, larger body size, greater competitive ability, delayed reproduction, repeated reproduction
Length of life	Longer (greater than one year)
Leads to:	Efficiency

1. Compare the typical population sizes and curves of r- and K-selected species. Include reference to carrying capacity:

2. Why are r-selected species able to adapt to changes in environment more quickly than K-selected species?

68 Population Age Structure

Key Idea: The age structure of a population refers to the relative proportion of individuals in each age group in the population.

Population age structure shows how many individuals are in each age group in a **population**. Populations can be classified according to specific age categories, e.g. years, life stage, e.g. egg, larvae, pupae, or size class, e.g. height or diameter in plants. A higher proportion of reproductive and pre-reproductive individuals indicates greater growth potential than in a population dominated by older individuals. Age structures are often shown as pyramids (below). The proportions of individuals in each category are plotted with the youngest individuals at the pyramid's base. The number of individuals moving from one age class to the next influences the age structure of the population from year to year. The loss of an age class can influence a population's viability and can even lead to population collapse (next page).

Age structures in human populations 2023

Population pyramids are useful tools for visualizing the age structure of a population and the ratios of males to females. The graphs show at a glance if a population is growing, declining, or stationary, and the information can be used to predict trends and plan for services in the future. e.g. more aged-care facilities in countries with an ageing population. The population pyramids below show three different population structures.

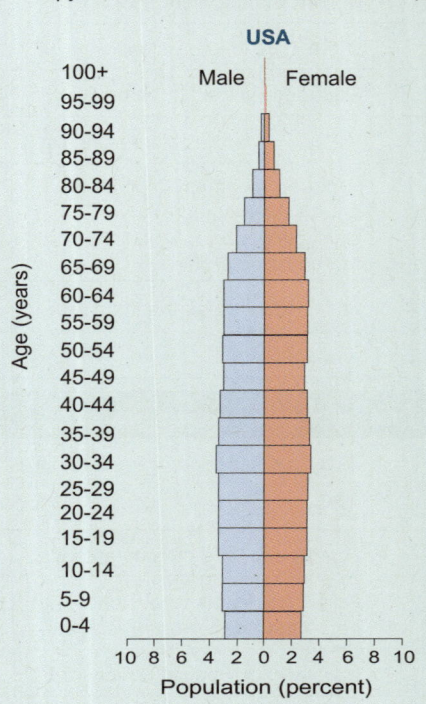

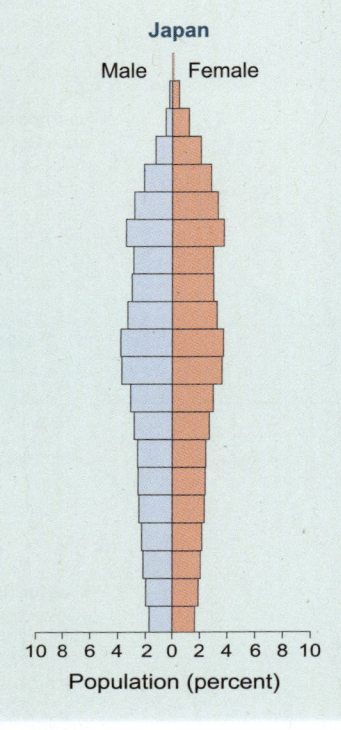

Stationary (USA):
Stable populations are characterized by an even, pillar shape pyramid, reflecting the population is neither growing or declining. There are relatively equal numbers of individuals at each age category and ratios of males and females remains fairly constant too. This pyramid is typical of developed countries where birth rates are low and the overall quality of life is high.

Expansive (Nigeria):
Rapidly growing populations are characterized by a classic triangle shape. The birth rate is high so the base of the pyramid is broad, reflecting the high number of young individuals within the population. Life expectancy is often lower, so fewer individuals live to old age. This pyramid structure is common in developing countries where access to health care and support services may be limited.

Constrictive (Japan):
Declining populations have a small base of young and are top heavy, reflecting a higher proportion of older individuals. This shape occurs because the birth rate is low and a high proportion of the people live to be very elderly. This pyramid is common in developed countries with high levels of education and where excellent health care and other services are available to a large proportion of the population.

1. How does population distribution differ between a stationary age pyramid and constrictive age pyramid?

2. Carry out some research to find out what age structure pyramid the country you live in shows. Write the answer here:

Age structures in animal populations

Age pyramids are especially useful for displaying age data about bird and mammal populations. In the hypothetical pyramids below, the young are represented by white bars and adults by blue bars. Growing populations are characterized by a high ratio of young to adult age classes. Aging populations with poor production are dominated by older individuals.

Population management

Analysis of the age structure of a population can assist in its management because it can indicate where most of the mortality occurs and whether or not reproductive individuals are being replaced. The age structure of both plant and animal populations can be examined; a common method is through an analysis of size which is often related to age in a predictable way.

Managed Fisheries

The graphs below show the age structure of a hypothetical fish population under different fishing pressures. The age structure of the population was determined by analysing the fish catch to determine the frequency of fish in each size (age) class.

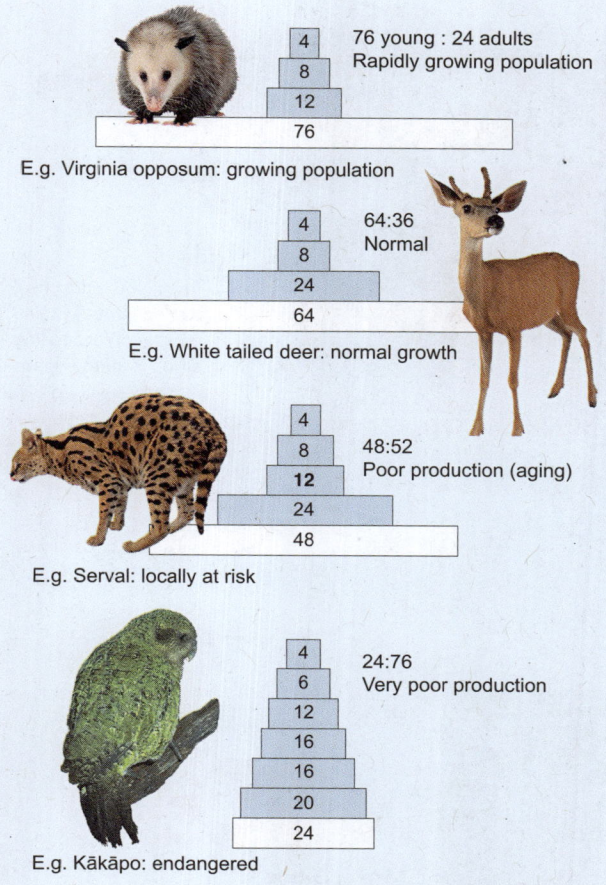

E.g. Virginia opposum: growing population
76 young : 24 adults
Rapidly growing population

E.g. White tailed deer: normal growth
64:36
Normal

E.g. Serval: locally at risk
48:52
Poor production (aging)

E.g. Kākāpō: endangered
24:76
Very poor production

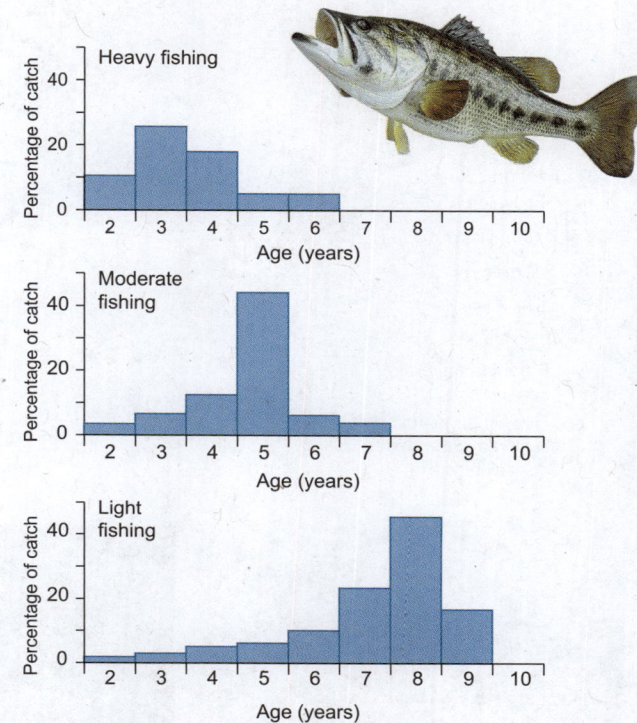

3. What characteristics of age pyramids identify populations at risk of being or becoming endangered?

_____
_____
_____

4. For the managed fish population above right, describe how the age structure changes when the fishing pressure increases from light to heavy levels:

_____
_____

5. State the most common age class for each of the above fish populations with different fishing pressures:

    (a) Heavy: _____ (b) Moderate: _____ (c) Light: _____

6. Why is knowledge of age structure important in managing a resource? _____
_____
_____

69 World Population Distribution

Key Idea: The human population is not evenly distributed across the globe. More people live in cities and towns than ever before.

In early 2024, the world's population was eight billion people. However, the human population is not evenly distributed across the globe, China and India account for a third of the world's population. Even within countries populations are unevenly dispersed. Population densities can be very high in certain areas, especially in regions with reliable water sources. Over the last few hundred years, **urbanization** (the move to living in cities or towns) has increased. 55% of people now live in urban centers, and this is predicted to increase.

World population distribution

Population size
- 1 billion+
- 100 million+
- 10 million+
- 1 million+
- 0-1 million
- Partial or severe coastal pollution
- Heavily populated delta regions vulnerable to sea level rise

China and India, the only two countries with a population exceeding 1 billion, together make up more than a third of the world's population.

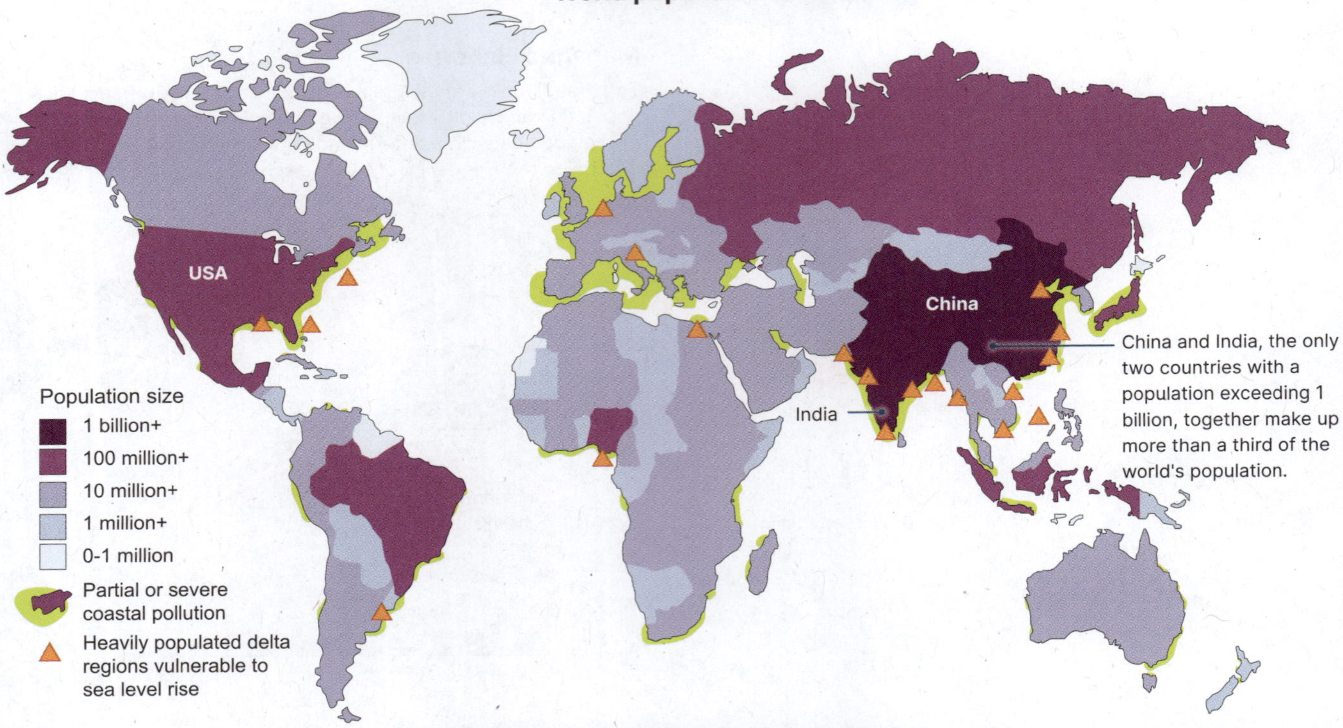

India 1.4 billion
The majority of India's population (66%) live rurally (above) and depend on agricultural activities for income and resources. India also has some very large cities, and they are the site of many important technology-based and manufacturing industries.

USA: 341 million
The night time image of mainland USA (above) shows the population is unevenly distributed: the majority live in the East. Coastal land accounts for 10% of the US footprint, but 40% of the population live in these coastal regions.

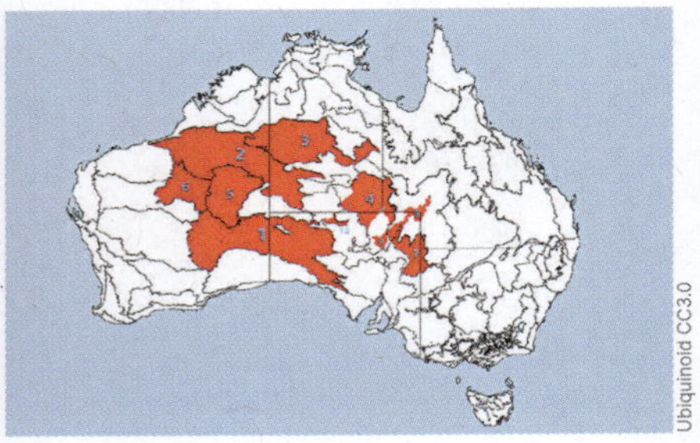

Australia: 26.6 million
Australia is the driest inhabited continent in the world. Its 10 deserts (red above) cover 18% of the mainland. The majority of Australia's population lives on or near the coast, near reliable water sources. Only 3% of the population live in desert regions.

1. What factor is an important driver for determining the location of human settlements? _____

2. Calculate how many people in Australia live in a desert region: _____

3. Asia has a large number of heavily populated delta regions vulnerable to sea level rise. Predict how sea level rise would drive the migration patterns of people in affected areas:

_____
_____
_____
_____

The Valeriepieris circle

▸ In 2013, a US teacher called Ken Myers constructed a map showing that more than half of Earth's population lived within an 8,000 kilometer diameter circle centered over the South China Sea. The circle was called the Valeriepieris circle (after Mr Myers' online user name).

▸ In 2015, Singaporean professor Danny Quah presented a revised model. His circle was smaller, it had a 6,600 kilometer diameter and was centered over the small town of Mong Khet in Mayanmar (right). In 2022, it was estimated that 4.2 billion people lived within this circle, more people live inside the circle than outside of it! This seems hard to believe when you see the small area the circle encompasses, and how much of the Earth sits outside it.

▸ The circle includes large areas of ocean and includes sparsely populated regions such as Mongolia and Serbia. However, it does also include India and China and several other populous Asian countries. The circle also contains 22 of the world's 37 megacities (cities with a population exceeding 10 million people).

▸ Mathematical calculations show Quah's Valeriepieris circle model is the smallest circle possible able to contain half of the world's population.

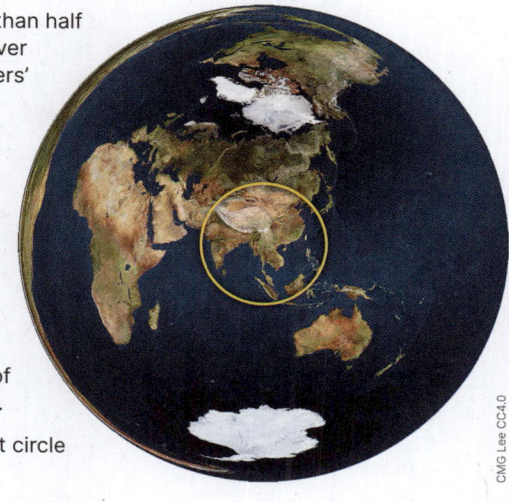

Population density: Two extremes

Macau

Kowloon walled city

Population density measures how many people live in a specified area, usually a square kilometre (km^2). It is calculated by dividing the total population of an area by its total land area. Globally, the average population density is 62 people per km^2. Island nations and independent territories can be some of the most densely populated countries.

Macau (a territory of China) is the most densely populated territory in the world. It has 21,000 people per km^2. Monaco (Europe) is second with 19,400 people per km^2 and Singapore has 8,430 people per km^2. The least densely populated country is Greenland (0.1 people per km^2). Australia's population density is 3.5 people per km^2.

At its peak, Kowloon walled city (Hong Kong), housed 50,000 people within a tiny 0.026 km^2 space. The population density was 1.9 million people per km^2. Living conditions and sanitation became intolerable, and it was demolished in 1994. Dharavi slum, India, is one of the most densely populated areas on Earth today (300,000 people per km^2).

4. Professor Quah's revised Valeriepieris circle contains more than half of the world's population, yet contains only a small proportion of Earth's surface. Explain how this can be:

_____
_____
_____
_____

5. (a) Define population density: _____
_____
_____

(b) The USA has a landmass of 9,147,590 km^2 and population of 334,914,895. Calculate its population density:
_____

(c) How does this compare to the global average of 62 people per km^2? _____
_____

(d) How many times more dense is the USA compared to Australia? _____
_____

Need help? See Activity 224

Urbanization: The shift to city living

Urbanization is the movement of people living in rural areas to urban environments (towns and cities). Humans traditionally lived in low density rural settings, and it was not until the last few hundred years that a significant shift to urban living occurred. In 1800, only 8.4% of the human population lived in towns and cities. In 2024, 55% of the population is considered urban. Many factors influence immigration to an urban environment (below). While there can be many positive effects, highly populated areas can also have negative effects, especially if the infrastructure and services cannot keep up with the increasing population growth.

Traditional villages in the rural populations of less economically developed countries have a close association with the land. Rural households depend directly on agriculture or harvesting natural resources for their livelihood. The residents are linked through family ties, culture, and economics.

Immigration push factors
- Rural overpopulation
- Lack of work or food
- Changing agricultural practices
- Desire for better education
- Racial or religious conflict
- Political instability

Immigration pull factors
- Opportunity for better jobs
- Chance of better housing
- More reliable food supply
- Opportunity for greater wealth
- Freedom from village traditions
- Government policy

In cities, the majority of the population does not depend directly on natural, resource-based occupations. Cities are hubs for commerce, education, and communication, but can suffer from overcrowding, pollution, and the rapid spread of infectious diseases.

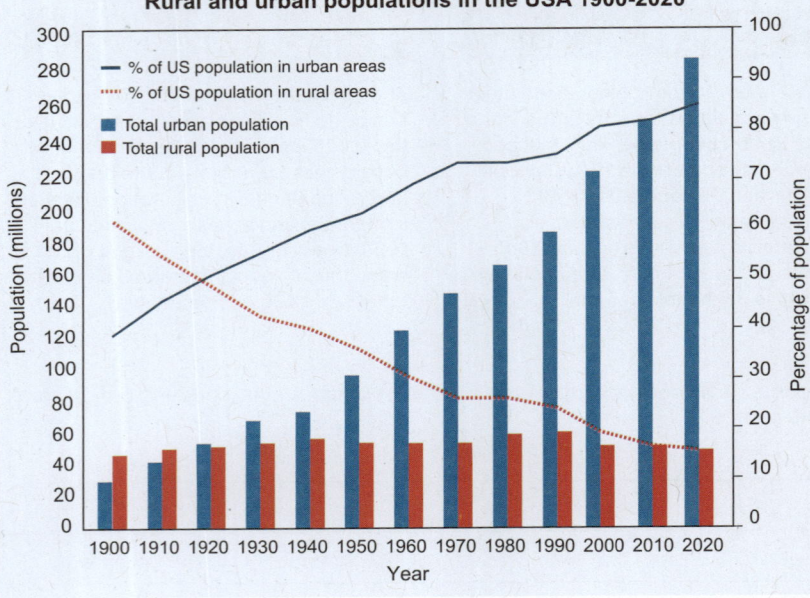

Cities, especially rapidly growing ones, face a range of problems associated with providing residents with adequate water, food, sanitation, housing, jobs, and basic services, such as health care. Slums or squatter settlements are found in many large cities in developing countries as more poor people migrate to urban areas.

6. Define urbanization: _____

7. Use the graph above to describe changes in urban populations in the US from 1900-2020: _____

8. List some negative effects of urbanization: _____

70 Population Growth Rate

Key Idea: The human population has grown substantially since the 1950s but growth is now slowing.

The human **population** reached one billion in 1804 and took a little more than 150 years to reach three billion in 1960. In early 2024, the world's population was eight billion and growing at a rate of 83 million per year. Although the global population is growing, the rate of increase has slowed since its peak in the 1960s. This is largely due to a drop in global **fertility rates**. In early 2024, total fertility rate was 2.4 births per woman; in the 1960s, the rate was five births per woman. While rapid **population growth** can put pressure on Earth's resources and have negative environmental effects, declining population rates can have negative economic and social effects, causing its own set of problems.

Differences in developing vs developed countries

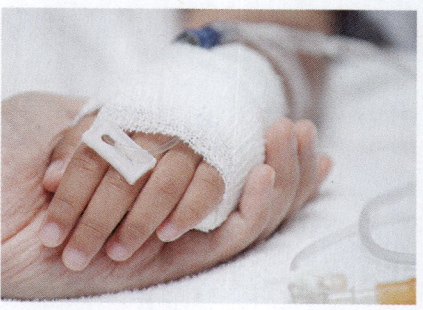

Families tend to be larger in **developing countries** than in **developed countries**. A major reason for large families in some developing countries is a societal belief that a large family is needed to help with work and to care for young or old family members. Children are often required to contribute to the family income.

In developing countries, child mortality can be high due to disease, malnutrition, poor healthcare, or poor sanitation, so parents may decide to have many children to ensure some survive to support the family. It may be harder to access contraception and family planning services which also contributes to large numbers of children.

In many developed countries, smaller families are common. This is influenced by having access to education, contraception, and family planning services. Families do not typically depend on children to contribute to family income. Having a large family can be a financial burden and this acts as a disincentive for high birth rates.

Case study: South Korea

South Korea has the lowest birth rate in the world and is no longer able to sustain its population without people moving (immigrating) to South Korea. For a population to hold steady, women need to have, on average, 2.1 children during their lifetime. In South Korea this number is 0.8 and is predicted to keep falling.

The situation in South Korea is extremely serious and a national emergency has been declared. In 50 years' time, the number of working-age people will have halved and half the population will be 65 years or older. The consequences of this include a severely reduced workforce, a reduced population for mandatory military service, and a significant drop in economic growth. A reduced workforce also means there is less tax collected, reducing money available for infrastructure and services like healthcare. There are fewer working-aged people to care for an aging population and public pension costs become extremely high.

The main drivers for the falling fertility rates in South Korea include women concentrating on career advancement and the increasing financial cost of raising children. Incentives to reverse the trend have so far failed.

1. Explain why family sizes tend to be larger in developing countries compared to developed countries:

   _____
   _____
   _____

2. Outline the social and economic affects of South Korea's low fertility rate: _____
   _____
   _____
   _____
   _____

World population growth and fertility rates

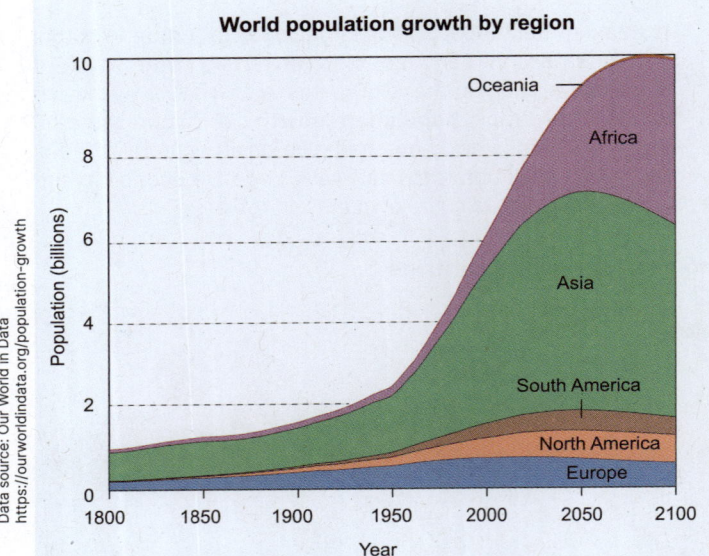

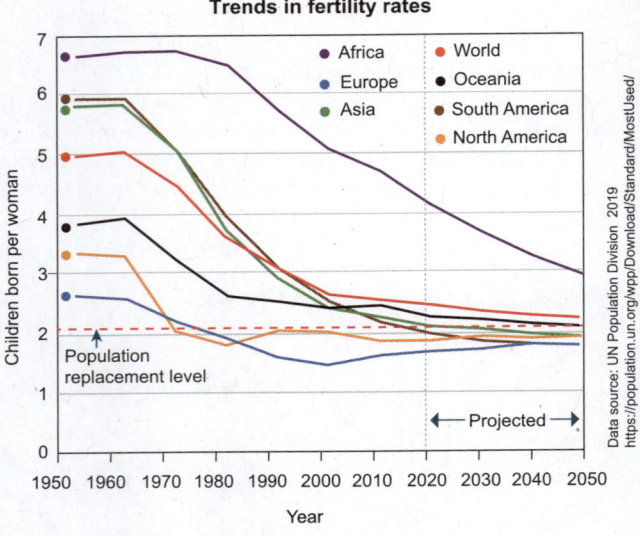

Africa is the only world region projected to have strong population growth for the rest of this century. Six countries are projected to account for more than half of the world's population growth through to the end of this century and five are in Africa (Nigeria, the Democratic Republic of Congo, Tanzania, Ethiopia, and Angola). The sixth country is Pakistan. Lagos is the most populous state in Nigeria and one of the fastest growing cities in the world. Nigeria is a lower middle-income economy but over 89 million of its more than 200 million people live in poverty.

The slowing of population growth is due largely to falling global fertility rates. Total fertility rate (TFR) is the number of children born to a woman during her lifetime. The replacement fertility rate to maintain the population is 2.1 births per woman. The global TFR is expected to be 1.8 in 2050 and 1.6 by 2100, down from 2.4 in 2024. This decline is associated with education of women, postponement of marriage, and improved access to contraception. Australia, Canada, China, Japan, New Zealand, the UK, and USA all have fertility rates below 2.1. In contrast, the fertility rates in sub-Sharahan Africa are nearly twice the global average.

3. Study the world population growth graph above.

   (a) Which geographic region is predicted to have continuing strong population growth? _____

   (b) Identify your geographic region and outline the projected population changes expected between 2000-2100:
   _____
   _____
   _____
   _____

4. Define total fertility rate: _____
   _____

5. Study the graph showing the trends in fertility rates above.

   (a) Using the world data, describe the general trend in fertility rate since 1950: _____
   _____
   _____

   (b) Which region has the highest fertility rate? _____

   (c) Estimate which regions are predicted to fall below the replacement fertility rate (2.1 in 2050)?
   _____
   _____

   (d) Describe some of the factors contributing to falling fertility rates: _____
   _____

71 Human Demography

Key Idea: The demographic transition model shows the relationship between births and deaths, and population growth over time. Its five stages are used to classify a country's industrial and economic development.

The **demographic transition model** (DTM) is used to visually show the relationship of birth rate and death rate over time. It shows how population growth rates cycle through stages as a country develops economically. Societies with minimal technology, education, and economic growth tend to have high birth rates and death rates. Countries which are more advanced and have better economic growth tend to have low birth rates and low death rates. This transition is usually associated with a shift to an industrialized economy. There are five stages (below). Most developed countries are beyond stage three, while the majority of developing countries are in stages two to three.

Population structure by DTM stage

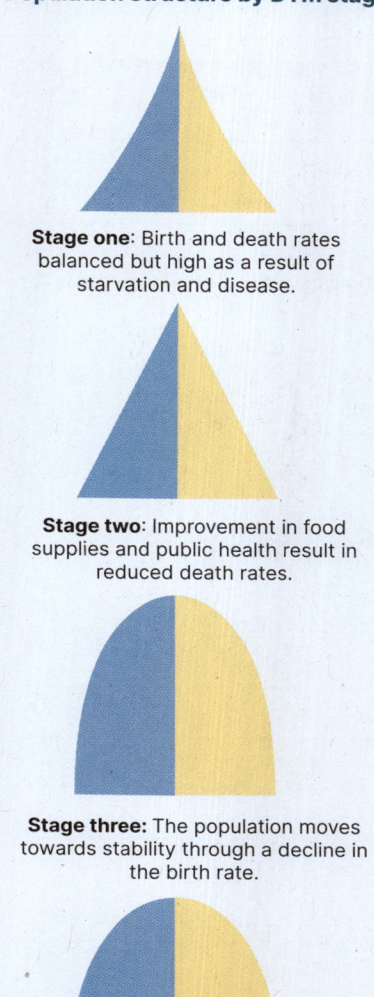

Stage one: Birth and death rates balanced but high as a result of starvation and disease.

Stage two: Improvement in food supplies and public health result in reduced death rates.

Stage three: The population moves towards stability through a decline in the birth rate.

Stage four: Birth and death rates are both low and the total population is high and stable.

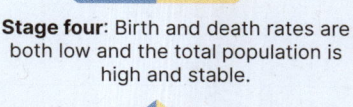

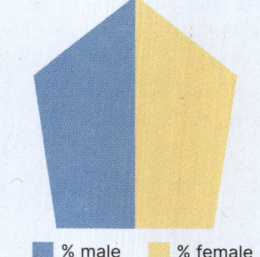

Stage five: Low birth rates and an ageing population. This stage has been added to the original model.

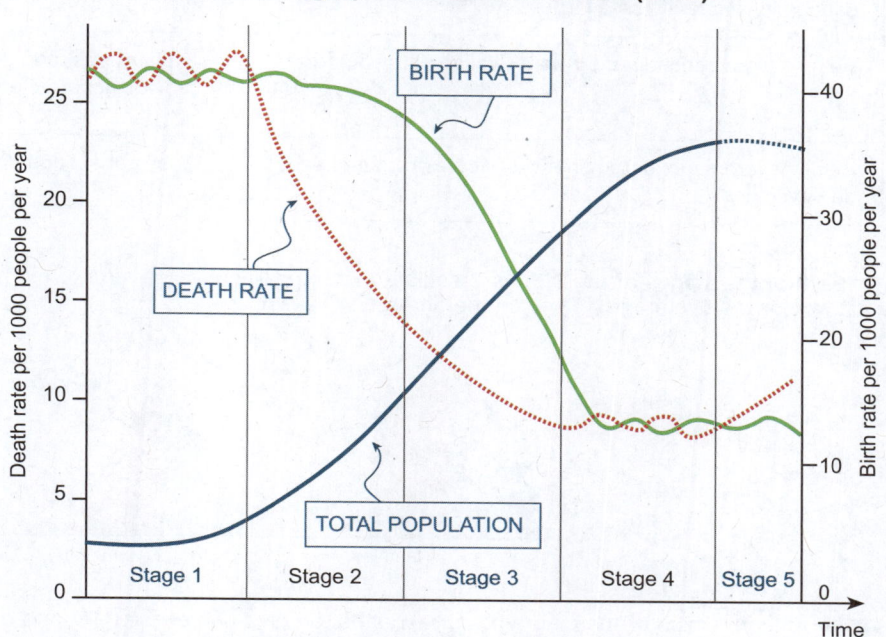

Stage one (high stationary): A balance between birth and death rates was true of all populations until the late 18th Century. Children are important contributors to the household economy. Losses as a result of starvation and disease are high. Population growth is typically very slow in this stage because it is constrained by the available food supply. Unless the society develops new technologies to increase food production, any fluctuations in birth rates are quickly matched by death rates. No countries are currently categorized as stage 1.

Stage two (early expanding): Rapid population expansion as a result of high birth rates and a decline in death rates. The changes leading to this stage in Europe were initiated in the agricultural revolution of the 18th century but have been more rapid in developing countries since then. Stage two is associated with more reliable food supplies and improvements in public health. Several sub-Saharan African nations and some Middle Eastern countries are in stage 2.

Stage three (late expanding): The population moves towards stability through a decline in the birth rate. This stage is associated with increasing urbanization and a decreased reliance on children as a source of family wealth. Family planning in nations such as Malaysia has been instrumental in their move to stage three. India, Indonesia, Mexico, and Kenya are examples of countries in stage 3.

Stage four (low stationary): Birth and death rates are both low and the total population is high and stable. The population is ageing and in some cases the fertility rate falls below replacement levels. Examples of countries in stage 4 include the USA, Argentina, Brazil, China, and most European countries.

Stage five (declining): Proposed by some theorists as representing countries that have undergone the economic transition from manufacturing based industries into service and information-based industries. Birth and death rates are low but the population is ageing and in decline because the population reproduces well below replacement levels. Countries in stage five include the United Kingdom, Germany, and Japan. Stage 5 models are largely theoretical because population outcomes can be altered by immigration.

Strengths and weaknesses of the DTM	
Strengths	**Weaknesses**
Provides a generalized tool for understanding and analysing population trends.	Oversimplifies complex population dynamics. Does not consider cultural, political, and environmental factors.
The DTM provides a simple model for comparing different countries.	DTM was originally developed for European countries and may not fully reflect non-Western countries. May not fully account for regional or country specific variations.
Helps governments make informed decisions on social, economic, and healthcare issues.	Assumes a linear transition between stages, does not necessarily apply to all countries.
Helps governments decide where to allocate resources.	Explains past and present demographic trends but may not provide accurate predictions for the future.
Helps governments determine where infrastructure needs to be developed.	Does not account for migration.

1. Each of the stages of the DTM is associated with a particular age structure. The diagrams below show structures for three of the five stages. They are not in order.

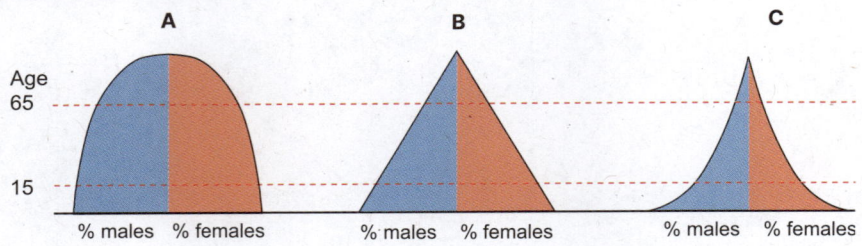

(a) Identify which of the diagrams corresponds to stage one of the DTM and explain your choice:

_____
_____
_____
_____

(b) Identify which of the diagrams corresponds to stage three of the DTM and explain your choice:

_____
_____
_____

2. Describe how population growth rates, fertility rates, and age structure differ between stages 4 and 5 of the DTM:

_____
_____
_____
_____

3. How might advancements in technology, education, and healthcare influence the progression of a country's demographic transition?

_____
_____
_____
_____

72 Humans and Resources

Key Idea: As the human population grows, the demand for resources also increases.

The expanding human population puts increasing strain on the world's resources, creates problems of pollution and waste disposal, and often places other species at risk. Even when resources are potentially sufficient to meet demands, problems with distribution and supply create inequalities between regions.

Air pollution: this contributes to global warming, ozone depletion, and acid rain, and is set to increase markedly in coming years.

Water use: global water consumption is rising rapidly. The availability of water is likely to become the most important resource issue of the 21st century.

Car emissions: combustion engine emissions are increasing rapidly, especially in Asia as their economies develop and their populations become more affluent.

Fossil fuel supplies: a dwindling supply of fossil fuels provides about 85% of the world's commercial energy. Most of this is consumed by the richest countries.

Global climate change: as a result of climate change, a rise in sea levels will threaten to inundate (flood) low-lying island nations and many coastal populations.

Waste: in industrialized societies, one person consumes many tonnes of raw materials each year, which must be extracted, processed, and then disposed of as waste.

Aquatic environments: coral reefs, lakes, rivers, and wetlands, and the species in them, are at risk from human activities, e.g. polluting waterways and overfishing.

Overfishing: unsustainable overfishing of the world's fish stocks has occurred in many fishing grounds. Many of these fish populations are unlikely to recover.

Natural resource use: consumption of natural resources, including fuels, water, and minerals, to power vehicles, supply homes, and meet technology and industrial needs are rapidly increasing.

Deforestation: forest fires and logging continue to cause shrinkage of the world's tropical and temperate forests. Almost 20% of the Amazon rainforest has been destroyed by human activity.

Biodiversity: threats to biodiversity are reaching a critical level. Extinction rates have increased 100 to 1000 times due to human activity affecting the biodiversity of natural environments.

Food security: although the world's food production is theoretically adequate to meet human needs, there are problems with distribution. 1/10 people do not get enough to eat.

1. The observable effects of human activities on the environment and resources vary depending on location. Identify one issue affecting your region or country and outline the consequences of it below:

_____
_____
_____
_____
_____
_____
_____

How many Earths are needed to sustain humans?

An environmental footprint (eco footprint) measures human demand for natural resources. It provides an indication of the quantity of natural resources required to support people and their activities. It can be applied to individuals, regions, countries, or the world. Globally, our current use of natural resources is unsustainable. Humans are using natural resources at a rate which is 1.7 times higher than our planet can regenerate them at. At this rate, we need 1.7 Earths to meet human demands.

The graphic below shows how many Earths would be needed if the world population lived like the residents in these countries. For example, if everyone on Earth lived like Americans, we would need 5.1 Earths. If everyone on the planet lived like the French, we would need 2.8 Earths.

There is only one Earth, but we need 1.7 Earths to meet our current resource demands. By 2030, it is predicted we will need two Earths to meet demand.

Country	Number of Earths each country requires to sustain resource use in 2022
Qatar (DV)	9.0
USA (D)	5.1
Australia (D)	4.5
Russian Federation	3.4
New Zealand (D)	3.3
Japan (D)	2.9
France (D)	2.8
United Kingdom (D)	2.6
China (DV)	2.4
Mexico (DV)	1.5
India (DV)	0.8
Kenya (DV)	0.6

D = Developed Country   DV = Developing Country   D/DV classification by International Monetary Fund (IMF), 2023

Data source: Earth overshoot day, 2022
https://overshoot.footprintnetwork.org/how-many-earths-or-countries-do-we-need/

2. Study the table above.

   (a) In general, what trend do you observe in terms of a country's developmental classification and resource use?

   _____

   _____

   _____

   (b) Identify the outlier country to the trend: _____

3. The ecological footprint for Canada is 7.9 gha per person. Global biocapacity (ability to generate resources) is 1.6 gha per person. Calculate how many Earths are required to sustain human resource use at Canada's consumption levels:

   _____

   _____

73 Did You Get It?

1. Populations are regulated by density dependent and density independent factors. Explain the difference between them and provide an example of each:

   _____
   _____
   _____
   _____
   _____
   _____

2. A population started with a total number of 100 individuals. Over the following year, population data were collected. Calculate birth rates, death rates, net migration rate, and rate of population change for the data below (as percentages):

   (a) Births = 14: Birth rate = _____   (b) Net migration = +2: Net migration rate = _____

   (c) Deaths = 20: Death rate = _____   (d) Rate of population change = _____

   (e) State whether the population is increasing or declining: _____

3. The diagrams (right) depict the age structure of two populations (A and B). The white bars represent young individuals.

   (a) Is population A expanding, declining, or stationary?

   _____

   (b) Is population B expanding, declining, or stationary?

   _____

4. The table below shows world population in 2024 by geographical region.

   (a) Convert the population numbers in row two to billions and write the answers in row three. The first one has been completed for you:

   (b) Convert the data into percentages for each region and write the answers in row four:

Region	Oceania	South America	North America	Europe	Africa	Asia	Total
Number of people	46.11 million	442.86 million	608.16 million	742.92 million	1.49 billion	4.78 billion	
Number of people (billions)	0.046						
Percentage of total population							

   (c) Plot the percentage for each region as a pie graph (right). Provide the title of your pie graph and the key below to save space:

   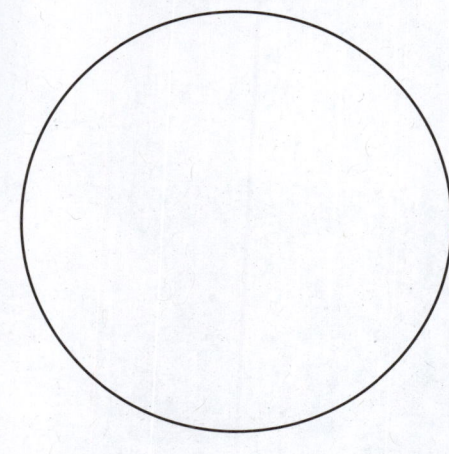

Chapter 4 The Living World
Investigating Ecosystems

Resource Hub
bit.ly/4cNYGsQ

Key Terms

- abundance
- classification keys
- datalogger
- DNA barcode
- ecosystem
- GPS tracking
- indirect sampling
- kite graph
- Lincoln index
- mark and recapture
- point sampling
- population
- quadrats
- radio tracking
- random sampling
- sample
- sampling
- sampling error
- satellite tracking
- Simpson's index of diversity
- species
- transect

Key Concepts

- Populations are typically too large to assess comprehensively, necessitating the use of sampling methods
- Common sampling methods include quadrats, transects, mark and recapture, and netting and trapping.
- When investigating ecosystems, it is crucial to gather information about the environment.
- Fieldwork should be conducted with careful consideration and respect for organisms and the environment.

Sampling Populations

Learning Objectives:

		Activity Number
☐ 1	Explain why sampling is necessary and what types of information samples can give about an ecosystem. Explain what is meant by sampling error and describe the difference between random and non-random sampling.	74-75
☐ 2	Describe the types of information that can be collected by data loggers. Describe techniques for monitoring water quality in the field and laboratory.	76-77
☐ 3	Compare different methods used to sample animal populations and explain why each method is suitable for sampling a particular organism.	78-79
☐ 4	Explain how sampling using quadrats can be used to estimate population abundance.	80-81
☐ 5	Describe how transect sampling is used and what information it gives about a species in a community. Use given data to draw kite graphs.	82
☐ 6	Explain the reason for using mark and recapture sampling and the type of organisms for which this method of sampling is appropriate. Use the Lincoln Index to estimate population sizes.	83
☐ 7	Describe some technologies that are used to monitor animals in their habitat. Describe how environmental DNA sampling is used in identification of organisms that may not be visible.	84, 86
☐ 8	Describe some indirect sampling methods and the reasons for their use.	84

Measuring Diversity and Using Classification Keys

☐ 9	Use Simpson's index of diversity to measure species diversity and compare biodiversity in different ecosystems from given data..	87
☐ 10	Explain how different classification keys can be used to identify organisms based on physical characteristics.	88

74 Why Do We Sample?

Key Idea: Sampling an ecosystem provides information about its composition and structure, its health, and the likelihood it will be able to resist change.

Would it be possible for you to count every individual organism in your backyard or local park? Could you reliably indicate their location, either on paper or digitally? Most likely not, because there are too many individuals and not enough time or resources to count them all. To get around these problems, researchers **sample** the **ecosystem**. Sampling involves choosing a small area that represents the ecosystem and counting the organisms in that area. The information gathered from the sample is used to draw conclusions about that ecosystem. Different sampling methods are chosen, depending on the location, the type of organisms, and the data to be collected. The sample should as representative of the **population** or ecosystem as possible.

What can sampling tell us?

Recording observations of species

Community composition
Sampling reveals which **species** are present in an ecosystem and helps to build a picture of community structure or identify species of particular interest. For example, are there endangered species, or introduced, or pest species present?

Great grey shrike with mouse prey

Species interactions
Sample data can be used to construct models of species interactions, e.g. food webs or ecological pyramids. The information can be used to predict the effect of a change in community structure, e.g. decrease in one species.

Gannets in breeding colonies

Species distribution
How is a particular species distributed in the ecosystem and does this change over time, e.g. seasonally? Sample data can tell us about the geographical range of the species and how this might be affected by environmental change.

Trevally school

Species abundance
Sampling reveals information about species abundance, i.e. how many of a particular species are present at the location. Species **abundance** is one measure for estimating biodiversity as well as ecosystem health and stability. The presence or absence of certain species can be used to indicate ecosystem health.

Brown-throated sloth, S. American jungle

Ecosystem stability
Data can be used to predict how likely it is that an ecosystem will remain unchanged in its characteristics. We know that low diversity systems are more likely to be negatively affected by disturbance than high diversity systems. The presence or absence of key indicator species are also used to monitor ecosystem changes.

Ocelot, mid-sized cat from central SA

Conservation management
Sampling provides a way to evaluate the success of conservation management strategies. For example, are the numbers of a threatened or endangered species increasing or decreasing? How are the numbers of an invasive species changing? If no progress is made towards conservation goals, the plan can be altered.

1. Suggest why it is important to select a sampling area that is a true representation of an area:

   _____
   _____
   _____

2. Why must scientists sample an ecosystem or population instead of studying it in its entirety?

   _____
   _____
   _____

75 Sampling Populations

Key Idea: Sampling a population provides information about its size, density, distribution, age structure, and more. The information gathered from **sampling** can be used to draw conclusions about **populations**, i.e. all the inhabitants of the same **species** in a singular location. Sampling can be used to investigate the size of a population and compare it to other populations. It can also show the distribution of populations around a geographical region, or the age structure, and whether the population is aging or being reinvigorated with an increased birth rate. Sampling a population always comes with a measure of error. Knowing the significance of that **sampling error** is important when calculating or using **sample** statistics. It can inform us about how well the sample represents the actual population.

What can sampling tell us?

Population size
How many individuals are in a population of organisms living in a certain area, such as a forest, or a pond? Counting individuals in a measured area and then applying specific equations to that sample allows us to estimate the size of the population over a larger area.

Species distribution
How does the distribution of **species** vary between different **ecosystem** and along environmental gradients? What factors influence species distribution patterns and how can sampling techniques help us understand these patterns?.

Age structure
What is the distribution of ages in a population? How many individuals are of breeding age, or are juveniles? Information like this is important when determining harvest quotas or breeding plans for endangered animals.

Sampling and sampling error

▸ Estimating population size can be difficult. Populations may be very spread out, making it difficult to find individuals, or they may be mobile, or occupy a vast area. In any sample area there may be more or fewer individuals than any other area simply by chance.

▸ Any statistic, e.g. the estimated population, or the average height of students, will have an error based simply on the fact that some of the population has been sampled and some has not. As a result, our sample statistic will never be exactly representative of the population. The difference between these is called the sampling error. The larger the sample, the smaller the sample error will be.

▸ The sampling error is the difference between the actual population number and the estimated population number. Because we almost never account for every individual making up the population (the reason why we have to sample), we can estimate the sampling error using a confidence interval.

4	3	2	3	2	0
2	3	4	1	2	1
4	6	4	4	3	3
3	2	5	3	6	5
0	4	3	4	3	4
2	3	1	1	0	2

1. (a) The box above (right) represents a small field divided into 36 squares. The number in each square represents the number of dandelions found in that area. Roll a six sided dice to randomly select six of the squares above (first roll for row, second roll for column). Record the numbers in these squares below (not repeating any squares):

   _____

   (b) Calculate the mean number of dandelions in the six squares and then estimate the total population of dandelions.

   Mean dandelions per square: _____ Population estimate: _____ Sampling error: _____

2. Repeat 1. above two more times:

   (a) Mean dandelions per square: _____ Population estimate: _____ Sampling error: _____

   (b) Mean dandelions per square: _____ Population estimate: _____ Sampling error: _____

3. What can you say about the estimated population of dandelions based on your sampling?

   _____
   _____
   _____

Sampling strategies

▶ **Random sampling:** In most ecological studies, it is not possible to measure or count all the members of a population. Instead, information is obtained through sampling in a manner that provides a fair (unbiased) representation of the organisms present and their distribution. This is usually achieved through **random sampling**.

▶ **Non-random sampling:** Sometimes researchers collect information by non-random sampling, a process that does not give all the individuals in the population an equal chance of being selected. While faster and cheaper to carry out than random sampling, non-random sampling may not give a true representation of the population.

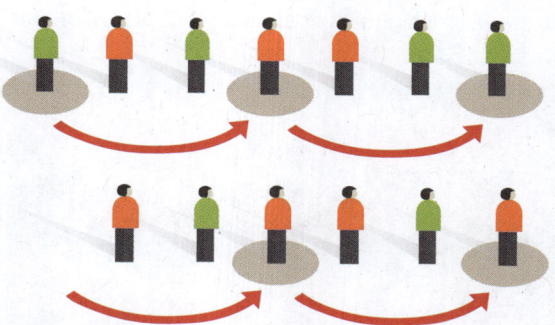

Systematic sampling

Samples from a larger population are selected according to a random starting point and a fixed, periodic sampling interval. For the example above, the sampling period is every third individual. Systematic sampling is a random sampling method, provided the periodic interval is determined beforehand and the starting point is random.

Example: Selecting individuals from a patient list.

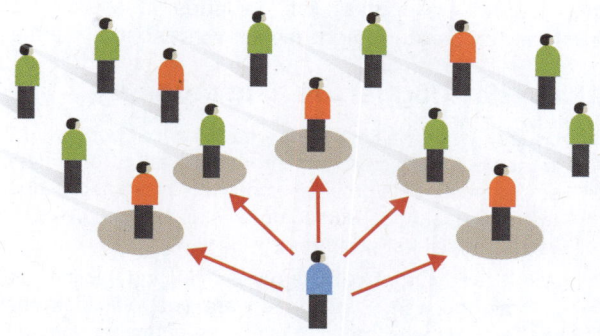

Opportunistic or convenience sampling

A non-random sampling technique in which subjects are selected because they are easily accessible to the researcher. Opportunistic sampling excludes a large proportion of the population and is usually not representative of the population. It is sometimes used in pilot studies to gather data quickly and with little cost.

Example: Selecting 5 closest people at a cafe at lunchtime.

Stratified sampling in environmental science

▶ Many study areas are not uniform. Instead, they include a variety of distinct habitats, especially if the study site is large. In stratified sampling, the various habitats are sampled separately in proportion to their representation in the total area. This ensures that the sampling fairly represents the entire habitat.

▶ The sample area is usually divided into groups (strata) based on biophysical features, e.g. landform, soil type, elevation etc., and then by vegetative structure, e.g. forest, woodland, grassland etc.

▶ Proportional sampling is an essential feature of stratified sampling. For example, the ecosystem on the right contained 30% mixed woodland and 70% grass. The researcher decided to place 20 random quadrat samples in total. To ensure proportional sampling, they placed six quadrats in the mixed woodland and 14 in the grass (Note: not to scale and not all quadrats visible).

4. A student wants to investigate the incidence of asthma in their school. Describe how they might select samples from the school population using:

   (a) Systematic sampling: _____

   _____

   (b) Stratified sampling: _____

   _____

   (c) Opportunistic sampling: _____

   _____

5. Distinguish between random and non-random sampling: _____

   _____

   _____

76 Sampling and Sensors

Key Idea: Sensors can be used to collect data about both biotic and abiotic factors in an environment during sampling. When studying **ecosystems**, it is important to gather information not only about the **population** numbers and distribution, but also about the environment in which those populations are found. This includes measuring factors such as temperature, light levels, wind speed, rainfall, and the levels of dissolved oxygen and solids in water. While visual or audio observations can be used to collect certain data, sensors and cameras offer a more accurate and efficient means of measurement. These instruments have the capability to provide single measurements at the time of sampling or continuous measurements for longer-term analysis, allowing insights over an extended period of time.

Using data loggers in field studies

▸ Usually, when we collect information about populations in the field, we also collect information about the physical environment. This provides important information about the local habitat of studied organisms and can be useful in assessing habitat preference.

▸ With the advent of **data loggers**, collecting this information is straightforward. Data loggers are electronic instruments that record measurements over time. They are equipped with a microprocessor, data storage facility, and sensor. Different sensors are used to measure a range of variables in water or air.

▸ The limits of the data logger's operation can be set before use, e.g. the sampling interval, and used remotely to record and store data. Data loggers in remote environments can be linked to computers at a base of operations via cellphone networks or satellite making data collection quick, accurate, and simple. They also enable data collection from multiple sites without the need for researchers to continually travel to them and allow prompt data analysis from those sites.

Drone-based cameras are used for estimating Galápagos marine iguana population abundance.

Acoustic monitoring, supported by AI enhancement, can detect presence and density of different bird species.

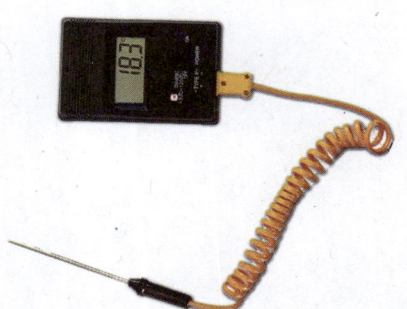 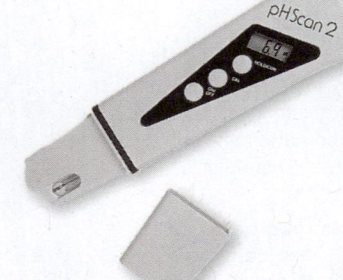 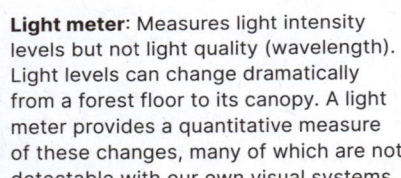

Digital thermometers: Digital thermometers fitted with probes give very precise measurements of temperature for water, air, soil etc. Simple meters and probes are relatively cheap and easy to use in multiple environments, providing useful information of changes over time or between locations.

pH meter: Measuring pH of water gives important information about water quality and the tolerance of organisms in it. Pure water has a pH of 7 but bodies of water generally have a pH range from 6 to 8. Modern digital meters can give very precise readings to measure small changes in pH.

Light meter: Measures light intensity levels but not light quality (wavelength). Light levels can change dramatically from a forest floor to its canopy. A light meter provides a quantitative measure of these changes, many of which are not detectable with our own visual systems.

1. Explain how data loggers make sampling abiotic factors a much simpler process than it might otherwise be:

   _____
   _____

2. Explain the need to sample abiotic factors when sampling animal or plant populations:

   _____
   _____
   _____

3. Imagine sampling a rocky shore habitat, from the water's edge to one meter above the high water mark. List the types of meters you might want to use to build up a picture of the environment:

   _____
   _____

77 Monitoring Water Quality

Key Idea: A range of sampling techniques and indicators can be used to sample water and to gauge its quality.

There is no single measure to objectively describe the water quality of a stream, river, or lake. Rather, it is defined in terms of various chemical, physical, and biological characteristics. Together, these factors define the health of the aquatic **ecosystem** and its suitability for various, desirable uses. It is normally not feasible to monitor for all contaminants potentially in water. For example, analysis for pesticides, dioxins, and other trace chemicals can be a costly, ongoing expense. Water quality is determined by making measurements on-site or by taking **samples** back to a laboratory for physical, chemical, or microbiological analysis. The method of abiotic monitoring is determined by the factor that is being measured and the reason for collecting the data.

Techniques for Monitoring Water Quality

Some aspects of water quality, such as black disk clarity measurements (above), must be made in the field.

The collection of water samples allows many quality measurements to be carried out in the laboratory.

Telemetry stations transmit continuous measurements of the water level of a lake or river to a central control office.

Temperature and dissolved oxygen measurements must be carried out directly in the flowing water.

1. What water quality variable observations would you expect from a high water quality stream in a forested area?

   _____
   _____
   _____
   _____
   _____

2. What water quality variable observations would you expect from a stream that had runoff from an urban area?

   _____
   _____
   _____
   _____
   _____

3. Explain why dissolved oxygen, temperature, and clarity measurements are made in the field rather than in the laboratory:

   _____
   _____
   _____
   _____
   _____

Water Quality Standards in Aquatic Ecosystems

Water quality variable	Why measured
Dissolved oxygen	• A requirement for most aquatic life • Indicator of organic pollution • Indicator of photosynthesis (plant growth)
Temperature	• Organisms have specific temperature needs • Indicator of mixing processes • Computer modeling examining the uptake and release of nutrients
Conductivity	• Indicator of total salts dissolved in water • Indicator for geothermal input
pH (acidity)	• Aquatic life protection • Indicator of industrial discharges, mining
Clarity - turbidity - black disk	• Esthetic appearance • Aquatic life protection • Indicator of catchment condition, land use
Color - light absorption	• Esthetic appearance • Light availability for excessive plant growth • Indicator of presence of organic matter
Nutrients (Nitrogen, phosphorus)	• Enrichment, excessive plant growth • Limiting factor for plant and algal growth
Major ions (Mg^{2+}, Ca^{2+}, Na^+, K^+, Cl^-, HCO_3^-, SO_4^{2-})	• Baseline water quality characteristics • Indicator for catchment soil types, geology • Water hardness (magnesium/calcium) • Buffering capacity for pH change (HCO_3^-)
Organic carbon	• Indicator of organic pollution • Catchment characteristics
Fecal bacteria	• Indicator of pollution with fecal matter • Disease risk for swimming etc.

78 Sampling Animal Populations

Key Idea: Sampling motile animals requires special techniques to ensure a representative sample is collected.

Unlike plants, most animals are highly mobile and present special challenges in terms of **sampling** them quantitatively to estimate their distribution and **abundance**. The equipment available for sampling animals ranges from various types of nets and traps (below), to more complex electronic devices, such as those used for radio-tracking large mobile **species**.

Plankton net
- Tow rope
- Bridle
- Metal hoop
- Cone of bolting silk
- Canvas sleeve
- Plastic container for collecting plankton sample
- Tie cord
- Direction of current

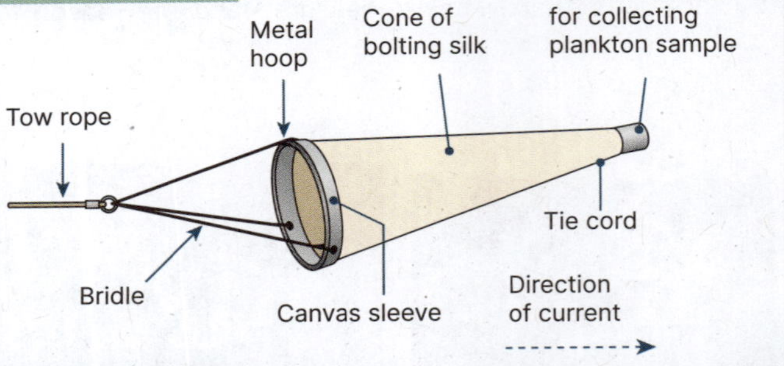

Beating tray
- Tree branch is shaken or beaten with a stick
- Insects and other invertebrates fall
- Canvas stretched over frame

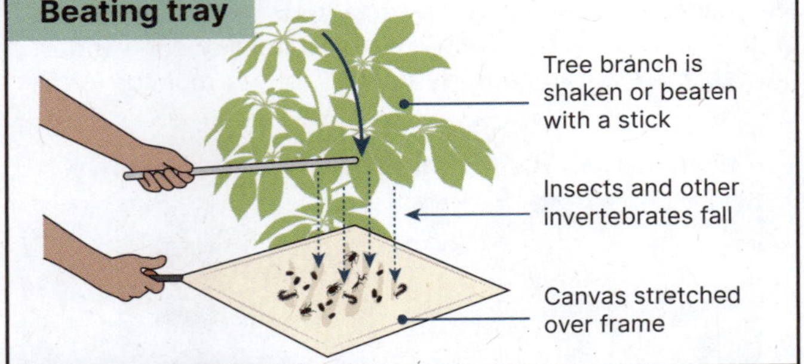

Kick sampling
- Direction of current
- Rocks upstream of the net are disturbed
- Small aquatic invertebrates are dislodged and collect in the net

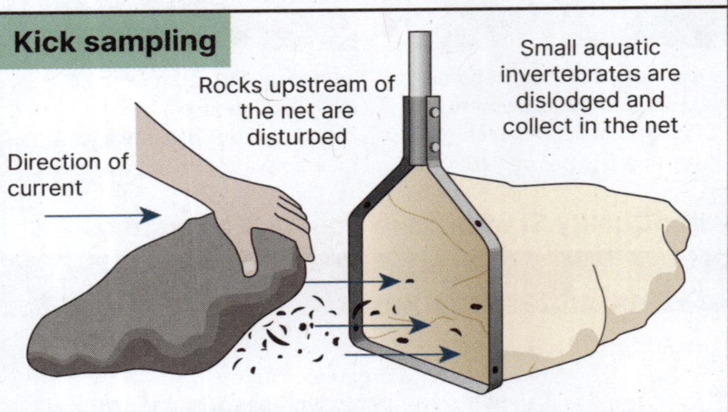

Tullgren funnel
- Light from a battery operated lamp drives the invertebrates down through the soil or litter.
- Large diameter funnel containing soil or leaf litter resting on a gauze platform.
- Gauze allows invertebrates of a certain size to move down the funnel.
- Collecting jar traps the invertebrates that fall through the gauze and prevents their escape.

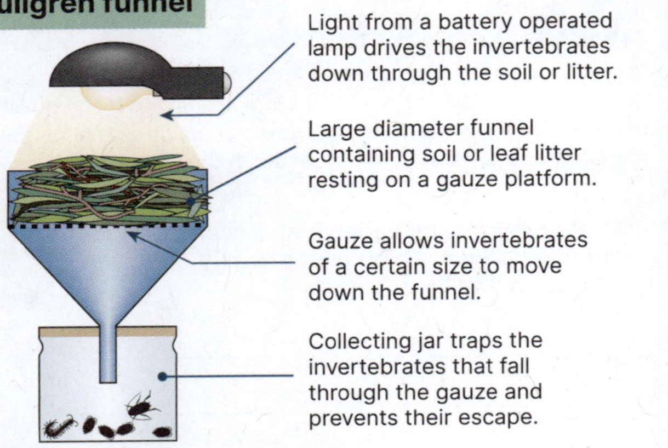

Pooter (aspirator)
- Glass collecting tube that sucks up small animals
- Clear plastic tube
- Rubber bung
- Gauze
- Specimen tube
- Glass mouthpiece through which operator sucks

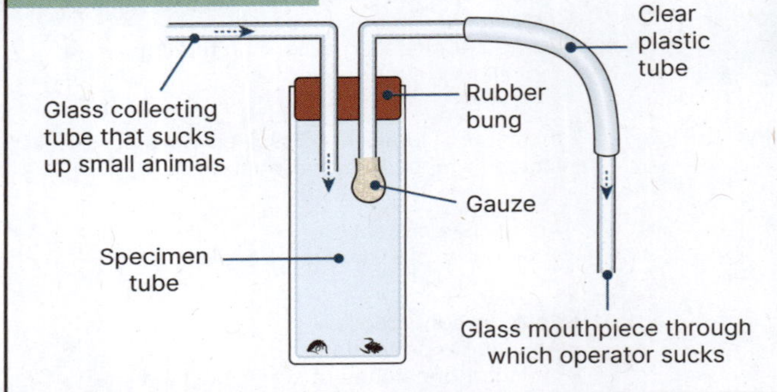

Pitfall trap
- Flat rock
- Support made of small stones or sticks
- Ground slopes away from trap to assist drainage
- Jam jar sunk into ground
- 3 cm of water or 50% ethanol may be added as immobiliser

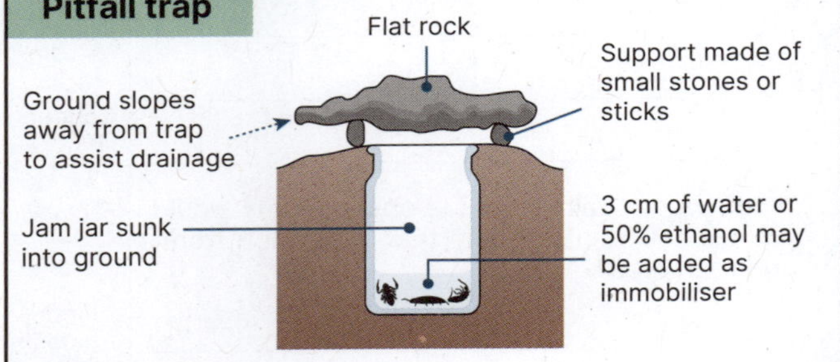

1. What are some challenges in sampling motile (moving) animals?

   _____

   _____

2. Pitfall traps are sunk into the ground and covered with a rock. What type of invertebrates would likely not be captured?

   _____

   _____

   _____

3. Plankton nets can be constructed from different mesh size, with smaller or larger holes in the fabric. How does this choice of mesh size affect the species captured when sampling animals in a pond?

   _____

   _____

   _____

   _____

79 Introducing Sampling Techniques

Key Idea: Choice of sampling method and design should be based on suitability to the populations being sampled, the environment, and the time and resources available.

During **sampling**, different options for collecting data can be selected depending on patterns of distribution and **abundance** observed of the biotic factors. Each sampling method is appropriate for different environments or organisms and has advantages and drawbacks. You must take several factors into account when sampling to make sure the data you collect accurately and impartially represents the **ecosystem** being investigated. Sampling should provide data that are unbiased and accurate.

Sampling designs and techniques

Point sampling
Individual points are chosen using a grid reference or random numbers applied to a map grid. The organisms at each point are recorded. Point sampling is often used to collect data about vegetation distribution.

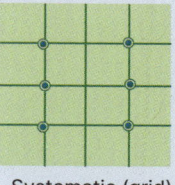
Systematic (grid)

Pros: Point sampling is efficient if time is limited. It is a good method for determining **species** abundance and community composition.
Cons: May miss organisms in low abundance.

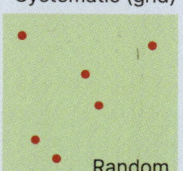
Random

Quadrats (see activity 80)
A **quadrat** provides a known unit area of sample, e.g. 1 m². Quadrats are placed randomly or in a grid pattern on the sample area. The presence and abundance of organisms in each square is noted. Quadrat sampling is appropriate for plants and slow moving animals and can be used to evaluate community composition.

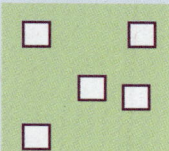

Line and belt transects (see activity 82)
In a line **transect**, a tape or rope marks the line. The species occurring on the line are recorded (all along the line or at regular points). Lines can be chosen randomly (right) or may follow an environmental gradient.
Pros: Low environmental impact and good for assessing the presence/absence of plant species.
Cons: Rare species may be missed.

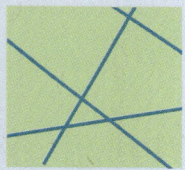

In a belt transect, quadrats are used to sample organisms at regular intervals along a measured strip.
Pros: Provide a lot of information on abundance and distribution as well as presence/absence.
Cons: Time consuming to do properly.

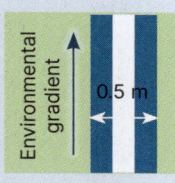

Mark and recapture sampling (see activity 83)
1. Animals are captured, marked, and then released back into the population (right).
2. After a suitable time to allow the marked animals to remix with the population, the population is resampled. The number of marked animals recaptured in a second sample is recorded as a proportion of the total.

1: All marked.

Pros: Useful for highly mobile species which are otherwise difficult to record.
Cons: Time consuming to do well.

2: Proportion recaptured

Sampling considerations

▸ **Random sampling** methods should be used to avoid bias in the data. In random sampling, every possible sample of a given size has the same chance of selection.

▸ The methods used to sample communities and their populations must be appropriate to the ecosystem being investigated. Communities in which the **populations** are at low density and have a random or clumped distribution will require a different sampling strategy from those where the populations are uniformly distributed and at higher density.

▸ The sample size, e.g. the number of quadrats, must be large enough to provide data to enable us to make inferences about aspects of the whole population.

1. Name a sampling technique that would be appropriate for determining:

   (a) Percentage cover of a plant species in pasture:
   _____
   _____

   (b) Change in community composition from low to high altitude on a mountain:
   _____
   _____

   (c) Association of plant species with particular soil types in a nature reserve:
   _____
   _____

   (d) The population size of a fish in a lake:
   _____
   _____

2. What are the benefits of collecting information about the physical environment when sampling populations?
   _____
   _____
   _____
   _____
   _____
   _____
   _____

80 Quadrat Sampling

Key Idea: Quadrat sampling uses a frame of known size randomly placed multiple times over an area of habitat to evaluate the abundance or diversity of organisms present in that area.

Quadrat sampling is a method of directly counting organisms in a representative portion of a habitat, typically used when the total number of organisms is too large to count individually. It can be used to estimate **population abundance** (number), density, frequency of occurrence, and distribution. Quadrats may be used without a **transect** when studying a relatively uniform habitat. In this case, the quadrat positions are chosen randomly using a random number table. The general procedure is to count all the individuals (or estimate their percentage cover) in a number of quadrats of known size and to use this information to work out the abundance or percentage cover value for the whole area.

Quadrat-based estimates

- The simplest description of a plant community in a habitat is a list of the **species** that are present. This qualitative assessment of the community has the limitation of not providing any information about the relative abundance of the species present.
- Quick estimates can be made using abundance scales, such as the ACFOR scale described below. Estimates of plant species percentage cover provide similar information. These methods require the use of quadrats which are used extensively in plant environmental studies.

Quadrat sizes

Quadrats are usually square and cover 0.25 m^2 (0.5 m x 0.5 m) or 1 m^2, but they can be of any size or shape. The quadrat size needs to be adjusted to habitat type. It must be large enough to be representative of the community but not so large as to be cumbersome to use. The quadrats used to sample plant communities are often 0.25 m^2. This size is ideal for low-growing vegetation.

A quadrat covering an area of 0.25 m^2 is suitable for most low growing plant communities such as this alpine meadow, fields, and grasslands.

Larger quadrats, e.g. 1 m^2 are needed for communities with shrubs and trees. Quadrats as large as 4 m x 4 m may be needed in woodlands.

Small quadrats (0.01 m^2 or 100 mm x 100 mm) are appropriate for lichens and mosses on rock faces and tree trunks.

How Many Quadrats?

As well as deciding on a suitable quadrat size, the other consideration is how many quadrats to take (the sample size). In species-poor or very homogeneous habitats, a small number of quadrats will be sufficient. In species-rich or heterogeneous habitats, more quadrats will be needed to ensure that all species are represented.

Determining the number of quadrats needed

- Plot the cumulative number of species recorded (on the y axis) against the number of quadrats already taken (on the x axis).
- The point at which the curve levels off indicates the suitable number of quadrats required.

Fewer quadrats are needed in species-poor or very uniform habitats, such as this bluebell woodland, below.

Describing Vegetation

Density (number of individuals per unit area) is a useful measure of abundance for animal populations but can be problematic in plant communities where it can be difficult to determine where one plant ends, and another begins. For this reason, plant abundance is often assessed using percentage cover. Here, the percentage of each quadrat covered by each species is recorded, either as a numerical value or using an abundance scale such as the ACFOR scale.

The ACFOR Abundance Scale

A = Abundant (30% +)
C = Common (20-29%)
F = Frequent (10-19%)
O = Occasional (5-9%)
R = Rare (1-4%)

The ACFOR scale could be used to assess the abundance of species in this wildflower meadow. Abundance scales are subjective, but it is not difficult to determine which abundance category each species falls into.

$$\text{Estimated average density} = \frac{\text{Total number of individuals counted}}{\text{Number of quadrats} \times \text{area of each quadrat}}$$

Guidelines for quadrat use:

1. The area of each quadrat must be known. Quadrats should be the same shape, but not necessarily square.
2. Enough quadrat samples must be taken to provide results that are representative of the total population.
3. The population of each quadrat must be known. Species must be distinguishable from each other, even if they have to be identified at a later date. It has to be decided beforehand what the count procedure will be and how organisms over the quadrat boundary will be counted.
4. The size of the quadrat should be appropriate to the organisms and habitat, e.g. a large size quadrat for trees.
5. The quadrats must be representative of the whole area. This is usually achieved by random sampling (right).

Quadrats are applied to the predetermined grid on a random basis. This can be achieved by using a random number table.

The area to be sampled is divided up into a grid pattern with indexed coordinates.

Sampling a centipede population

A researcher by the name of Lloyd (1967) sampled centipedes in Wytham Woods near Oxford in England. A total of 37 hexagon–shaped quadrats were used, each with a diameter of 30 cm (see diagram on right). These were arranged in a pattern so that they were all touching each other.

Use the data in the diagram to answer the following questions.

1. (a) Determine the average number of centipedes captured per quadrat:
   _____

   (b) Calculate the estimated average density of centipedes per square meter (remember that each quadrat is 0.08 square meters in area):
   _____

   (c) Looking at the data for individual quadrats, describe in general terms the distribution of the centipedes in the sample area:
   _____
   _____

   (d) Describe one factor that might account for the distribution pattern:
   _____

Each quadrat was a hexagon with a diameter of 30 cm and an area of 0.08 square meters.

The number in each hexagon indicates how many centipedes were caught in that quadrat.

2. Describe one difference between the methods used to assess species abundance in plant and in animal communities:
   _____
   _____

3. Identify the main consideration when determining appropriate quadrat size: _____
   _____

4. Identify the main consideration when determining number of quadrats: _____
   _____
   _____

81 Sampling a Leaf Litter Population

Key Idea: The estimates of a population obtained through quadrat sampling may vary depending on the placement. Using larger samples can help account for this variation. The diagram (next page) represents an area of forest floor with its resident organisms. The distribution of leaves and four animal **species** are shown. This exercise is designed to prepare you for planning and carrying out a similar procedure to practically investigate a natural community.

1. **Decide on the sampling method**
   For the purpose of this exercise, it has been decided that the populations to be investigated are too large to be counted directly. A quadrat sampling method is to be used to estimate the average density of the four animal species as well as that of the leaf litter.

2. **Mark out a grid pattern**
   Use a ruler to mark out 3 cm intervals along each side of the sampling area (area of quadrat = 0.03 × 0.03 m). Draw lines between these marks to create a 6 × 6 grid pattern (total area = 0.18 × 0.18 m). This will provide a total of 36 quadrats that can be investigated.

3. **Number the axes of the grid**
   Only a small proportion of the possible quadrat positions will be sampled. It is necessary to select the quadrats in a random manner. It is not sufficient to simply guess or choose your own on a 'gut feeling'. The best way to choose the quadrats randomly is to create a numbering system for the grid pattern and then select the quadrats from a random number table. Starting at the top left hand corner, number the columns and rows from 1 to 6 on each axis.

4. **Choose quadrats randomly**
   To select the required number of quadrats randomly, use random numbers from a random number table. The random numbers are used as an index to the grid coordinates. Choose 6 quadrats from the total of 36 using the table of random numbers provided for you at the bottom of the next page. Make a note of which column of random numbers you choose. Each member of your group should choose a different set of random numbers, i.e. different column: A–D, so that you can compare the effectiveness of the sampling method.

   Column of random numbers chosen: _____

   NOTE: Highlight the boundary of each selected quadrat with coloured pen/highlighter.

5. **Decide on the counting criteria**
   Before the counting of the individuals for each species is carried out, the criteria for counting need to be established. There may be some problems here. You must decide before sampling begins as to what to do about individuals that are only partly inside the quadrat. Possible answers include:

   (a) Only counting individuals that are completely inside the quadrat.
   (b) Only counting individuals with a clearly defined part of their body inside the quadrat (such as the head).
   (c) Allowing for 'half individuals' e.g. 3.5 centipedes).
   (d) Counting an individual that is inside the quadrat by half or more as one complete individual.

   Discuss the merits and problems of the suggestions above with other members of the class (or group). You may even have counting criteria of your own. Think about other factors that could cause problems with your counting.

6. **Carry out the sampling**
   Carefully examine each selected quadrat and count the number of individuals of each species present. Record your data in the spaces provided on the next page.

7. **Calculate the population density**
   Use the combined data TOTALS for the sampled quadrats to estimate the average density for each species by using the formula:

$$\text{Density} = \frac{\text{Total number in all quadrats sampled}}{\text{Number of quadrats sampled} \times \text{area of a quadrat}}$$

   Remember that a total of 6 quadrats are sampled and each has an area of 0.0009 m^2. The density should be expressed as the number of individuals *per square meter* (no./m^2).

   Woodlouse: ☐   False scorpion: ☐

   Centipede: ☐   Leaf: ☐

   Springtail: ☐

8. (a) In this example, the animals are moving. Describe the problems associated with sampling moving organisms. Explain how you would cope with sampling these same animals if they were particularly active:

   _____
   _____
   _____

   (b) Carry out a direct count of all 4 animal species and the leaf litter for the whole sample area (all 36 quadrats). Apply the data from your direct count to the equation given in (7) above to calculate the actual population density (remember that the number of quadrats in this case = 36):

   Woodlouse: ☐   Centipede: ☐   False scorpion: ☐   Springtail: ☐   Leaf: ☐

   Compare your estimated population density to the actual population density for each species:

   _____
   _____
   _____

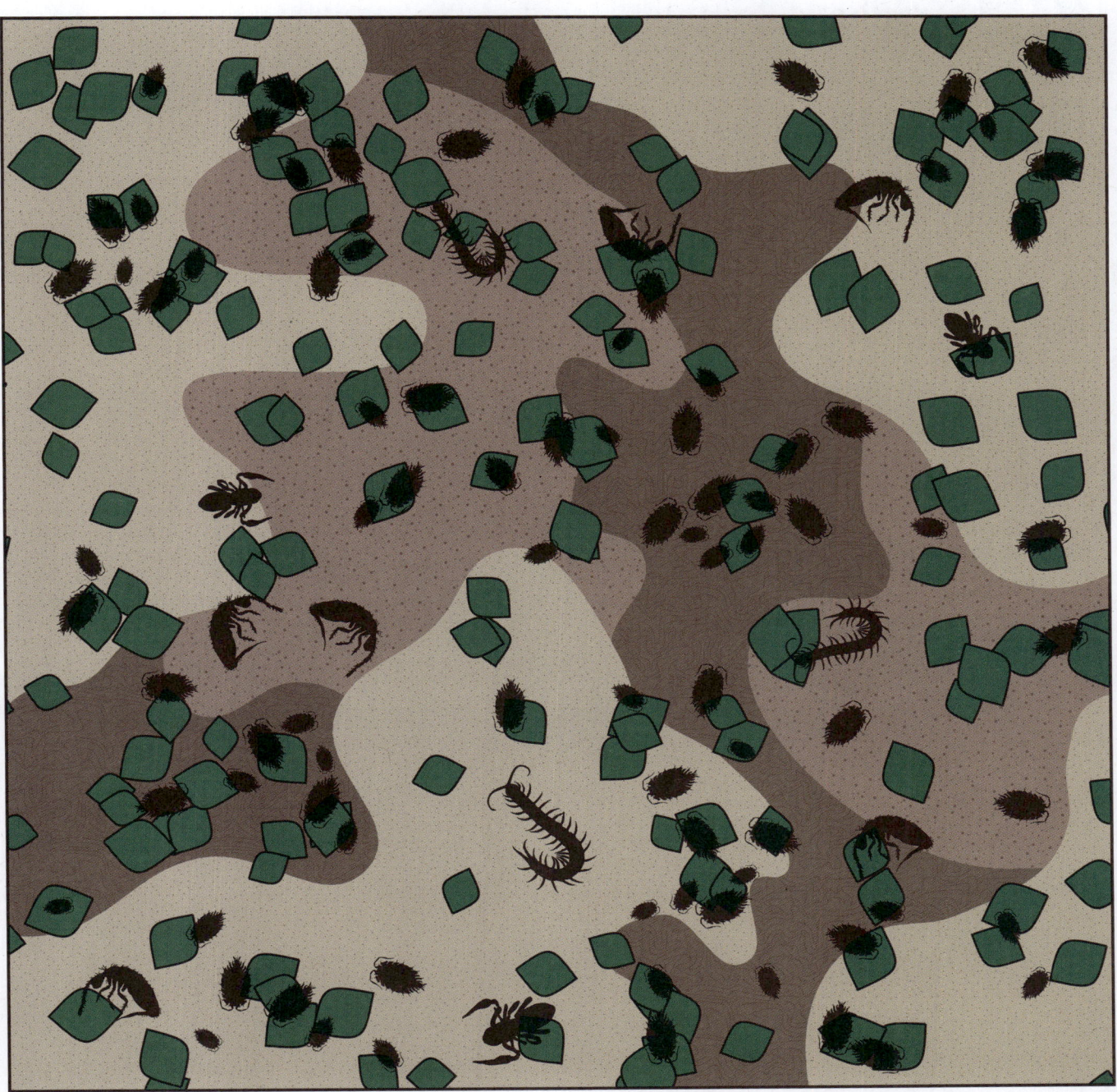

Coordinates for each quadrat	Woodlouse	Centipede	False scorpion	Springtail	Leaf
1:					
2:					
3:					
4:					
5:					
6:					
TOTAL					

Table of random numbers

A	B	C	D
2 2	3 1	6 2	2 2
3 2	1 5	6 3	4 3
3 1	5 6	3 6	6 4
4 6	3 6	1 3	4 5
4 3	4 2	4 5	3 5
5 6	1 4	3 1	1 4

The table above has been adapted from a table of random numbers from a statistics book. Use this table to select quadrats randomly from the grid above. Choose one of the columns (A to D) and use the numbers in that column as an index to the grid. The first digit refers to the row number and the second digit refers to the column number. To locate each of the 6 quadrats, find where the row and column intersect, as shown below:

Example: 5 2 refers to the 5th row and the 2nd column

82 Transect Sampling

Key Idea: Transect sampling is useful for providing information about species distribution along an environmental gradient.

A **transect** is a line placed across a community of organisms. Transects provide information on the distribution of **species** in the community. They are particularly valuable when the transect records community composition along an environmental gradient, e.g. up a mountain or across a seashore. The usual practice for small transects is to stretch a string between two markers. The string is marked off in measured distance intervals and the species at each marked point are noted. The **sampling** points along the transect may also be used for the siting of quadrats, so that changes in density and community composition can be recorded. Belt transects are essentially a form of continuous **quadrat** sampling. They provide more information on community composition but can be difficult to carry out. Some transects provide information on the vertical, as well as horizontal, distribution of species, e.g. tree canopies in a forest.

Point sampling

Sample points

Continuous belt transect

Some sampling methods require the vertical distribution of each species to be recorded.

Continuous sampling

Quadrats are placed adjacent to each other in a continuous belt

Interrupted belt transect

4 quadrats across each sample point — Line of transect

1. Belt transect sampling uses quadrats placed along a line at marked intervals. In contrast, point sampling transects record only the species that are touched or covered by the line at the marked points.

   (a) Describe one disadvantage of belt transects: _____
   _____

   (b) Why might line transects give an unrealistic sample of the community in question? _____
   _____
   _____

   (c) How do belt transects overcome this problem? _____
   _____

   (d) When would it not be appropriate to use transects to sample a community? _____
   _____

2. How could you test whether or not a transect sampling interval was sufficient to accurately sample a community?
   _____
   _____

Kite graphs

A **kite graph** is a good way to show the distribution of organisms sampled using a belt transect. Data may be expressed as abundance or percentage cover along an environmental gradient. Several species can be shown together on the same plot so that the distributions can be easily compared.

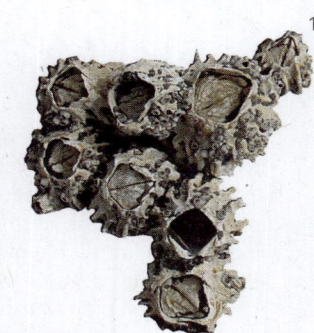

3. The data in the table were collected from a rocky shore field trip. Four common species of barnacle were sampled in a continuous belt transect from the low water mark to a height of 10 m above that level. The number of each of the four species in a 1 m² quadrat was recorded.

Plot a kite graph of the data for all four species below. Be sure to choose a scale that takes account of the maximum number found at any one point and allows you to include all the species on the one plot. Include a scale on the diagram so that the number at each point on the kite can be calculated.

Distribution of 4 common barnacle species on a rocky shore

Height above low water (m)	Barnacle species			
	Plicate barnacle	Columnar barnacle	Brown barnacle	Sheet barnacle
0	0	0	0	65
1	10	0	0	12
2	32	0	0	0
3	55	0	0	0
4	100	18	0	0
5	50	124	0	0
6	30	69	2	0
7	0	40	11	0
8	0	0	47	0
9	0	0	59	0
10	0	0	65	0

An example of a kite graph

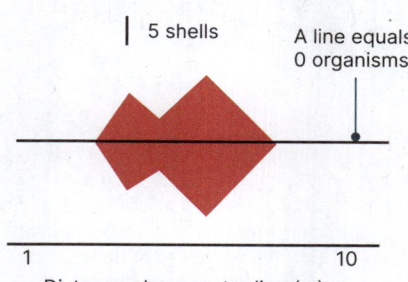

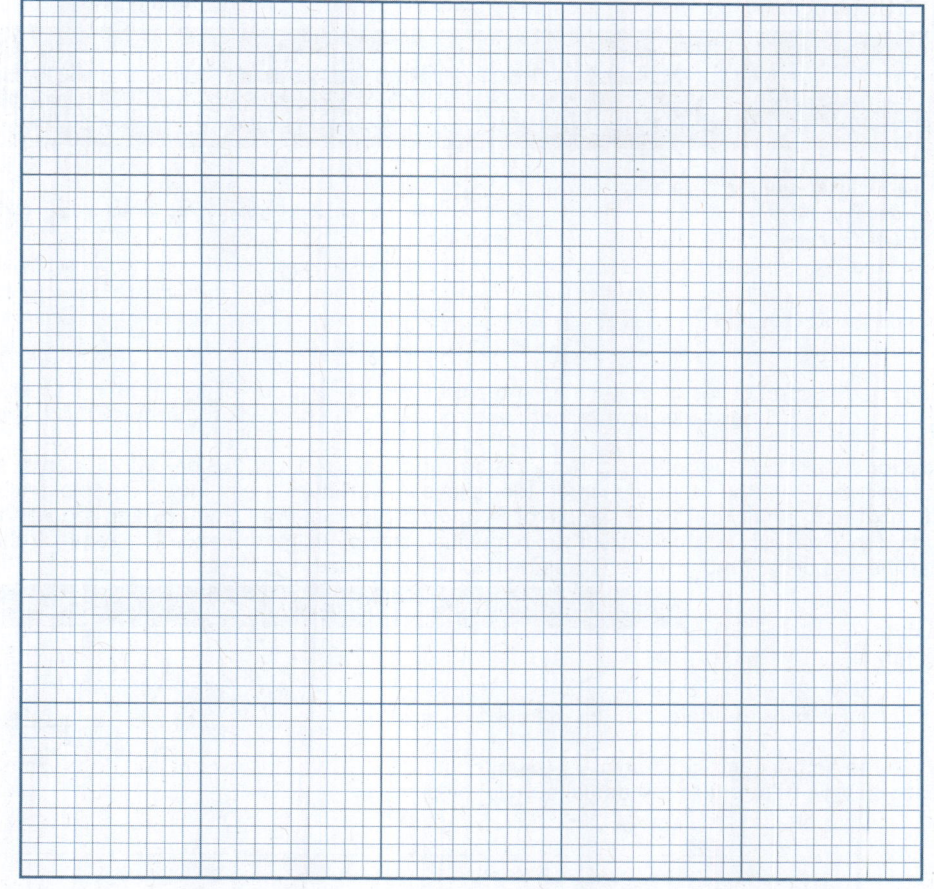

83 Mark and Recapture Sampling

Key Idea: Mark and recapture sampling enables estimates of the population size of highly mobile organisms.

The **mark and recapture** method of estimating **population** size is used in the study of animal populations in which the individuals are highly mobile. It is of no value where animals do not move or move very little. Once tagged, the animals need enough time to disperse before recapture, so **sampling error** does not occur from capturing clusters. Likewise, if the animals are left too long between tagging and recapture, some individuals may have died or left the population, also creating sampling errors. The number of animals caught in each sample must be large enough to be valid. This **sampling** method is suitable for estimating fish population numbers, providing both commercial and recreational fishermen report tagged and caught fish. The **Lincoln index** is a statistical strategy to estimate the total population size, given the size of the first **sample**, and number of tagged individuals in the second sample.

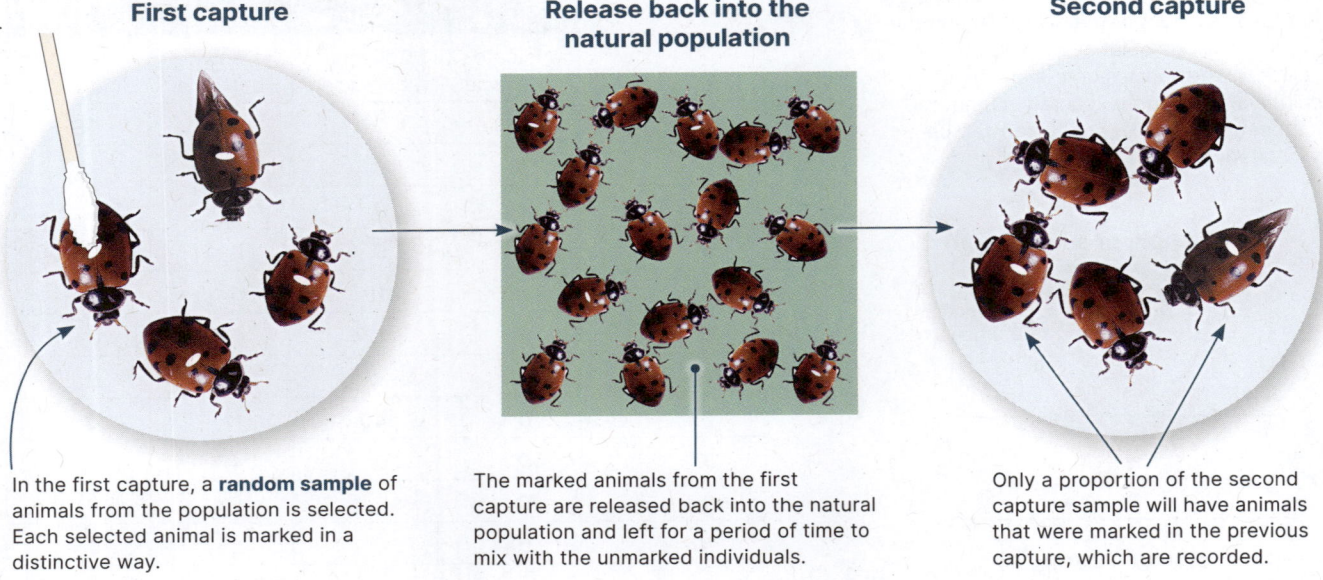

First capture — In the first capture, a **random sample** of animals from the population is selected. Each selected animal is marked in a distinctive way.

Release back into the natural population — The marked animals from the first capture are released back into the natural population and left for a period of time to mix with the unmarked individuals.

Second capture — Only a proportion of the second capture sample will have animals that were marked in the previous capture, which are recorded.

The Lincoln Index

Data from the mark and recapture sampling techniques can be used to estimate population size. If repeated over successive cycles of mark and recapture, population trends can be used by conservationists to understand the stability of the **species**.

$$\text{Total population (N)} = \frac{\text{Number in 1st sample (all marked) (M)} \times \text{Number in 2nd sample (n)}}{\text{Number of marked individuals recaptured in 2nd sample (R)}}$$

Steps in the mark and recapture technique:
1. Sample the population by capturing as many individuals as is possible and practical. Capture technique will depend on the animal.
2. Mark the captured animals to distinguish them from unmarked animals.
3. Return the marked animals to their habitat and leave them for an extended period to allow them to redistribute themselves in the population.
4. Sample the population again (the sample must be large enough to provide valid data but the sample size can be different to the first).
5. Determine the numbers of marked to unmarked animals in the second sample. Use the equation above to estimate the population size.

1. (a) You will need several boxes of toothpicks and a pen. Work in a group of 2-3 students to 'sample' the population of toothpicks in the full box by using the mark and recapture method. Each toothpick will represent one animal.

    Instructions: Take out 10 toothpicks from the box and mark them on 4 sides with a pen so that you will be able to recognize them from the other unmarked toothpicks later. Return the marked toothpicks to the box and shake the box to mix the toothpicks. Take a sample of 20 toothpicks from the same box and record the number of marked and unmarked toothpicks. Determine the total population size by using the equation above and record in table below. Repeat the sampling 4 more times (steps b–d above) and record your results:

	Sample 1	Sample 2	Sample 3	Sample 4	Sample 5
Estimated population					

    Count the actual number of toothpicks in the matchbox: _____

    (b) Compare the actual number of 'animals' to your estimates, state by how much it differs, and explain why the sample estimate may be different to the actual population size:

    _____

    _____

2. In 1919 a researcher by the name of Dahl wanted to estimate the number of trout in a Norwegian lake. The trout were subject to fishing so it was important to know how big the population was in order to manage the fish stock. He captured and marked 109 trout in his first sample. A few days later, he caught 177 trout in his second sample, of which 57 were marked.

    (a) Use the Lincoln index (above) to estimate the total population size:

    Size of first sample: _____

    Size of second sample: _____

    Number marked in second sample: _____

    Estimated total population: _____

    (b) What factors could have caused sampling errors in Dahl's research?

    _____
    _____
    _____
    _____

3. Discuss some of the problems with the mark and recapture method if the second sampling is:

    (a) Left too long a time before being repeated: _____
    _____
    _____
    _____

    (b) Too soon after the first sampling: _____
    _____
    _____
    _____

4. Describe two important assumptions being made in this method of sampling, which would cause the method to fail if they were not true:

    (a) _____
    _____
    _____

    (b) _____
    _____
    _____

5. Some types of animal would be unsuitable for this method of population estimation, i.e. it would not work.

    (a) Name an animal for which this method of sampling would not be effective: _____

    (b) Explain your above answer: _____
    _____
    _____
    _____
    _____

84 Indirect Sampling

Key Idea: Signs and calls made by animals can be used to sample for population numbers. This is indirect sampling.

If **populations** are small and easily recognized, they may be directly monitored quite easily. However, direct measurement of elusive, easily disturbed, or widely dispersed populations is not always feasible. In these cases, indirect methods can be used to assess population **abundance**, provide information on habitat use and range, and enable biologists to link habitat quality to species presence or absence. **Indirect sampling** methods provide less reliable measures of abundance than direct **sampling** methods, such as **mark and recapture** but are widely used nevertheless. They rely on recording the signs of a **species**, e.g. scat, calls, tracks, and markings on vegetation, and using these to assess population abundance. In Australia, the Environmental Protection Agency (EPA) provides a Frog Census Datasheet (below) on which volunteers record details about frog populations and habitat quality in their area. This programme enables the EPA to gather information across Australia.

Recording a date and accurate map reference is important.

Population estimates are based on the number of frog calls recorded by the observer.

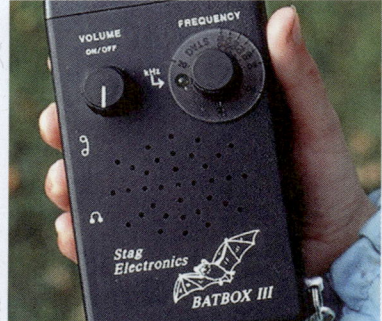

Electronic devices, such as the bat detector pictured above, can be used to estimate population density of nocturnal, highly mobile species such as bats. In this case, the detector is tuned to the particular frequency of the hunting clicks emitted by specific bat species. The number of calls recorded at a time can be used to estimate numbers per area.

The analysis of animal tracks allows wildlife biologists to identify habitats in which animals live and to conduct population surveys. Interpreting tracks accurately requires considerable skill as tracks may vary in appearance even when from the same individual. Tracks are particularly useful as a way to determine habitat use and preference.

All animals leave scat (feces) which is species specific and readily identifiable. Scat can be a valuable tool by which to gather data from elusive, nocturnal, easily disturbed, or highly mobile species. Fecal analyses can provide information on diet, movements, population density, sex ratios, age structure, and even genetic diversity.

1. (a) Describe the kind of information that could be gathered from the Frog Census Datasheet:

   _____

   (b) Identify the benefits of linking a measure of abundance to habitat assessment:

   _____
   _____

2. Describe one other indirect method of population sampling and outline its advantages and drawbacks:

   _____
   _____

85 Radio, Satellite, and GPS Animal Tracking

Key Idea: Technology can allow monitoring of sampled animals whilst producing minimal disruption.

Fieldwork involving difficult terrain, aquatic environments, or highly mobile, secretive, or easily disturbed **species**, has been greatly assisted in recent years by the use of radio-transmitter, and more recently, **satellite** and **GPS tracking** technology. Tracking technology can be used to quickly obtain accurate information about an animal's home range and can provide information about dispersal, distribution, habitat use, and competitive relationships. Radio-tracking is particularly suited to **population** studies of threatened species (because it is relatively non-invasive) and of pests (because their dispersal and habitat use can be monitored). The information can be used to manage an endangered species effectively or to plan more efficient pest control operations. Satellite transmitters, sending data to a computer via satellite, can be used to study migratory movements of large animals and marine species which are more difficult to follow. GPS tracking attaches a receiver, rather than transmitter, to an animal and provides real-time movement.

Radio-tracking
A tracking antenna and receiver can locate animals by accurately fixing their position. They can be mounted on light aircraft or off-road vehicles for mobile **radio-tracking** over large areas. Portable hand-held antennae are used for work in inaccessible or difficult terrain. It is important for the antenna to be pointed in the direction of the animals to avoid signal loss.

Satellite-tracking
A transmitter attached to the animal sends location information to a satellite, which then transmits the data to land-based computers for analysis. Satellite tracking is advantageous for aquatic mobile species like turtles and whales because scientists do not need to be physically near the animals. The commonly used ARGOS system provides an accuracy range of around 150m, but data refreshes can be delayed.

Tagged American condor

GPS (Global Positioning System)
GPS technology provides precise and continuous movement data. Small receiver devices with built-in computers calculate the location of tagged animals. Four orbiting satellites enable almost instantaneous data streaming across most of the Earth. Some data is stored on tags that need to be retrieved, but new technology allows for remote data transmission, similar to airtags.

Record-breaking flight of the Godwit

▸ The Bar-tailed Godwit (*Limosa lapponica*) sets the record for the longest, non-stop migration of all animals. The 2020-21 flights of a male were tracked using a satellite transmitter that was solar-powered (shown in the map, right). In 2022, the record was broken once more by another male called 'B6', only 4 months old and wearing a 5G satellite tag.

▸ Unlike the albatross and other seabirds, the godwit needs to continuously flap its wings to stay aloft. It does this for around 11 days continuously on the spring flight from its breeding grounds in Alaska to its feeding grounds in the North Island of New Zealand.

▸ Although radio-tracking is used in the nesting grounds of Alaska where the godwits remain in the local area for half a year, satellite tracking must be used for migration. The godwit is a small bird, just 300 g, so the tracker needs to be light (5 g) to ensure it does not add undue energy burdens during the long flight.

▸ Scientists use the data to understand bird population and migration path changes due to climate, ecosystem damage, and pollution.

1. Why would satellite tracking rather than radio-tracking be the most suitable choice for locating migrating birds?

___

2. Radio-tracking equipment tends to be much cheaper than the other satellite based methods but requires heavier transmitters and predictable home ranges for animals. Which species would suit this form of tracking and why?

___

86 Environmental DNA Sampling

Key Idea: Organisms that inhabit a particular area but cannot be directly observed or found can be identified using environmental DNA.

The advent of rapid genome sequencing and the analysis of biological data using computers (bioinformatics) has allowed scientists to identify **species** based on sequences of DNA. Each species is given a **DNA barcode** based on short sections of conserved DNA that are unique to each species.

DNA can be shed into the environment by an organism, e.g. in droppings, or from skin. This can be sampled from water or soil for some time after the organism is gone. Scientists can identify the presence of the organism via its DNA barcode without ever seeing the organism. This method was (and still is) used to monitor outbreaks of Covid-19 in a **population** without having to test every person through sampling of wastewater in populated areas.

The DNA barcode

▸ DNA barcoding uses short, highly conserved (unchanging through evolution) sequences of DNA to produce species-specific information. The sequence of DNA chosen for analysis depends on the type of organism, e.g. plant, animal, fungus, bacterium.

▸ The goal of DNA barcoding is to enable identification of individual species from short sequences of DNA from samples. This information can then be applied whenever species-specific knowledge is important. Applications include evolutionary biology, conservation, detection of invasive species, dietary analysis (to help describe food webs), and food safety.

Plant barcoding uses up to three genes in the chloroplasts and one in the nuclear DNA.

The barcode from animals is taken from the commonly-shared respiratory gene.

Fungal barcoding uses the ITS (internal transcribed spacer) region, commonly present in fungi.

Barcoding and environmental sampling

DNA barcoding works on the assumption that each species' DNA is different so it can be used to identify specific species. Closely related species will have similar DNA barcodes.

Collect DNA from environment.

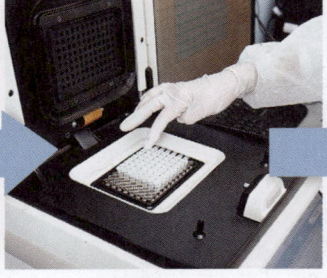
Isolate DNA sequences in sample.

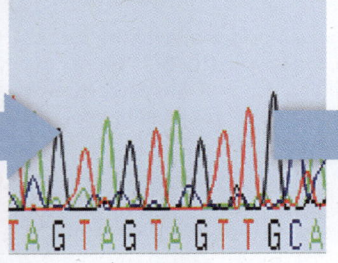
Sequence the DNA found.

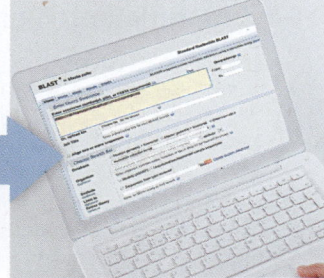
Compare sequence with database to identify species present.

1. Why is DNA barcoding considered a type of indirect sampling? _____

2. Explain how sampling DNA barcodes provides a useful way to:

   (a) Monitor species that are difficult to sample directly: _____

   (b) Monitor disease in a population: _____

   (c) Identify the range of species in an ecosystem: _____

87 Measuring Diversity in an Ecosystem

Key Idea: Diversity indices incorporate both species richness and relative abundance so are a good measure of diversity.

A diversity index, like **Simpson's index of diversity**, is a mathematical measure of **species** diversity in a community. It takes into account the number of species present as well as their relative **abundance**. Diversity indices can also be used to assess the health of an **ecosystem** when examining collected **sample** data. A change in species composition can indicate changes in an ecosystem's status, e.g. in response to pollution or climate change. Certain 'sensitive' species are associated only with specific conditions, e.g. clean, cold water. The presence (or absence) of these indicator species tells us about the health of an ecosystem.

High diversity

Low diversity

Using diversity indices and the role of indicator species

▶ The health of an ecosystem can be assessed by measuring the variety of organisms living in it. Certain species, called indicator species, are typically found in ecosystems that are either polluted or pristine. A change in the species assemblage can therefore indicate deterioration or recovery of an ecosystem.

▶ To be properly interpreted, indices are often evaluated with reference to earlier measurement or a standard ecosystem measure.

▶ The images above show samples from two streams: a high diversity community with a large number of invertebrate species (left) and a low diversity community (right) with fewer species in large numbers. These images also show typical stream indicator species.

▶ The left photograph shows a stonefly (1) and an alderfly larva (2). These species (and mayfly larvae) are typical of high water quality.

▶ The right photograph shows a dominance of snails (3) which are tolerant of a wide range of conditions, including degraded environments.

Simpson's index of diversity

Simpson's index of diversity is a method of quantitatively measuring diversity.

In the equation, n = number of individuals of each species in the sample and N = total number of individuals (of all species) in the sample. It generates a value between 0 and 1. The closer the value is to 1, the higher the biodiversity.

Simpson's index of diversity (D) is easily calculated using the simple formula below. Communities with a wide range of species produce a higher score than communities dominated by larger numbers of only a few species.

$$D = 1 - (\sum (n/N)^2)$$

D = Simpson's index of diversity

N = Total number of individuals (of all species) in the sample

n = Number of individuals of each species in the sample

1. Why might it be useful to have baseline data (prior knowledge of a system) before interpreting a diversity index?

_____

_____

2. (a) How might you monitor the recovery of a stream ecosystem following an ecological restoration project?

_____

_____

_____

   (b) What role could indicator species play in the monitoring programme? _____

_____

_____

Comparing the biodiversity of different ecosystems

A student observed that in a conifer plantation there seemed to be only a few different invertebrate species and wondered if more would be found in a nearby oak woodland. They carried out an investigation and the results are tabled below. The invertebrates are not drawn to scale.

Site 1: Oak woodland

Species	Number of animals (n)	n/N	$(n/N)^2$
Species 1	35		
Species 2	14		
Species 3	13		
Species 4	12		
Species 5	8		
Species 6	6		
Species 7	6		
Species 8	4		
	∑n = 98		∑(n/N)² =

Site 2: Conifer plantation

Species	Number of animals (n)	n/N	$(n/N)^2$
Species 1	74		
Species 2	20		
Species 3	3		
Species 4	3		
Species 5	1		
Species 6	0		
Species 7	0		
Species 8	0		
	∑n = 101		∑(n/N)² =

Species 1 Mite | Species 2 Ant | Species 3 Earwig | Species 4 Woodlice | Species 5 Centipede | Species 6 Longhorn beetle | Species 7 Small beetle | Species 8 Pseudoscorpion

3. (a) Complete the two tables above by calculating the values for n/N and $(n/N)^2$ for the student's two sampling sites:

   (b) Calculate the Simpson's Index of diversity for site 1: _____

   (c) Calculate the Simpson's Index of diversity for site 2: _____

   (d) Compare the diversity of the two sites and suggest any reasons for it: _____

4. (a) Species richness is a measure of the number of different species in an area. Which of the two areas sampled above has the greatest species richness?

   (b) Why would measuring species richness not be as informative as measuring species diversity?

88 Classification Keys

Key Idea: Classification keys are used to identify an organism based on its distinguishing features and assign it to a species. Dichotomous keys are a useful tool in biology and can enable identification to the **species** level provided the characteristics chosen are appropriate for separating species. An organism's classification should include a clear, unambiguous description, an accurate diagram, and its unique name, denoted by the genus and species. **Classification keys** are used to identify an organism and assign it to the correct species (assuming that the organism has already been formally classified previously). Typically, keys involve a series of linked steps. At each step, a choice is made between two features (dichotomous key). Each alternative leads to another question until an identification is made. If the organism cannot be identified, it may be a new species or the key may need revision. This activity describes two examples of dichotomous keys. The first type is a branching tree, used with larvae of caddisflies. The second type is a nested style, used for aquatic insect orders, and *Acer* plants.

Caddisfly larvae

Branching tree: classification key for caddisfly larvae
The key shown here is a simplified version of one commonly used to identify caddisfly larvae. It identifies the organisms to genus level only. To use the key to identify the larvae pictured below, start at the top and branch at each feature until you reach the bottom.

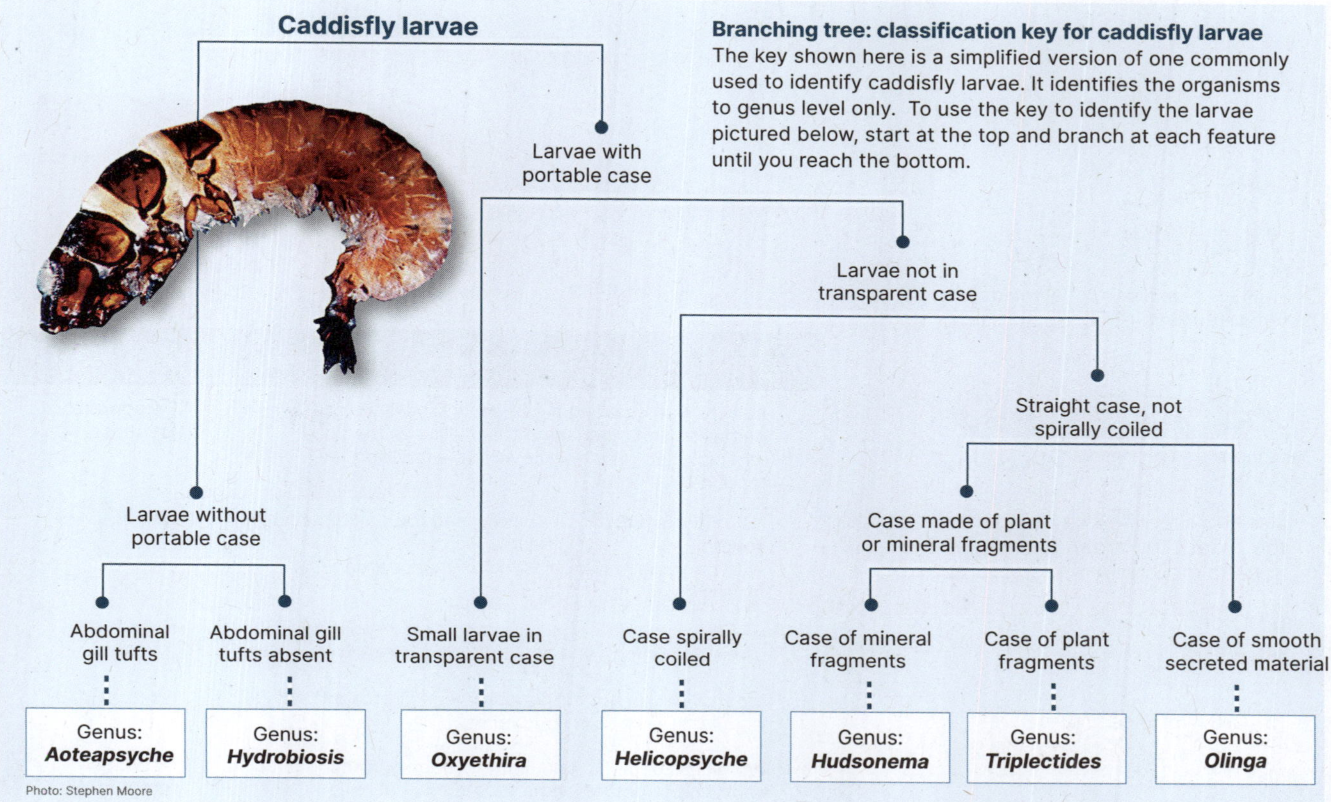

Larvae without portable case
- Abdominal gill tufts → Genus: *Aoteapsyche*
- Abdominal gill tufts absent → Genus: *Hydrobiosis*

Larvae with portable case
- Small larvae in transparent case → Genus: *Oxyethira*
- Larvae not in transparent case
  - Case spirally coiled → Genus: *Helicopsyche*
  - Straight case, not spirally coiled
    - Case made of plant or mineral fragments
      - Case of mineral fragments → Genus: *Hudsonema*
      - Case of plant fragments → Genus: *Triplectides*
    - Case of smooth secreted material → Genus: *Olinga*

Photo: Stephen Moore

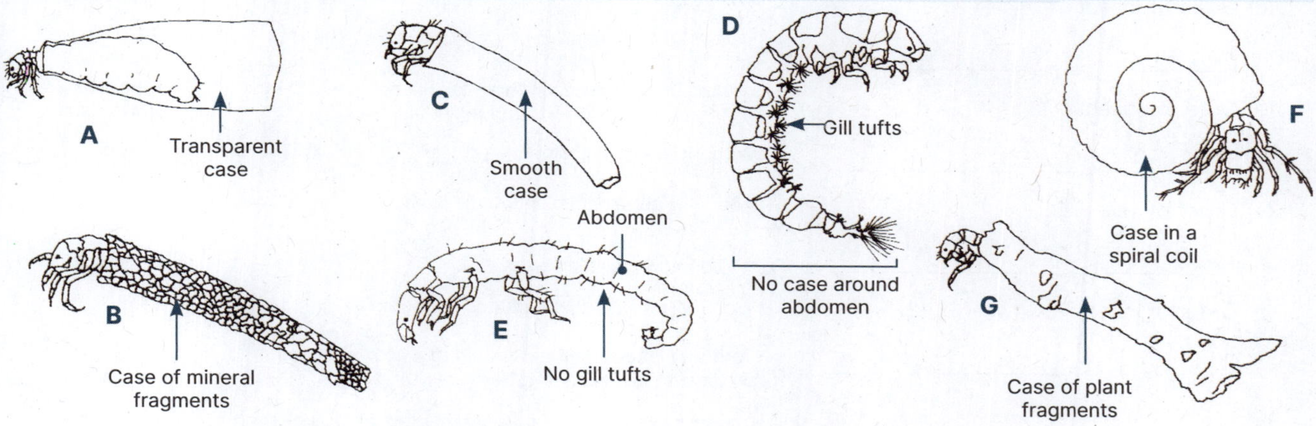

Specimens:
- A: Transparent case
- B: Case of mineral fragments
- C: Smooth case
- D: Gill tufts; No case around abdomen
- E: Abdomen; No gill tufts
- F: Case in a spiral coil
- G: Case of plant fragments

1. Describe the main feature used to distinguish the genera in the key above: _____
   _____

2. Use the key above to assign each of the caddisfly larvae (A-G) to its correct genus:

   A: _____   D: _____   G: _____

   B: _____   E: _____

   C: _____   F: _____

©2025 BIOZONE International
ISBN: 978-1-99-101409-2
Photocopying prohibited

Nested style: classification key for aquatic insect larvae

The key shown here identifies the organisms to order level only. To use the key to identify the larvae pictured below, start at the top question and move to the next question matching your response. The last statement will indicate the classification name.

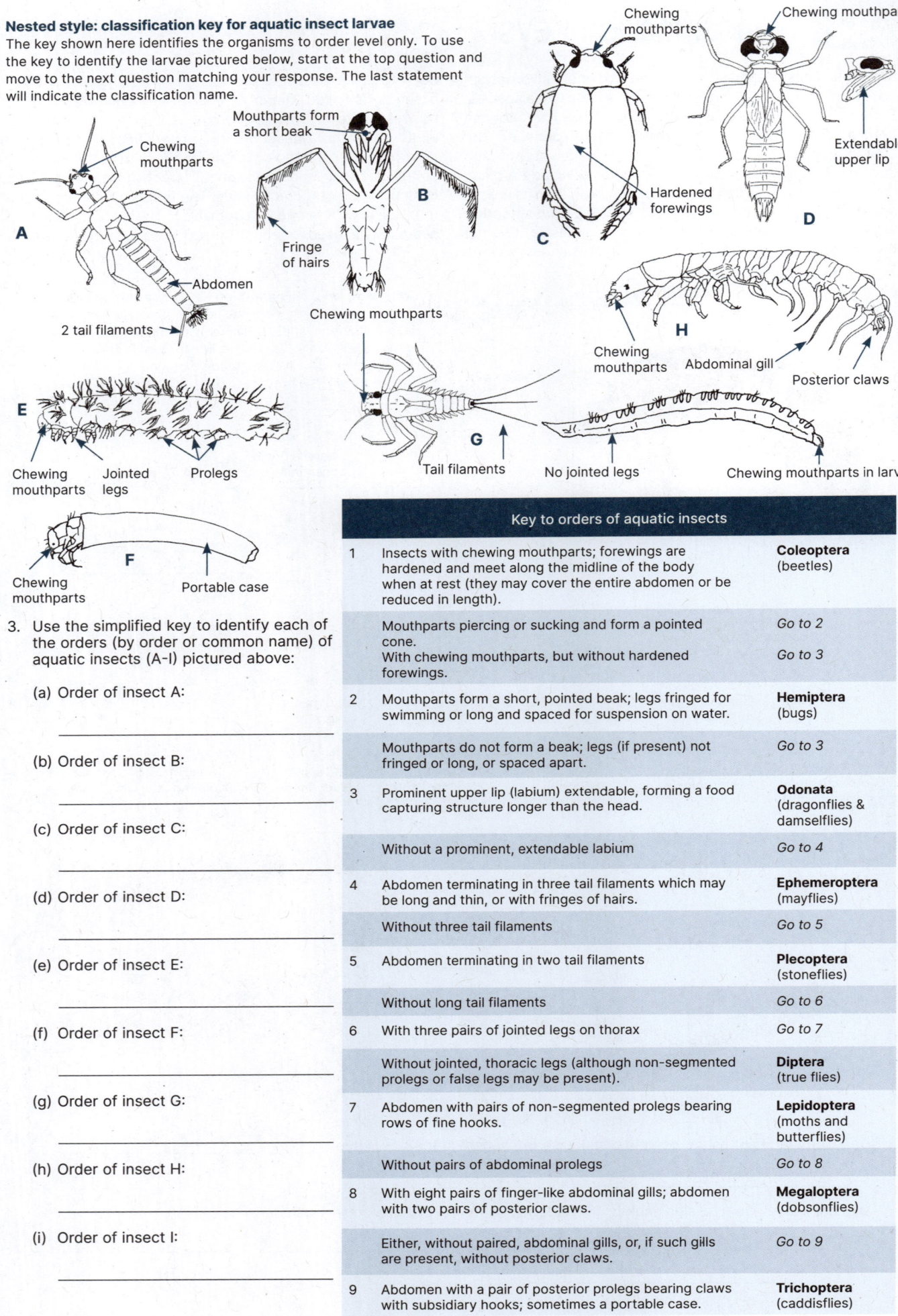

3. Use the simplified key to identify each of the orders (by order or common name) of aquatic insects (A-I) pictured above:

(a) Order of insect A: _____

(b) Order of insect B: _____

(c) Order of insect C: _____

(d) Order of insect D: _____

(e) Order of insect E: _____

(f) Order of insect F: _____

(g) Order of insect G: _____

(h) Order of insect H: _____

(i) Order of insect I: _____

Key to orders of aquatic insects		
1	Insects with chewing mouthparts; forewings are hardened and meet along the midline of the body when at rest (they may cover the entire abdomen or be reduced in length).	**Coleoptera** (beetles)
	Mouthparts piercing or sucking and form a pointed cone.	Go to 2
	With chewing mouthparts, but without hardened forewings.	Go to 3
2	Mouthparts form a short, pointed beak; legs fringed for swimming or long and spaced for suspension on water.	**Hemiptera** (bugs)
	Mouthparts do not form a beak; legs (if present) not fringed or long, or spaced apart.	Go to 3
3	Prominent upper lip (labium) extendable, forming a food capturing structure longer than the head.	**Odonata** (dragonflies & damselflies)
	Without a prominent, extendable labium	Go to 4
4	Abdomen terminating in three tail filaments which may be long and thin, or with fringes of hairs.	**Ephemeroptera** (mayflies)
	Without three tail filaments	Go to 5
5	Abdomen terminating in two tail filaments	**Plecoptera** (stoneflies)
	Without long tail filaments	Go to 6
6	With three pairs of jointed legs on thorax	Go to 7
	Without jointed, thoracic legs (although non-segmented prolegs or false legs may be present).	**Diptera** (true flies)
7	Abdomen with pairs of non-segmented prolegs bearing rows of fine hooks.	**Lepidoptera** (moths and butterflies)
	Without pairs of abdominal prolegs	Go to 8
8	With eight pairs of finger-like abdominal gills; abdomen with two pairs of posterior claws.	**Megaloptera** (dobsonflies)
	Either, without paired, abdominal gills, or, if such gills are present, without posterior claws.	Go to 9
9	Abdomen with a pair of posterior prolegs bearing claws with subsidiary hooks; sometimes a portable case.	**Trichoptera** (caddisflies)

Keying out plant species

▶ Keys are also extensively used to identify plants as they are quick and easy to use in the field, although they sometimes rely on the presence of particular plant parts such as fruits or flowers. Some also require some specialist knowledge of plant biology. The following simple activity requires you to identify five species of the genus *Acer* from illustrations of the leaves using a nested style, dichotomous key. It provides valuable practice in using characteristic features to identify plants to species level.

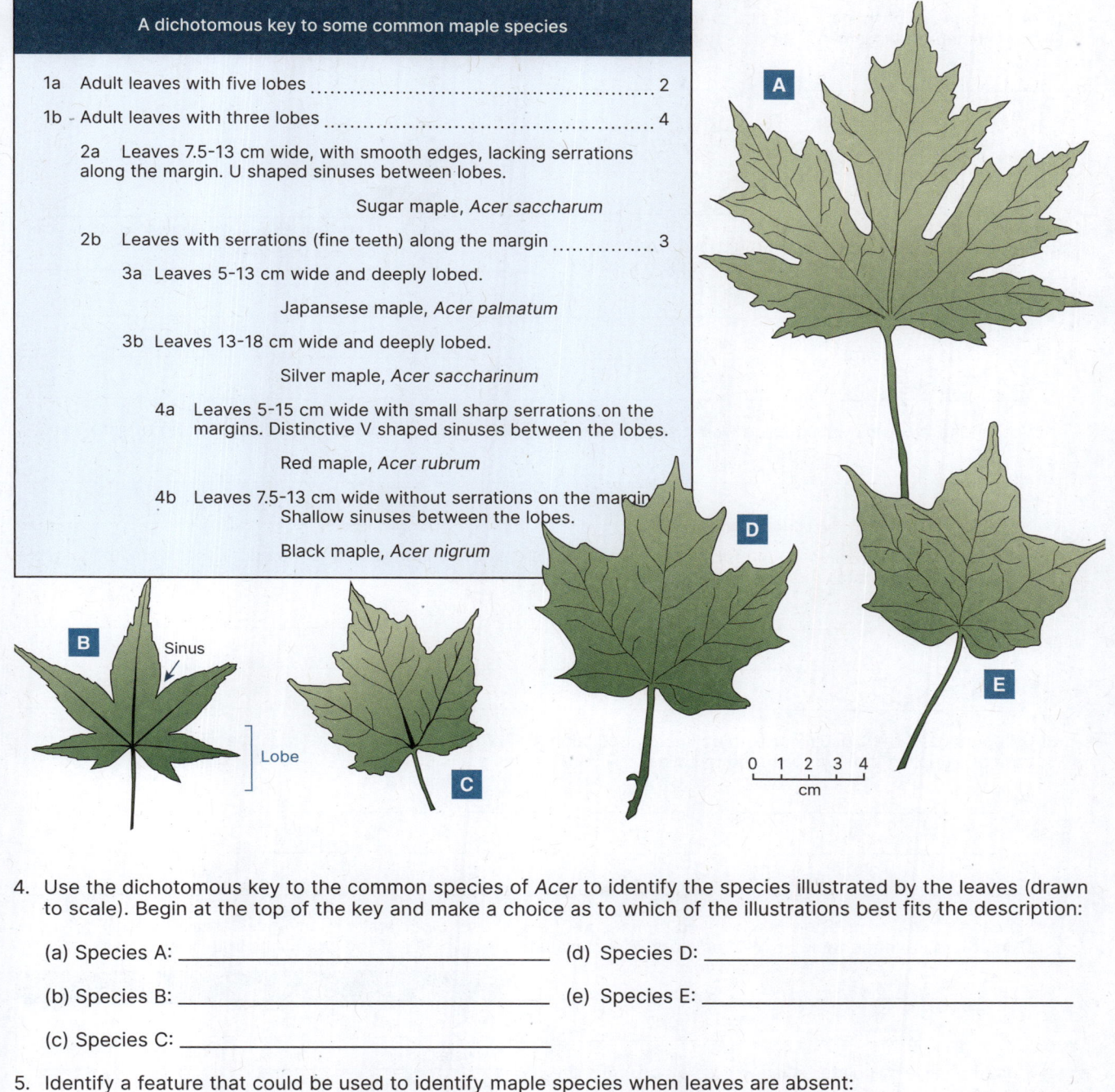

A dichotomous key to some common maple species

1a Adult leaves with five lobes 2
1b Adult leaves with three lobes 4
    2a Leaves 7.5-13 cm wide, with smooth edges, lacking serrations along the margin. U shaped sinuses between lobes.
                    Sugar maple, *Acer saccharum*
    2b Leaves with serrations (fine teeth) along the margin 3
        3a Leaves 5-13 cm wide and deeply lobed.
                  Japansese maple, *Acer palmatum*
        3b Leaves 13-18 cm wide and deeply lobed.
                  Silver maple, *Acer saccharinum*
        4a Leaves 5-15 cm wide with small sharp serrations on the margins. Distinctive V shaped sinuses between the lobes.
                  Red maple, *Acer rubrum*
        4b Leaves 7.5-13 cm wide without serrations on the margin. Shallow sinuses between the lobes.
                  Black maple, *Acer nigrum*

4. Use the dichotomous key to the common species of *Acer* to identify the species illustrated by the leaves (drawn to scale). Begin at the top of the key and make a choice as to which of the illustrations best fits the description:

(a) Species A: _____   (d) Species D: _____

(b) Species B: _____   (e) Species E: _____

(c) Species C: _____

5. Identify a feature that could be used to identify maple species when leaves are absent:
_____

6. When identifying a plant, suggest what you should be sure of before using a key to classify it to species level:
_____
_____

7. Explain the importance of correctly identifying species when carrying out environmental science:
_____
_____
_____

89 Did You Get It?

1. Mangrove crabs live in estuaries that support specially adapted mangrove plants. Mangrove crabs are omnivorous, eating both mangrove plants, detritus, and small mud invertebrates, sheltering in burrows to avoid predators.

   (a) Abundance of both mangrove and crab populations can be estimated by sampling. Why do scientists sample?

   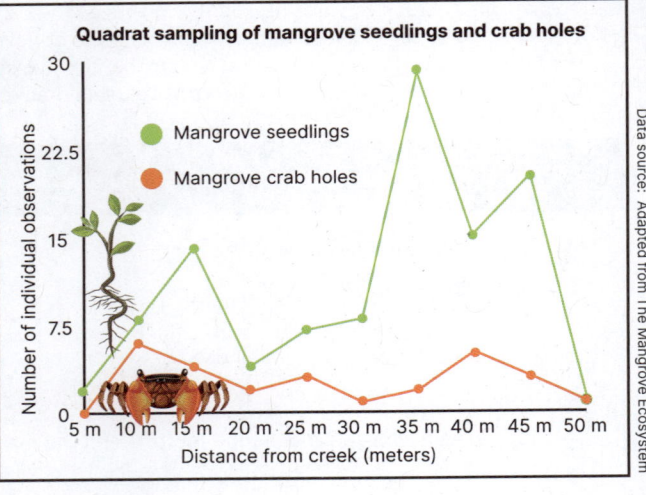

   Quadrat sampling of mangrove seedlings and crab holes

   Data source: Adapted from The Mangrove Ecosystem

   (b) Quadrat sampling by students (right) was used to estimate abundance. Describe this sampling method:

   (c) Both indirect and direct sampling were used (above, right). Distinguish between the types, referring to data given:

   (d) Different plant species are found further away from the creek. Describe the type of sampling that would be suitable for collecting this data, and how it differs from quadrat sampling:

   (e) Students used the Lincoln index to estimate population size of the mangrove mud crabs. Describe and explain the type of sampling that was required to collect the data for this statistical tool:

2. (a) Simpson's Index of diversity can be used to measure the extent of biodiversity in any given area: $D = 1 - (\Sigma(n/N)^2)$. Using the data in the table below, complete, then calculate the diversity of the organisms in this rocky seashore:

   (b) From the value calculated, how would you describe the biodiversity in this area? Give a reason for your answer:

Species	Number	n/N	$(n/N)^2$
i. Plicate barnacle	114		
ii. Oyster borer	8		
iii. Snakeskin chiton	12		
iv. Limpets	34		
v. Dog whelks	128		
vi. Muscles	98		
vii. Cockles	57		

   (c) One of the seabird species that visits the rockpools is migratory. Explain a method that could be used to track the bird during its migration, providing reasons for your choice:

©2025 BIOZONE International
ISBN: 978-1-99-101409-2
Photocopying prohibited

Food Production

Land use	• Crop uses • The green revolution • Food security
Production systems	• Intensive production systems • Sustainable production systems
Managing pests	• Chemical pesticides • Integrated pest management

Earth's Resources

Soils	• Soil degradation • Soil protection
Land resources	• Mining • Forestry practices
Water Resources	• Freshwater management • Freshwater conflict
Fisheries	• Managing fish stocks • Aquaculture

A huge proportion of land on Earth is used for the purpose of feeding the human population.

Sustainable use of Earth's resources is essential if they are to be available for future generations.

Global Resources

The growing human population is putting pressure on the Earth's limited resources.

Global resources include land, water, minerals, and energy.

Sustainable and environmentally sound use of the Earth's resources can help reduce habitat loss and resource depletion.

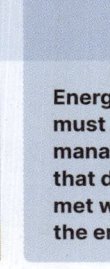

Energy supplies must be carefully managed to ensure that demands are met while protecting the environment.

Energy is necessary for modern living and new technology must be developed to meet human requirements.

Precision Seafood Harvesting

Energy Sources

Energy	• Energy transformations • Energy consumption
Renewable energy	• Wind and water • Solar and geothermal • Biofuels
Non-renewable energy	• Fossil fuels • Nuclear energy

Energy Security

Energy conservation	• Energy efficiency • Comparing fuel efficiencies
Energy threats	• Space weather • Natural disasters
Energy storage	• Energy recovery systems • Batteries

ON THE FARM

Humans have been farming for around twelve thousand years. It is estimated that possibly a quarter of a million plant species are available for agricultural use. However, just 150 make up the majority of crops plants used by humans. Of these, just 12 plant species provide 75% of all the world's plant based food. More than half is provided by the three megacrops rice, wheat, and maize.

The vast majority of meat and animal products used by humans come from five animals species. 70% of all bird mass on Earth is domesticated, the majority of which are chickens. Domesticated mammals make up about 60% of the mass of all mammals (with humans making up over 33%).

Commercial farming, whether conventional or organic, uses an enormous amount of energy. This energy use is from the production of fertilizers and sprays, and the use of on-farm machinery such as harvesters and irrigators. Much of this energy comes from fossil fuels, although there are moves to try to electrify much of this machinery to reduce fuel input.

Power hungry

Agriculture uses large amounts of energy. While much of this energy comes from fuel, used for machinery such as tractors and harvesters, a large proportion is electricity energy that comes directly from the local grid. This is to power pumps, lighting, heating etc. Much of agriculture's energy requirements could be based on renewable sources if electricity was generated from resources such as wind.

A combine harvester harvests a field of ripe wheat.

Animal farm

Animal farming is land intensive. It requires large areas of land for the animals to live and graze on and more land to grow extra feed crops for times of low growth, e.g. winter or very dry summers.

This can be reduced by raising animals in small feed lots and stalls but this has ethical issues.

Monocultural agriculture

Only about 20% of farmland is used to grow plants exclusively for human consumption. The rest is used to raise animals or grow crops to make feed for them, e.g. maize. The genetic diversity of plant crops is reducing every year as fewer farmers grow heritage crops and opt for newer, commercially produced, but more productive varieties.

🔍 Take a Deeper Look

- Investigate the feasibility of electrifying the farming process. Discuss this with your classmates. What parts of farming are simpler to electrify than others?
- Farming uses many different chemicals. Investigate the various types of chemical used in farming, their purpose, the volume used and any effects on the environment.
- In recent decades, meat production has expanded and intensified. What effects might this have had on the global environment?

Chapter 5 — Global Resources
Land and Water

Resource Hub
bit.ly/3VXllfq

Key Terms
- agriculture
- bycatch
- eutrophication
- fertilizer
- globalization
- green revolution
- insecticide
- integrated pest management
- mineral
- monoculture
- overfishing
- rangeland
- vertical farming
- water footprint

Key Concepts
- Agriculture and forestry require careful management to ensure their sustainability.
- Modern farming practices have intense energy requirements.
- Careful land management and use of Earth's resources are essential to reduce environmental disturbance.
- Water is an essential resource that must be carefully managed.
- The oceans provide fish as a food source but fish stocks must be maintained carefully.

Food Production

Learning Outcomes: | **Activity Number**

- [ ] 1 Describe the extent of land required for agriculture and the importance of plants for human survival. — 90-91
- [ ] 2 Explain what the green revolution was and describe the impacts of different methods of farming on the environment. — 92-95
- [ ] 3 Describe worldwide trends in cereal and meat production. Compare and contrast growing conditions of different cereal crops and different methods of meat production. — 96-97
- [ ] 4 Explain the term 'food security' and describe the relationship between food insecurity and poverty. — 98
- [ ] 5 Describe the types of pests that can affect food production systems and how they are controlled. Explain how pesticide resistance arises. Explain how an integrated pest management approach manages pests in a cropping system. — 99-101

Land as a Resource

- [ ] 6 Explain how soils become degraded and describe some practices that reduce the likelihood of soil erosion. — 102-103
- [ ] 7 Compare different ways of growing and harvesting trees in a forestry system and describe how the danger of fire can be lessened. — 104
- [ ] 8 Compare the effects of grazing in a rangeland area and describe how good rangeland management leads to a healthy ecosystem. Explain the role of national parks and conservation areas. — 105-106
- [ ] 9 Compare three models of city development and how describe how a city can become more sustainable. Compare different modes of transportation. — 107-108
- [ ] 10 Explain why metals and minerals are mined, compare the quantities of different mined metals, and describe some uses of metals and minerals. — 109
- [ ] 11 Explain globalization and compare its advantages and disadvantages. — 110

Water as a Resource

- [ ] 12 Locate the world's main freshwater resources and describe how human activities can affect water supplies. Describe why water is needed for survival and how water can lead to conflict situations. Describe uses of water in industrial settings. — 111-113
- [ ] 13 Describe the ecological impacts of fishing and explain how and why fish stocks must be managed responsibly. — 114

90 Land for Agriculture

Key Idea: Currently, just under half of the world's available land is used for agriculture.

A finite amount of land is available on Earth for food production. Around 48 million square kilometers of Earth is used for **agriculture**, of which croplands uses one third, and land for grazing uses the rest. As the human population continues to grow, other ways of economically producing food will need to be found. Alternative food production systems such as vertical farms may provide a way to feed humans without the need for more land. Human dietary changes that rely less on pasture raised animals could help reduce the amount of land needed for agriculture.

Feeding the world

As the human population grows, so too does the need for land on which to grow enough food for everyone. Currently, almost half of Earth's habitable land has been turned over to agriculture to feed an ever increasing human population. Land covered in ice or desert cannot be used for agriculture. As seen below, most agricultural land is used to raise livestock for meat and dairy products; only 16% of cultivated land is used to produce crops to feed humans directly.

Global land use

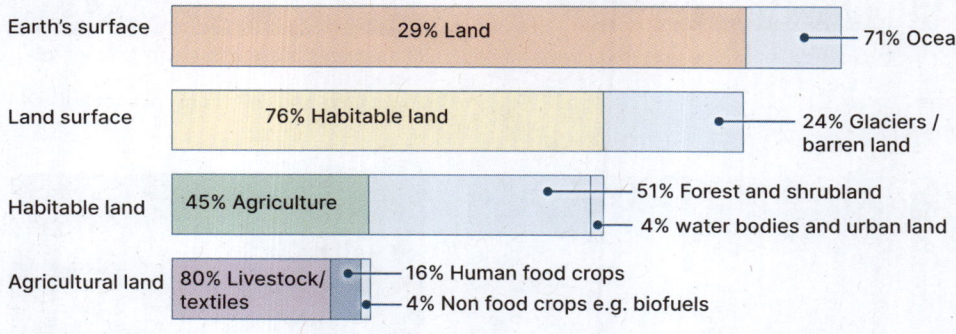

Earth's surface: 29% Land — 71% Ocean
Land surface: 76% Habitable land — 24% Glaciers / barren land
Habitable land: 45% Agriculture — 51% Forest and shrubland — 4% water bodies and urban land
Agricultural land: 80% Livestock/textiles — 16% Human food crops — 4% Non food crops e.g. biofuels

Data source: adapted from Richie and Roser, 2023. Our World in Data CC-BY

Some research suggests that a change in human dietary habits could result in less land being required to feed the human population. A diet more reliant on plant rather than animal protein would require less pastureland to raise animals, in particular cattle and sheep.

▶ As urban populations continue to increase, more food has to be brought into cities to feed everyone. Large transport vehicles, some of which have to travel long distances from where food is produced, contribute to pollution problems.

▶ One solution to these so called food miles is the concept of **vertical farming**. Crops are grown in specially designed rooms under controlled lighting conditions. They are precisely supplied with water and nutrients and harvested at peak selling condition.

▶ This has shown promise for small, leafy vegetables, particularly salad greens and fresh herbs which are perishable, with short shelf lives.

▶ While expensive to set up, vertical farms in cities can provide a ready supply of fresh greens in peak condition to local buyers.

1. What type of land is not suitable for agricultural purposes? _____

2. What is agricultural land used for other than growing crops to feed humans directly? _____

3. Why would a change in human diet to include more plant based products reduce land requirements for agriculture? _____

4. Describe both the benefits and drawbacks of vertical farming: _____

91 The Importance of Plants

Key Idea: We rely on plants and plant products for survival. Plants produce oxygen via the process of photosynthesis and are also the ultimate source of food and metabolic energy for nearly all animals. Besides foods, e.g. grains, fruits, nuts, and vegetables, plants also provide people with shelter (wood for construction), clothing (textile fibers), medicines, fuels, and raw materials from which innumerable other products are made. Without plants, life on Earth could not be sustained. They are found on every continent, even in extreme environments such as Antarctica.

Plant tissues provide the energy for almost all heterotrophic life. Many plants produce highly nutritious fruits in order facilitate the spread of their seeds by animals and birds.

Plant tissues can be used to provide shelter in the form of framing, cladding, and roofing.

Many plants provide fibers for a range of materials including cotton (above), linen (from flax), and coir (from coconut husks).

Plant extracts, including rubber from rubber trees (above), have many uses, including manufacturing materials.

Coal, petroleum, and natural gas are fossil fuels which were formed from the dead remains of plants and other organisms. Together with wood, they provide important sources of fuel.

Over 25% of all modern medicines are derived from plant extracts. Not all plant extracts are beneficial. Many recreational drugs, e.g. opium and cannabis (above) are derived from plants.

1. Using examples, describe how plant species are used by people for each of the following:

   (a) Food: _____

   (b) Fuel: _____

   (c) Clothing: _____

   (d) Building materials: _____

   (e) Aesthetic value: _____

   (f) Therapeutics: _____

2. Suggest three reasons why forests should be protected: _____

92 The Green Revolution

Key Idea: The Green Revolution enhanced global food production but the requirement for water and fertilizer introduced problems.

The development of high yielding dwarf varieties of food crops by targeted cross breeding programs was known as the **Green Revolution**. Norman Borlaug in the USA developed shorter stemmed varieties of wheat that gave greater yields of grain per head. Their shorter stems made them more resistant to toppling over in bad weather. These new varieties were introduced to the grain growing areas of Mexico and resulted in vastly improved wheat yields from the same land area. Similar programmes for rice were later developed in Asia, notably India. Again, the new varieties developed by cross-breeding gave higher yields on shorter stems. Grain production increased widely. However, the new varieties required increased inputs of fertilizer and water which was costly for farmers and introduced the problems associated with high input **agriculture**.

The **first agricultural revolution** saw human society change from one of hunter gatherers to one of farming.
Humans settled in one place, developed villages and social systems and worked together to produce food for themselves. They began domesticating specific plants and animals and developing ways of preserving food to feed themselves year round. Sections of land were fenced off to protect them from damage by both wild and domesticated animals. Food was produced locally and exchanged at small, local markets.

The **second development in agriculture** took place around or just after the Industrial Revolution.
Development of better machinery such as plows and seed drills on farms allowed for more intensive agriculture.
Instead of growing food for a small, local community, food was grown on a larger scale and transported away from where it was being grown into the towns and cities to feed the populations of people now living in urban environments.

In the 1940s, a US scientist called Normal Borlaug began to develop new strains of wheat as part of an initiative in Mexico. By selectively cross-breeding specific varieties, he produced a wheat plant that grew on shorter stems and gave a high yield of grain. Its strong, short stem prevented it from falling over (lodging), even after heavy rainfall.
Mexican yields of wheat began to dramatically improve. Without having to convert more land to agriculture, wheat output went from half a million tonnes in 1940 to 5.5 million tonnes in 1985. Borlaug won the Nobel Peace Prize in 1970 for his efforts in increasing world food production to prevent starvation.
This **green revolution** of better varieties, coupled with developments in fertilizer manufacture increased yields dramatically.

1. What was the first agricultural revolution? _____
_____
_____
_____

2. Explain how the green revolution helped prevent human hunger: _____
_____
_____
_____
_____
_____
_____

Wheat production in Mexico over 25 years

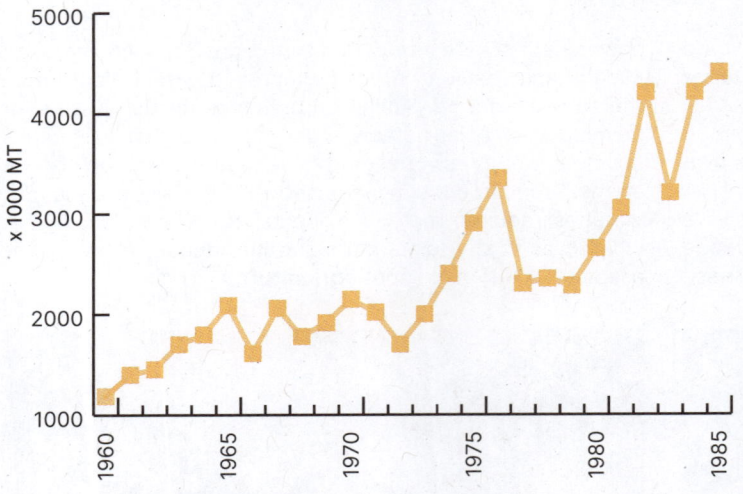

Data source: Adapted from https://www.indexmundi.com/agriculture/

- The graph (left) shows how wheat production in Mexico improved during the Green Revolution: 1960s to 1980s. Borlaug's varieties with shorter stems and higher grain yields caused a rise in the country's wheat output.
- The new, higher yielding varieties required high inputs of fertilizer to maintain yields and had high water requirements. Although there was a drop in production in some years, notably the mid 1970s and early 1980s, the overall trend was upwards.
- One downside of such increases in production is the need for more fertilizer, particularly nitrogen, which uses fossil fuels in its manufacture.

Improvements to rice

- Rice composes a large percentage of the diet in many parts of the world.
- In the 1950s, the population of Asia was increasing but food production was not. Rice was a staple for most people but not enough was being grown to sustain the growing population.
- The International Rice Research Institute (IRRI) was founded in the Philippines, with the goal of developing new varieties of rice that would give higher yields.
- A high-yielding strain from Indonesia was crossed with a short stemmed variety from China giving a high yielding dwarf form, named IR8. It was suggested that, under optimum conditions, it could increase rice yields ten fold. Its use rapidly spread throughout India and averted the possibility of famine.
- Some critics say that, as a result of use of IR8, many older varieties have been lost. These varieties could have provided genes for better resistance to climate change.
- Golden rice was developed in the 1990s and is genetically engineered to contain high levels of beta-carotene, which is converted in the body to vitamin A. This allows better nutrient delivery to people in developing countries where rice is the staple food.
- Other areas of research include improving the resistance of rice to insects, bacteria, and herbicides, improving yields, and delivering edible hepatitis and cholera vaccines in the rice.

3. Examine the graph of wheat production in Mexico on this page:

   (a) Between which years did wheat production show the greatest decreases? _____

   (b) Suggest some possible reasons for these large drops in production: _____

   (c) Between which years was the highest increase in production? _____

4. Discuss the difference between the development of IR8 rice and golden rice:

93 Impacts of Farming

Key Idea: Farming greatly affects the Earth. It requires huge amounts of land and contributes to global pollution.

Human requirements for food and other goods means that almost half of the Earth's habitable land has been altered to produce these. Farming has large freshwater requirements and is a source of pollution both in the form of greenhouse gases and animal waste products. It also causes soil erosion and pollution of waterways from fertilizer runoff. Without application of nitrogen fertilizers, we would not be able to produce food in the quantities needed to sustain the growing human population. The production of fertilizer from ammonia was a breakthrough in chemistry. However, the process is dependent on methane, a fossil fuel, to provide one of the key raw materials for its production. Its energy and therefore fuel requirements are high, and the carbon dioxide it generates as a waste product is also a pollutant.

Impactor	Impact from agriculture
Greenhouse gas emissions	26% is from food production
Land use	50% of habitable land used for agriculture
Freshwater withdrawals	70% of freshwater withdrawals used for agriculture
Ocean and freshwater pollution	78% of ocean and freshwater pollution is from agriculture
Mammal biodiversity	94% of global mammal biomass (excluding humans) is livestock
Bird biodiversity	71% of global bird biomass is poultry livestock

Data source: Adapted from Richie 2022. Our World in Data CC-4.0

▶ According to the United Nations Food and Agricultural Organization, almost 50% of the world's habitable land is used for agriculture.

▶ The table (left) indicates the impact of agriculture on the environment. Agriculture is a key contributor to global greenhouse gas emissions and also has large water requirements for both plant crop and animal production.

▶ Farming has had enormous influences on the animal and bird populations on Earth. Most bird and mammal biomass in the world exists because humans breed them as a food source.

▶ Farming is our source of food and also provides non-food materials used in many other human activities, e.g. timber for construction and manufacturing; other plant raw materials such as rubber, turpentine, fiber crops such as cotton or linen for clothing, essential oils such as lavender, and plants used for production of pharmaceuticals.

Large scale fertilizer production - the Haber process

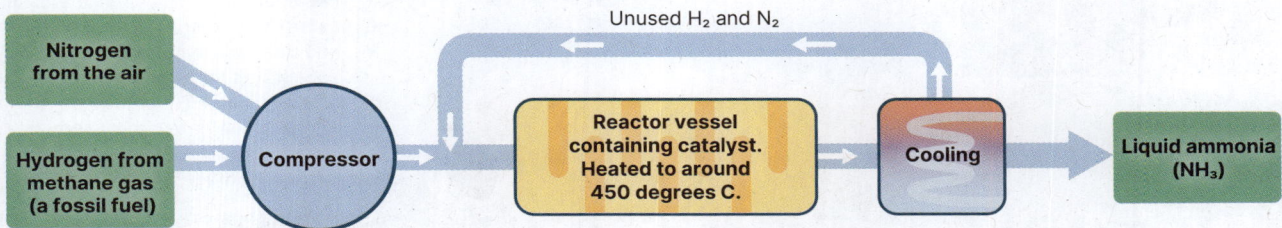

▶ Modern, intensive agricultural practices require high inputs of fertilizer, particularly nitrogen. As nutrients are removed from the soil by harvesting crops year after year, these must be returned to the soil artificially.

▶ A process developed by German scientist Fritz Haber found a way to convert the nitrogen from the air into ammonia, a chemical which can be turned into fertilizer.

▶ A source of hydrogen, which comes from natural gas (methane) is reacted with nitrogen sourced from the air.

▶ The reaction chamber containing an iron catalyst is heated to 450 degrees Celsius. Once cooled, liquid ammonia is produced that can be used to make fertilizer and other products.

▶ Not only does the Haber process use fossil fuels, it also has high energy requirements due to the heat needed for the reaction to occur. It also generates CO_2 as a waste product, contributing to greenhouse gases.

1. Discuss the effect of farming on the Earth's biodiversity: _____

2. What are the negative environmental effects of industrial fertilizer production: _____

94 Intensive Farming Practices

Key Idea: Intensive farming practices can harm the environment in many different ways.

Human demand for food requires techniques of farming that can give the maximum possible yield from the land available. Land must be cleared for farming if it has the potential to grow useful products but this can damage the land if it is not carefully managed. Pests and diseases must be controlled in both animals and plants through the use of chemical pesticides which can harm beneficial organisms as well as killing harmful ones. In a cropping system, weed cover must be reduced as much as possible to provide the desired crop with maximum resources to give as good a yield as possible.

▸ Producing food from a limited amount of land presents several challenges: to maximize yield while minimizing losses to pests and disease, to ensure sustainability of production, and (in the case of animals) to meet certain standards of welfare and safety.

▸ By maximizing the efficiency of resource use and energy conversion into product, industrialized intensive agricultural systems meet these demands by using high inputs of energy to obtain high yields per unit of land farmed. Such systems apply not just to crop plants, but to animals too, which are raised to slaughter weight at high densities in confined areas (a technique called factory farming).

Some features of industrialized agriculture

Antibiotics are used to treat diseases such as mastitis in dairy cattle. This improves herd health and reduces milk losses, due to disease.

Fertilizers are applied to maintain soil fertility and replace nutrients extracted by growing plants. Crop residue is plowed into the ground, reusing nutrients and reducing the amount of fertilizer required.

Fertilizers can be sprayed using aerial topdressing in inaccessible areas, increasing the area that can be farmed.

Pesticides and fungicides are used extensively to control crop pests and diseases in industrialized agriculture. Indiscriminate use of these leads to increased resistance to commonly used chemicals and contamination of land and water.

Clearing land of trees for agriculture increases available land but can lead to slope instability, soil erosion, and land degradation.

Antibiotics are used in the intensive farming of poultry for egg and meat production. Proponents regard antibiotics as an important management tool to prevent, control, and treat disease, allowing farmers to raise healthy animals and produce safe food.

Feedlots are a type of confined animal feeding operation which is used for rapidly feeding livestock, notably cattle, to slaughter weight. Diet for stock in feedlots are very dense in energy to encourage rapid growth and deposition of fat in the meat (marbling). As in many forms of factory farming, antibiotics are used to combat disease in the crowded environment.

Maximizing productivity

Maximizing agricultural productivity is essential if the human population is to be fed using the land available. Several ways exist that allow for maximum productivity:

Large scale farming uses enormous fields and huge machines to maximize production. Larger machines are more efficient: more seeds can be sown and harvested using fewer people and less fuel.

Crops that can withstand more (or less) rainfall can be grown in response to changing climate. New research into factors that affect flood resistance in staple crops could be used to produce better yields.

Precision agriculture using technology such as drones can pinpoint, for example, which plants need water in a particular field, or where pests are beginning to affect a crop. Action can be taken early.

Modern technology can help farm managers manage livestock numbers, monitor weather, schedule tasks, and help with financial management. This can reduce the burden of administrative work.

By breeding or raising healthier livestock that reach marketable weight quickly and are less prone to disease, costs for veterinary treatment will be reduced, producing a greater financial return.

Educating farm staff and managers to ensure they are aware of the latest research and technology and are mindful of health and safety practices ensures that farms are safe as well as profitable places to work.

1. Explain the need for each of the following in industrialized intensive agriculture:

    (a) Pesticides: _____
    _____

    (b) Fertilizers: _____
    _____

    (c) Antibiotics: _____
    _____

2. Explain where the energy in intensive agriculture is used: _____
    _____
    _____
    _____
    _____

3. Describe some of the issues that arise when land is cleared for the purpose of agriculture:
    _____
    _____
    _____
    _____

95 Sustainable Farming

Key Idea: Changes to agricultural practices can help to provide a more sustainable way of farming.

Sustainable agriculture refers to farming practices that maximize the net benefit to society by meeting current and future food and material demands while maintaining ecosystem health and services. Two key issues in sustainable agriculture are biophysical and socio-economic. Biophysical issues center on soil health and the biological processes essential to crop productivity. Socio-economic issues center on the long-term ability of farmers to manage resources, such as labor, and obtain inputs, such as seed. Sustainable agricultural practices aim to maintain yields and improve environmental health. Crops are often grown as polycultures (more than one crop type per area), which reduces pest damage by providing a trap crop or pest confuser, e.g. planting onions in a carrot crop masks the carrots' odor and reduces damage by carrot sawfly. However, yields obtained using sustainable practices can be up to 25% lower than those obtained using intensive practices. Food needs are projected to be 50% greater by 2050 than today, so this is a major disadvantage that must be overcome, either in the management of **agriculture** or by society as a whole.

Factors affecting sustainable agriculture

Fertilizers
Adding artificial **fertilizer** into an agricultural system can increase crop production. However, excess amounts can kill soil bacteria and reduce the amount of nitrogen naturally produced.

Soil erosion
Terracing along the slope, instead of downwards reduces soil erosion by breaking long slopes into a series of shorter ones. Terraces protect water quality by intercepting agricultural runoff.

Soil
Agriculture requires healthy soils. Soil health can be maintained by growing crops that naturally produce soil nitrogen (legumes) and adding organic matter by recycling crop waste and manure.

Water and pollution
Agriculture uses water for irrigation and watering stock. Sustainable practices for water use include increasing irrigation efficiency, protecting catchments, e.g. by riverside planting, storing excess rainwater, and decreasing runoff. These practices maintain and improve water quality.

Biodiversity
Biodiversity in agriculture is important for soil, plant, and animal health. Using many different agricultural crops (rotation) or grasses in a paddock decreases the risks of pests and diseases spreading in the soil and affecting crop yield. It also reduces the need for pesticides.

Natural cycles
Sustainable agriculture matches crops with natural cycles and systems. Legumes fix nitrogen and reduce the need for applied fertilizer. Crops are grown in suitable climates, reducing the need for irrigation or pest management. Materials are recycled as much as possible to promote environmental health.

Nutrient leaching
Excess nutrients from fertilizer and plant waste can leach into the waterways and result in **eutrophication**. Excess algal growth can reduce dissolved oxygen and damage the ecosystem.

The carbon footprint and sustainable agriculture

▸ Areas where carbon is stored are known as sinks or carbon pools. In agriculture, soil, plant biomass, and microbial biomass can act as sinks.

▸ Photosynthesis removes carbon from the atmosphere and converts it into organic molecules, stored in crops.

▸ This organic carbon may eventually be returned to the atmosphere as CO_2 through plant respiration, or respiration of stock or humans who eat the crops. The microbial biomass also respires. Additionally, carbon can be released from frozen soil when it thaws. Agricultural machinery is nearly always run on fossil fuels which, when combusted, add CO_2 to the atmosphere.

▸ Stable ecosystems rely on a balanced carbon cycle. When carbon outputs in an agricultural system equal the inputs, the system has a net zero carbon footprint (see Activity 216). However, most unsustainable agricultural systems add to the atmospheric carbon pool.

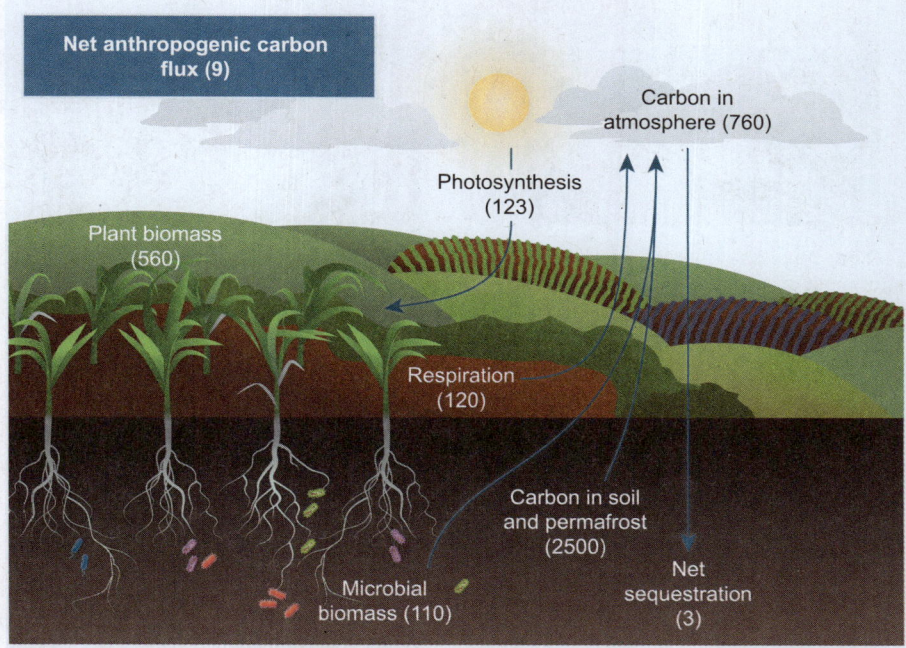

Carbon stocks are shown in gigatons. Anthropogenic carbon flux is human-induced carbon from all sources. full Oceanic contributions to atmospheric pool not shown.

Data source: Adapted from Jansson et al. 2021. Plant Sci. https://www.frontiersin.org/journals/plant-science/articles/10.3389/fpls.2021.636709/

How can the carbon footprint be reduced?

A reduction in the carbon footprint equates to reducing greenhouse gas emissions, specifically CO_2. Agricultural practices that decrease the use of fossil fuel powered machinery can be utilized. These include low tillage methods that keep the soil structure intact, reduction of bare soil between crops by use of a cover crop, and accurate application of fertilizer that can be reduced with rotated crops and nitrogen fixing plants. New technology is leading to greater use of electric farm vehicles and renewable energy to reduce the reliance on fossil fuelled machinery.

1. What is a carbon footprint and why is the reduction of it relevant to agriculture? _____

_____

_____

_____

_____

2. Explain how sustainable agriculture manages each of the following resources to meet its goals of long term sustainability:

   (a) Biodiversity: _____

   _____

   _____

   (b) Water: _____

   _____

   _____

   (c) Soil: _____

   _____

   _____

   _____

96 Cereal Crop Production

Key Idea: Cereals are extremely important crops required to feed the enormous human population.

Humans are reliant on just a few of the world's vast numbers of different plant species for food, of which many are cereals. These are an excellent source of carbohydrates, many proteins, as well as essential vitamins and minerals. Maize has overtaken wheat as the most widely grown crop, mainly because it is used heavily for animal feed and is also used to make biofuels which are alternatives to fossil fuels. In order to grow the amount of cereals required, inputs of fertilizer and water are high, which can damage the environment and contribute to world climate problems.

Maize (corn), *Zea mays*, is the most widely produced cereal crop globally, making it the most important cereal crop. It is used for both human and animal feed and for biofuels. Approximately 30% of the world's maize crop is genetically modified. These modifications include herbicide, insect and disease resistance, higher yields, and improved nutritional quality.

Rice, *Oryza sativa*, is the basic food crop of most of Asia as well as being a staple in other parts of the world. The most common varieties are aquatic and are often grown under irrigation. Upland varieties are also grown which have similar requirements to other cereal crops. China is the largest producer, mainly for its own use but rice is also extensively grown in India, Pakistan, Japan, Thailand, and Vietnam.

Wheat, *Triticum* spp., is the most widely grown cereal crop after maize, mainly in temperate regions. Bread wheat is soft, with a high gluten content. Durum wheat is a harder, low gluten variety used mainly for making pasta. Key areas for wheat production include Canada, USA, and Europe. The economic stability of many nations is affected by the trade in wheat and related commodity prices.

Barley, *Hordeum vulgare L.*, is the fourth most cultivated cereal. It is able to grow in high altitude regions with short growing seasons and prefers lower temperatures than many other cereals. It is cultivated in temperate regions of the world and is widely used as animal feed. It is commonly malted (partially germinated) as a basis for the production of many alcoholic beverages, including beer and whisky.

Sorghum, *Sorghum bicolor*, is a frost-tender, warm climate plant, well adapted to arid conditions. It has low soil nutrient and water requirements, reflecting its origin in the sub-Saharan Sudan region of Africa. It is widely cultivated across the world in warm regions. It is nutritious and used as human food in Asia and Africa. In other regions it is used as animal feed and as industrial raw materials.

Rye, *Secale cereale*, is a cool climate crop and is mainly grown in North Eastern Europe and Russia. It is tolerant of poor soil and will withstand very low temperatures and snow cover. Rye can be affected by a fungus called ergot in damp, humid conditions. It is mainly used to make rye bread, popular in northern European countries, and also alcoholic beverages such as rye whiskey.

1. Why is maize such an important crop? _____

2. Why is wheat such an important crop? _____

3. Suggest why sorghum is cultivated and used as a food source in Africa: _____

Cereal crop production

Cereals are the most widely cultivated crops. Other main crops grown worldwide are sugar, leafy vegetables, oil crops, roots and tubers, and fruits. Because cereals are used both as direct food for humans and for feeding livestock raised for human consumption, huge quantities are grown annually. As shown below, maize production has now surpassed rice and wheat production, mainly because of its use in biofuels and animal feeds.

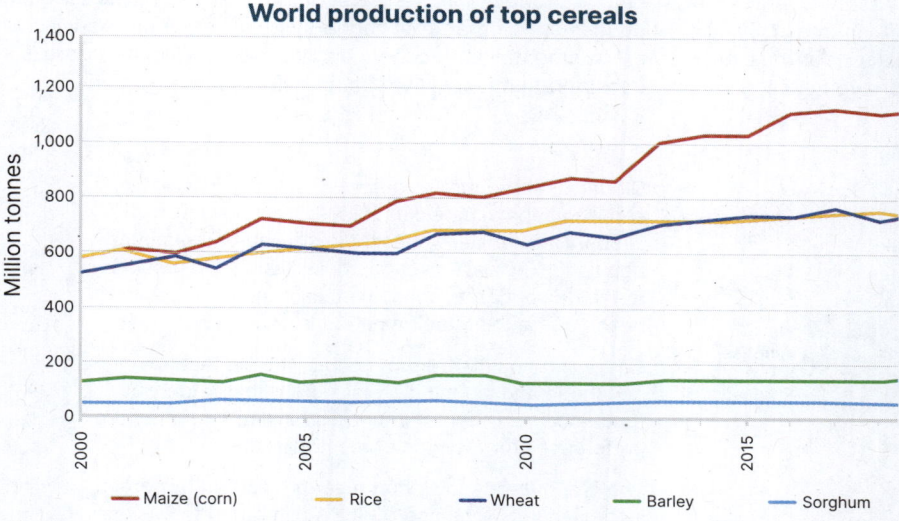

Cropping potential of major cereals		
Crop	Yield (kg grain/ha)	Specific growth requirements
Maize	1000-4000	Warm, frost free climate. Fertile soil. Drought intolerant.
Wheat	1000-14,500	Adapted to wide range of temperate climates and soils.
Rice	4500 (lowland aquatic varieties)	Tropical varieties grown in paddy fields are drought intolerant
Sorghum	300-2000 Can be up to 6500 for irrigated hybrids	Wide range of soils. Drought tolerant. Grown in regions that are too dry for maize.

▸ As the human population grows, so too does the need for cereals. A great deal of scientific research has gone into producing high yielding, pest resistant varieties but many require high **fertilizer** and **pesticide** input.

▸ High inputs of nitrogen fertilizer are not good for the environment. Much of what is applied will run off into waterways, causing pollution.

▸ Producing nitrogen fertilizer uses fossil fuels as a raw material (methane supplies the source of hydrogen). The chemical process used to make it needs very high temperatures, and therefore energy inputs.

4. Examine the graph above showing world production of major cereals:

    (a) Which crops have not increased in production over the time period shown? _____

    (b) Which crop has had the biggest increase in production? _____

    (c) Which crops have shown approximately the same increase in production? _____

5. Why are barley and rye more commonly grown in Northern/Eastern parts of Europe than in other parts of the world?

6. It is estimated that there are 350-400 thousand species of plants on Earth and yet humans are reliant on only around 12 of these for food. Write a paragraph below explaining why this could be a potential problem for human existence. You may wish to research case studies to support your argument:

97 Meat Production

Key Idea: Global meat production has increased dramatically over the past 60 years.

Meat is an important source of many essential nutrients. Global meat production has been steadily increasing since the 1960s, partly because people in many countries have become wealthier. Meat, which was previously a luxury item, is now an everyday staple food. Cattle, poultry, pigs, and sheep are the most commonly farmed animals. Meat can be produced in intensive, high density systems that use up less space than free range systems; both methods have environmental and economic advantages. Animal welfare is a consideration for many meat consumers.

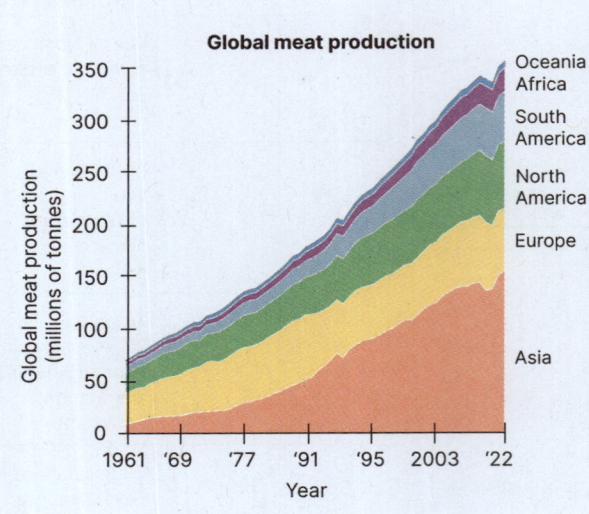

Global meat production has been steadily increasing, showing approximately a four fold increase since the 1960s. This has partly been in response to the global population increase over this time, but also in response to the demand from consumers. In general, as people become wealthier, their meat consumption increases.

Cattle and poultry are produced in the highest numbers, but sheep and pig meat are also produced widely as food sources. Different countries have different cultural preferences for the type of meat they eat

The graph (left) shows that meat production has greatly increased in Europe and the USA compared to 60 years ago. The increase in production in Asia over the same time period is much higher.

In **intensive production systems**, meat animals are grown to slaughter weight in carefully controlled spaces. Their diets are primarily manufactured feedstuffs. Some people believe that animals have a poor quality of life in such systems.

▸ Advantages: Faster to raise animals to slaughter weight and less land is needed. For beef, more meat is produced per animal than in free range systems.

▸ Disadvantages: Animals housed in close proximity to one another have a high risk of disease so antibiotics are given routinely. Although the animals live in a small area, large areas of agricultural crop land that could otherwise be used to grow food directly for humans are required to produce corn and soy as feedstuffs.

In **free range production systems**, meat animals are grown to slaughter weight in open spaces where they have room to move naturally. Their diets are primarily forage. Some people believe that animals have a better life when out on pasture.

▸ Advantages: Animals are less likely to contract diseases from each other. Land that is unsuitable for crops because of its topography or soil type can be used for animal production.

▸ Disadvantages: It takes longer for animals to reach slaughter weight and, in beef production, less meat is obtained per animal. For example, eight US citizens can be fed per animal in contrast to ten per animal in an intensive system.

1. Compare the production of animals in intensive systems and free range systems:

   _____

   _____

   _____

   _____

2. Examine the graph of global meat production above.

   (a) Describe the overall trend displayed by the graph: _____

   _____

   (b) Which region has shown the highest increase in meat production? _____

98 Food Security

Key Idea: Food security is linked to human wellbeing and prosperity.

Having enough food to eat is vital for human survival. Many factors lead to food insecurity, some natural and some man-made. The increase in world population has made the requirement for food ever greater. Vast areas of land have been altered to grow crops to feed humans and the animals that are farmed for meat and dairy produce. Extreme weather events can cause crops to fail, which can affect people at both a local and global level. War zones can lead to severe food insecurity for the local population, those having to flee the area, and the wider world. The health and economic prosperity of humans is linked to their food security, as a lack of food affects both physical and mental health.

Climate extremes. Extreme weather such as floods, fires, extreme storms, and natural disasters, e.g. earthquakes and tsunamis, can lead to catastrophic crop failures. This can lead to severe shortages of certain foods which makes prices rise. Foodstuffs may have to be imported to places that were previously self-sufficient in certain crops, making them too expensive for some people to buy.

World population. Producing enough food to feed an increasing world population is an ongoing challenge that has led to more and more land being converted to agricultural land. This has resulted in deforestation of key areas of the world's forests, particularly the Amazon rainforest. When land use is changed and attempts are later made to change it back to its original use, many challenges arise.

World conflict and instability. One of the biggest drivers of food insecurity in the world is conflict. Wars greatly affect the everyday lives of people, often driving them from their homes at short notice. Farms are often abandoned or have reduced productivity. This can affect world supply of certain crops, as demonstrated by the increase in grain prices as a result of the war in Ukraine, a leading wheat producer.

The vicious circle of poverty and food insecurity

- The FAO (Food and Agricultural Organization) describes a vicious circle of food insecurity, health issues, low productivity, and poverty.
- People who are malnourished often fail to develop properly physically and may have physical and mental impairment.
- Parents who struggle to feed their children and worry about the cost of food may suffer from mental health issues such as depression or anxiety.
- Any physical or mental impairment generally leads to poor workplace productivity. This prevents people from advancing in the workplace, preventing them from earning more. This contributes to ongoing poverty.

Adapted from FAO 2008

1. Explain why human conflict situations contribute to a lack of food security at both a local and global level:

   _____
   _____
   _____

2. Why does food insecurity contribute to poverty? _____
   _____
   _____
   _____
   _____

©2025 BIOZONE International
ISBN: 978-1-99-101409-2
Photocopying prohibited

99 Pest Control

Key Idea: Pests of agricultural products must be managed in a way that minimizes damage to non-pest species.

The fight against pests is ongoing for any food production system. Pests attack and diminish the quality of crops and can ruin them altogether, causing huge financial losses for farmers. Pests attack both plants and animals and can be insects, bacteria, fungi, or parasites. Weeds are also considered a pest and many farmers spray crops with selective herbicides to control them. Chemicals used to destroy or reduce pests can build up in the food chain and cause health issues for other animals.

Fungal diseases: Fungal diseases affect many food crops, including tomatoes. Fungicides prevent damage by rusts, molds and mildew. Like all pesticides, they can work either through contact or be systemic. Contact sprays kill where they touch. Systemic sprays are circulated in the plant tissues and kill from within.

Insect damage: Insecticides work in various ways and vary from those used in large-scale agricultural operations to fly sprays used inside buildings. **Insecticides** can work by over-stimulating the nervous system of the target organism. Pyrethroids are natural insecticides commonly used in homes.

Weed control: Glyphosate is currently the most commonly used herbicide. It works only on growing plants, affecting metabolic pathways. Its use has become controversial in recent years, resulting in a ban on use in some but not all countries. Many chemicals target different types of weeds, e.g. they will kill pasture weeds but not grasses.

Lungworm in cattle

Parasites in animals can cause ill health or death, and animals must be protected against several dangerous organisms. Lungworm (*Dictyocaulus viviparous*) is a parasite that affects cattle. Higher stocking rates and changes in weather patterns have caused lungworm to become an increasing issue for farmers in the UK and Europe. Treatments include vaccination of calves over eight weeks' old and oral worm treatment (drenches).

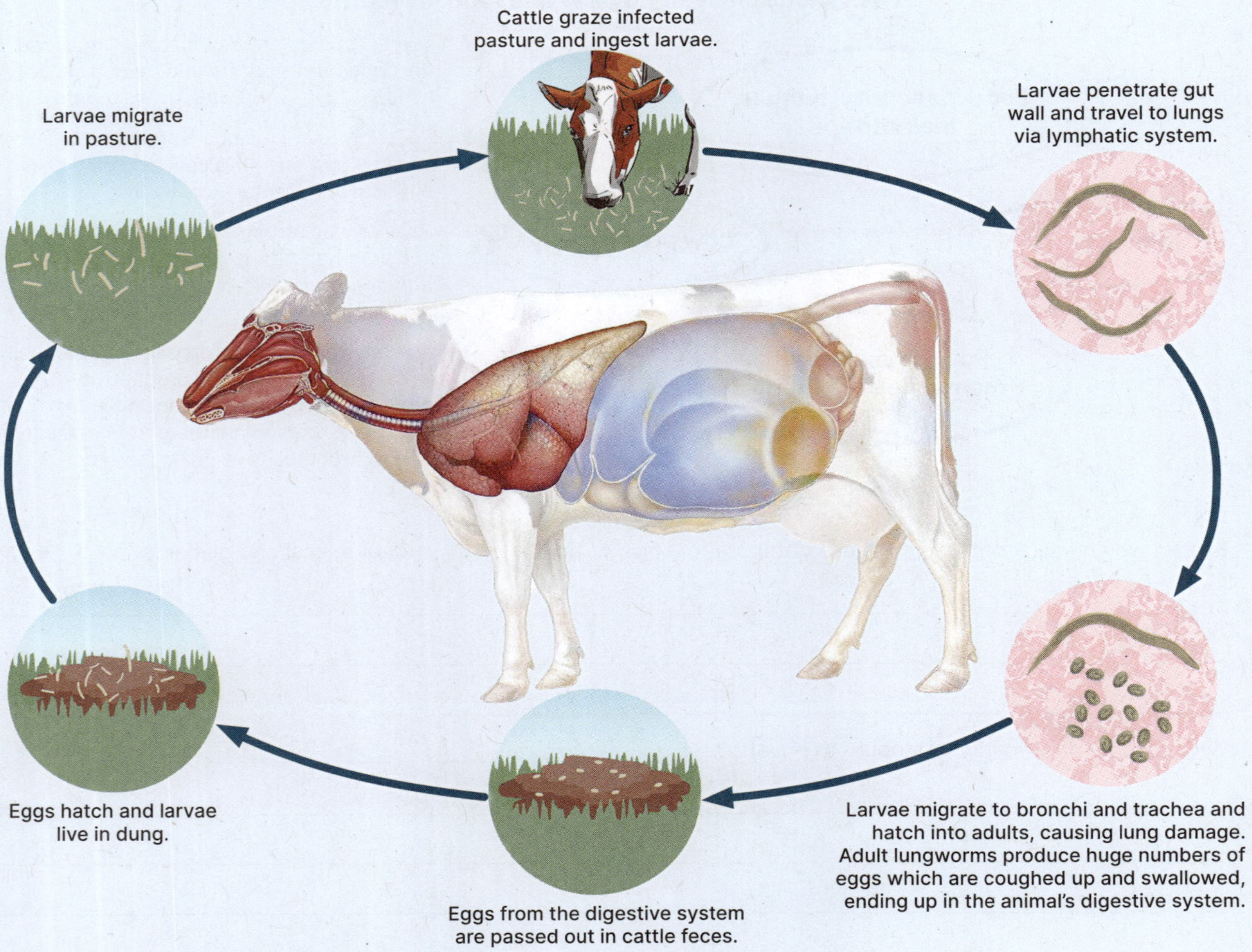

- Larvae migrate in pasture.
- Cattle graze infected pasture and ingest larvae.
- Larvae penetrate gut wall and travel to lungs via lymphatic system.
- Larvae migrate to bronchi and trachea and hatch into adults, causing lung damage. Adult lungworms produce huge numbers of eggs which are coughed up and swallowed, ending up in the animal's digestive system.
- Eggs from the digestive system are passed out in cattle feces.
- Eggs hatch and larvae live in dung.

Protecting bees from pesticides

Bees are important pollinators for many crops. Orchards that are not pollinated will not produce adequate amounts of fruit. Intensive use of pesticides can have a detrimental effect on bees, often leading to death. This presents a problem for farmers who rely on their activity as pollinators.

Bees can be protected from the harmful effects of pesticides if these are not sprayed on plants that are in full bloom, as this is usually when bees visit.

By using pesticides that break down quickly, for example, within a few hours, bees are less likely to be harmed.

Case study: the white rumped vulture

▸ The white-rumped vulture, *Gyps bengalensis*, was once widespread in India. Today, only one bird remains for every 1000 recorded in 1992.

▸ These vultures are found throughout India, Pakistan, and Nepal, where they play an important ecological (and cultural) role in cleaning up and recycling nutrients from animal carcasses and human bodies.

▸ The reason for their decline was found to be the anti-inflammatory drug Diclofenac, widely used to treat illness and injury in livestock. Vultures ingest the drug and die from renal failure after feeding at carcass dumps.

▸ In 2006, the governments of India, Nepal, and Pakistan banned the use of veterinary diclofenac and a vulture-safe alternative is now available.

▸ A number of breeding programs were established to help increase bird numbers. These birds were released into the wild. Analysis of dead birds confirmed that death was from natural causes and not poisoning.

1. Why are pests an issue for any agricultural production system? _____

2. Research some steps farmers can take to reduce the incidence of lungworm in cattle, other than vaccination:

3. How can bees be protected from pesticides when these need to be applied to crops that they pollinate?

4. Research a pest that threatens the area where you live. It may be an insect, mammal, disease or invasive plant. Write a paragraph below explaining why it is a pest, and what steps are being taken to manage it:

100 Pesticide Resistance

Key Idea: Pesticide resistance develops through several different mechanisms.

Insecticides are **pesticides** used to control insects considered harmful to humans, their livelihood, or environment, and their use has proliferated since the advent of synthetic insecticides in the 1940s. Resistance develops when the target species becomes adapted to the effects of the pesticide and it no longer controls the population effectively. Resistance can arise through a combination of mechanisms but the underlying process is a form of natural selection in which the most resistant organisms survive to pass on their genes to their offspring. Insecticides can also kill useful insects and birds, reducing biodiversity and leading to bioaccumulation in food chains. Some pest species develop multiple resistance against many different pesticides. Because insecticides are used in medical, agricultural, and environmental applications, the development of resistance has serious environmental and economic consequences.

Mechanisms of resistance in houseflies

INSECTICIDE

1. **BEHAVIOR** — avoidance of contact
2. **MECHANICAL RESISTANCE** — resistance to entry lowers uptake levels
3. **DESTRUCTION** — detoxification methods
4. **BIOCHEMICAL RESISTANCE** — enzymatic changes at points of contact

Insecticide resistance in houseflies can arise through a combination of mechanisms:

(1) Increased sensitivity to an insecticide will cause the pest to avoid a treated area.

(2) The *Pen* gene confers stronger physical barriers, decreasing the rate at which the chemical penetrates the fly's cuticle.

(3) Chemical changes within the fly's body can render the pesticide harmless.

(4) Structural changes to the target enzymes make the pesticide ineffective. No single mechanism provides total immunity but together they transform the effect from potentially lethal to insignificant.

Development of resistance

Susceptible
Resistant

A small proportion of the population will have the genetic makeup to survive the first application of a pesticide.

The genetic information for pesticide resistance is passed to the next generation.

The proportion of resistant individuals increases following subsequent applications of insecticide. Eventually, almost all of the population is resistant.

The application of an insecticide can act as a potent selection pressure for resistance in pest insects. The insecticide acts as a selective agent, and only individuals with greater natural resistance survive the application to pass on their genes to the next generation. These genes (or combination of genes) may spread through all subsequent populations.

1. Give two reasons why widespread insecticide resistance can develop very rapidly in insect populations:

   (a) _____

   (b) _____

2. Explain how repeated insecticide applications acts as a selective agent for evolutionary change in insect populations:

   _____

©2025 **BIOZONE** International
ISBN: 978-1-99-101409-2
Photocopying prohibited

101 Integrated Pest Management

Key Idea: Integrated pest management uses multiple approaches to manage pests.

Integrated pest management (IPM) uses a variety of methods to carefully monitor pests and take appropriate, targeted action quickly. This approach to pest management can cut down on pesticide use but requires expert knowledge of crop and pest ecology. It is more labor intensive than other methods of pest management as it combines chemical, biological, and mechanical means to reduce pests and relies on regular examination of crops in the field. Often, pests are not completely eradicated but their numbers are significantly reduced such that crops are not adversely affected and can still be sold. When pesticides are required, they are targeted and ecologically sensitive such that they minimize effects on non target species. IPM is less likely to induce resistance in the pest population than traditional approaches.

Step one in an efficient IPM program is **prevention**. A grower should use pest resistant crop varieties. Soil should be carefully managed to give plants the best growing environment possible as healthy plants are less susceptible to pest damage. Crops should be rotated to ensure that the same crops are not grown in the same place year after year.

Step two is to decide on an **action threshold** based on economic factors (graph, right).
How much will it cost to buy pesticides and treat the pest? Will that cost be recouped through sales of produce? Will there be a cost in cleanup of soil/waterways after treatment? Will there be health and safety issues for staff? At what point will it cost more to treat the pest than the farmer can recoup in sales?

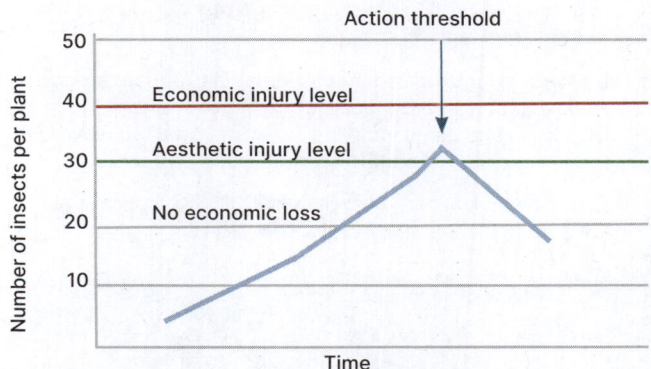

Data source: adapted from https://extension.usu.edu/planthealth/research/eil-concept

Step three is to **monitor** crops in their growing environment. Yellow and/or blue sticky traps should be used to identify pests as they appear. The different colored traps attract different insect pests. Blue sticky traps are attractive to pests such as thrips, whereas yellow sticky traps attract a wider range of pest insects.

Step four is to take **action** by use of appropriate biological and/or chemical controls. These should be used only when necessary to avoid economic losses. Biological controls are predators that eat pests, e.g. ladybugs eat aphids; pheromone traps attract pests away from the crop plant. Chemical controls should be targeted to specific pests.

Step five is to **evaluate** whether actions taken in step four have been effective. Monitoring, action, and evaluation should be ongoing throughout the lifecycle of the crop. Action taken can be targeted to different pest species which may differ depending on the time of year and weather conditions.

1. Describe the main steps in an IPM program, briefly stating the importance of each:

   _____
   _____
   _____
   _____
   _____
   _____

2. What economic factors should be taken into account before deciding to treat a crop for pests?

   _____
   _____
   _____
   _____

IPM case study: Japanese Beetle, *Popillia japonica*.

Japanese beetles are a pest of many economically important horticulture crops in the USA, including vegetable and fruitcrops, landscape plants, trees, and grass grown as turf. An IPM approach to managing these pests is outlined below:

▸ If possible, grow plants that aren't attacked by beetles, e.g. holly, spruce, rhododendron, white oak, red maple, lilac.

▸ Learn to identify the beetles by their metallic green heads, metallic, tan colored wing casings and, in adults, distinctive white rear tufts and lateral white tufts of hair (image, right).

▸ If pest numbers are manageable, remove adults by hand and destroy. Netting small areas of crops will prevent adult beetles from landing on them.

Japanese beetles on leaf. Note distinctive white rear tufts.

▸ **Biological controls** can be applied at appropriate stages of the Japanese beetle's life cycle. Parasitic nematodes, such as *Heterorhabditis bacteriophora* will target the larve in the soil. *Bacillus thuringiensis* is a bacteria that affects both the adults and the larvae and can be applied in a water spray to leaves where adults are present and to the soil when larvae are underground. It does not affect beneficial insects such as bees.

▸ Often, biological controls will not kill all of the pest organisms but will reduce numbers to levels where damage to plants is lowered to manageable levels.

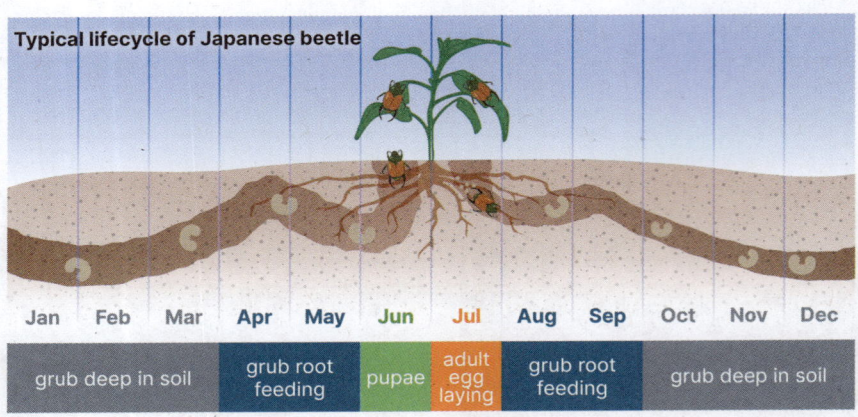

▸ If damage is great enough to warrant **chemical treatment**, use an appropriate treatment for the lifecycle stage of the pest, shown left.

▸ Larvae live in soil and destroy roots. Pesticides that target larvae must be applied to the soil in May-June when larvae are underground. In June, beetles begin to emerge and climb up plants.

▸ In mid-July to mid-Sept, adult are on leaves. This is the time to apply a pesticide, before the adult beetles cause too much damage, spread to neighboring plants, or begin egg laying.

3. Why is Japanese beetle a pest of crops in many countries? _____

4. Suggest why Japanese beetle is not a pest in its native country, Japan: _____

5. Identify any four steps that could be taken to implement an IPM strategy for Japanese beetle. Refer to the steps on the previous page to guide you:

   (a) _____

   (b) _____

   (c) _____

   (d) _____

6. Compare and contrast a chemical control treatment with a biological control treatment: _____

7. Why would a biological control such as a parasitic nematode need to be applied to the soil and not to the adult beetles on leaves?

102 Soil Degradation

Key Idea: Soils can be severely damaged by poor agricultural practices and become unusable.

Soil is a fragile resource and can be easily damaged by inappropriate farming practices. Some soils, such as those under the Amazon rainforest, are very vulnerable to human interference. These soils are thin and nutrient poor; after only a few years of farming they may be abandoned due to poor production. Overgrazing and deforestation may cause desertification. Chemically intensive agricultural practices call for ever-increasing doses of herbicides, **insecticides**, fungicides, and **fertilizers**. These often result in high crop yields, but produce soils that are compacted and lacking structure. Healthy soils are alive, with a diverse community of organisms, including bacteria, fungi, and invertebrates. These organisms improve soil structure and help to create humus. Repeated chemical applications kill soil organisms and eventually result in a soil that is hard, lacking in organic material, and unproductive.

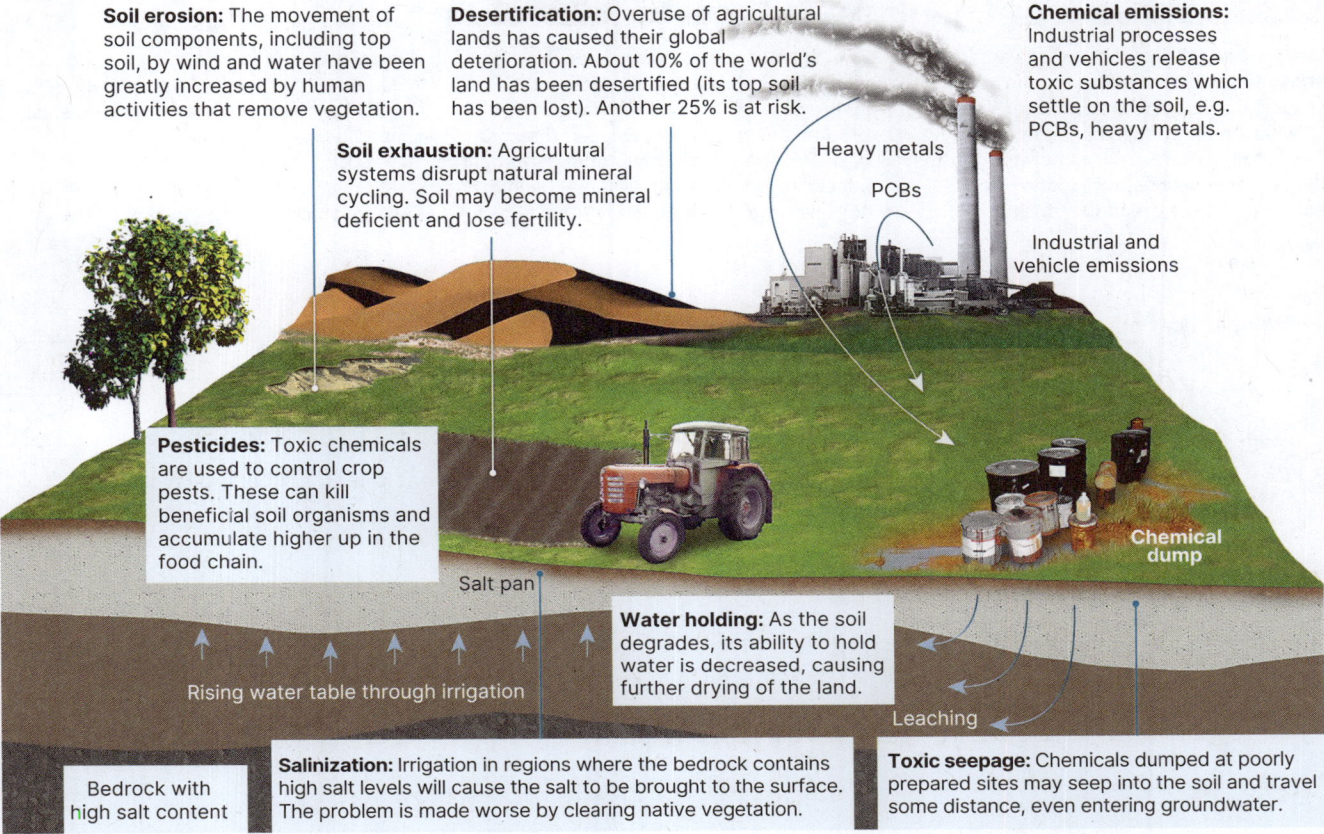

Soil erosion: The movement of soil components, including top soil, by wind and water have been greatly increased by human activities that remove vegetation.

Desertification: Overuse of agricultural lands has caused their global deterioration. About 10% of the world's land has been desertified (its top soil has been lost). Another 25% is at risk.

Chemical emissions: Industrial processes and vehicles release toxic substances which settle on the soil, e.g. PCBs, heavy metals.

Soil exhaustion: Agricultural systems disrupt natural mineral cycling. Soil may become mineral deficient and lose fertility.

Pesticides: Toxic chemicals are used to control crop pests. These can kill beneficial soil organisms and accumulate higher up in the food chain.

Water holding: As the soil degrades, its ability to hold water is decreased, causing further drying of the land.

Salinization: Irrigation in regions where the bedrock contains high salt levels will cause the salt to be brought to the surface. The problem is made worse by clearing native vegetation.

Toxic seepage: Chemicals dumped at poorly prepared sites may seep into the soil and travel some distance, even entering groundwater.

Soil is a living material that takes thousands of years to develop into a fertile substrate that farmers can use to grow crops. A healthy soil is teeming with organisms; some, such as earthworms and other invertebrates are visible, but many are invisible to the naked eye. Soil contains bacteria and fungi, some of which are essential for plants to grow. Mycorrhizal fungi have symbiotic relationships with plant roots and some bacteria grow in the root nodules of legumes, allowing them to obtain nitrogen in a usable form. Once destroyed, a soil is lost for use in our lifetime as it can take thousands of years to regenerate.

Salinization of soil

Salinization is most common in arid, semi-arid, or coastal regions. All soils contain salts, many of which are water soluble and can build up in the soil. In arid areas, salts can slowly seep upwards into the soil by capillary action. They accumulate in the top layers if no rain falls to wash them away or take them back downwards.

- Irrigation and over-fertilization add water soluble salts to the soil. Over time, these accumulate and diminish soil quality for agricultural use.
- Flood plains are also susceptible to salinization as salts build up over time with continual flooding and drying out.
- It is estimated that salinization is responsible for 0.3-1.5 million hectares of land being abandoned per year.

The problem of disposing of unwanted agricultural chemicals has reached major proportions in developed countries. Chemical dumps, such as the one illustrated, suffer from deterioration, with the contents spilling from rusting drums and entering the groundwater system.

The high use of pesticides in developed countries is claimed to be necessary by growers to maintain high levels of production. This often comes at the cost of destroying the natural predators of pest species. Pesticides can accumulate in the soil and enter the groundwater system.

Human activities such as overgrazing livestock on pasture and deforestation may cause regional climate changes and a marked reduction in rainfall. This in turn may lead to the formation of a desert environment, or the encroachment of an existing desert onto formerly arable land.

1. What effect does pesticide application have on soil? _____
   _____
   _____

2. Describe how salinization of soil occurs and the steps that can be taken to prevent it in an agricultural system:
   _____
   _____
   _____

3. What are the main factors that lead to soil degradation in modern agricultural practices?
   _____
   _____
   _____
   _____

4. Research and discuss some problems associated with the use of agricultural chemicals throughout their lifecycle:
   _____
   _____
   _____
   _____
   _____

5. It is sometimes said that, 'human civilizations were built on soil'. Discuss what you understand by this statement and argue whether you agree or disagree with it:
   _____
   _____
   _____
   _____
   _____
   _____

103 Reducing Soil Erosion

Key Idea: Soil is very vulnerable if left bare. Once lost, it cannot be recovered and takes a long time to regenerate. Good soil is vital for productive agriculture. Most modern cropping techniques use heavy machinery to turn over the remnants of harvested crops, break up the soil, and smooth it flat to form a planting surface. This leaves the soil exposed, and large volumes of topsoil can be lost through wind or rain **erosion** before there is sufficient crop cover to protect it. Even when a crop is fully established, there may still be exposed ground from which soil can be lost. Alternative planting techniques such as minimum tillage farming, terracing, contour plowing, windbreaks, and intercropping reduce the exposure of soil to the elements. Protecting the soil is vital for agricultural practices.

Crops are often planted parallel to the slope of the land so that machinery can move through them easily. This orientation produces channels down which water can easily flow, taking valuable topsoil with it.

Plowing and planting across, rather than down, slopes produces contours that slow water runoff and reduce soil loss by up to 50%. Water has more time to settle into the soil, reducing the amount of irrigation required.

Terracing converts a slope into broad strips, slowing or preventing water and soil runoff and reducing erosion. This technique is commonly used in rice paddy fields. Terraces also help to control flooding downstream.

Windbreaks reduce soil erosion by reducing wind speed close to the ground. They also reduce water loss, and so lower irrigation requirements. Windbreaks placed near drainage ditches help to reduce erosion because the tree roots stabilize soil at the edge of the ditch.

Agroforestry is a combination of agriculture and forestry. Crops or stock are raised on the same land as a stand of woody perennials. Biodiversity levels are often higher than in conventional agricultural systems, and soil loss is much reduced.

Cropping System	Average annual soil loss t/ha	% rain runoff
Bare soil	41.0	30
Continuous corn	19.7	29
Continuous wheat	10.1	23
Rotation: corn, wheat, clover	2.7	14
Continuous grass	0.3	12

Soil erosion is significantly reduced when the vegetative cover over the soil is maintained (above). Continuous cover can be achieved by using machinery to plant crops directly into the soil, along with fertilizers and pesticides, beneath the existing ground cover.

1. Explain how terracing and contour plowing reduce soil loss compared to plowing parallel to the slope:

_____
_____

2. Explain why maintaining vegetative cover reduces erosion: _____

_____

3. Discuss the effects of traditional intensive cropping versus alternative cropping systems on soil erosion:

_____
_____
_____
_____

104 Forestry

Key Idea: Forestry species should be grown and harvested in a manner that affects the environment as little as possible. For forestry to be sustainable, demand for timber must be balanced with the regrowth of seedlings. Sustainable forestry allows timber demands to be met without over-exploiting the timber-producing trees. Different methods for logging are used depending on the type of forest being logged. This allows the various services provided by forests to remain undisrupted. These services include providing shelter for wildlife, acting to reduce water runoff by absorbing excess rainwater, and moderating the local climate. Careful selection of species and means of cultivation will ensure that damage to the environment during growing and harvesting is kept to a minimum.

Harvesting methods

Clear cutting

A section of a mature forest is selected (based on tree height, girth, or species), and all the trees are removed. During this process the understory is destroyed. A new forest of economically desirable trees may be planted. In plantation forests, the trees are generally of a single species and may even be clones. Clear cutting is a very productive and economical method of managing a forest; however, it is also the most damaging to the natural environment. In plantation forests, this may not be of concern and may not affect sustainability, but clear-cutting of old growth forests causes enormous ecological damage.

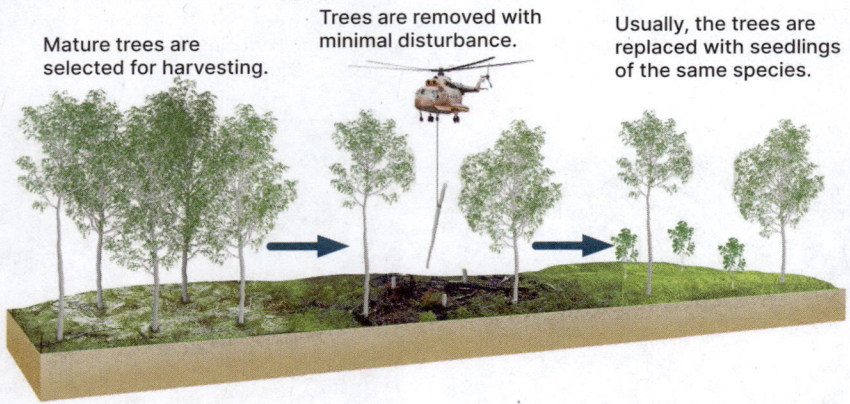

A selection of mature plantation forest is selected for harvesting. → All the trees are removed. → Seedlings of the same species are replanted.

Selective logging

A mature forest is examined, and trees are selected for removal based on height, girth, or species. These trees are felled individually and directed to fall in such a way as to minimize the damage to the surrounding younger trees. The forest is managed in such a way as to ensure continual regeneration of young seedlings and provide a balance of tree ages that mirrors the natural age structure.

Mature trees are selected for harvesting. → Trees are removed with minimal disturbance. → Usually, the trees are replaced with seedlings of the same species.

Strip cutting

Strip cutting is a variation of clear cutting. Trees are clear cut out of a forest in strips which are narrow enough that the forest on either side is able to reclaim the cleared land. As the cleared forest reestablishes (3-5 years), the next strip is cut. This allows the forest to be logged with minimal effort and damage to forest on either side of the cutting zone. At the same time, the original forest can reestablish.
Each strip is not cut again for around 30 years, depending on regeneration time.

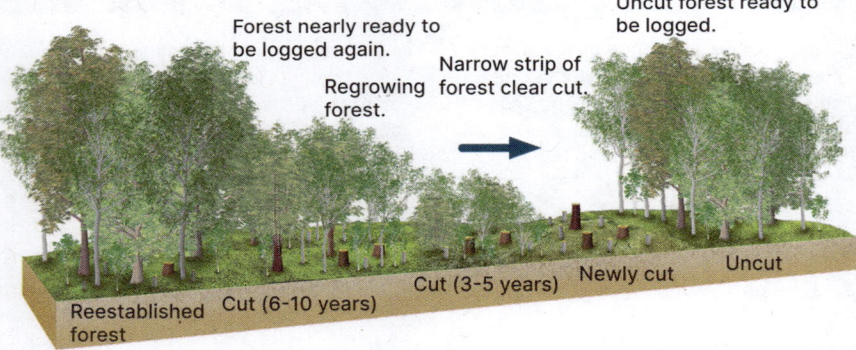

Reestablished forest | Cut (6-10 years) | Forest nearly ready to be logged again. Regrowing forest. | Cut (3-5 years) | Narrow strip of forest clear cut. | Newly cut | Uncut forest ready to be logged. | Uncut

1. Suggest which logging method best suits the ideas of sustainability and explain your answer: _____
   _____
   _____
   _____

2. Why might selective logging not be chosen as a harvesting method on a large scale? _____
   _____
   _____
   _____

Forest fires

Fires are a part of natural forest development and may occur as a result of lighting strikes, the concentration of heat on dry tinder, or by human activities. For many years, these natural fires were extinguished by fire services or prevented by education campaigns, but this led to a build up of flammable material in forested areas.

Forest management aims to reduce the occurrence of large, serious fires. Controlled burns are designed to remove excess flammable material in a section of forest to significantly reduce the risk of a wildfire. This is done in colder seasons where the risk of the burn becoming uncontrollable is reduced. The controlled burns also help to stimulate new growth.

Effects of forestry

Forestry harvesting operations lead to large tracts of bare land which causes loss of habitat for organisms living in the forested environment. Trees which were sequestering carbon are no longer there to do this. Rainfall can lead to severe erosion until newly planted trees grow, washing away valuable soil and nutrients.

Large forestry **monocultures** such as this gum plantation in Australia provide a very limited ecosystem for other organisms to inhabit. There is little biodiversity. Pests and disease can spread rapidly because of this. Because all of the trees are the same, they are all removing the same nutrients from the soil, leading to soil degradation.

When mature forests are harvested, it can be uneconomical to remove wood other than valuable logs. All other branches and small trunks are left as 'slash'. In severe storm events, this waste wood can wash down hills into rivers and onto beaches. In extreme storms, the slash can destroy bridges and block transport routes.

3. Describe the advantages and disadvantages of different logging methods: _____
_____
_____
_____
_____
_____
_____
_____

4. Describe what is meant by a controlled burn and why are these used in a managed forestry system:
_____
_____
_____

5. Identify some issues from the planting of monocultures: _____
_____
_____

6. What is forestry 'slash' and what problems result from it? _____
_____
_____
_____

105 Managing Rangelands

Key Idea: Careful management of grazing animals on rangelands is necessary to preserve the natural environment and reduce environmental damage.

Rangelands are large, relatively undeveloped areas populated by grasses, grass-like plants, and scrub. They are usually semi-arid to arid areas and include grasslands, tundra, scrublands, coastal scrub, alpine areas, and savanna. Globally, rangelands cover around 50% of the Earth's land surface. The US has about 3.1 million km^2 of rangeland, of which 1.6 million km^2 is privately owned. Rangelands cover 80% of Australia, mostly as the outback, but only 3% of Australia's population live in rangeland areas. Rangelands are often used to graze livestock such as sheep and cattle but, because they occur in low-rainfall areas, they do not regenerate rapidly. Careful management is required to prevent damage and soil loss as a result of overgrazing.

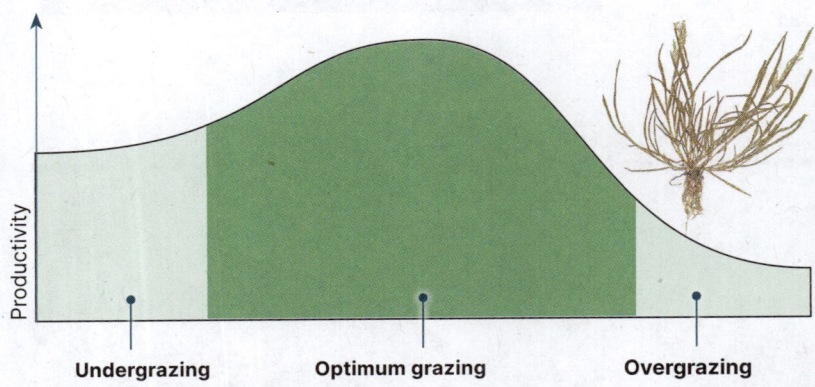

Undergrazing: Net productivity is reduced because standing dead material leaves little room for new growth to come through.

Optimum grazing: Old material is removed so new growth can come through, but enough growing material is left to allow recovery.

Overgrazing: Too much material is removed and new growth cannot become established. Plants die and erosion occurs.

Grasses (left) grow continuously from a meristem close to the ground, so the leaf can be cropped without causing growth to stop. This allows a field to be grazed in a near-continuous fashion. Grazing by animals stimulates grass to grow and removes dead material. Grasslands cropped at their optimum capacity can be much more productive than if left uncropped (left).

Overgrazing occurs when too many animals are grazed for too long on a section of pasture and the grass does not have enough time between cropping to regrow. Overgrazing may destroy the meristem, in which case plant regeneration stops. Exposed soils may become colonized by invasive species (below) or eroded by wind and rain.

Total net primary production and efficiency of grazed and ungrazed grasslands			
		Net production (kcal/m^2)	Efficiency %
Grazed	Desert	1081	0.13
	Shortgrass plains	3761	0.80
	Mixed grasslands	2254	0.51
	Prairie	3749	0.77
Ungrazed	Desert	1177	0.16
	Shortgrass plains	2721	0.57
	Mixed grasslands	2052	0.47
	Prairie	2220	0.44

Effect of grazing on plant species composition

Intensive grazing causes changes in the species composition. Species that perform better under grazing will increase their range, while others will reduce their range. Grazing also opens gaps in plant distribution which allows invasive species to establish or increase in range.

Ungrazed / Moderate grazing / Overgrazed

Decreasers   Increasers   Invaders

1. Explain how carefully managed grazing on a rangeland can increase its productivity: _____

2. Describe the effect of grazing on the diversity of rangeland plants: _____

Managing grazing

The key to successful rangeland management is allowing only the numbers of animals that the grasses and other plants can support. Rotational grazing allows for pasture recovery. Calculating optimum stocking rates is important to ensure that the land is not overgrazed and stock rotations are sustainable. Livestock are selective grazers, so care must be taken when grazing only one livestock species that particular grasses or shrubs are not eaten to local extinction.

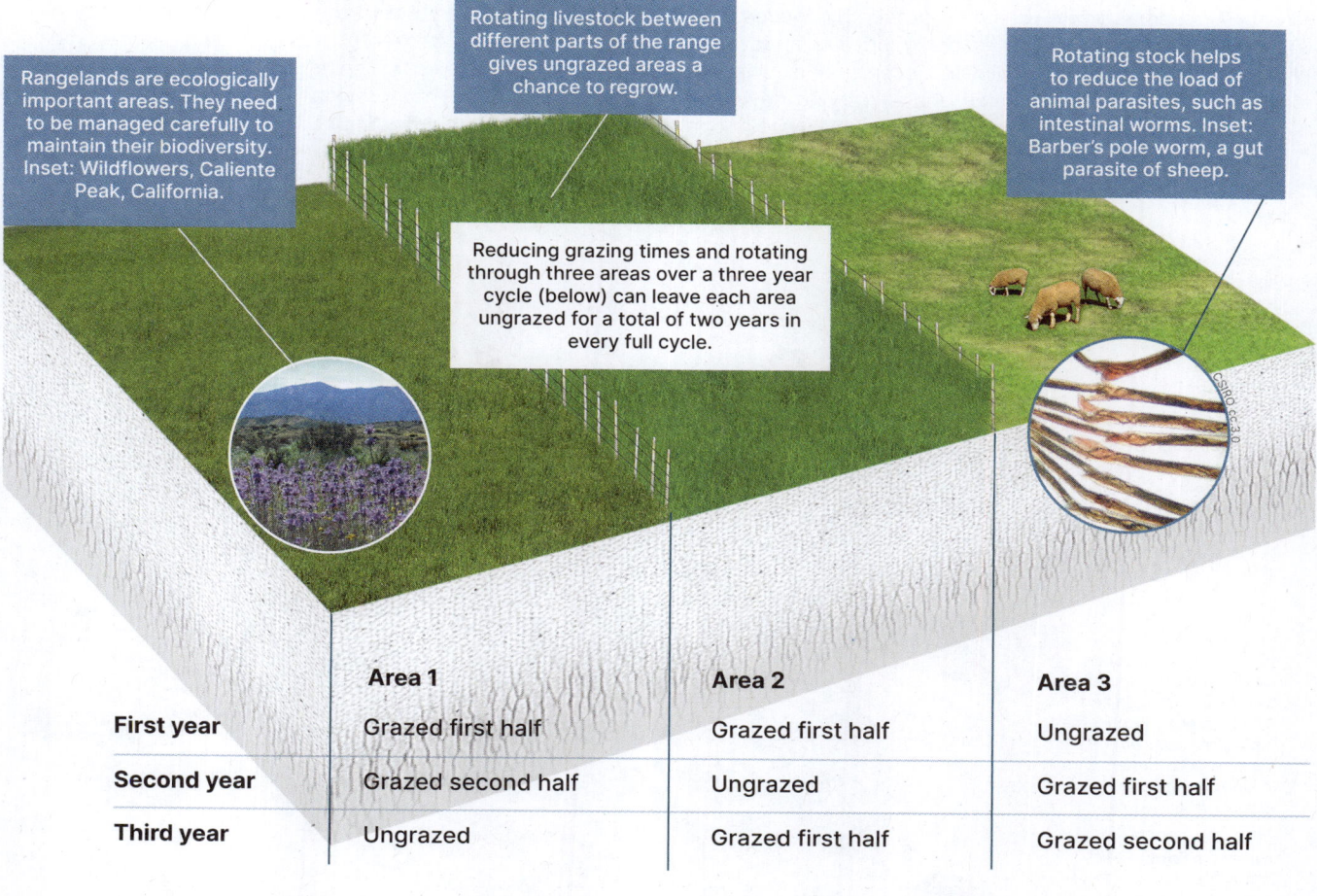

Rangelands are ecologically important areas. They need to be managed carefully to maintain their biodiversity. Inset: Wildflowers, Caliente Peak, California.

Rotating livestock between different parts of the range gives ungrazed areas a chance to regrow.

Rotating stock helps to reduce the load of animal parasites, such as intestinal worms. Inset: Barber's pole worm, a gut parasite of sheep.

Reducing grazing times and rotating through three areas over a three year cycle (below) can leave each area ungrazed for a total of two years in every full cycle.

	Area 1	Area 2	Area 3
First year	Grazed first half	Grazed first half	Ungrazed
Second year	Grazed second half	Ungrazed	Grazed first half
Third year	Ungrazed	Grazed first half	Grazed second half

3. Discuss the relationship between a rangeland's productivity and the number of animals grazed on the land:

4. Describe the effect of grazing on the diversity of rangeland species:

5. Explain why rangelands can easily become overgrazed:

106 Reserve Lands

Key Idea: Reserves lands play an important role both in conserving natural environments and supporting local economies through careful management of tourism.

Nature reserves may be designated by government institutions in some countries or by private landowners. Different reserves, e.g. wildlife, scenic and scientific reserves, and National Parks, have varying levels of protection depending upon country and local laws. National parks are usually located in places which have been largely undeveloped, and often feature areas with exceptional ecosystems such as those with endangered species, high biodiversity, or unusual geological features. They are protected by law for public understanding, appreciation, and enjoyment, while being maintained for future generations. Reserves provide habitats for endangered species, migratory birds, and big game.

Parks and reserves in North America

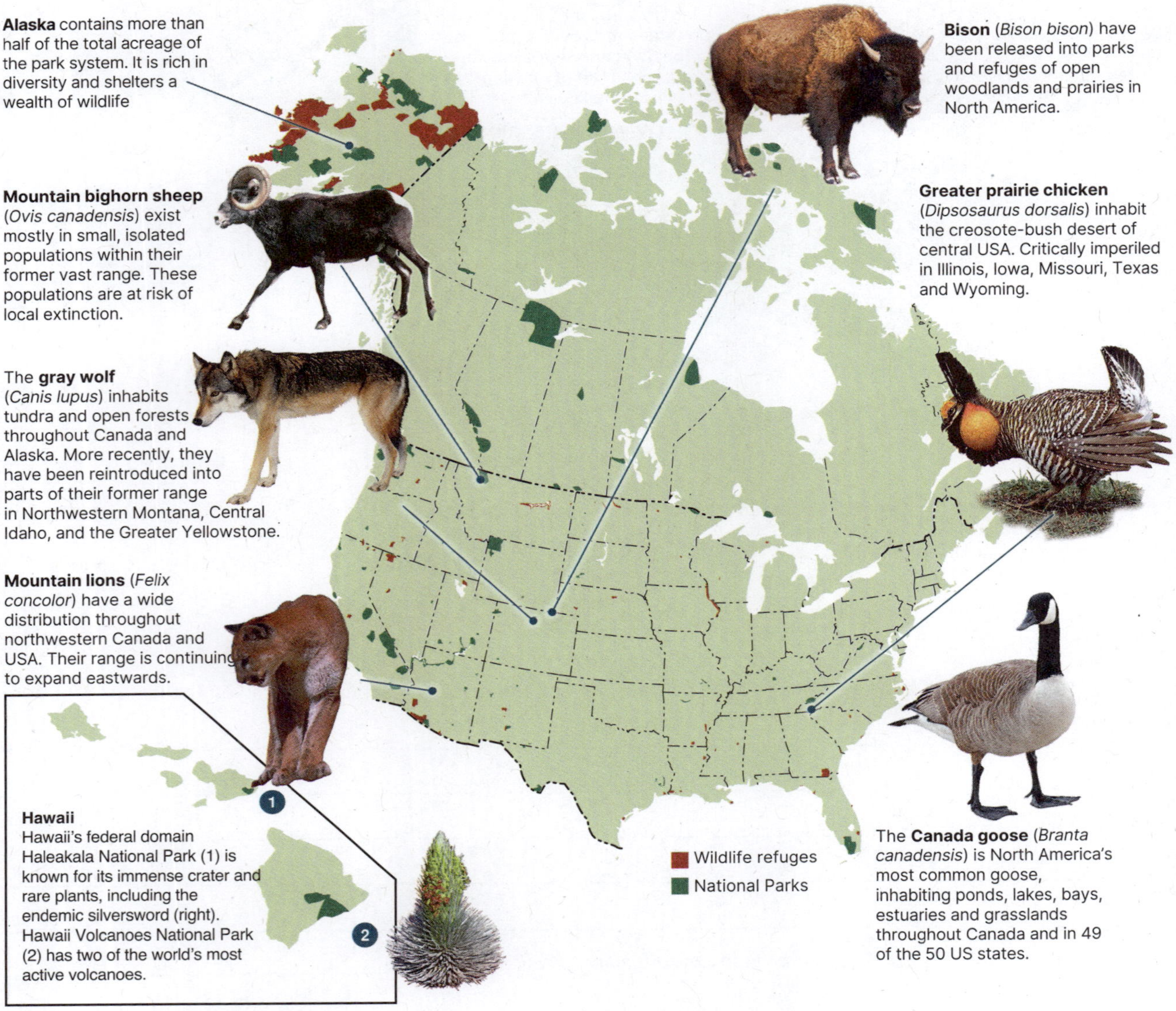

Alaska contains more than half of the total acreage of the park system. It is rich in diversity and shelters a wealth of wildlife

Mountain bighorn sheep (*Ovis canadensis*) exist mostly in small, isolated populations within their former vast range. These populations are at risk of local extinction.

The **gray wolf** (*Canis lupus*) inhabits tundra and open forests throughout Canada and Alaska. More recently, they have been reintroduced into parts of their former range in Northwestern Montana, Central Idaho, and the Greater Yellowstone.

Mountain lions (*Felix concolor*) have a wide distribution throughout northwestern Canada and USA. Their range is continuing to expand eastwards.

Hawaii
Hawaii's federal domain Haleakala National Park (1) is known for its immense crater and rare plants, including the endemic silversword (right). Hawaii Volcanoes National Park (2) has two of the world's most active volcanoes.

Bison (*Bison bison*) have been released into parks and refuges of open woodlands and prairies in North America.

Greater prairie chicken (*Dipsosaurus dorsalis*) inhabit the creosote-bush desert of central USA. Critically imperiled in Illinois, Iowa, Missouri, Texas and Wyoming.

The **Canada goose** (*Branta canadensis*) is North America's most common goose, inhabiting ponds, lakes, bays, estuaries and grasslands throughout Canada and in 49 of the 50 US states.

- Wildlife refuges
- National Parks

1. Discuss the importance of National Parks and other reserves with reference to conservation and tourism. Carry out further research, if necessary:

_____
_____
_____
_____
_____
_____

Australia's National Reserve system administers over 9300 protected areas including national parks, indigenous lands, and privately protected land. In total, they cover more than 980,000 km^2 or 13% of the country.

In New Zealand, the Department of Conservation administers 14 national parks, but many private organizations fund and administer other restoration projects or protected areas. In all, more than 80,000 km^2, nearly one third of the country, is protected.

The UK's natural environment has a long history of human exploitation and is highly modified as a result. In England 9.3%, Wales 19.9%, and Scotland 7.2% of the land is protected by national parks, covering 22,660 km^2.

Internationally there are 113,000 national parks. Together, they cover 149 million km^2. Although this sounds large, it equates to just 6% of the Earth's land surface. The world's first national park was Yellowstone (USA) (above), established in 1872.

The role of non-governmental organizations

Private organizations and charities play an important role in protecting natural areas. They supplement the role of public departments and increase the area under protection while keeping public costs down.

The Nature Conservancy is one such organization. The mission of the Conservancy is to preserve the plants, animals, and natural communities that represent the diversity of life on Earth, by protecting the lands and waters they need to survive. With donations from over a million members, the Conservancy has purchased over 69,000 km^2 in the USA and a further 470,000 km^2 outside the USA (an area greater than the combined size of Costa Rica, Honduras, and Panama). Larger nature reserves usually promote conservation of biodiversity more effectively than smaller ones, with habitat corridors for wildlife and edge effects also playing a part.

The Great Sand Dunes National Park in Colorado, USA, was in part created by the influence and funds of the Nature Conservancy. It covers 340 km^2.

2. Choose either a national park or reserve close to where you live, or in a country that interests you. Highlight its importance in the conservation of specific species, its importance to tourism, its significance in scientific research or other contributions. Write your findings below:

107 City Planning

Key Idea: Cities evolve in different ways. Modern cities can be altered to be more sustainable and provide pleasant living environments for residents.

Cities are urban centers of population and commerce. The population that defines a city varies from country to country. Cities can be modeled along three main plans: concentric, sector, or multi-nuclei. However, both history and geography of an area results in variations of these three models. Local planning regulations prevent urban sprawl and dictate where certain developments can occur. These regulations also prevent mixing of incompatible activities such as heavy industry and housing. Modern city planning must integrate many factors including sustainable development, aesthetics (how something looks), safety, transport, and environmental protection. A well designed city can enhance the economic and social aspects of a community.

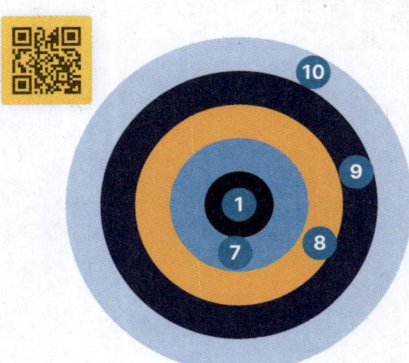
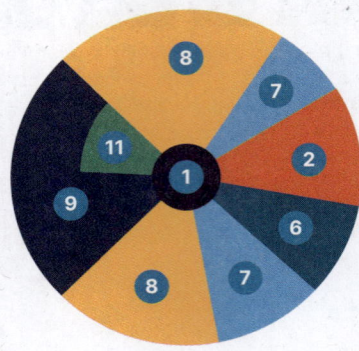
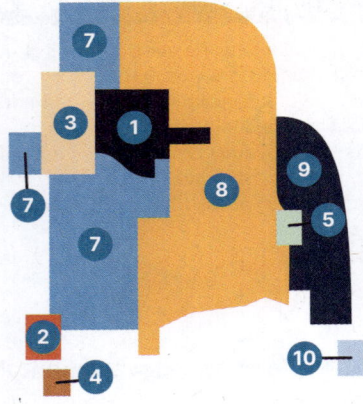

The **concentric ring** model of a city, developed by Ernest Burgess in the 1920s, consists of a system of circles. Land values in the center are very high, hence why buildings are very tall: it is cheaper to build upwards than outwards. The CBD often develops close to the intersection of major transport routes. Outside the CBD is an industrial zone, followed by a working-class housing zone, a more elite residential zone, and a commuter zone. This model was applicable to cities that were developing in the 1920s and Chicago is a classic example of a concentric ring city. In many developing countries the pattern of housing is reversed, with more affluent areas close to the city center and poorer areas on the outskirts.

The **sector model** of urban design was developed by Homer Hoyt in the 1930s. This model describes cities that develop around a CBD in distinct wedges or sectors. An industrial sector is likely to be aligned along major transport lines so that materials can be obtained for manufacturing etc. This design often occurs because wealthier areas are built around desirable features (such as scenic areas or waterfronts). Industry and high density housing are built close to transportation routes for easy transport of goods and workers. Philadelphia matched this model well in years gone by. The industrial part of the city developed along the main railway line for easy transport of manufactured goods.

The **multi nuclei model** was developed by Harris and Ullman in the 1940s. Multi-nuclei cities are probably the closest representation to the majority of the world's cities. They are built around multiple centers or satellite towns. As the city expands, the distance between the CBD and outlying areas becomes greater, and workers often have to commute long distances to their place of work. As the population grows, satellite areas become larger and more influential. They develop their own commercial and industrial areas. Business may shift from the CBD to the satellite areas because there is more land for expansion, lower rents, and often better access to the workforce.

Key

- 1 Central Business District
- 2 Heavy Manufacturing
- 3 Wholesale / Light Manufacturing
- 4 Industrial Suburb
- 5 Outlying Business District
- 6 Transportation
- 7 Low Class Residential
- 8 Middle Class Residential
- 9 High Class Residential
- 10 Commuter Residential
- 11 Education / Recreation

1. Explain why few cities completely match the models described above:

2. Suggest why the pattern of housing in cities in developing countries is reversed, i.e. with poorer areas on the outskirts:

Building sustainable cities

Rooftop gardens are becoming common in many cities. They contribute to a city's food supply and also help to regulate a building's temperature.

Farms may no longer be able to supply the food requirements of an increasing population. High-rise buildings may be converted into greenhouses to grow crops, or into indoor stock raising areas.

High-speed rail linking all parts of a city provides cost effective and convenient transport. It also reduces traffic congestion and smog.

New suburbs and housing developments are being designed in clusters to fit into the natural landscape. Cluster housing leaves areas for the original plants and animals to remain. The design also reduces traffic noise pollution.

Green belts and green spaces are an important part of city development. They provide habitats for animals and recreational areas for people. Areas of undeveloped land act as barriers to development, limiting urban sprawl.

Greenways are walkways or bicycle lanes connecting areas. They separate pedestrians from traffic and provide quick pedestrian access to different parts of the city. Greenways also provide habitats for wildlife.

Solar panels will power street lights, transport information systems, and many public amenities. It will be mandatory for houses to use solar energy to supplement heating and energy requirements. More efficient waste and recycling systems will also be developed.

3. Explain how a sustainable city might achieve:

   (a) Efficient transport: _____

   _____

   _____

   (b) Ecological balance: _____

   _____

   _____

4. Suggest some difficulties that might arise in converting an existing city into a more sustainable one:

   _____

   _____

108 Transportation

Key Idea: Different means of transport are required for different purposes. Reliance on fossil fuels for transportation can be minimized through alternative transport options. Efficient movement around and between cities is often limited by the geography, design, and planning of the transport system. To be efficient, public transport systems need a high population density and must be able to transport people to within a short walk of their destination. In many cities, extensive light rail systems are used to achieve this. Roads, especially busy highways, have a significant impact on local ecosystems, contributing to pollution and forming dangerous barriers for animals to cross.

Air travel comes with heavy environmental costs. Developments in electric aeroplanes and sustainable aviation fuels are looking to reduce reliance on carbon-based fuels.

Electrically powered trains are efficient both for short and long distance travel.

Buses, including electric buses provide a convenient way of transporting large numbers of people around cities.

Rideshare schemes that allow people to pick up and drop off a car for short periods of time cut down on the need for individual car ownership. These could work out cheaper for many people in terms of maintenance and other expenses.

Electric bike and scooter pick up and drop off facilities allow for ease of transport for individuals over short distances.

Electric vehicle charging stations allow people to quickly and conveniently recharge vehicles.

Many towns and cities have invested in retrofitting their existing road networks with cycle lanes to encourage more people to cycle.

Traditional Transport System	Private petrol/diesel car	Diesel/petrol bus	Non electric train	Plane
Advantages	• Convenient, all-hours transport • Personalized • Can access most locations • Especially useful in areas with low populations	• Flexible access to most locations • Useful for short to medium length journeys • Low proportional operating costs if used by large numbers	• Fast transport over medium to long distances • Low per-person pollution • Can haul large volumes of people and freight	• Fast transport over long distances • Able to cross all geographic barriers
Disadvantages	• Expensive to run and maintain • Produce large amounts of pollution per person • Contributes to road congestion • High risk of accident/injury	• Timetables may be subject to traffic congestion • Can cost more to maintain than they earn • Can be slow to reach destinations because of the many stops	• Run to inflexible timetables • Expensive to run and maintain • Produce heavy vibration and some noise	• Fixed and inflexible schedules • Can be uncomfortable over long distances • Produce large amounts of localized noise • Expensive to run and maintain

1. Explain how effective transport systems can reduce congestion and pollution: _____
_____
_____
_____

2. Explain why all four transport systems described in the table are needed: _____
_____
_____
_____

109 Mining and Minerals

Key Idea: The metals and minerals available in the Earth have long been mined for human use. Modern inventions require the extraction of previously unused resources.

Humans have long utilized many of the **minerals** contained within the Earth. Minerals are naturally occurring, solid chemical substances formed through geological processes. Metals conduct heat and electricity and often have a shiny appearance. When minerals are concentrated in an area that makes them economically viable to extract, they are called ore deposits. Most of the Earth's high quality, easily accessible ore deposits have been mined, leaving only more inaccessible deposits left to extract. Some low quality ore deposits, previously uneconomical to extract, may require mining in the future as mineral resources become depleted. As the cost of extraction and processing becomes higher and the mineral becomes more scarce, the value of the ore increases, which in turn increases the cost of products containing the mineral.

Metals, minerals and ores

- Metals: components of ores that can be extracted and used. Sometimes found in native, elementary form, e.g. gold.
- Minerals are naturally occurring products found on Earth. Some are crystalline structures, e.g. sapphires, diamonds, pyrite.
- Ores are minerals occurring in nature that contain extractable metals, e.g. bauxite is an ore containing aluminum.

The formation of minerals and metals deposits varies, but generally involves three stages: formation, transport, and concentration. For example, the formation of gold and silver is associated with volcanic activity. The metals are transported in geothermal water (above) and precipitate out as deposits when the water cools.

Mineral deposits may be eroded and the minerals washed into streams, where they are sorted by sedimentary processes. Dense particles fall to the streambed first, forming placer deposits. These deposits are important sources of valuable minerals such as gold, platinum, tin, and iron.

The evaporation of mineral-rich waters concentrates mineral salts. Many common salts, including NaCl (table salt) are harvested by evaporative mining (above). The sun and wind evaporate water from shallow pools, leaving mounds of salt crystals behind.

Metals have many applications. Precious metals, e.g. gold, platinum, diamonds, are relatively rare and have a high economic value. They can be made into jewellery or traded as currency. The strength of many metals makes them useful material for construction, e.g. iron. Conductive properties make them essential in many high-tech devices.

Lithium mine, Chile

Mining is associated with many environmental problems including erosion, environmental contamination from chemicals, and habitat destruction. Lithium is extensively used in rechargeable batteries in electric vehicles.

1. What is the difference between an ore and a mineral? _____

2. Why are some metals called 'precious metals'? _____

Demand for mined products

Metals Mined in 2019

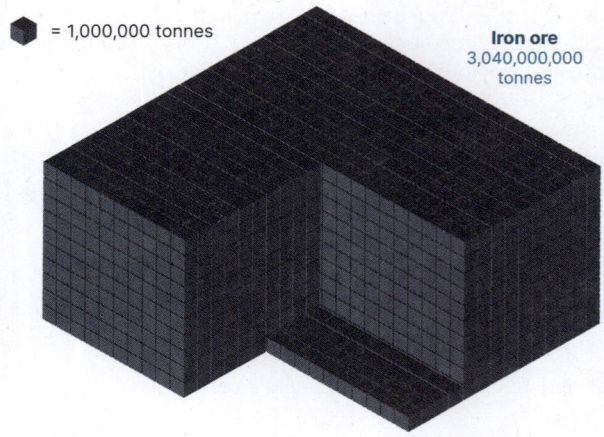

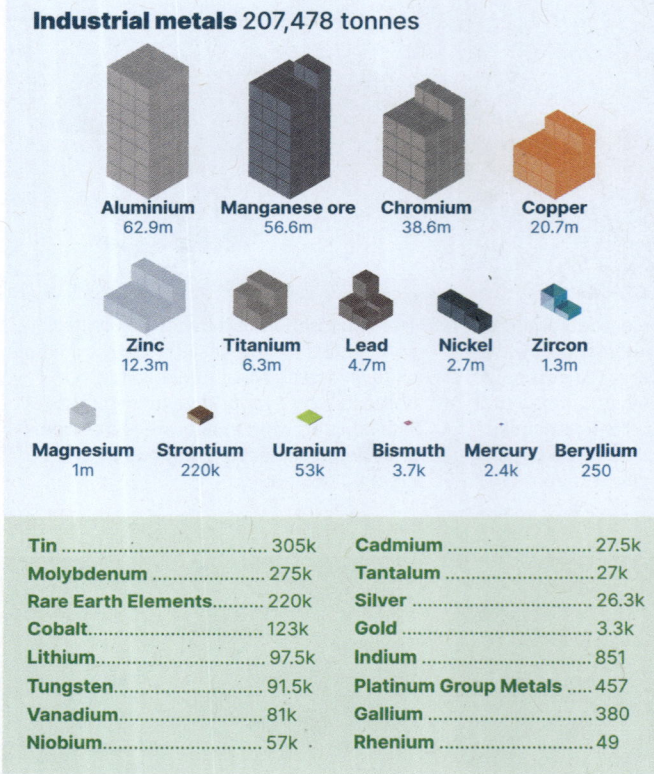

The diagram on the left gives a snapshot of metals mined in one year.

- Iron is used to make steel and is used in construction and manufacturing, from houses to skyscrapers, bridges, roads, cars and anything where its strength is needed. Iron is the most abundant metal, by mass, on Earth.

- Aluminum is strong and lightweight and, unlike iron, resists corrosion. It has numerous uses, including aircraft construction, building materials, car parts and others. High speed trains, such as Japan's shinkansen, use aluminum in their construction as it is much lighter than steel, thereby reducing friction between the train and the track. This allows for higher speeds.

- Chromium and manganese are both used in steelmaking, hence why they are so extensively mined.

- Copper is highly conductive and used in all forms of electrical wiring.

- Zinc is used as a protective coating called galvanizing. It coats many iron based components that are used outside as it prevents them from rusting.

- Titanium is as strong as steel. However, it is much lighter and is used as an alloy (a metal made by combining other metals). When alloyed with steel, it is extremely strong and is used in aircraft and spacecraft construction.

- Lead is a dense metal and is used extensively in battery manufacture, weights, and in some glassware.

- Nickel is resistant to corrosion. It is used in many alloys. Zircon sand is used in heat resistant linings for furnaces and other items that need to withstand extreme temperatures.

- Other metals are used in a wide range of infrastructure and manufactured goods, including technological products such as computers, smartphones, batteries for electric vehicles, and computer monitors.

- As the human population grows and as more people move to urban areas, housing and infrastructure needs to grow and develop to meet the needs of growing populations. Demand for metals is unlikely to decrease in the near future.

Source: Adapted from https://elements.visualcapitalist.com/all-the-metals-we-mined-in-one-visualization/

3. What property of iron makes it ideal as a construction material? _____

4. Identify the three most commonly mined metals:

    (a) _____ (b) _____ (c) _____

5. What common everyday items can you think of that are made of metal? _____

6. Choose one of the items you listed in the previous question. Research its construction and make a list of all the metals it includes, where on Earth this metal is found, and what is involved in its extraction:

_____
_____
_____

110 Globalization

Key Idea: Globalization is the changing of a local market for a good or service into a global one and has both advantages and disadvantages.

Globalization refers to the transformation of local or regional markets or societies into global ones. It involves the migration of people and spread of trade, information, and technology past their traditional national borders. Globalization is not new. In the past, culture and ideas were distributed together with goods along trade routes. Modern globalization began around the same time as the Industrial Revolution, but is usually associated with the rise of large, multinational corporations and the reduction of trade restrictions between many countries over the last 30 years. The global export of Western ideas and culture through mass media (film, radio, television, and the internet) has increased the rate of globalization. Some powerful groups have been set up to advance globalization and development. The advantages and disadvantages of these are discussed below.

The Tragedy of the Commons

Garrett Hardin proposed the Tragedy of the Commons in 1968 to describe how over-exploitation of a resource may occur. When there is no clear ownership of a resource, users overexploit it until it is depleted. Hardin's original example was based on a group of herders sharing land to graze cows. Each herder continues to place cows on the land until the carrying capacity is exceeded. Each individual herder initially benefits, but eventually the land is damaged from overgrazing and the whole group suffers. The tragedy of the commons can be applied to modern issues of sustainability including overfishing, deforestation, human population growth, pollution, habitat destruction, and resource depletion.

Anti globalization protests at 31st G8 summit

Globalization was driven partly by measures taken after World War II, in which large organizations were set up to reconstruct war-torn countries. These organizations now work to develop trade between countries. They include:

- The World Bank provides financial and technical assistance to developing countries to promote social and economic progress. 189 countries are members.
- The International Monetary Fund (IMF) is a global organization with 190 member countries. Its goal is to aid global growth and economic stability by providing financial aid to member countries.
- The World Trade Organization (WTO) helps countries negotiate trade deals so producers, exporters, and importers can conduct their business efficiently on a global scale. 164 countries are members.
- The United Nations (UN) is a global organization committed to maintaining international peace and security, economic development, and human rights. It has 193 member states.

Advantages of globalization	Disadvantages of globalization
Access to new resources and ways of thinking	Increased consumerism and consumption of goods and resources
Increased awareness of global environmental issues	Exploitation of the workforce in developing countries
Greater access to global markets with fewer restrictions, e.g. reduced tariffs, more free trade agreements	Exploitation of resources, especially in developing countries, to meet global demand
Greater understanding of cultural differences	Financial disasters occur on a larger scale, i.e. global rather than regional
Greater travel opportunities allows exposure to other societies and cultures	Potential loss of regional cultural beliefs and customs

1. Describe the impact of globalization on the environment: _____
_____
_____
_____
_____

111 Global Water Resources

Key Idea: The world's supplies of usable freshwater must be carefully managed to avoid over-exploitation.

Earth is an aqueous planet: 71% of its surface is covered by water. Only about 0.0071% of the world's water exists as usable freshwater at the Earth's surface (in lakes, rivers, and wetlands). Rivers and lakes provide sources of water for human use, especially the irrigation of crops. Some of the largest and most important water courses are shown below and opposite. Over-exploitation of water for irrigation and other human uses can lead to irreversible damage and this has already occurred in certain parts of the world. Over-extraction of water can lead to ground subsidence.

The Ogallala aquifer is a vast water-table aquifer located beneath the Great Plains in the US. It covers portions of eight states and is extensively used for irrigation. 30% of it has already been used and 70% of it is expected to be gone by 2065. The aquifer is essentially non-renewable as it will take thousands of years to recharge.

The Colorado River runs through seven states of Southwestern United States. It has several large dams including the Hoover dam and the Glen Canyon dam. The river is so heavily used for irrigation purposes that its flow rarely reaches the sea any more.

The Amazon River accounts for 20% of the world's total river flow and drains 40% of South America. The Amazon is the largest rainforest in the world and has the world's highest biodiversity.

The North American Great Lakes are the largest group of freshwater lakes on Earth, containing 22% of the world's fresh surface water.

The Mississippi River drains most of the area between the Rocky Mountains and the Appalachians. A series of locks and dams provide for barge traffic.

Rivers have been vital transport routes since ancient times. Cities and towns could transport goods and resources along rivers for trade. Large rivers such as the Mississippi, above, are able to accommodate large ships for much of their length. Adding locks to the river can extend the distance a ship can travel.

Rivers provide vital water for irrigation of crops and act as reservoirs for town supplies. However, overuse of a river for irrigation or building dams can lower its level and reduce the downstream flow. This can cause warming of downstream river waters which can lead to degradation of the riverine habitat.

The flow of water in a river provides a way of generating electricity by driving turbines in dams. The damming of rivers severely affects flow rates downstream. Dams are often used for flood control, but this often means floodplains no longer receive a supply of vital nutrients during floods, lowering the fertility of the soils.

©2025 **BIOZONE** International
ISBN: 978-1-99-101409-2
Photocopying prohibited

The Volga River and its many tributaries form an extensive river system, which drains an area of about 1.35 million km^2 in the most heavily populated part of Russia. High levels of chemical pollution currently give cause for environmental concern. The Volga River receives the majority of its water supply from snow melt, and the water is extracted by many urban areas along its path. Water loss, evaporating from dams, or diverted for agricultural irrigation, have greatly reduced the flow released at the river mouth, and therefore, the volume of the Caspian Sea it enters.

The fertile Ganges Basin is central to the agricultural economies of India and Bangladesh. The Ganges and its tributaries currently provide irrigation to a large and populous region. A a recent UN climate report indicates that the glaciers feeding the Ganges may disappear by 2030, as the temperature rise from climate change can no longer sustain them, leaving a seasonal system fed by the monsoon rains.

From glacial origins, the Yangtze River flows 6300 km eastwards into the East China Sea. The Yangtze is subject to extensive flooding, which is only partly controlled by the massive Three Gorges Dam, and it is heavily polluted. The polluted water has degraded the aquatic ecosystems, as well as reducing drinkable water supplies. Glacial melt that supplies the headwater is likely to reduce due to climate change impacts, and available water resources will decline significantly.

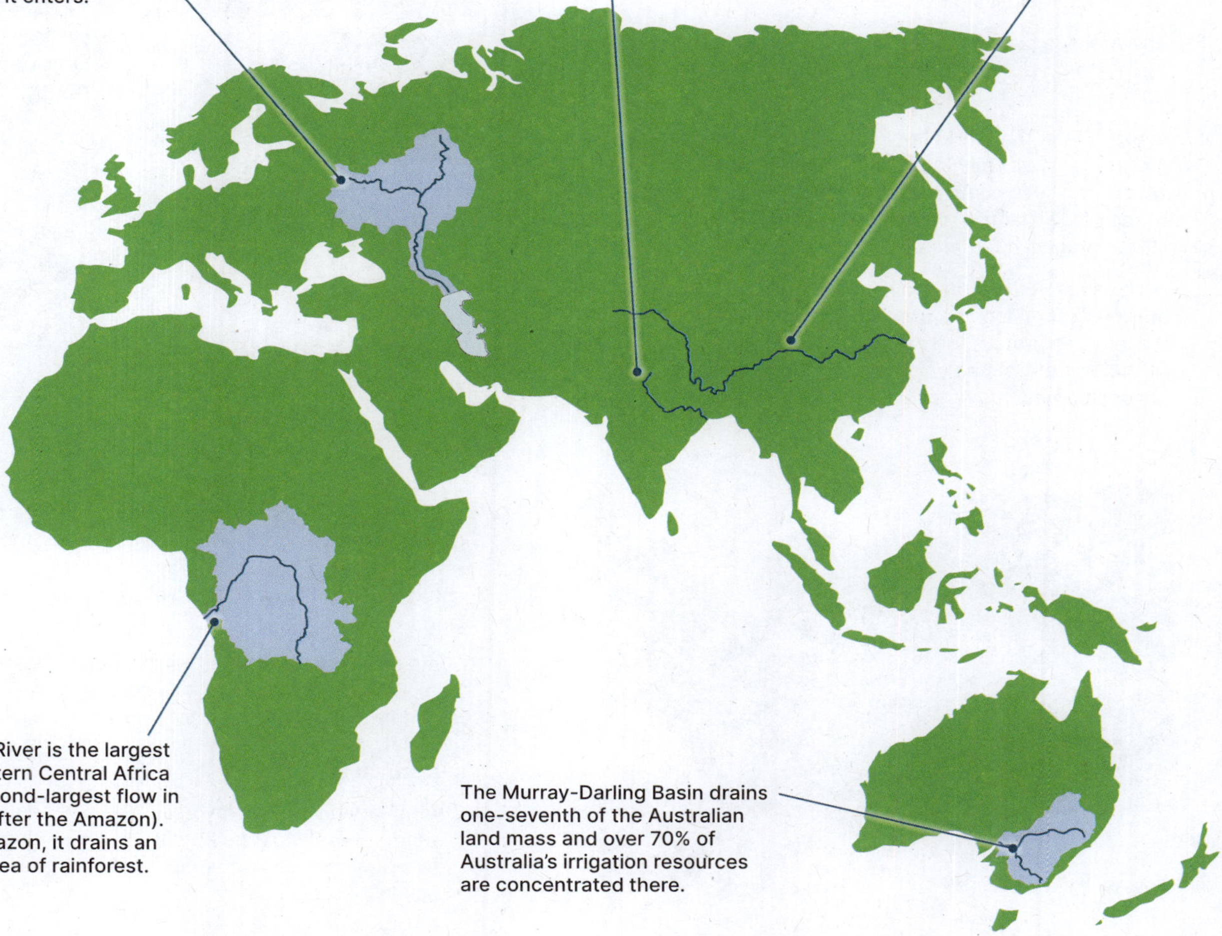

The Congo River is the largest river in Western Central Africa with the second-largest flow in the world (after the Amazon). Like the Amazon, it drains an extensive area of rainforest.

The Murray-Darling Basin drains one-seventh of the Australian land mass and over 70% of Australia's irrigation resources are concentrated there.

1. Explain why some deep but extensive aquifers such as the Ogallala are considered to be non-renewable:
   _____
   _____

2. Identify three uses of rivers by humans:

   (a) _____

   (b) _____

   (c) _____

3. Describe some of the negative aspects of human use of global water resources:
   _____
   _____
   _____

Human effects on water supplies: case studies

Aral sea in Central Asia

- The shallow Aral Sea in central Asia was once the world's fourth largest freshwater lake. It provided a livelihood for hundreds of people who fished its abundant supplies of species such as bream and carp.

- The sea is fed mainly by two rivers that flow from mountainous regions far away. In the 1960s, the Soviet Union diverted water from the rivers to irrigate arid land and create cottonfields. This had enormous implications for the Aral sea and for those who made their living from fishing.

- The Aral sea has now effectively been split in two, with the northern part in Kazakhstan and the southern part in Uzbekistan.

- In the 1990s, the World Bank stepped in to help Kazakhstan. A dam was built to help prevent the flow of water to the South which was successful in restoring water to the northern part of the sea. Fishing is once more a viable livelihood for people in the villages in the area.

- In Uzbekistan the situation is different. The government is more reliant on income from the cotton and there is less incentive to protect the Aral sea. In 2015, the eastern basin of the South Aral Sea completely dried up and the water never returned.

Aral sea 2001

Aral sea 2018

Eastern USA

- Underground aquifers provide water supplies to the numerous large settlements that run down the eastern seaboard of the USA. Withdrawal of water for human activities from these aquifers over time is causing the land to slowly subside.

- Some places are sinking at rates of up to 2mm per year. This includes major cities such as New York and Baltimore. While, this may not appear to be large, over time it is likely to have major effects on city infrastructure.

- Compounded with sea level rise and a changing climate, this subsidence puts the region at high risk of flooding such as happened as a result of Hurricane Sandy in 2012.

4. What is the main reason for the rapid disappearance of the Aral Sea?

_____
_____
_____

5. Compare and contrast the economic incentives for restoring the Aral Sea between the two countries that border it:

_____
_____
_____
_____

6. What combination of factors is leading to increased flooding risks on the East coast of the USA?

_____
_____
_____
_____

112 Water and People

Key Idea: Water is essential for the everyday survival of people, and sources of essential freshwater do not respect international boundaries.

Water is the most important substance on the planet. Life could not exist without it. There are approximately 1.4 billion trillion liters of water on Earth and Earth is the only planet in the solar system where large volumes of water are found on the surface in liquid form. However, water is not evenly distributed throughout the globe. Deserts receive very little rainfall, whereas other places experience large volumes of seasonal or daily rainfall. Despite the enormous amount of water on this planet, wasteful usage and poor management of treatment and supply has reduced the amount of fresh water that is available to much of humanity.

Water's unique properties make it an unusual substance. It has no taste or odor, it is less dense in its solid form than its liquid form (allowing ice to float), and has an extremely high boiling point compared to other similar molecules (such as H_2S). It is polar and can dissolve ionic solids and conduct electricity. It also has an extremely high surface tension caused by the strong bonds formed between water molecules.

The human body is nearly 70% water and, to keep it healthy and functioning correctly, health authorities recommend drinking between 1.8 and 2 liters of water per day. In some countries, access even to this small amount is difficult. In some areas people may use just 15 liters of water per day for all of their domestic uses including drinking water (compared to over 570 liters per person per day in the United States).

Nearly half of the water supplied by municipal water systems in the US is used to flush toilets and water lawns, and another 20-35% is lost through leakages. Treatment of wastewater places major demands on cities yet there are few incentives to reduce water use and recycle. Providing reliable, clean water remains a major public health issue in many poorer regions of the world.

Water use in the home

Water is used in many ways in domestic situations. The pie chart (right) demonstrates average water use for a number of activities. This does not include the water used outside the home, for example, watering the garden, washing the car, cleaning windows etc.

People can take individual actions to reduce their use of water such as turning off running taps while cleaning teeth, washing clothes less often, taking shorter showers, or installing dual flush toilets.

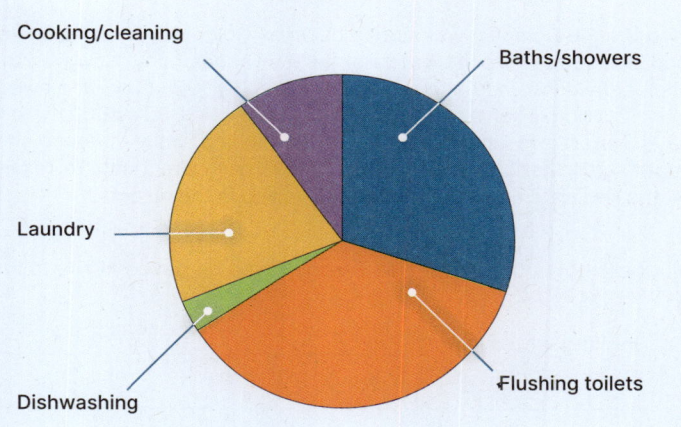

1. (a) Describe some of the properties of water: _____

   (b) Explain why water is so important in living organisms: _____

2. What factors limit water supply to some countries? _____

Freshwater crosses borders

Water is a strategic community resource. Its scarcity has led to disputes throughout the world and throughout history. Conflicts over water can occur both within and between countries. Water conflicts between countries often occur when a water course flows through or is bounded by more than one country. Water polluted in one country may affect the next, or water removed for irrigation in one country will reduce supply to another. Water conflicts within countries often occur because of the differing demands and interests of industrial, agricultural, domestic, and recreational users. Government policies need to ensure fair water access for all parties.

An international conflict

The Colorado River and the Rio Grande are sources of water conflict between the USA and Mexico. A treaty signed in 1944 requires Mexico to deliver 431.7 million m^3 of water from the Rio Grande per year. The USA must deliver 1.85 billion m^3 of water from the Colorado to Mexico per year. In 2004, 17 irrigation districts, the North Alamo Water Supply Corporation, and 29 farmers threatened to sue Mexico over its failure to supply the agreed volume from the Rio Grande. According to the claim, Mexico owed the group more than 1.2 billion m^3 of water.

The Colorado river runs dry in places due to over-extraction of water.

A national conflict

The Rio Grande Compact was signed between Colorado, Texas, and New Mexico in 1938 to share the waters of the Rio Grande fairly. However, the agreement failed to ensure the supply of water. By 1966, Colorado owed New Mexico 1.2 billion m^3 of water while New Mexico owed Texas 600 million m^3 of water. New Mexico and Texas sued Colorado and, while Colorado cleared its debt to New Mexico, New Mexico must still pay water debts to Texas. In 2024, the Federal Government became involved in the dispute.

3. Suggest how both water and energy could be saved in the following domestic situations. You may have to carry out some research:

   (a) Drinking water supply: _____

   (b) Cooking: _____

   (c) Sanitation: _____

   c) Laundry: _____

4. Explain why water is a common source of conflict: _____

113 Water and Industry

Key Idea: Immense volumes of water are used in industrial processes to satisfy consumer demand for food, clothes, and other products.

Water is an essential commodity in communities. Not only is it essential for domestic supply, it is also important in industry and agriculture. Water is most often used by industry as a solvent, a coolant, and for cleaning purposes. A product's **water footprint** is defined as the volume of fresh water required for its production. Most water in agriculture is used for irrigation of crops, although it is also drunk by stock and used as a solvent for sprays. Every year, millions of cubic meters of water are diverted or pumped out of rivers and aquifers to meet the demands for industry or crops such as avocados. The fashion industry has a high water footprint.

Agricultural use of water
Intensive agriculture uses large volumes of water. Crop irrigation accounts for 65% of the world's water use, yet it is largely inefficient, with only 40% of the water reaching the crops and the rest being lost through evaporation, seepage, and runoff. Improved irrigation practices, such as drip irrigation, could double the volume of water delivered to crops and free large amounts of water for other uses.

Industrial use of water
Industrial related water usage increases as the human population increases. The processing of food, and the manufacture of metal, wood and paper products, gasoline and oils all consume large amounts of water. High consumer demand and low levels of recycling exacerbate the problem, but there are large savings to be made by using recycled water and improving the efficiency of water use throughout a product's lifecycle.

Average water footprint of some common products
The water footprint of a product is the total of all the water used in its production. It includes groundwater, rainwater, irrigation, and freshwater used to dilute any wastewater from manufacturing to maintain water quality determined by local standards.

Product	Water footprint (liters)
Pair of jeans	10850
Pair of leather shoes	8000
Car	52000-83000
Smartphone	12760
Cotton T-shirt	2720
One sheet of A4 paper	5.1

Avocado production has increased in recent years in response to consumer demand. It is a warm climate tree crop and often requires irrigation to supplement rainwater and natural groundwater. Its water footprint is an average of 744.3 m³ per tonne. Production is causing water stress conditions in some countries and water extraction for avocados from aquifers has been blamed for small earthquakes in Mexico.

1. Explain what is meant by the term water footprint: _____
   _____
   _____

2. Research an industrial use of water in your local area. What is the main use of water in this industry and how could this particular industry be more water efficient?
   _____
   _____
   _____
   _____

3. How much water is required to produce one average cotton T-shirt? _____
   _____

Fast fashion

According to the UN, the fashion industry is the second most polluting on the planet.

- 93 billion cubic metres of water used annually with large amounts of wastewater generated.
- Estimates range from 6-10% of all global CO_2 emissions being generated by textile production.
- Discharge of dyes and fixatives in wastewater cause pollution of waterways.
- Plastics and microfibres are released into water when synthetic fabrics are washed.
- Only 15–20% of textiles are recycled annually; the rest are deposited in landfills or incinerated.

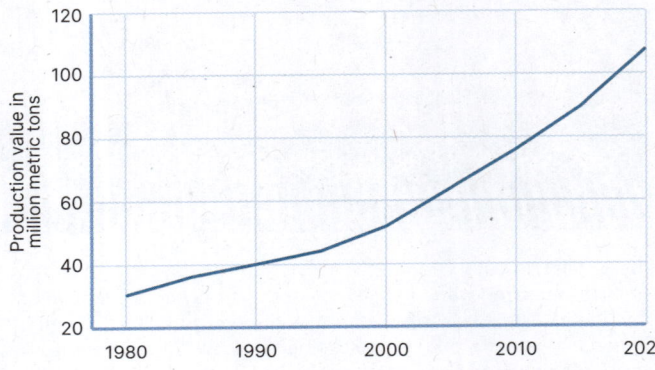

- The graph, left, shows the rise in production of textiles worldwide since 1975. Approximately 60% of the textiles produced are used for clothing.
- Part of this production increase is driven by demand for fast fashion.
- People have more disposable income to spend on clothes, many of which are only worn a few times before being discarded.

4. Research and write your own definition for the term fast fashion: _____
   _____
   _____
   _____
   _____

5. The graph above displays the increase in textile production between 1975 and 2020:

   (a) Approximately what weight of textiles were produced in 2020? _____

   (b) What weight of textiles for clothes were produced in this year? _____

   (c) How many years did it take for production to double from 1975? _____

   (d) How many years did it take production to double from the year 2000? _____

6. Describe some of the ways in which fast fashion contributes to environmental issues affecting the Earth:
   _____
   _____
   _____
   _____

7. Identify some steps that students at your school could take to reduce the environmental impact of fast fashion:
   _____
   _____
   _____
   _____
   _____

114 Ecological Impacts of Fishing

Key Idea: Fishing and overfishing have had severe environmental impacts on the world's oceans. Many fish have been considerably over-exploited.

Fishing is an ancient human tradition and is economically, socially, and culturally important. Today, it is a worldwide resource extraction industry. Several decades of overfishing in all of the world's oceans have pushed commercially important species such as cod into steep decline. The United Nations Food and Agriculture Organization (FAO) reports that almost seven out of ten of the ocean's commercially targeted marine fish stocks are either fully or heavily exploited, over-exploited, depleted, or very slowly recovering from previous **overfishing**. The maximum sustainable yield has been exceeded by too many fishing vessels catching too many fish, often using wasteful and destructive methods. Organisms not intended for food are wasted as **bycatch**.

Lost fishing gear (particularly drift nets) threatens marine life. Comprehensive data on ghost fishing impacts is not available, but entanglement in, and or ingestion of, fishing debris has been reported for over 250 marine species.

Overfishing has resulted in many fish stocks at historic lows and fishing effort at unprecedented highs.

Over-capitalization of the fishing industry has led to too many fleets, and too many large scale vessels. The fishing activities of these large vessels are unsustainable and, for every calorie of fish caught, a fishing vessel uses 15 calories of fuel.

Bottom trawls and dredges cause large scale physical damage to the seafloor. Non-commercial, bottom-dwelling species in the path of the net can be uprooted, damaged, or killed, turning the seafloor into a barren, unproductive wasteland unable to sustain marine life. An area equal to half the world's continental shelves is now trawled every year. In other words, the world's seabed is being scraped 150 times faster than the world's forests are being clear-cut.

The limited selectivity of fishing gear results in millions of marine organisms being discarded for economic, legal, or personal reasons. These organisms are defined as by-catch and include fish, invertebrates, protected marine mammals, sea turtles, and sea birds. Depending on the gear and handling techniques, some or all of the discarded organisms die. A recent estimation of the worldwide bycatch is approximately 30 million tons per year, which is about one third of the estimated 85 million tons of catch that is retained each year.

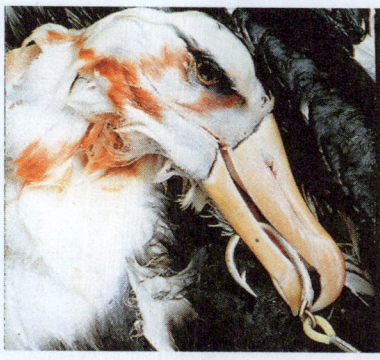

Longline fishing (mainly for tuna) results in the death of 100,000 albatrosses and petrels every year in the southern Pacific alone. Six of the world's twenty albatross species are in serious decline and longline fishing is implicated in each case.

Over-harvesting of abundant species, or removal of too many reproductive individuals from a population, can have far reaching ecological effects. Modern boats, with their sophisticated fish-finding equipment, have the ability to catch entire schools of fish.

Fish farming, once thought to be the solution to the world's overfishing problems, actually accelerates the decline of wild fish stocks. Many farmed fish are fed meal made from wild fish, but it takes about one kilo of wild fish to grow 300 g of farmed fish. Some forms of fish farming destroy natural fish habitat and produce large scale effluent flows.

Towards more sustainable fishing

Part of the problem with fishing is bycatch, i.e. the fish that are not wanted. Even if thrown back, these fish or other marine organisms often don't survive. Techniques to reduce bycatch include changing hook design and attaching devices called pingers to the line or nets, that frighten away non-target organisms.

Different fish species are fished with different kinds of lines or nets. In particular, the mesh size of the net can be changed so that small fish can swim through it and larger fish are caught. This can help to ensure that young fish survive to breeding age or that the wrong species of fish are not caught in the net.

New types of net designs are constantly being tested. One of the newest net designs is called Precision Seafood Harvesting. It consists of a PVC liner towed by a trawler and forms a tunnel of water that reduces stress and damage to target fish, increasing catch efficiency. Holes allow unwanted fish to escape.

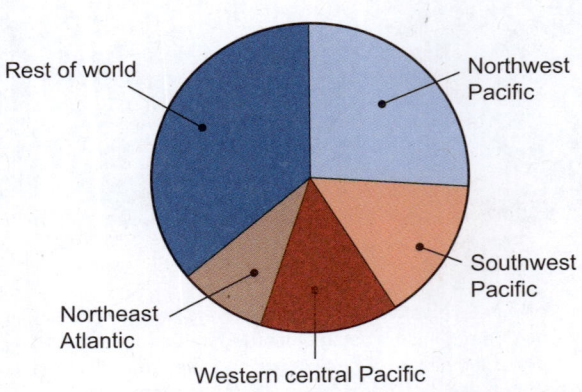

The world's largest fishery is the Northwest Pacific, taking 28% of the global catch

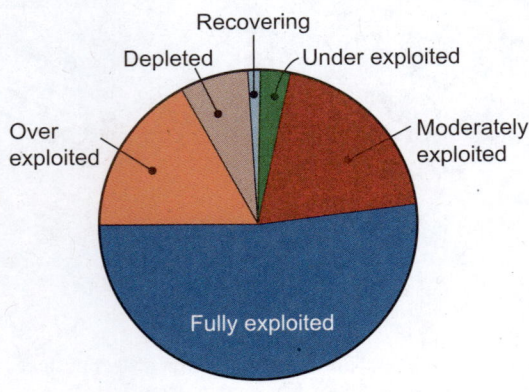

Of the world's fish species, 52% are already fully exploited. Any increase in catch of these species would result in over-exploitation. Seven percent of fish species are already depleted and 17% are over exploited

1. Define the term bycatch: _____

2. Describe two ways that bycatch can be reduced: _____

3. What percentage of fish stocks in total are depleted, over exploited or fully exploited? _____

4. Explain what is meant by over-exploitation in relation to commercial fisheries management: _____

5. Describe some environmental issues related to the fishing industry, other than bycatch and over-exploitation: _____

115 Fisheries Management

Key Idea: Scientists use data to inform management of fisheries operations.

For a fisheries program to be sustainable, rates of harvest must be carefully controlled. Fish populations must be monitored on an ongoing basis and careful data analysis must take place to protect socks from overfishing. Overall populations are calculated using a biomass indicator which gives an estimate of the weight of all the fish in a particular area. Most countries publish quotas regularly for various species to ensure that more fish are not harvested than can be replaced by natural reproductive rates. These quotas can vary from year to year and from season to season.

▸ The sustainable harvesting of any food resource requires that its rate of harvest is no more than its replacement rate. If the harvest rate is higher than the replacement rate, then it follows that the food resource will continually reduce at ever increasing percentages (assuming a constant harvest rate), and eventually be lost. Scientists can collect data and use mathematical calculations, such as maximum sustainable yield (MSY), to establish how many fish can be harvested without affecting future populations.

▸ The maximum sustainable yield (MSY) is the maximum amount of fish that can be taken without affecting the stock biomass and replacement rate. Calculating an MSY relies on obtaining precise data about a population's age structure, size, and growth rate. If the MSY is incorrectly established, unsustainable quotas may be set, and the fish stock may become depleted.

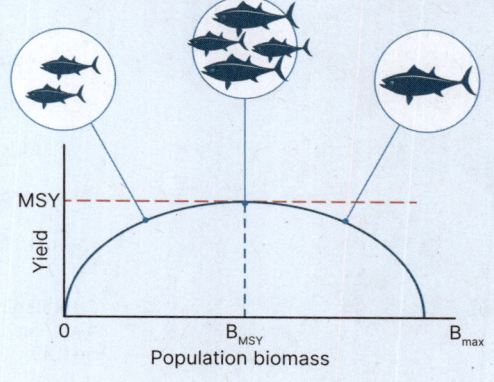

Maximum Sustainable Yield and fishing levels

1. Fishing below B_{MSY}
Fewer available fish. Reduced catch rates and average fish size due to fewer fish in the water. Fish size profile is altered.

2. Fishing above B_{MSY}
Larger and older fish dominate, and therefore result in less productive fish stock.

▸ Scientists often use biomass, the total weight of all fish stock, as a useful measure for the 'amount' of fish present. Biomass increases due to fish reproduction and growth rates, which can differ for each fish species. Biomass decreases due to fish death, both natural and caused by fishing.

▸ MSY biomass (**BMSY**) is calculated at 50% of the maximum (unfished) biomass of an ecosystem and identifies the most effective amount of fish harvesting / fishing.

▸ Under ideal conditions, harvesting at this rate (BMSY) should be able to continue indefinitely. However, the growth rate of a population is likely to fluctuate from year to year.

▸ If a population has below-average growth for several years while the take remains the same, there is a high risk of population collapse because an ever-increasing proportion of the population will be taken with each harvest.

1. What is the maximum sustainable yield and why is that an important indicator of sustainable harvesting?

_____

_____

2. A fish population consists of about 3.5 million individuals. A study shows that about 1.8 million are of breeding age.

   (a) Researchers want to know the maximum sustainable yield for the population so that it can be fished sustainably. What factors will they need to know to accurately determine the MSY?

   _____

   _____

   (b) Should these smaller, non-breeding individuals be included in the catch? Explain your reasoning:

   _____

   (c) It is found that the larger a breeding individual is, the more fertile it is. What implications might this have on the harvesting method for these fish and the viability of the fishery?

   _____

   _____

   _____

   _____

©2025 **BIOZONE** International
ISBN: 978-1-99-101409-2
Photocopying prohibited

Monitoring fish stocks

▸ Marine ecosystems are just as much at risk as those on land and removal of fish for human consumption is a risk to marine biodiversity. Fish is the main source of protein nutrition for a large percentage of the world's population and fishing provides a livelihood for around 60 million people worldwide.

▸ Fish do not respect international boundaries and countries must therefore work together to agree on how many fish from a particular species can be caught each year to provide an ongoing, sustainable harvest.

▸ Individual nations have organisations that monitor fish stocks within their jurisdictions. For example, in the US, NOAA, the National Oceanic and Atmospheric Administration, in association with regional fishery management councils monitors stocks. In Australia, the Australian Fisheries Management Authority is the regulatory body. In the UK, a number of different regulatory organizations operate, depending on location and species.

▸ An exclusive economic zone (EEZ) is a border surrounding a country and its territories, generally 200 nautical miles, over which it has jurisdiction and is therefore allowed to fish according to quotas set nationally and internationally.

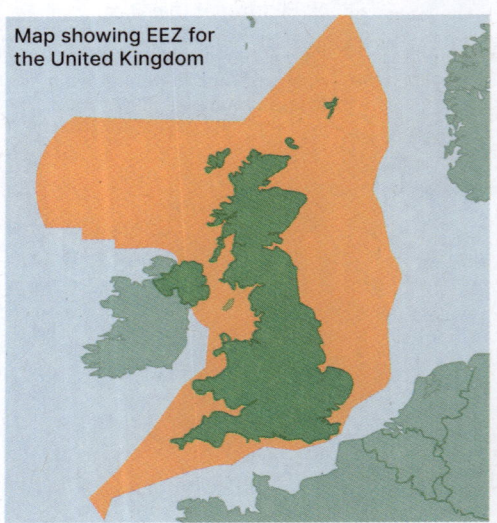

Map showing EEZ for the United Kingdom

Illegal fishing

Industrial fishing 2017-2021 (number of vessels)

- Asia: 19,900
- Europe: 2700
- N. America: 2000
- Africa: 1800
- S. America: 1000
- Australia: 500

■ Not publicly tracked
■ Publicly tracked

Dark fishing is a worldwide problem. Satellite tracking can help identify 'dark vessels' which operate with no registration, no flag, and no license to operate. The extent of the problem is widespread but is a particular issue in Asian waters.

Data sources: Paolo et al. 2024. Nature; Park et al. 2020. Science Advances.

International cooperation

▸ The International Commission for the Exploration of the Seas (ICES) comprises 5000 scientists from 20 member countries and aims to share scientific knowledge across borders to assist in fish stock management and conservation.

▸ The United Nations Fish Stocks Agreement aims to ensure the conservation of fish that straddle international boundaries and/or are migratory.

Western Atlantic bluefin tuna, *Thunnus thynnus*, is the largest of the tuna species and can reach almost 4m in length. It is a sustainably managed fish and, according to NOAA, equipment used to catch it does not affect the ocean bed and allows non-target species to escape. Regulations disallow targeted catching of bluefin tuna in the Gulf of Mexico, as this is an important spawning area.

Yellowfin sole, *Limanda aspera*, is found around the coast of Alaska. It can grow up to 45cm in length and live for up to 39 years. Females do not start to reproduce until they are 10.5 years old so it is important that young fish are not caught. They live on soft, sandy ocean bottoms. According to stock assessments of 2022, it is not overfished and equipment used to catch them does not damage the seafloor.

Atlantic cod, *Gadhus morua*, used to be a plentiful fish and Cape Cod in the US was named for the abundant fishing there. Due to high levels of fishing in the 20th century, stocks of cod plummeted and populations are now carefully monitored.
A stock rebuilding plan is in place to help the population recover; this includes seasonal and sometime year round area closures to protect stock numbers.

3. What is an exclusive economic zone (EEZ)? _____

_____

_____

4. Why is 'dark fishing' an international issue? _____

_____

_____

Aquaculture

Aquaculture is the breeding, growing and harvesting of seafood in specially designed tanks or ponds in water and is a form of farming. It is used to raise fish, including shellfish, used as animal or human food and can also be a means of raising youngstock to help rebuild numbers of threatened and/or endangered species.

- In Norway, several varieties of fish, including salmon, are farmed and production has been increasing steadily since the early 2000s to meet demand.
- It is a major contributor to the Norwegian economy and provides a livelihood to many people, particularly in rural areas.
- Sea lice are a major health issue for farmed salmon as these can impact fish health.
- The environmental effects of such intense farming has been criticized as escaped fish from sea cages allow farmed fish to interact with wild fish, possibly spreading disease.

- New Zealand Greenshell mussels, also called greenlip mussels, *Perna canaliculus*, is an indigenous species that is a popular food, both wild caught and farmed.
- Farming them takes place on long-line systems, with a taught line strung across a line of surface floats and anchored to the seafloor at each end.
- The mussels for cropping are grown along a series of dropper ropes that are attached to the backbone.
- They are harvested at 8-10cm shell length, around 18-30 months after seeding onto the crop lines. The farming of greenshell mussels does not impose heavily on the local environment.

- Farming of lobster species has not proved to be as simple as farming other organisms.
- They have a complex life cycle, including multiple larval stages. They can be difficult to feed and will be cannibalistic if enough of their preferred food is not always available.
- They are expensive to produce as it takes 5-7 years before they are big enough to sell.
- Many countries have ongoing research programs into raising lobster species on farms as it could potentially be a lucrative business with considerable inputs to the economy. In Australia, a 10 year project is underway to assess the viability of farming the tropical lob lobster *Panulirus ornatus*.

5. How does aquaculture help reduce the problems of overfishing? _____

6. (a) What environmental benefits might an aquaculture program offer? _____

   (b) What are some of the negative environmental impacts of aquaculture? _____

7. From the information given on the three farmed species above, outline criteria for a successful aquaculture operation: _____

116 Did You Get It?

1. Describe some issues that can arise by applying chemical pesticides to crop plants and soils:
   _____
   _____
   _____

2. What actions taken by individuals would have the greatest impact on reducing in-home water consumption?
   _____
   _____

3. Describe some ways in which a city could develop in sustainability:
   _____
   _____
   _____
   _____

4. Examine the graph on the right:

   (a) What oil crop uses the most land area to produce one metric ton?
       _____

   (b) What oil crop uses the least land area to produce one metric ton and suggest what might be the reason for this?
       _____
       _____
       _____

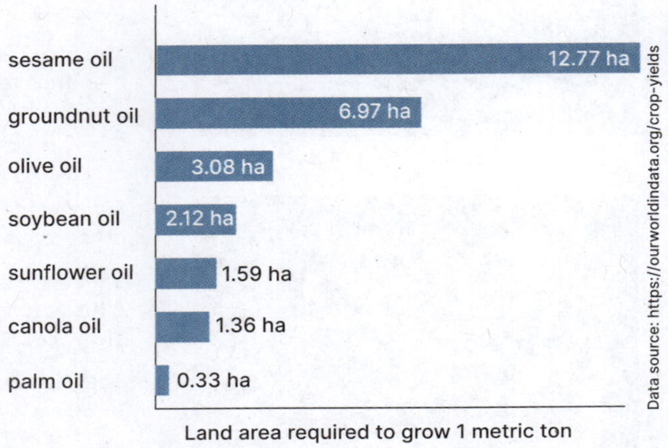

   Land area required to grow 1 metric ton
   - sesame oil: 12.77 ha
   - groundnut oil: 6.97 ha
   - olive oil: 3.08 ha
   - soybean oil: 2.12 ha
   - sunflower oil: 1.59 ha
   - canola oil: 1.36 ha
   - palm oil: 0.33 ha

   Data source: https://ourworldindata.org/crop-yields

5. Suggest what has taken place to affect the landscape shown in the photo (right). What are some of the environmental consequences of this land use:
   _____
   _____
   _____
   _____
   _____

6. Suggest why the development of resistance to insecticides is a problem for humans:
   _____
   _____

7. Why was the change from a hunter gatherer society to an agricultural society an important development in human/environment interactions?
   _____
   _____

Chapter 6 Global Resources
Energy

Resource Hub
bit.ly/3LhnHRm

Key Terms

- biofuel
- coal
- energy conservation
- fossil fuel
- geothermal energy
- hydroelectric power
- hydrogen fuel cell
- natural gas
- non-renewable energy
- nuclear fission
- oil
- photovoltaic cell
- renewable energy
- solar energy
- watt
- wind energy
- wind turbine

Key Concepts

▸ Energy cannot be created or destroyed, it is simply transformed from one form to another.
▸ Non-renewable resources provide immediate low cost power in the short term.
▸ Renewable energy technology is rapidly becoming more efficient and more reliable.
▸ Increasing the efficiency of energy usage can dramatically reduce energy demands.

Energy

Learning Outcomes:	**Activity Number**
☐ 1 Describe basic energy concepts, including the laws of thermodynamics and energy transformation.	117
☐ 2 Describe the general and specific methods of electricity generation.	118
☐ 3 Identify trends in human energy use, including historic, current, and future energy demands.	119

Sources of non-renewable energy

☐ 4 Describe the methods of coal and oil extraction and the effects of these on the environment.	120-123
☐ 5 Discuss the advantages and disadvantages of using coal and oil as a source of energy. Include considerations of economic and environmental impacts.	120-123
☐ 6 Describe methods of nuclear power generation via fission reactors. Discuss its advantages and disadvantages, including short and long term environmental effects.	124

Sources of renewable energy

☐ 7 Define the term 'renewable energy' and explain why this source of energy is becoming increasingly important for global energy production.	125
☐ 8 Discuss the production of electricity using renewable resources, including their advantages and disadvantages, capacity to produce electricity, and their effect on the environment. Renewable sources include: wind, solar, hydroelectric, geothermal, ocean power, hydrogen, and biofuels.	126-131
☐ 9 Compare gasoline, hydrogen, and electricity based transport systems.	132-133

Energy conservation and security

☐ 10 Describe the concept of energy conservation. Explain how energy conservation helps reduce energy demand and CO_2 emissions.	134
☐ 11 Discuss the importance of energy security and the effect of conflict and natural disasters on energy security.	135
☐ 12 Describe methods of energy storage and explain the need and use for different methods of energy storage, including batteries.	136-137

117 Using Energy Transformations

Key Idea: Most commercial electricity is generated by transforming kinetic energy into electrical energy.

Producing electricity is usually achieved by using kinetic energy to turn a turbine attached to a magnet or electromagnet housed inside a large set of wire coils (or vice versa) (the generator). Moving the magnet through the coils produces electricity by a process called electromagnetic induction. The difference between most forms of electricity generation is the method employed to turn the turbine. Energy comes in many forms, from potential (stored) energy to kinetic (movement) energy. Energy can be transformed easily between these forms. A rock at the top of a hill has gravitational potential energy. Giving it a push so that it rolls down the hill converts the gravitational potential energy into kinetic energy, along with some sound and heat. Energy is lost from a system, normally as heat due to friction, whenever energy is transformed from one form into another. Removing causes of energy loss improves the efficiency of the device being used. Generally, the fewer steps involved in energy transformation, the less energy will be lost from the system.

Photovoltaic cells (or solar cells) are increasingly being used in the production of electricity. The solar cell is able to produce electricity directly from the Sun's energy without the need for a turbine.

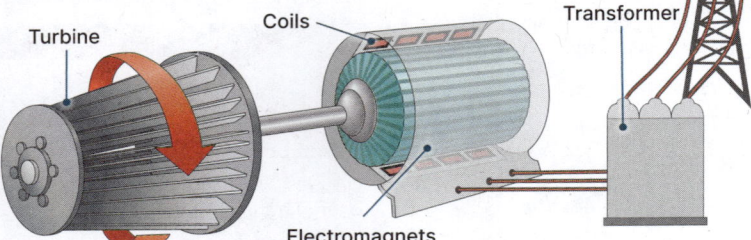

The method used to turn or drive the turbine is what distinguishes the power plant. **Hydroelectric power** plants use falling water, thermal plants use steam, and **wind turbine**s use wind. The capacity of an electricity generation plant refers to its instantaneous power output, e.g a 1000 MW power plants can produce up to 1000 megajoules of electricity per second.

An energy chain can be used to describe where the energy used to generate electricity comes from, and goes to. The number of steps in the chain depends on the form of energy being used and the type of energy generation.

1. Describe the process by which electricity is commercially generated: _____
   _____
   _____
   _____

2. Explain why no form of electricity generation can ever be 100% efficient: _____
   _____
   _____
   _____

3. Research each of the following. Create an energy chain showing energy transformations in the generation of electricity:

   (a) Geothermal power generation: _____
   _____

   (b) Hydroelectric power station: _____
   _____

   (c) Nuclear power station: _____
   _____

118 Global Energy Consumption

Key Idea: Global energy consumption has increased over the course of human history.

There is an argument that the history of human social, technological, and cultural development is linked directly to our ability to harness and transform energy. One of the earliest uses of energy was the invention of ways to provide reliable fire. Energy could then be used to cook food, produce weapons to hunt animals, and provide light. Domesticating animals and using them to provide mechanical power increased the work that could be done by one person. Using **coal** and charcoal to provide heat enabled the smelting of iron and metal alloys, expanding the range of tools that could be used. Steam and internal combustion engines provided ways for heat to be transformed into reliable and powerful mechanical work. Harnessing other types of mechanical work to drive electrical turbines, e.g. water turbines, produced reliable electrical energy. Developing ways to harness the fission of atoms in nuclear power plants and to capture the energy in sunlight are the latest ways to produce electrical power. However, despite these advances, by far most energy production is based on the **fossil fuels**, coal, oil, and **natural gas**, which are used in roughly equal proportions.

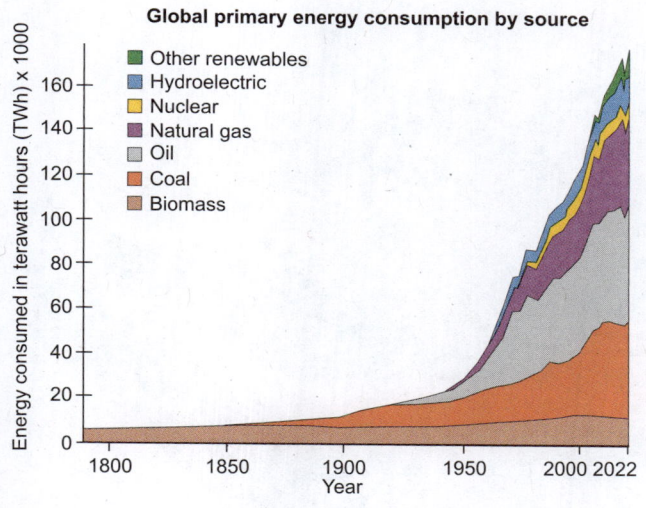

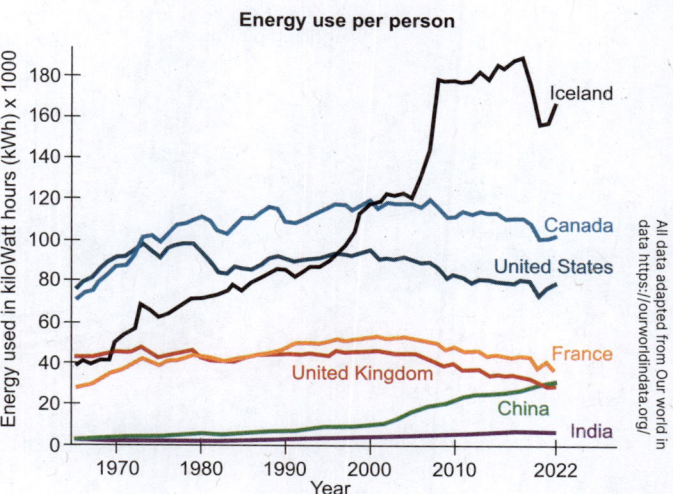

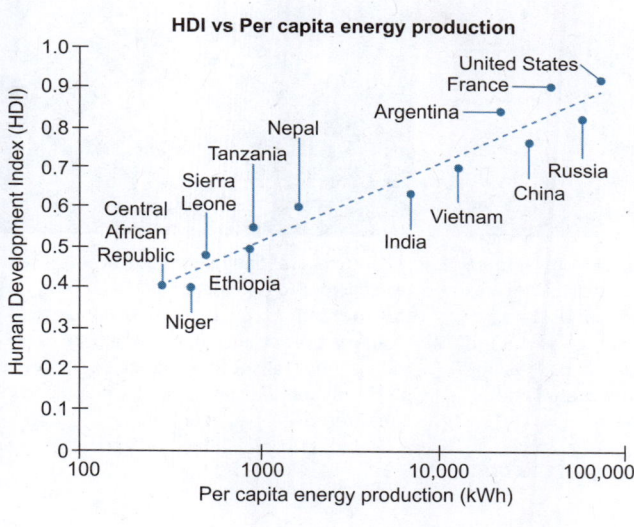

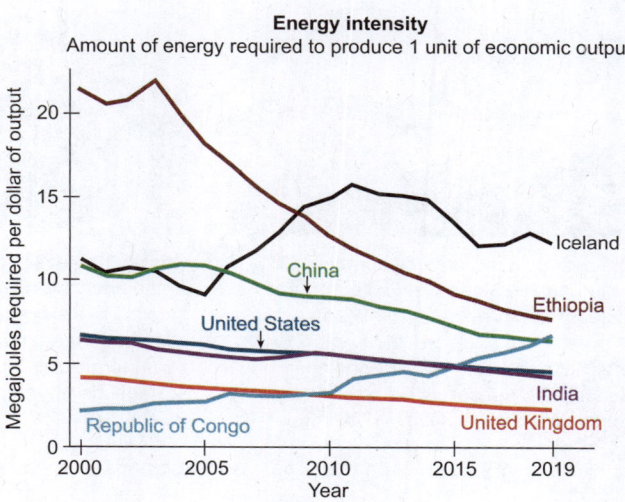

1. Describe how global energy production has changed over the last 200 years: _____

2. Describe the relationship between HDI and energy production: _____

3. What is the general trend in energy intensity over the last 30 years? _____

119 Non-Renewable Resources

Key Idea: Non-renewable resources are finite and cannot be replaced once they have been used.

The Earth contains enormous mineral resources which can be obtained and used with relative ease to produce usable energy. The most commonly used of these are the **fossil fuels**, i.e. **coal**, **oil**, and **natural gas**. These can be burnt immediately to produce heat energy, or they can be refined to provide for a variety of energy or material needs.

As well as fossil fuels, radioactive minerals can be mined and concentrated, and the nuclear energy they produce harnessed to provide electrical energy. Around 80% of the world's energy needs comes from burning fossil fuels, with around 5% coming from nuclear energy. Recently, minerals involved in the production of photovoltaic cells have become important. Although the cells produce **renewable energy**, the minerals in them are **non-renewable**.

Non-renewable energy resources from the Earth's crust

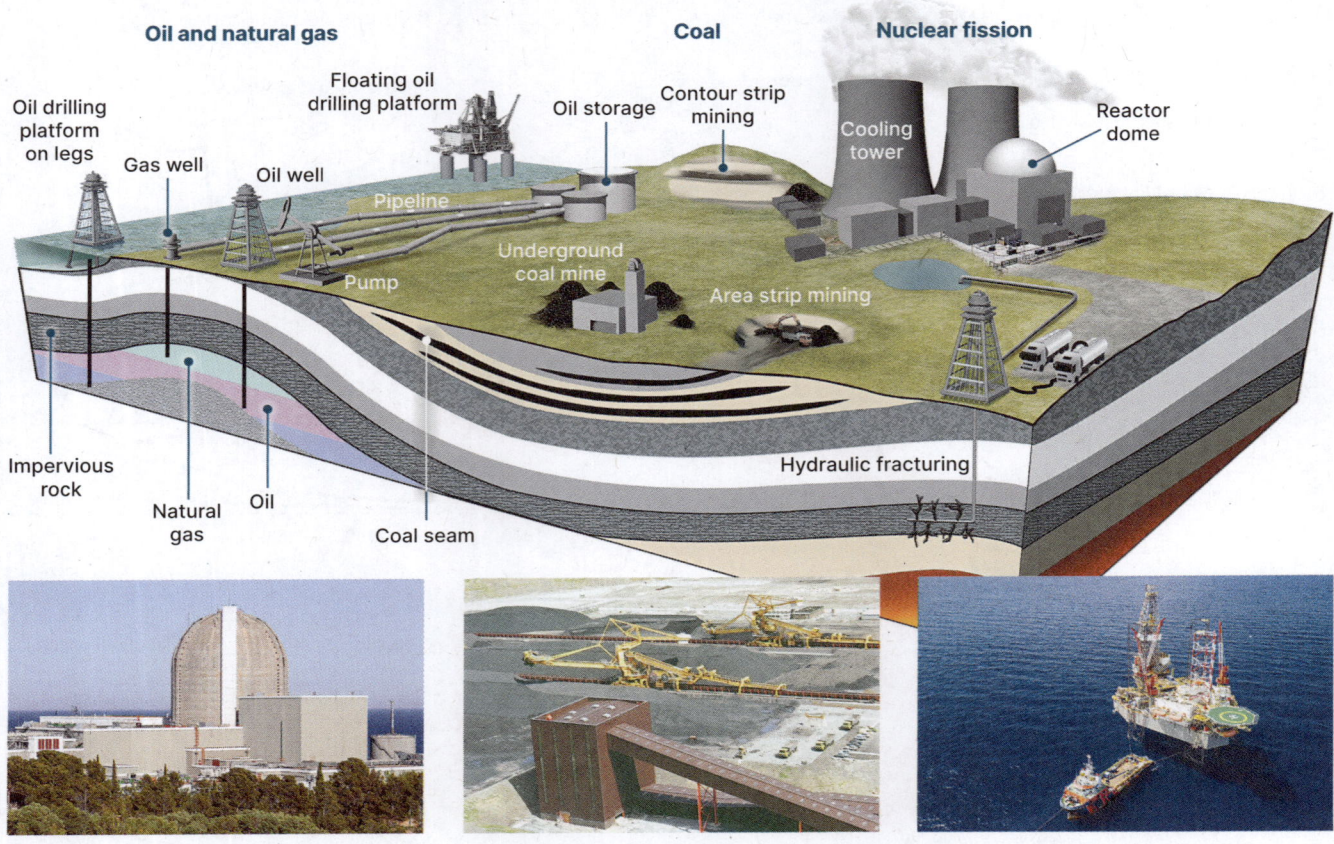

A nuclear power plant uses uranium-235 or plutonium-239 as fuel in a controlled **nuclear fission** reaction to release energy for propulsion, heat, and electricity generation. Nuclear power does not release CO_2, but safe storage and disposal of nuclear waste remains a challenge.

Coal can be easily extracted from seams found near the surface. This causes a large amount of disruption to the landscape. Coal from deeper seams can be extracted by underground mining, which causes little surface disruption provided there is no land subsidence.

Oil and natural gas can be extracted by drilling into a reservoir and pumping the contents to the surface. Many large reservoirs are found offshore, along the continental shelves. Special drilling platforms can be towed out by ships and anchored over the reservoir.

1. Explain why coal, oil, natural gas, and nuclear fuels such as uranium are non-renewable:

   _____
   _____
   _____
   _____

2. Describe some of the issues associated with extracting energy resources from the Earth's crust:

   _____
   _____
   _____
   _____

120　Coal

Key Idea: Coal is one of the most commonly used fuels.

Coal is formed from the remains of terrestrial plant material buried in vast, shallow swamps during the Carboniferous Period (359 to 299 mya) and subsequently compacted under sediments to form a hard black material. Burning coal accounts for about one third of the world's energy production and it is used for domestic and industrial purposes. Coal reserves are estimated at 1.1 trillion tonnes, but removing it from the ground requires large amounts of energy and causes immense disturbance of the surrounding landscape. Burning coal produces vast quantities of greenhouse gases and pollutants, contributing to smog and global warming.

Surface mining

- Surface mining, e.g. strip mining, is used when the coal bed is found close to the surface. The layers of soil above the coal bed (overburden) are removed by machinery and may be stored for later remediation.
- Area strip mining removes coal in long strips. Overburden from the new strip is used to fill the previous one. Contour strip mining is similar but is used on steeper terrain, following the contours of the land.
- Surface mining causes a vast amount of disturbance to the local environment. The land itself is disturbed and the machinery used contributes to noise and visual pollution. Land mined in this way can be quickly eroded by heavy rains, and sediments (often laden with toxic substances) can be washed into streams and rivers.
- Surface mines can be restored once the mining operation is complete, although this does not always happen. Restoration does not always return the mine to its former state either. Open pit mines may be left to fill with water to create a lake.

Subsurface mining

- Subsurface mining uses two main methods. Room and pillar mining remove large areas of coal but leave behind coal pillars that help support the roof of the mine. Long wall mining uses machines that move along the coal face. Coal falls on to a conveyor as it is cut from the seam and is taken to the surface. As the machine moves forward, the space behind is allowed to collapse.
- Subsurface mining causes far less land disturbance than surface mining. However, it is far more dangerous and much of the coal is left in the ground. Risks include roof collapse, build up of explosive or toxic gases, and lung diseases from inhaling fine dust. Although less land is disturbed by digging, land subsidence can leave ripples or holes in the land above.

Uses of coal

Coal as a direct fuel source: Coal is pulverized and used to fuel thermal power stations. In developing countries it is often used for home heating and cooking. This can lead to health problems if furnaces or stoves are not properly vented and coal ash not handled carefully, as coal can contain many toxic substances.

Coal for steel production: Unlike thermal coal, metallurgical coal (coking coal) is very low in ash, sulfur, and phosphorus content. Coking coal is used to produce coke by heating in the absence of air. Coke is used to produce iron and steel. It produces heat for the process and reduces iron oxide to iron at the same time.

Coal as a carbon source: Coal is not only used for heat production. It is a useful source of carbon for many industries. For example, activated carbon made from coal is used in industrial air purifiers, chemical scrubbers, and waste water treatment. Coal-tar is used in some asphalts and medical treatments.

1. How is coal formed? _____

2. Give two important industrial uses of coal: _____

3. What is the main gaseous waste product of burning coal? _____

Major world coal reserves

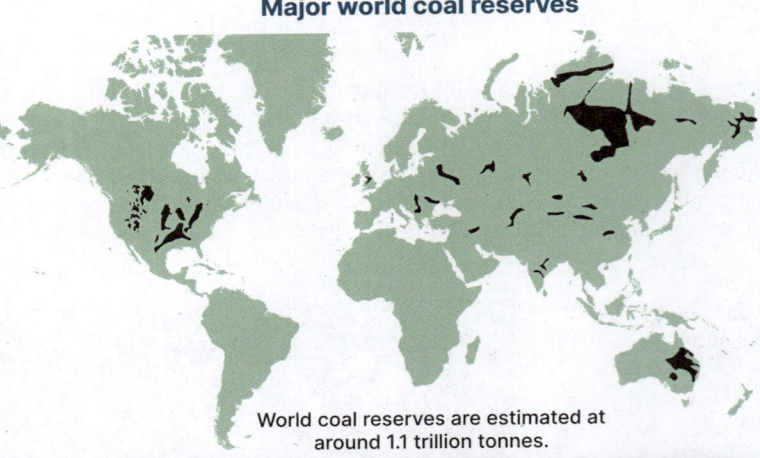

World coal reserves are estimated at around 1.1 trillion tonnes.

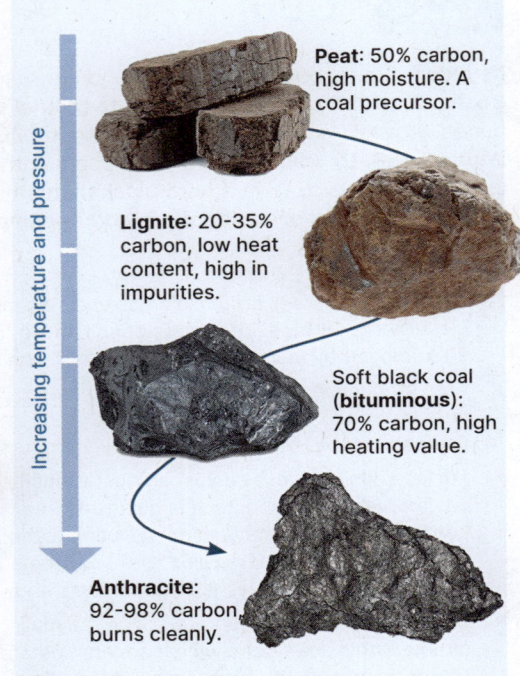

Peat: 50% carbon, high moisture. A coal precursor.

Lignite: 20-35% carbon, low heat content, high in impurities.

Soft black coal (bituminous): 70% carbon, high heating value.

Anthracite: 92-98% carbon, burns cleanly.

Coal is a sedimentary rock. The longer it has been buried and compressed, the better it is as a fuel. Peat still has a high moisture content. It is not coal, but is part of the coal 'family.'

Coal	
Advantages	**Disadvantages**
• Important in industry as coke	• High CO_2 production when burned
• High net energy yields.	• High particle pollution from soot
• Can be used to produce syngas and converted to other fuels, e.g. hydrogen.	• Sulfur dioxide produced as a pollutant when burned
• Relatively easy to extract when near to surface	• Low grade coals produce high pollution and contribute to acid rain
• Huge supplies (at least 500 years' worth)	• High land disturbance through mining

4. Describe the advantages and disadvantages of using coal: _____

5. Explain why anthracite is a better fuel than peat: _____

6. Suggest one reason to account for the non-uniform distribution of coal globally: _____

7. In 1988, the United States' EPA conducted a detailed study of 20 potentially toxic substances from coal-burning electric utilities. It concluded that, with the possible exception of mercury, there were no compelling health risks from burning coal. In many developing countries, emissions from burning coal are a serious problem for health due to exposure to arsenic, fluorine, radioactive particles (uranium and thorium), and carcinogenic organic compounds.

   Discuss the EPA's statement above. Suggest why burning coal in the United States presents little or no health risk, whereas developing countries experience many health problems from burning coal. Use more paper if you wish.

121 Oil and Natural Gas

Key Idea: Oil and natural gas have multiple fuel applications. The world's **oil** reserves formed from the remains of algae and zooplankton (animal plankton) that settled to the bottom of shallow seas and lakes about the same time as the **coal** forming swamps existed. These remains were buried and compressed under layers of non-porous sediment. The process, although continuous, occurs so slowly that oil (like coal) is essentially **non-renewable**. Crude oil can be refined and used for an extensive array of applications including fuel, road tar, plastics, and cosmetics.

Oil and natural gas

▸ Oil and **natural gas** are both composed of a mixture of hydrocarbons and are generally found in the same underground reservoirs. Natural gas is generally defined as a mixture of hydrocarbons with four or fewer carbon atoms in the chain (as these are gaseous at standard temperatures and pressures). Oil is defined as the mixture of hydrocarbons with five or more carbon atoms in the chain.

Peak oil

▸ Peak oil once centred around the idea that at some point oil **production** would reach its maximum potential and then begin an irreversible decline. This idea has been challenged in the last few decades as new technologies and discoveries have continually increased oil production.

▸ With **renewable energy** technologies becoming more efficient and increasing concern over CO_2 levels in the atmosphere, the idea of peak oil is now about when the **demand** for oil will peak and then decline as energy production moves away from **fossil fuels**.

▸ At this stage, fossil fuel demand is predicted by the International Energy Agency (IEA) to peak some time around 2030 and then begin a slow decline.

▸ The current reality is that fossil fuels will remain an important part of the world's energy mix until at least 2050. However, as history has shown, predicting the future of oil is extremely difficult.

According to BP in 2021 proven world oil reserves are estimated at around 270 billion m^3 of oil and 200 trillion m^3 of natural gas.

Oil	
Advantages	**Disadvantages**
• Large supply	• Many reserves are offshore and difficult to extract
• High net energy gain	• High CO_2 production
• Can be refined to produce many different fuel types	• Potential for large, widespread environmental damage if spilled
• Easy to transport	• Rate of use will use up reserves in the near future

1. Describe the difference in the composition of natural gas and oil: _____

2. How has the idea of peak oil changed over time and why is this significant? _____

3. Describe some of the advantages and disadvantages of using oil as a fuel: _____

Oil and natural gas production and transport

Natural gas is often found in the same reservoirs as oil. Drilling rigs require specialized facilities to store the gas. Because of this, much natural gas is either vented, or reinjected to maintain pressure in the reservoir.

Transport of natural gas requires specialized equipment. Liquid Natural Gas (LNG) tankers are able to cool the gas to -162°C and transport it as a liquid (saving space). Gas can also be piped to shore if facilities are nearby.

Oil can be found in materials that make conventional extraction methods impossible. These **unconventional oils**, e.g. oil shale, are often mined in the same way as coal and then crushed and heated to release the oil.

Crude and heavy oils require refining before use. Crude oil is separated into different sized fractions by a distillation tower. Heavy oils may be heated under pressure to break them into smaller, more usable molecules.

Oil is refined in a distillation tower by fractional distillation. The tower is around 400°C at the bottom, but cools towards the top to less than 100°C. Crude oil is pumped into the bottom of the tower and evaporates. The oil vapor cools and condenses as it travels up the tower. Long chain hydrocarbons condense near the bottom while short chain hydrocarbons condense near the top.

Short chain hydrocarbons find use in portable lighters. **Butane** is commonly used in cigarette lighters and camp stoves. **Propane** is commonly used for larger barbecue grills.

Petrol and **diesel** are formed from hydrocarbons with between 6 and 12 C atoms. They provide a high energy, easily combustible fuel that, being a liquid, is easily stored and transported.

Mid length hydrocarbon chains ~15 C atoms, are used as **jet fuel**. They are less volatile and less flammable than shorter chain hydrocarbon fuels while providing high energy per unit volume.

Long chain hydrocarbons may be heated to split them into shorter chains to boost the fractions of petrol and diesel produced, or used in **lubricating oil**, **heavy fuel oil**, **waxes**, and **tar**.

4. Explain how crude oil is refined: _____

5. Investigate the use of refined fuels and then explain why short chain hydrocarbons, such as propane and butane, are used in gas stoves and portable lighters whereas longer chained hydrocarbons are used in vehicles:

Extreme oil

Increases in the price of oil and new developments in technology has made oil fields that were once thought to be uneconomic or unreachable, viable. In addition to the estimated 900 billion barrels of conventional oil that is still relatively easily obtained, there are at least 1500 billion barrels of unconventional oil or offshore oil that could still be extracted.

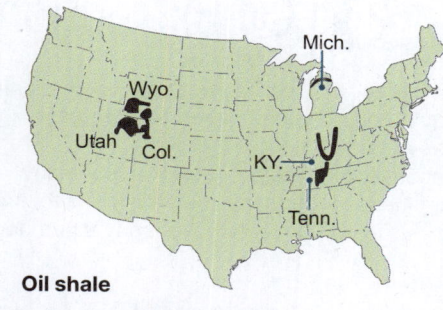

Oil shale
- Shale containing kerogen.
- Rock is mined and heated to extract oil.
- Mining and processing are expensive and have a massive environmental impact.
- Reserves could be up to 800 billion barrels at a cost of US$100 a barrel.

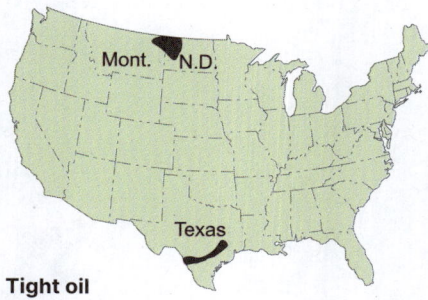

Tight oil
- Light crude tightly bound in formations of impermeable shale.
- Extracted by hydraulic fracturing.
- Possible risks of ground water contamination and triggering earthquakes.
- Reserves of up to 300 billion barrels at US$50 a barrel to produce.

Arctic offshore
- The recession of the Arctic icesheet has opened up the possibility of drilling for oil in the Arctic Ocean.
- Conditions would be treacherous.
- Reserves could be up to 90 billion barrels at a cost of US$100 a barrel.

Oil sands
- Loose sand or sandstone saturated with bitumen.
- Mined in vast open pits.
- The extraction process leaves huge piles of tailings and produces far more CO_2 than the equivalent amount of conventional oil.
- Reserves are up to 169 billion barrels and may cost US$75 a barrel to produce.

Presalt deepwater
- Ultra-deepwater drilling is now relatively common. The presalt reservoirs offshore of Brazil are under nearly 3000 m of water and 1500 m of rock.
- Any blowout would be extremely difficult to contain.
- Reserves are estimated at 100 billion barrels at a cost of US$65 a barrel.

Data source: Time magazine 2012

6. What are some difficulties transporting and storing natural gas? _____

7. How is oil and natural gas formed? How is this different from the formation of coal?

8. Why might offshore Arctic oil fields may soon be exploitable? Explain the consequence of this exploitation:

122 Oil Extraction

Key Idea: Oil extraction requires the drilling of wells and equipment to pump, separate, store, and transport the oil and gas obtained.

Because **oil** can be found in so many different kinds of sediments or deposits, many different techniques are required to extract it. Conventional oil can be pumped directly from the reservoir, but other unconventional oils, such as heavy crude, oil sands, and oil shales, require quite different extraction techniques. These are often energy intensive, reducing the net energy gain. Estimates put global conventional oil reserves at 1.5 trillion barrels and unconventional oil reserves at around 3 trillion barrels.

Oil and natural gas extraction

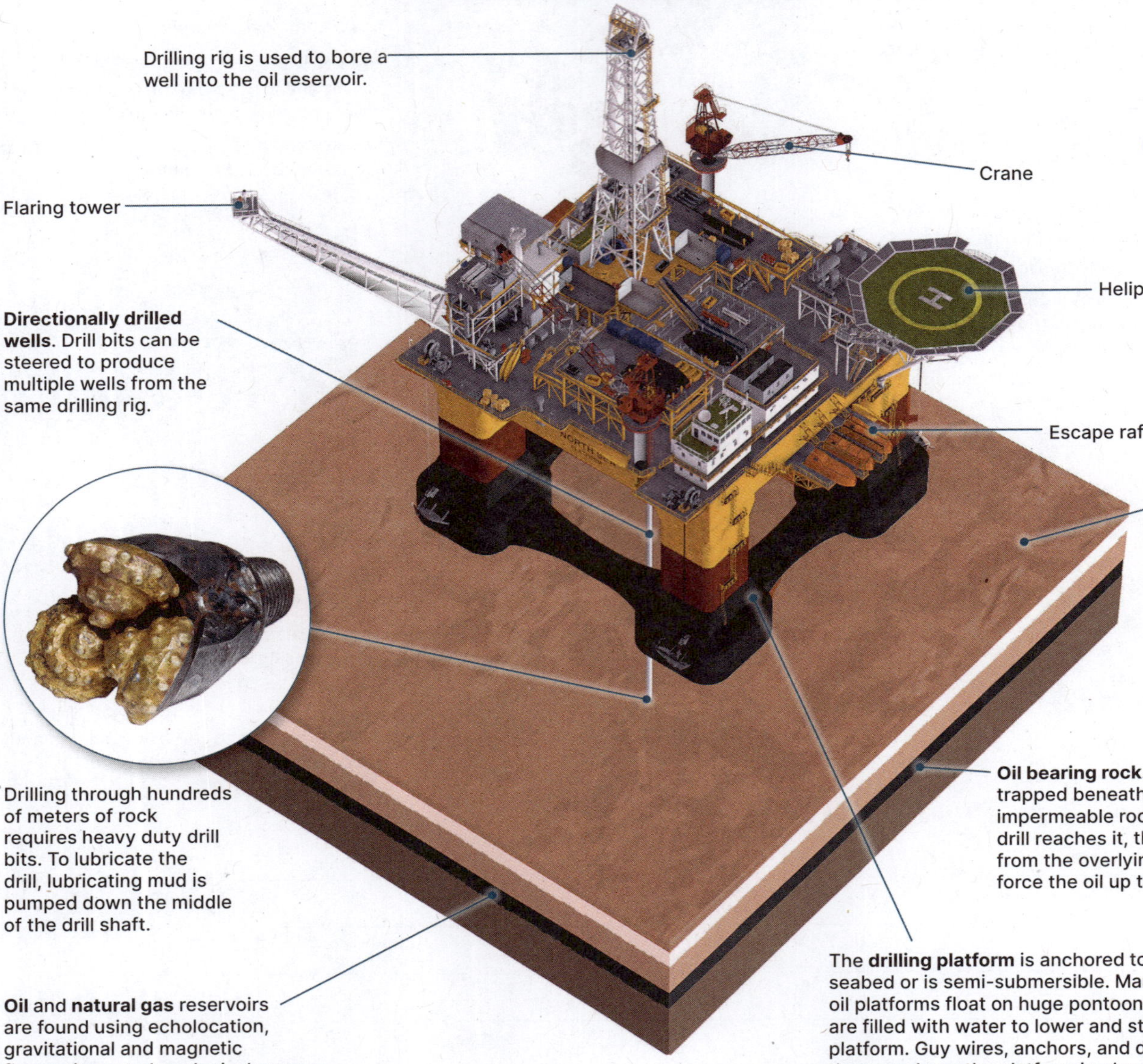

- Drilling rig is used to bore a well into the oil reservoir.
- Crane
- Flaring tower
- Helipad
- **Directionally drilled wells**. Drill bits can be steered to produce multiple wells from the same drilling rig.
- Escape rafts
- Sea bed
- Drilling through hundreds of meters of rock requires heavy duty drill bits. To lubricate the drill, lubricating mud is pumped down the middle of the drill shaft.
- **Oil bearing rock**. Oil is trapped beneath layers of impermeable rock. Once the drill reaches it, the pressure from the overlying rock will force the oil up the well.
- **Oil** and **natural gas** reservoirs are found using echolocation, gravitational and magnetic fluctuations, and geological surveys.
- The **drilling platform** is anchored to the seabed or is semi-submersible. Many deep sea oil platforms float on huge pontoons. These are filled with water to lower and stabilize the platform. Guy wires, anchors, and directional thrusters keep the platform in place.

Oil shales and oil sands

- **Oil shale** contains a solid, bituminous material called kerogen. The rock is mined by open pit mining and the oil extracted by heating the shale and washing the oil out. This process is very energy intensive and its carbon footprint is much larger than that of conventional oil extraction. However, global reserves are large, at ~1.6 trillion barrels (proven). Potential reserves may be several time more than this.
- **Oil sands** are loose sands and sandstones saturated with bitumen. The sands are mined by open pit mining and the oil is extracted in a similar way to shale oil. Again, the process has a large carbon footprint but the reserves are substantial (more than 170 billion barrels in the US).

Oil sand mine, Canada

©2025 **BIOZONE** International
ISBN: 978-1-99-101409-2
Photocopying prohibited

On-shore oil fields

On-shore or land based oil fields are often iconically characterized by fields of pumpjacks (far right) slowly moving up and down to bring oil to the surface. Unlike off-shore rigs, land based drilling (near right) can be done by small, mobile rigs that can be set up relatively quickly. The rig drills the well and inserts steel pipes (casing) into the shaft. Cement is pumped into the shaft and rises on the outside of the casing, aiding structural integrity. If the oil is under pressure, it will flow to up the well to the surface. Pumpjacks are used when the pressure of the well is reduced and more effort is needed to bring the oil to the surface.

Fields of pump jacks are mainly the result of drilling many single vertical wells. With modern equipment being able to steer the drill head multiple horizontal wells can be drilled by one platform. Less pumpjacks are needed where this has been done.

Hydraulic fracturing

The increasing price of a barrel of oil at the beginning of this century created more interest in extracting oil and **natural gas** from sediments that were once uneconomic to mine. Some geological formations may contain large amounts of oil or natural gas, but due to low pressure or impermeable rock, e.g. shales, these have poor flow rates. A process called **hydraulic fracturing** (fracking) is used in order to increase the flow rate in these environments.

1. A well is bored into the layer containing the oil or gas.

2. The well is lined with concrete and steel casings.

3. The well is then drilled horizontally, up to 1500 m from the vertical well.

4. A perforation tool is inserted into the well and explosives are used to fracture the rock.

5. Fluid, which is 99.5% water and 0.5% additives, is pumped into the well to increase flow and keep the fracture open.

6. Gas and oil then flow back up the well. Water (flowback) is recovered and stored in lined pits for later treatment.

Hydraulic fracturing causes fractures in the rock. Opponents of hydraulic fracturing claim the rock fractures allow methane and flowback to leak into groundwater, contaminating it.

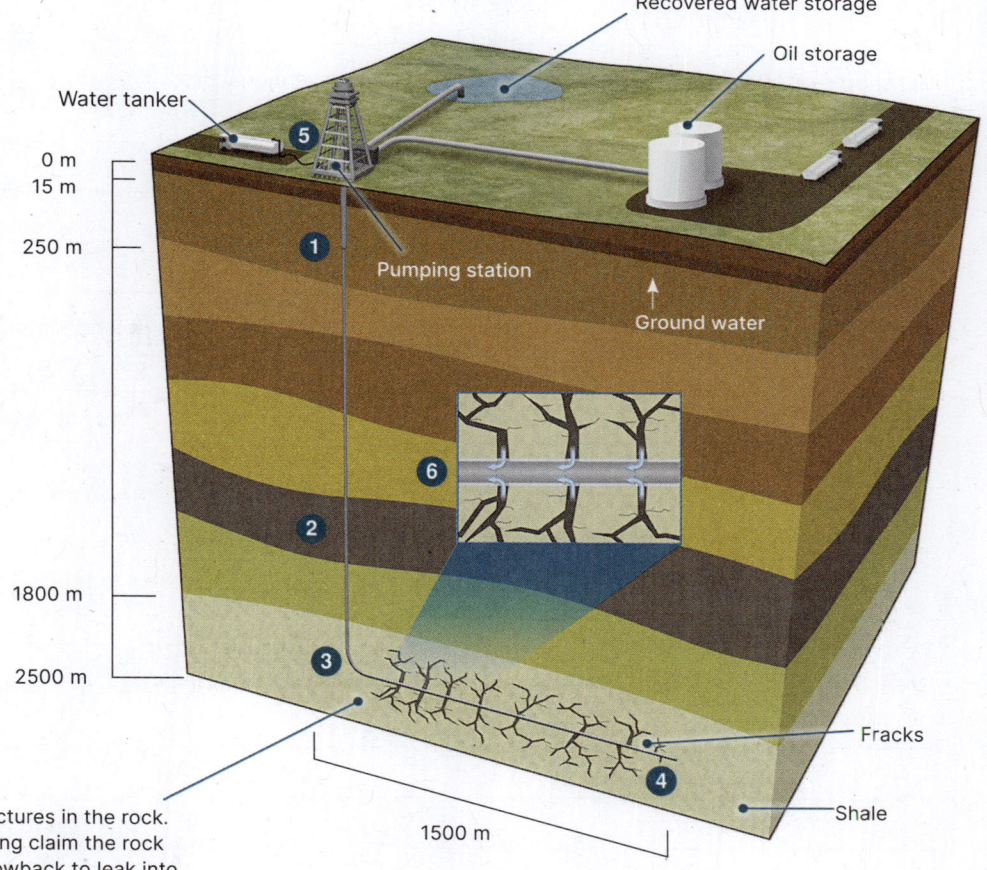

1. The world uses about 100 million barrels of oil per day. Assuming no more oil reserves are located, calculate:

   (a) How many years conventional oil supplies will last based on current consumption? _____

   (b) How many years non-conventional oil supplies will last based on current consumption? _____

The cost of extraction

The energy invested to produce a barrel of oil can be expressed as a ratio of energy expended to energy gained.

- In the early 1900s the ratio was around 1:100 (1 barrel of oil was needed to produce one hundred barrels of oil).
- This ratio has reduced over time to between 1:30 and 1:10 today, depending on the region the oil is extracted from.
- The ratio of energy input to energy gained, or the energy return on energy invested can also be expressed as:

Energy used in locating oil
Energy used in extracting oil
Energy used in refining oil
Energy used in transporting oil
Net energy available for use

Crude oil: 1:10
Open pit oil shale: 1:7.2
In situ oil sands: 1:5

Ratio: barrels of oil input/output

$$ERoEI = \frac{\text{Usable acquired energy}}{\text{Energy expended}}$$

2. Briefly describe the process of extracting crude oil: _____

3. Describe the difference between conventional and unconventional oils: _____

4. Why is it important to account for the energy cost of oil extraction when determining whether to extract oil?

5. Explain how hydraulic fracturing increases the flow rate of natural gas or oil: _____

6. Describe the environmental concerns associated with hydraulic fracturing: _____

7. How do steerable drill heads increase the efficiency of oil extraction? _____

123 Environmental Issues of Oil Extraction

Key Idea: Oil extraction is associated with a number of environmental issues, even before considering its transportation and use.

Drilling for **oil** on land risks groundwater pollution, oil spills from drilling offshore affect vast areas, and the mining of oil sands and oil shales destroys thousands of hectares of boreal forests. However, oil is still the most important fuel for the world's transportation industry and is integral to our daily lives. Some of the issues associated with oil extraction are described below.

Land based oil platform
- Disruption of land to construct pads for pumps, storage, and pipelines. Runoff and leaks from wells contaminates ground and surface water.
- Accidental release of air pollutants, such as methane, contributes to global warming.
- Land subsidence above oil or gas fields.
- Abandoned wells can still contaminate water supplies and emit methane.

Offshore oil platform
- Disruption of seabed in laying down pipelines and wellheads.
- Potential for oil spills to affect vast areas of ocean and shorelines.
- Flaring and release of gases contributes to global warming.

Oil sands and oil shale mining
- Removes and destroys vast tracts of forest.
- Produces large volumes of solid and liquid toxic tailings.
- Storage of tailings uses huge tracts of land.
- Leakages of tailings ponds contaminates water.

In situ extraction
- Uses enormous amounts of energy and water.
- Millions of liters of water must be stored after use in massive tailings ponds.
- Extraction produces up to three times as much CO_2 as the same quantity of conventional oil.
- Oil requires upgrading/cracking which uses large quantities of energy.

Hydraulic fracturing well
- Uses enormous volumes of water, up to 16 million liters per horizontal well and 240 thousand liters of additives.
- Up to 75% of water is recovered as flowback and must be stored.
- Leaks in storage ponds may contaminate groundwater.
- Groundwater can be contaminated by additives and oil if the well casings are not sealed correctly.
- Unknown fissures from fractured layers to groundwater could allow contaminants to rise up into groundwater.
- Fracks and high pressure injection of water in areas of unknown faultlines could trigger minor earthquakes.

1. What environmental issues do hydraulic fracturing and the mining of oil sands and oil shales have in common?

2. Explain why an oil spill from offshore oil extraction can potentially affect huge marine areas:

3. Explain why the production of oil from in situ extraction produces three times more CO_2 per barrel of conventional oil:

124 Nuclear Power

Key Idea: Nuclear power offers low-carbon energy if the wastes are dealt with safely.

Nuclear power plants produce about 5% of the world's usable energy supply but 14% of the world's electricity supply because virtually all of them are used for electricity production. **Nuclear fission** reactors are currently the only reactor type used to produce commercial electricity, although there are a number of reactor designs. Nuclear power stations were first developed for commercial electricity production in the 1950s and there are now more than 400 reactors throughout the world. Some countries, e.g. France, have a high reliance on nuclear power stations for electricity. The popularity of nuclear power stations declined during the 1970s and 80s due to high costs and two high profile accidents. Until the Fukushima disaster in March 2011, in Japan, nuclear power was growing in popularity as a way of reducing greenhouse gas emissions by power stations. However, the disaster served to remind many of their potential hazards.

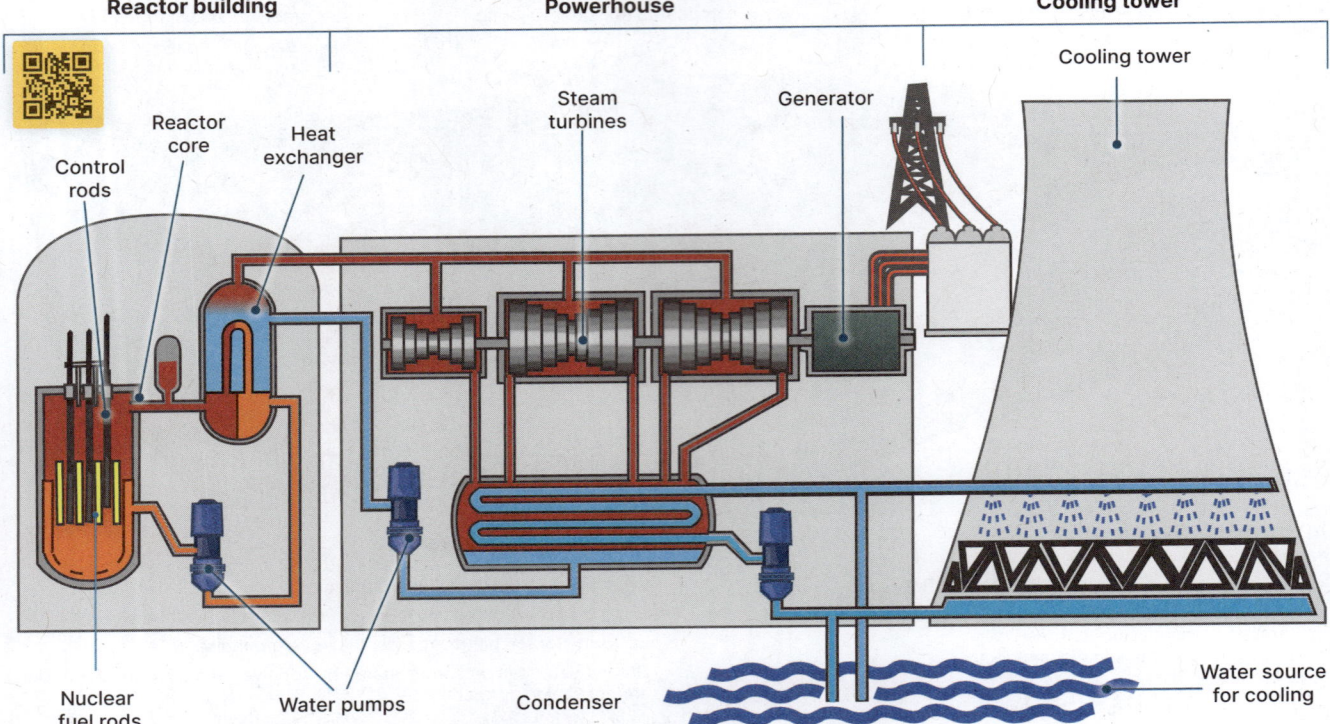

A nuclear power station consists of a reactor building, powerhouse, and cooling tower(s). The reactor building houses the reactor core, which consists of a series of nuclear fuel rods set between removable control rods. Heat produced in the reactor is passed through a heat exchanger to heat water to steam which drives the turbines and generator. Steam then passes into a condenser cooled by water pumped from the cooling tower.

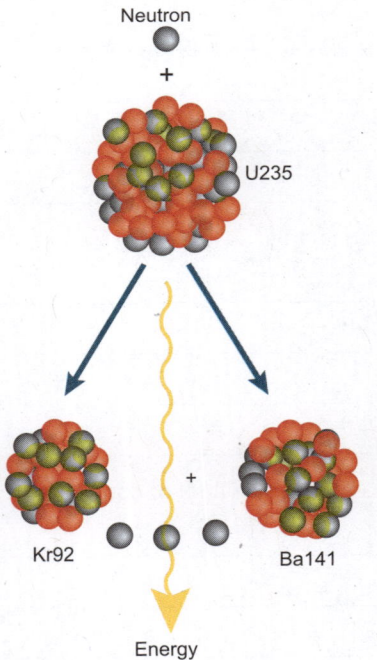

The energy in a nuclear reactor comes from the splitting of a **plutonium** or **uranium** atom. Part of the mass of the atom is converted into energy. The energy released is used to heat a heat sink (water or a metal or molten salt) surrounding the reactor. For a 1000 MW reactor, about 27 tonnes of uranium is needed every year. An equivalent coal fired power station uses about 2.5 million tonnes of coal per year.

Uranium subjected to fission produces around three million times as much energy as an equal mass of coal. U-235 makes up 0.7% of uranium ore. It is enriched to 3.5% for use in fuel rods. Once spent it makes up just 1% of the uranium in the fuel rod.

There have now been two level 7 nuclear disasters since the first use of nuclear power stations. Both level 7 events were caused by human error, faulty reactor design, and inadequate safety features and procedures.

Controlling nuclear fission

▸ Just as conventional power stations generate electricity by harnessing the thermal energy released from burning **fossil fuels**, the thermal energy released from nuclear fission can be used to produce electricity in a nuclear power station.

▸ When uranium-236 splits and releases neutrons, those neutrons can be absorbed by nearby uranium-235 atoms which then split and release neutrons and so on in a chain reaction. With each fission reaction, energy is released as heat. The reactor core generates heat in a number of ways.

- The kinetic energy of fission products is converted to thermal energy when the nuclei collide with nearby atoms.
- Some of the gamma rays produced during fission are absorbed by the reactor and their energy is converted to heat.
- Heat is produced by the radioactive decay of fission products and materials that have been activated by neutron absorption. This decay heat source will remain for some time even after the reactor is shut down.

▸ A nuclear reactor coolant – usually water, but sometimes a gas, liquid metal, or molten salt – is circulated past the reactor core to absorb the heat that it generates. The heat is carried away from the reactor and is then used to turn water into high pressure steam.

▸ So much heat is produced that poorly managed nuclear fuel can become hot enough to melt.

▸ The power output of the nuclear reactor depends on the rate at which the nuclear chain reaction proceeds. It is adjusted by controlling how many neutrons are able to produce more fission reactions. Control rods are used to absorb neutrons. Absorbing more neutrons in a control rod means that there are fewer neutrons available to cause fission, so pushing the control rod deeper into the reactor will reduce the reactor's power output and extracting the control rod will increase it.

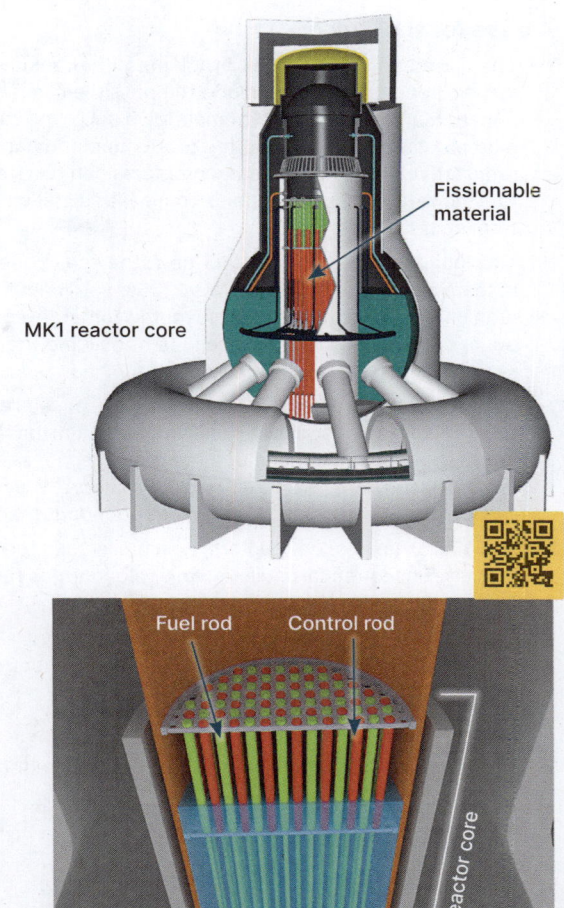

Nuclear power generation	
Advantages	**Disadvantages**
• Large potential fuel supply	• High start up costs
• Little fuel is needed so supplies last a long time	• Disposal of waste presents major technical and environmental problems
• Low air pollution (low CO_2 emissions)	• Risk of catastrophic environmental disaster if accident occurs
• Little land required	• Technology can be adapted to develop nuclear weapons
• Large amount of energy generated	• Potential terrorist target

1. (a) Explain where the energy released in a uranium based nuclear reaction comes from:

   _____
   _____
   _____
   _____

   (b) Explain how a nuclear power station harnesses this energy to produce electricity:

   _____
   _____
   _____
   _____
   _____

A case for thorium?

- There are several important problems with uranium-based nuclear reactors. These include the possible misuse of enriched uranium, the risk of meltdown and explosion of high pressure reactors, and the production and handling of radioactive waste. While many problems can be overcome, they none-the-less persist and come with significant, long term costs and risks.
- Thorium based reactors are not new; they have been researched and tested since the 1960s. For various reasons, uranium-based reactors were favoured for large-scale designs, but in the last few decades the idea of building thorium-based reactors has begun to gather momentum in many countries.
- One of the reasons for this is that thorium-based reactors are, theoretically at least, much safer than uranium-based reactors. There is almost no chance of a meltdown, almost no wastes are produced, and, like other nuclear reactors, they produce no carbon dioxide. Thorium is also more abundant than uranium.
- Unlike uranium-based reactors, which use solid fuel pellets, thorium-based reactors use molten salts both as fuel and a coolant.
- One of the major advantages of thorium reactors is that they recycle their own fuel and do not need to be shut down to be refuelled.

Thorium reactors	
Advantages	**Disadvantages**
• Large thorium supply	• High start up costs
• Thorium can be used in a variety of reactor types	• Lack of current infrastructure
• Reduces risk of nuclear proliferation	• High cost of producing thorium fuels.
• Less nuclear waste	• Emission of gamma rays
• Reactors are safer than current types	

2. Explain how the radiation produced by the reactor core is prevented from contaminating the rest of the power plant:

3. Explain the purpose of the water pumps in a nuclear power plant:

4. How does nuclear fission help to reduce greenhouse emissions?

5. *"Nuclear power provides clean, cheap, and safe electricity."* Discuss the merits of this statement:

6. How many times more energy is obtained from a nuclear reactor in terms of tonnes of fuel used, than a coal fired power station?

125 Renewable Energy

Key Idea: Technologies for harvesting renewable energy can be adapted to many kinds of environment.

A **renewable energy** resource is one where energy can be extracted repeatedly without its source ever being depleted. Renewable energy resources have been used by humans for centuries, the most common being water wheels and windmills providing rotational energy to mills and small factories. Both these technologies have been modernized and scaled up as hydroelectric dams and **wind turbines**. **Fossil fuels** are polluting, supplies are limited, and their extraction is environmentally damaging, so the development and improvement of renewable energy technologies is an increasingly high priority. Along with this, renewable energies can help countries become more energy independent, thus reducing the effect of outside economic influences. Many renewable energy technologies can be scaled down easily to provide portable energy, such as solar cookers or solar panels. However, renewable energies have their own problems to do with space, noise, and use of rare or toxic minerals, as well as the cost of operation and length of life.

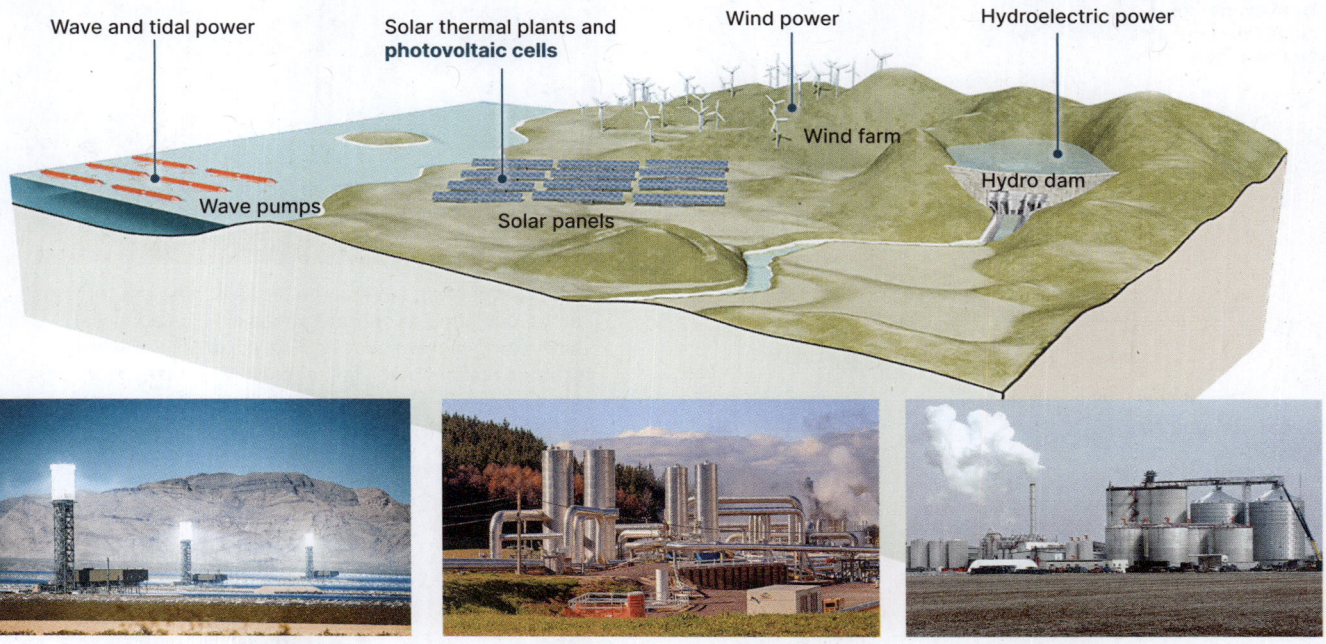

Solar thermal power plants   Geothermal power plants   Biofuel production

1. Describe the type of environment in which each of the following renewable energy resources would work best:

    (a) Solar: _____

    (b) Wave: _____

    (c) Wind: _____

    (d) Geothermal: _____

    (e) Hydro: _____

2. Explain why renewable energy is likely to become the predominant energy source in the future:

    _____

126 Wind Power

Key Idea: Wind power provides a relatively simple and scalable way to produce electricity.

Wind power has been used for centuries to provide the mechanical energy to pump water or run milling machinery. Today, it is mainly used to produce electricity. Wind power is becoming increasingly reliable and cost effective as the technology develops and turbines are able to operate in a range of conditions and wind speeds. In fact **wind energy** is one of the cheapest types of energy to build, maintain, and use. Globally, wind power is steadily increasing in generation capacity, but wind is a variable energy provider. There can be problems matching output to demand, such as during seasonal demands and low (or extremely high) winds. This means systems for managing and distributing electricity will be required as well as backup or base load electricity supplies, e.g. hydro or geothermal power.

Wind turbine

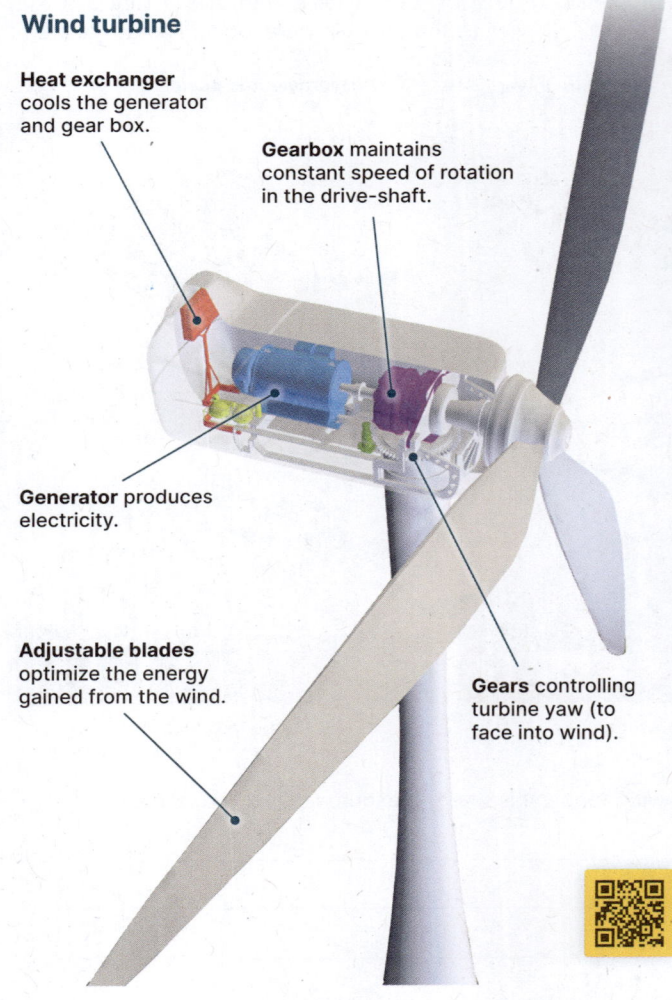

Heat exchanger cools the generator and gear box.

Gearbox maintains constant speed of rotation in the drive-shaft.

Generator produces electricity.

Adjustable blades optimize the energy gained from the wind.

Gears controlling turbine yaw (to face into wind).

Wind farms often cover large areas of land but turbines can be designed to operate at sea and, on a smaller scale, along highway edges. The scalability of **wind turbines** makes them simple to install in many locations, with turbine sizes ranging from a few metres to over 200 metres in diameter.

At the end of 2022, the power output from wind turbines was around 7% of global electricity production. Global installed capacity was more than 800 GW. Electricity generation from wind is rising every year.

1. A typical wind turbine produces around 2.3 MW. The average house uses 30 kWh of energy per day (a kilowatt hour is the equivalent of 1000 joules of energy per second (1kW) running for 1 hour). Calculate the following:

    (a) The minimum number of wind turbines required to power a town of 20,000 households:

    _____
    _____

    (b) Wind turbines cost around $1.3 million per MW of energy production to build. What will the be the cost of (a) above?

    _____

    (c) The cost of building, running, and maintaining wind turbines over their 20 year lifetimes is about $50 per MWh. What could the 20,000 households using the wind turbines above expect to pay in dollars per year for the use of electricity provided by the wind turbines?

    _____
    _____

    (d) Why can households actually expect to have to pay a lot more than this? _____
    _____

Environmental effect of wind turbines

▸ Over their entire life cycle, wind turbines produce some of the lowest greenhouse gas emissions of any electricity production facility. This includes mining of minerals, manufacture and installation, and lifetime use.

▸ Similar to solar power plants, wind farms require large, open spaces and therefore tend to be in rural areas. This has led to what is often called 'industrialization of the country' and 'energy sprawl'. This includes the spreading of access roads and transmission networks, all of which affect the visual appeal of the areas around wind farms.

▸ There is documented evidence of wind farms affecting flying animals, especially where they are placed at the top of hills. Birds and bats following the contours of these hills are struck by turbine blades and killed.

▸ The blades have a service life of between 10 and 20 years and are generally made of fiberglass or (increasingly) carbon fiber, and there is no easy way of recycling these. When their lifespan ends, the blades must be disposed of in landfills.

▸ Because wind turbines sit on single, tall towers well above the ground, the area underneath them is minimally affected. Therefore, if they are sited on agricultural ground, the land can continue to be used for agriculture (right).

Wind power	
Advantages	**Disadvantages**
• No emissions	• Production of visual and noise pollution
• Little ground disturbance during or after construction	• Requires steady winds
• Compact and transportable to most locations	• Can interfere with the flight paths of flying animals
• Can be located in many areas (even at sea)	• Much of actual cost to user is repaying start-up costs.
• Cost certainty. Operating cost is not affected by fuel prices	• Back up systems are required in low winds

2. A major problem with generating electricity is the effect of the facility on the environment. Describe how wind power solves some of these problems. What problems does it create?

_____
_____
_____
_____
_____
_____

3. Explain why increasing uptake of wind power will require better management of the electrical grid: _____
_____
_____
_____
_____
_____

4. (a) From the graph on the right, what is the optimum wind speed for wind turbine use?

   _____

   (b) What happens to efficiency as wind speed increases?

   _____

5. Wind turbine towers have become progressively taller. Why might this help increase reliability of output?

   _____
   _____

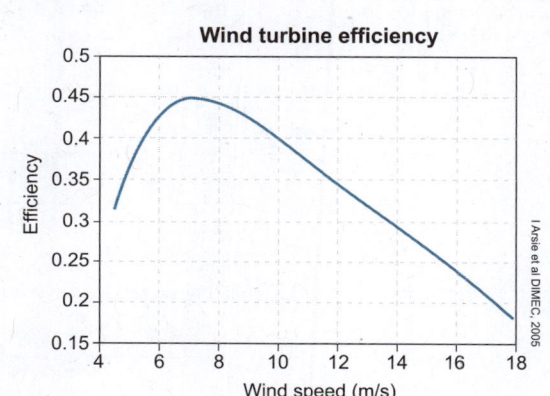

127 Hydroelectricity

Key Idea: Hydroelectricity dams have the dual usage of both producing electricity and storing water for domestic and agricultural use.

Hydroelectricity accounts for around 20% of global electricity production. Electricity is produced by utilizing the gravitational potential energy of water stored in reservoirs behind dams. As water falls, directed along pipes into the powerhouse, the potential energy is converted into kinetic energy, which turns turbines to generate electricity. The larger the volume of water and the further it has to fall, the greater the amount of energy it contains. Large dams can therefore produce large amounts of electricity. The generation of electricity itself produces no CO_2 emissions or other air pollution, but the construction of the dam requires massive amounts of energy and labor and often requires river diversions. Construction of large hydroelectric dams is highly controversial because creating a reservoir behind the dam often requires the submergence of towns and land. Dams constructed inefficiently can also fill up with silt and gradually decline in their generation capacity.

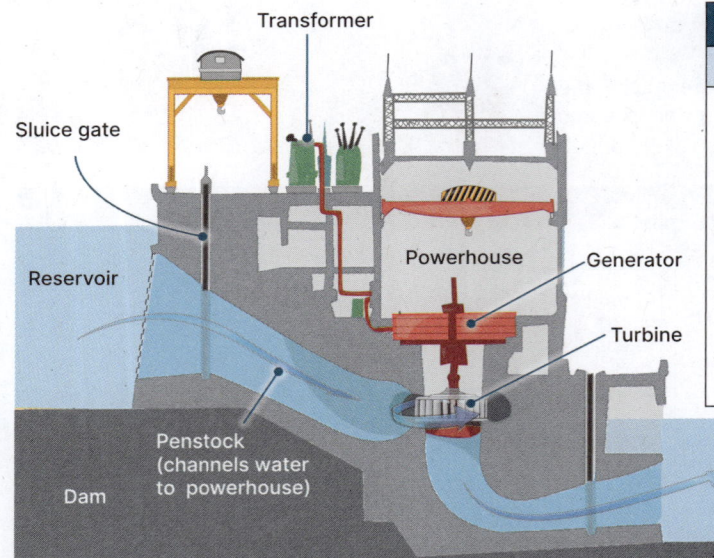

Hydroelectricity	
Advantages	**Disadvantages**
• High net energy gain	• High construction costs
• High efficiency (high percentage of energy converted into electricity)	• High initial CO_2 production from reservoir as drowned material rots
• Produce reservoirs that can be used for recreation and irrigation	• River diversions during and often after construction
• Provide flood control	• Dams Interfere with fish migration
• Long life spans	• Drown river valleys behind dam

Using hydroelectric power

Hoover Dam, Colorado River NV/AZ

Chief Joseph Dam, Columbia River, WA

Dlouhé stráně pumped storage plant, Czech Republic

The mass of water and the distance it falls are important in determining the amount of electricity that can be produced. The power (the energy produced per second) produced by a hydroelectric power plant can be approximated from the mass of water flowing past the turbine and the height of its fall.

Water doesn't have to be stored in a dam for a hydroelectric power plant to work. Water can be directed to flow past the turbine and simply use the force of the flowing water (called **run-of-the-river**). The dam is usually there to divert water towards the intake or powerhouse or to store water in case of lower river levels.

Pumped storage is a useful way of storing excess energy in hydroelectric plants. During off-peak times, water flowing through the plant is used to pump water to a higher storage pond. During high demand, this water can be run through a separate powerhouse to provide extra electricity to the local grid.

1. (a) Explain how hydroelectric dams are used to generate electricity: _____

_____

_____

_____

(b) Describe the relationship between water volume, height of the dam, and electricity production:

_____

_____

Issues with hydroelectric power

The Yangtze River

1987: before

2006: after

Silt from Yangtze River

Siberian Crane

The **Three Gorges Dam** (above) on the Yangtze river, China, is 2.3 km wide and 101 m high, with a reservoir 660 km long. It has a generation capacity of 22,500 MW. The construction of the Three Gorges Dam in China caused the river water level to rise by 100 m, and required 1.2 million people to be relocated.

Dams reduce flood damage by regulating water flow downstream. However, they also prevent deposition of fertile silts. Flooding land behind the dam to create a reservoir seriously disrupts the feeding areas of wading birds.

The Colorado River

A number of dams are found along the Colorado River, which runs from Colorado through to Mexico. The two largest hydroelectric dams on the river are the Glen Canyon Dam and the Hoover Dam. Together, they have a generation capacity of over 3000 MW and provide irrigation and recreation for thousands of people. Both dams control water flow through the Colorado River and were controversial even before their construction.

The construction of **Glen Canyon Dam** effectively ended the annual flooding of the Colorado River. This has allowed invasive plants to establish and has caused the loss of many camping beaches as new silt is trapped behind the dam. The reduced flow rate of the river has severely affected native fish stocks. Controlled floods held in 1996 and 2004 have had beneficial effects on the downstream ecosystems.

Hoover Dam, which impounds Lake Mead, has a generation capacity of over 2000 MW. Water from Lake Mead serves more than 8 million people in Arizona, Nevada, and California. The dam has had a major effect on the Colorado delta, which has reduced in size from around 800,000 hectares to barely 73,000 hectares. Native fish populations have also been reduced.

2. Explain why pumped-storage hydroelectric power is an efficient use of electricity resources:

_____
_____
_____
_____

3. Discuss the following statement: 'Hydroelectric power produces clean, environmentally-friendly electricity':

_____
_____
_____
_____
_____
_____
_____

4. Why are large dams suitable for providing base-load electricity generation (continual operation)?

_____
_____
_____

128 Solar Power

Key Idea: The Sun's energy can be used to produce electricity or direct heating.

The energy reaching the Earth from the Sun is in the order of trillions upon trillions of joules per day, far more than all of humanity uses in an entire year. This energy can be harnessed in many ways for heating or to create electricity. Solar thermal power plants, use concentrated sunlight to heat a fluid, which will turn water to steam to drive a turbine. These only work efficiently at large scales. **Photovoltaic cells** (solar panels) are becoming more popular as their cost goes down and their efficiency increases. Importantly, the solar panels can be used on almost any scale without a loss of efficiency, making them useful for home installation as well as large scale power plants.

Thermal solar power stations have several designs. The most commonly used are the central receiver system, which uses fields of mirrors (heliostats) to concentrate light on a central tower, and the distributed receiver system, which uses rows of parabolic troughs to concentrate light on a pipe running through the trough. Excess heat can be stored in tanks of molten salt.

Photovoltaic cells (PV cells) produce electricity directly from light. Electricity is produced when a photon of light hits a semiconducting material (such as silicon) and knocks an electron loose. The electron is captured and forced to travel in one direction around a circuit, and so produces direct current electricity. Excess power can be stored in a battery.

Solar power can be scaled up or down relatively easily. PV cells can be mounted on a roof for domestic use, or used for portable electricity generation. Solar water heaters can reduce electricity bills. Solar cookers (above) are useful in places where there are high sunshine hours. They are essentially a parabolic mirror with a platform for a pot for food in the middle.

Solar heating

- Passive solar heating is becoming more common in houses. It can efficiently heat a home without electrical input or equipment for moving heat around the house. The design and placement of the house is important when using **solar energy** for heating. Houses placed with their main windows facing South in Northern Hemisphere and North in the Southern Hemisphere gain large amounts of solar energy during the day. Double glazed windows and insulation help to store this energy to keep the house warm during the night.

- Active solar heating uses pumps to circulate heat gathered from a rooftop collector to various parts of a house. Pumps may circulate water through a tank to provide hot water or through a heat exchanger to feed radiators.

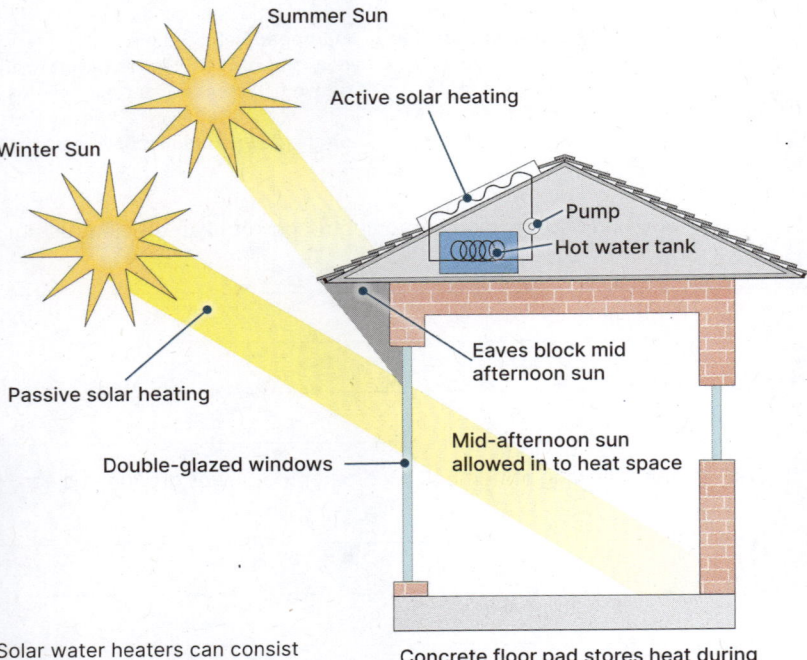

Solar water heaters can consist of simple coils of pipe containing fluid heated by the Sun, heating water via a heat exchanger.

Concrete floor pad stores heat during day, keeping floor warm at night.

1. Explain why solar energy could provide almost limitless energy for humans to use:

___

Developments in solar technology

▶ PV cells are becoming more popular as their efficiency increases and their cost decreases. Even many Antarctic research stations are now powered at least partially by PV cells.

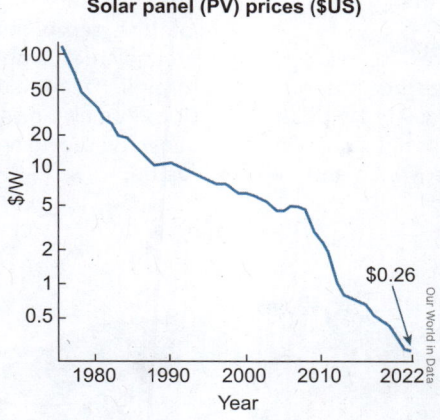

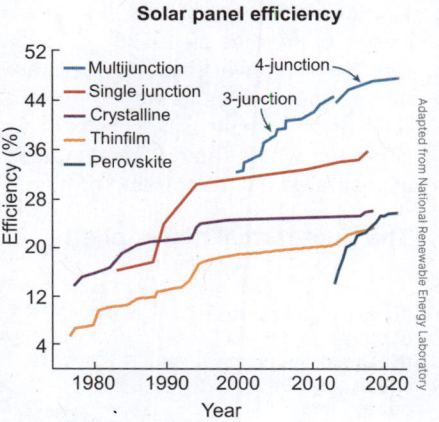

New kinds of PV cells are constantly being developed. Although silicon-based cells are the most familiar there are multiple types of various efficiencies, which are constantly being bettered. New technologies include bifacial PV cells, which capture light on both faces of the cell (above), perovskite cells, organic cells, dye-based cells, and quantum dots.

The cost of PV cells has dramatically decreased over time. This reduction in price is partially due to better production processes, better access to the minerals required, and new techniques and technologies for capturing photons.

The efficiency of PV cells has constantly increased over time. Theoretical limits of efficiency are about 80%. New designs such as coupling perovskite based cells with traditional silicon cells have rapidly increased PV efficiencies.

Solar power	
Advantages	**Disadvantages**
• Low or no CO_2 emissions	• Ground shaded by large solar panels
• Relatively high net energy gain	• Back up systems required
• Small photovoltaic cells are portable and can power many applications	• Large land area needed for commercial scale production
• Unlimited energy source during fine weather	• High sunshine hours required

2. Why have PV cells become a viable option for powering private residences? _____

3. Explain how solar energy can be used to provide electricity even at night: _____

4. Explain the difference between passive and active solar heating: _____

5. Discuss how a house could meet all its energy demands from solar energy: _____

129 Geothermal Power

Key Idea: The natural heat in the Earth can be used to heat homes or produce electricity.

Geothermal power stations operate where volcanic activity heats groundwater to steam. Bores drilled into the ground release this steam and transfer it via insulated pipes to a separator where the dry steam (steam without liquid water) is separated and directed to turbines. Wet steam and waste dry steam are then condensed to water and injected back into the geothermal reservoir to maintain the pressure and groundwater supply. This management practice is essential to prevent land subsidence and depletion of the reservoir. Geothermal power stations often operate close to full capacity, providing a base load, which other power sources top up.

The geothermal power plant

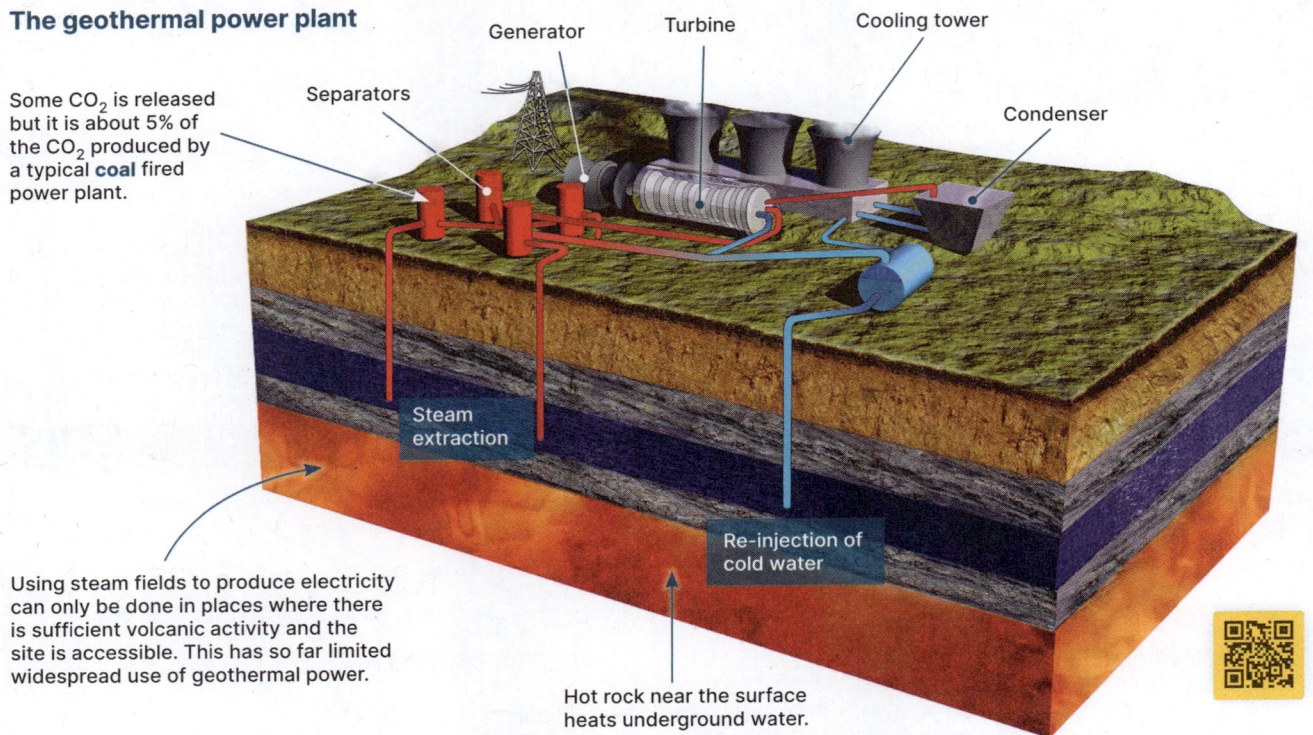

Some CO_2 is released but it is about 5% of the CO_2 produced by a typical coal fired power plant.

Using steam fields to produce electricity can only be done in places where there is sufficient volcanic activity and the site is accessible. This has so far limited widespread use of geothermal power.

Hot rock near the surface heats underground water.

Wairakei geothermal area, New Zealand

Flash steam geothermal plant, Iceland

Wairakei geothermal power station, NZ

Geothermal energy is produced by the fission of radioactive material deep in the Earth. This causes an enormous amount of heat, which heats groundwater close to the surface, producing a geothermal reservoir. The geothermal activity is usually visible at the surface as geysers and boiling mud (above left).

Geothermal power stations often provide base load supplies. This is the minimum continuous electrical supply for an area and is supplied by power stations that can operate constantly at near full capacity. A newer and now more common type of geothermal plant is the binary cycle plant (above center). Here low temperature water (~57°C) is passed by a secondary (binary) fluid with a very low boiling point. The fluid flash-vaporizes to gas and is used to drive a turbine. Geothermal fields (above right) are often large and steam must be sent along specially designed pipes that can expand and contract up to several meters in various weather conditions.

1. Explain how a geothermal power station works: _____

2. Explain why geothermal power plants can be used as baseload supplies: _____

Other uses of geothermal energy

▶ Geothermal energy is only around 20% efficient when used to produce electricity. However, waste hot water from the power plant can be used to heat other industrial operations, such as heating ponds for producing tropical shrimp in temperate environments (top right).

▶ Geothermal energy is useful for direct heating and only needs electricity to run pumps to move water or fluid about. As a result, this kind of heating can extract 4-5 times more energy from the ground that was used to run the pump.

▶ This kind of heating is often used in heating geothermal hot pools. Water can be pumped into a heat exchanger deep in the ground and heated by hot rocks, before being pumped into a pool. Other hot water pools simply pump hot water directly from the ground to the pool.

Heating the home

▶ Below about 3 m, the ground temperature is relatively stable and similar to the average air temperature. This stable temperature is exploited by geothermal heat pumps, which are now installed in many houses (bottom right).

▶ These use a small pump to circulate fluid inside pipes from the roof and floor space of a house into the ground. In summer, this transfers heat from the house to the ground, cooling the house. In winter, it transfers heat from the ground into the house. Geothermal heat pumps do not have to be used in geothermal areas, they simply use the relatively constant temperature of the ground.

| Geothermal power ||
Advantages	Disadvantages
• Moderate to high net production of usable energy	• Few suitable sites
• Low - moderate CO_2 emissions	• Easily depleted if not carefully managed
• Low cost (in suitable areas)	• Noise and odor pollution
• Low environmental impact if managed correctly	• Land subsidence possible

3. Explain why geothermal electricity is currently only viable in a few places on Earth: _____

4. Explain why geothermal reservoirs used for electricity production must be carefully managed: _____

5. Describe some uses for geothermal energy other than generating electricity: _____

130 Ocean Power

Key Idea: The energy in tides and waves can be used to provide a source of electricity, but designing equipment to withstand the sea is difficult.

An enormous amount of energy is stored in the world's oceans. Twice daily, tides move huge volumes of water up and down the coasts of the continents, while billions of joules of energy are transferred when waves meet the shore. Many of the world's energy problems could be solved if this energy could be harnessed, but many problems exist in doing this. Machinery to harness tidal or wave energy requires certain shoreline contours and seabed features, and regular swells. It must also be able to withstand constant immersion in seawater and the relentless and often unpredictable movement of the sea. Many designs have been proposed to exploit various types of seawater movement. While some have shown promise, most have not proved economically viable and concerns also exist over effects on marine life and shorelines. For these reasons, ocean power is unlikely to contribute much to future world energy needs.

Machines to harness tidal or wave energy

Underwater turbines use the simplest designs for harnessing tidal power use. These exploit the currents produced by tides. These operate in much the same way as wind turbines. The largest of these designs has been the SeaGen (below). This operated two turbines producing 1.2 MW between 2008 and 2019. A key feature of its design was the ability to operate in both ebb and flow tides.

Tidal barrages require several meters of tidal difference and may negatively affect the estuaries across which they are built. The largest is the Sihwa Lake Tidal Power Station powerplant in South Korea with a output of 254 MW. It cost over $500 million to build.

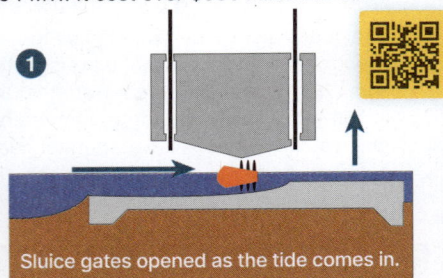
① Sluice gates opened as the tide comes in.

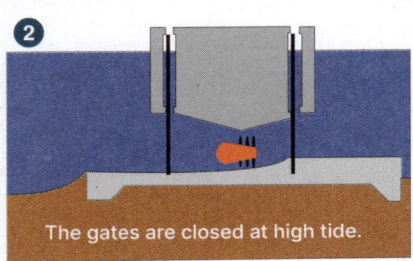
② The gates are closed at high tide.

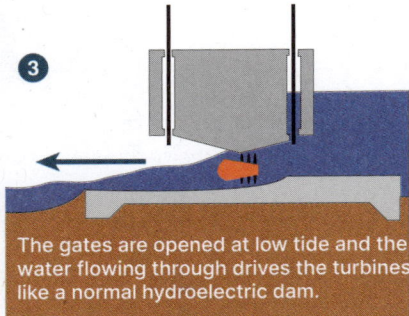
③ The gates are opened at low tide and the water flowing through drives the turbines like a normal hydroelectric dam.

Wave Dragon

Pelamis

- **Wave power** as a means of electricity production is very complex. Developers must take into account wave height and period. These vary almost continuously, sometimes subtly and sometimes by extremes. Unlike many tidal power stations, wave powered systems need to withstand continual pounding by waves.
- Very few designs have proved economically viable or been able to operate for more than a few years. Pelamis was a promising, but ultimately uneconomic, design. Wave Dragon is still undergoing testing.

1. Describe the technical problems associated with producing energy from the sea: _____

2. Describe the potential benefits of harnessing ocean power: _____

3. Explain why ocean power is unlikely to ever produce much of the world's energy: _____

131 Energy from Biomass

Key Idea: Energy from biomass is the oldest form of energy use by humans. Extracting efficient and clean energy from it is still problematic.

Biomass is any material derived from a living organism, e.g. animal waste or wood from plants. Fuels made from biological processes have been used for many years. In many regions of the world, dried animal dung is still used to fuel fires. More recently, a move to produce more commercial quantities of renewable **biofuels** for use in transport and industry has taken place. Biofuels include methanol, ethanol, gasohol (a blend of petrol and ethanol), and biodiesel made from a blend of plant oils and traditional diesel oil. Biogas (methane) is an important renewable gas fuel made by fermenting wastes in a digester.

Biomass

▸ Biomass fuels include almost any solid form of biomass, such as wood and wood pellets, crop residue, and even dried animal waste. It can include biomass grown specifically for fuel or, more commonly, waste products from other industries, such as corn husks or wood chips from the timber industry.

▸ The problem with biomass as a fuel is that it burns much faster than it is produced (it may take ten years for a tree to grow, but less than a day to burn it as fuel). As a result, biomass power plants are relatively small and limited to the waste material they can acquire. What's more, much of this material is not concentrated, e.g. compare the energy in 1 kg of wood compared to 1 kg of **coal**, so biomass power plants are not particularly efficient at producing electricity.

Wood waste

Biofuels

▸ Biofuels are made from processing biomass. The most common types are bioethanol (95% of the world's ethanol is produced biologically) and biodiesel. Like simple biomass, biofuels are renewable, but are limited by their rate of production and use.

▸ Biofuels can be divided into generations depending on how the fuel is produced.

Sugarcane

Legume crop residue

Photobioreactor for algal culture

Photoelectric cell

1st generation biofuels come from biomass that is also a food source, e.g. sugarcane and corn. Thus, they take up resources that could produce food.

2nd generation biofuels come from non-food biomass, e.g. crop residues. However, this removes nutrients normally returned to the soil after harvest.

3rd generation biofuels are still mostly experimental. They do not compete with food sources and are mostly based around algae, which contain lipid oils.

4th generation biofuels use living material to produce energy directly. The photoelectric cell above produces H_2 and O_2 directly. These are still at an experimental stage.

Producing methane

▸ Different types of organic material often need to be treated in specific ways to extract energy from them. Methane can be produced from the same material as ethanol, but it can also be produced from material that is not suitable for ethanol production, such as sewage and manure.

▸ The process of producing methane from organic material is done in conditions that cause anaerobic digestion.

▸ Methane-producing bacteria combine hydrogen and carbon dioxide to produce methane (a process called methanogenesis). The methane can then be used as a high energy fuel and as a starting product for the synthesis of various chemicals.

Roof is able to inflate as gas is recovered. Digester

1. (a) Is corn a 1st or 2nd generation biofuel? _____

   (b) Is the corn residue left after harvesting a 1st or 2nd generation biofuel? _____

2. Suggest why 3rd generation biofuels are still in the developmental stage, with virtually none in commercial production:

   _____
   _____
   _____

Ethanol as an alternative fuel

▶ Ethanol is an important industrial chemical. It has properties that make it useful in both food production and industry. Ethanol has been proposed as a replacement for **fossil fuels** such as gasoline because it burns well as a fuel, it can be stored easily, it is easy to produce in large quantities, and it can be produced from plant material.

▶ Many countries mandate that ethanol should be blended into gasoline (as E10) to reduce the carbon footprint of gasoline use. In the US, biofuel blending volumes are mandated to increase to 85 billion liters by 2025.

Producing ethanol from feedstock

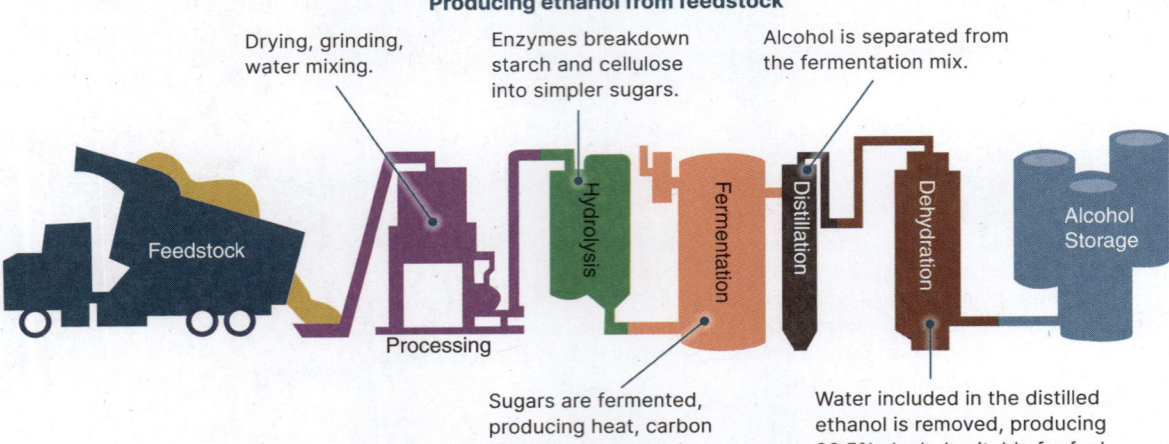

Drying, grinding, water mixing.

Enzymes breakdown starch and cellulose into simpler sugars.

Alcohol is separated from the fermentation mix.

Sugars are fermented, producing heat, carbon dioxide, and ethanol.

Water included in the distilled ethanol is removed, producing 99.5% alcohol suitable for fuel.

▶ Different types of feedstock can be used to produce ethanol, but starch and sugar-heavy feedstock is best. In the US, corn (maize) provides 98% of material used to produce ethanol whereas In Australia, wheat makes up about 60% of feedstock for ethanol production.

▶ Crops for ethanol production compete with arable land for food production. In the US, the corn ethanol industry requires around 13 million hectares of land. The amount of ethanol produced is limited to 60 billion liters, to reduce competition with food crops for land. 40% of all corn grown in the US is used for ethanol production.

▶ Although ethanol is often called a carbon neutral fuel, it requires large amounts of fuel to grow, harvest, transport, and distil the crop. Ethanol also contains only two thirds the energy of the equivalent volume of gasoline. It is therefore better called a carbon-reduced fuel, with lifetime emissions being about 46% lower than gasoline.

▶ As an alternative, corn waste after harvest could provide up to 1.27 billion tonnes of useful waste and produce 30% of the US transport fuel needs. However, this would use the organic material that is normally plowed back into the land as fertilizer. This would have implications for soil fertility and increase dependence on inorganic fertilizers, which are costly to produce and create problems of water contamination.

Ethanol plant

3. The grain required to produce 100 L of ethanol can feed a person for a year. Around 49 billion liters more ethanol was produced in the US from corn in 2018 than in 2001. How many people could this have fed?

_____

4. Explain the disadvantages of ethanol production from corn as an alternative fuel: _____

_____
_____
_____
_____
_____
_____
_____

5. Describe how biofuels can form part of the solution to reducing CO_2 emissions: _____

_____
_____
_____

132 Hydrogen Fuel Cells

Key Idea: A hydrogen fuel cell uses the oxidation of hydrogen to produce electricity.

Fuel cells have been in use for decades but recently, development has accelerated, especially for powering vehicles. **Hydrogen fuel cells** use hydrogen as a fuel and produce water by reacting the hydrogen with oxygen.

The hydrogen fuel cell

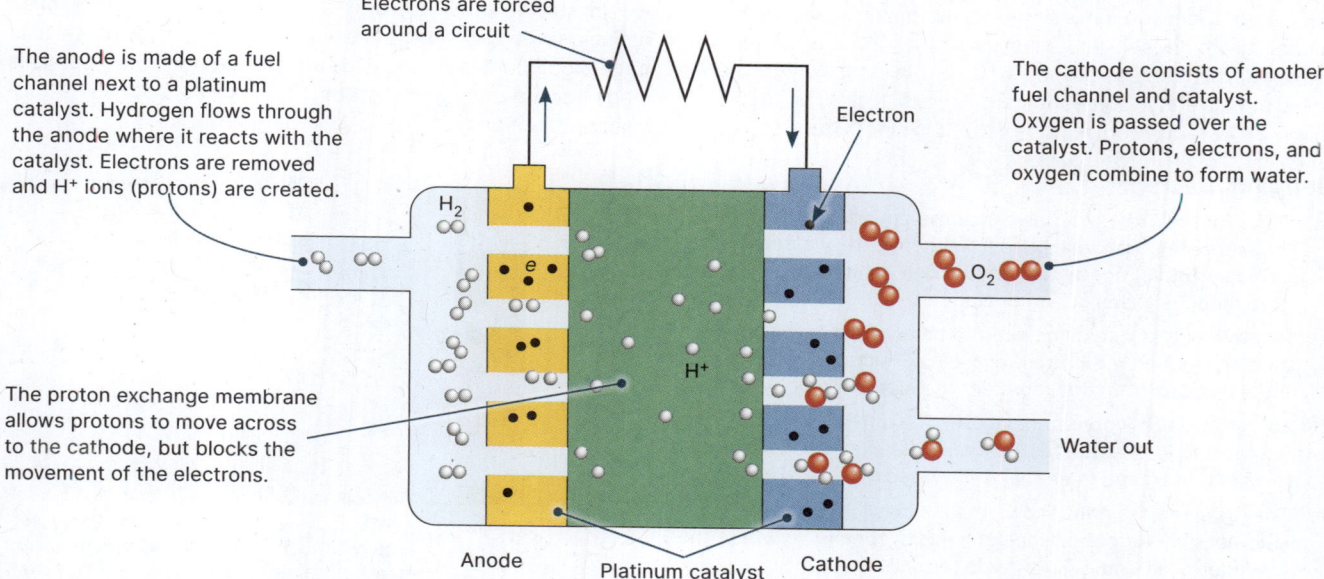

The anode is made of a fuel channel next to a platinum catalyst. Hydrogen flows through the anode where it reacts with the catalyst. Electrons are removed and H⁺ ions (protons) are created.

Electrons are forced around a circuit

The cathode consists of another fuel channel and catalyst. Oxygen is passed over the catalyst. Protons, electrons, and oxygen combine to form water.

The proton exchange membrane allows protons to move across to the cathode, but blocks the movement of the electrons.

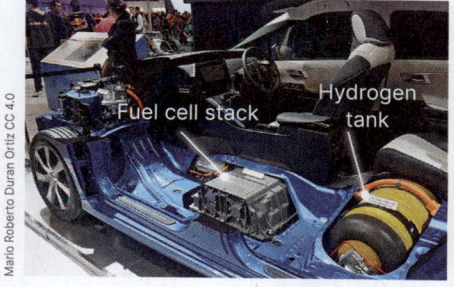

The only emission from hydrogen fuel cells is water. A fuel tank holding the hydrogen fuels the fuel cell which produces electricity for an electric motor. This makes fuel cell-based vehicles a good alternative to conventional combustion engine vehicles.

At the moment, most of the hydrogen for fuel cells is produced by steam reforming from **natural gas**. The rest is produced using electrolysis. This could be sustainable if the energy used to do this is produced using sustainable sources (such as solar PV).

Part of the problem with a hydrogen-based transport system is the amount of energy required to transport the hydrogen to fuelling stations. This adds steps into the fuelling chain and wastes energy, unlike charging stations for battery vehicles.

1. Explain how a hydrogen fuel cell produces electricity: _____

2. (a) What are some advantages of using hydrogen fuel cell vehicles? _____

   (b) Explain why a transport industry based on hydrogen fuel cells is not as efficient as one based on battery electric vehicles: _____

133 Comparing Fuel Choices

Key Idea: Vehicles can be powered by different sources of energy. The most viable option will be the one which delivers the greatest set of advantages and conveniences.

The transport industry accounts for about 20% of global carbon dioxide (CO_2) emissions. It is therefore an important sector to focus on emissions reduction in order to lower global greenhouse gas emissions. In terms of powering vehicles, this centers around what is the best way to fuel a vehicle. Currently, domestic cars, trucks, trains, planes, and ships all use different kinds of fuels refined from oil, each developed to store and release energy in a certain way based on the energy needs and use of the vehicle. Both battery and hydrogen power would serve the needs of all vehicle types without the need for different fuel types (but like other vehicles, different motor and drive train configurations would still be required) and would reduce the greenhouse emissions of the vehicles being used. However, both also have significant costs and disadvantages in their use. This leads to a debate over which fuel type is best for future transport options.

Judging fuel types

▸ Different fuel/energy types cause the emission of different amounts of greenhouse gases. Neither **battery electric vehicles** (BEVs) nor **hydrogen fuel cell vehicles** (HFCVs) emit greenhouses gases during driving.

▸ However, other questions need to be asked: Are there emissions when the battery or fuel cell is made? What about the emissions from the source of the electricity or hydrogen?

▸ To gain a better understanding of each of the three fuel types, we need to look at the full life cycle of each type of vehicle as it is built, used and fueled, and disposed of.

▸ The graph on the right shows the life cycle greenhouse emissions for different types of vehicle, specifying where the fuel/energy and emissions come form.

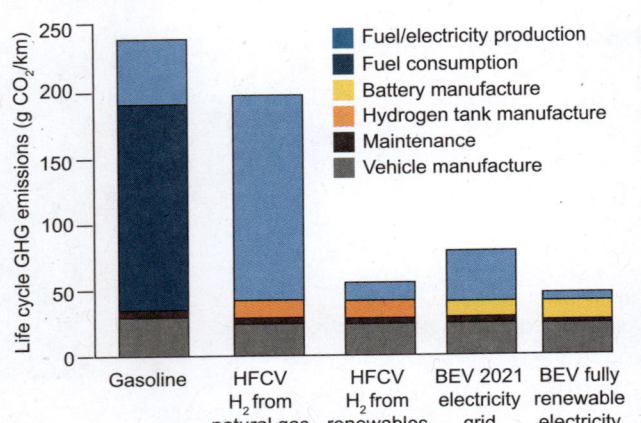

Adapted from: ICCT WHITE PAPER | GLOBAL COMPARISON OF THE LIFE-CYCLE GREENHOUSE GAS EMISSIONS OF PASSENGER CARS

Efficiency

▸ One of the most important aspects of transport is its efficiency, i.e. the energy lost from the fuel to the movement of the vehicle.

▸ The use of gasoline as a fuel is hampered by the laws of thermodynamics. No matter how well an internal combustion engine is engineered, it will never be 100% efficient. Energy is lost as heat from combustion, sound, and overcoming friction. Gasoline powered engines are, at best, about 38% efficient. This becomes even worse if the manufacture and transport of the fuel itself is taken into account.

▸ Battery vehicles are more efficient because the motors have fewer moving parts, and the electricity to charge the battery is transported in power lines.

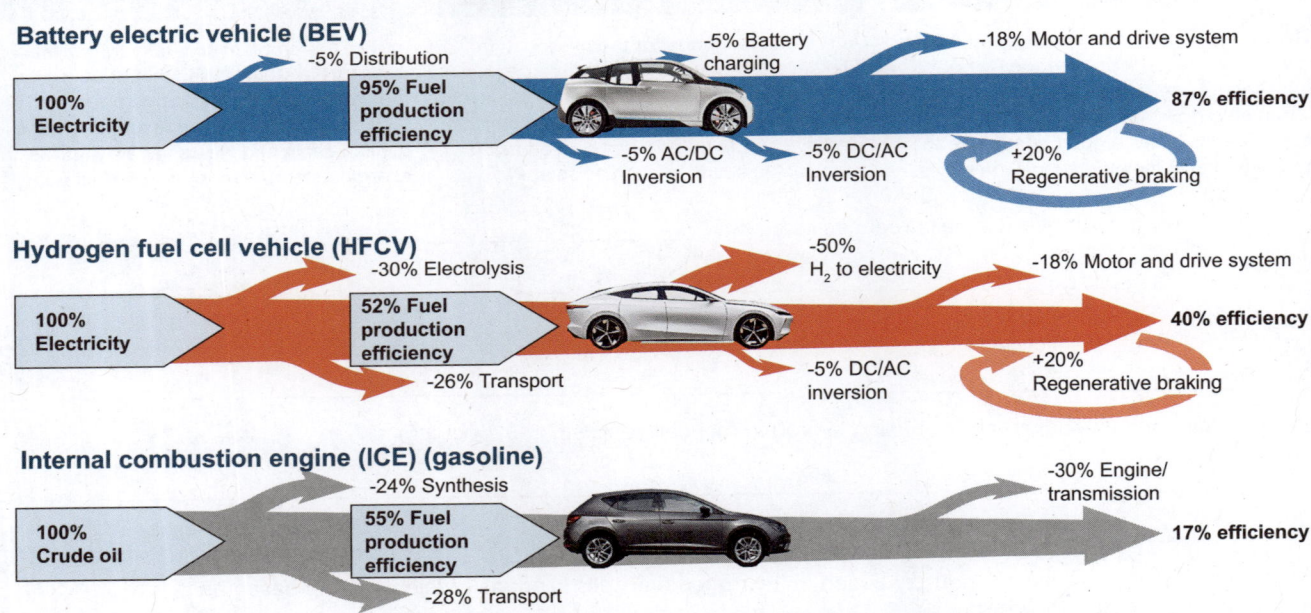

1. Describe what differences in viewpoint of efficiencies might arise between an end-user (owner/buyer) and governments and/or manufacturers in terms of personal vehicle development:

_____

_____

_____

134 Energy Conservation

Key Idea: Conserving energy by changing ways of use and increasing efficiency of appliances can help reduce the need for new energy production facilities.

Energy conservation makes better use of the energy resources we have. Large amounts of potential energy are lost during energy conversions. In the United States, around 63% of potential energy is lost simply in the conversion fuel, e.g. **coal**, into electricity. More is lost through poorly designed or insulated buildings. Traditionally, the solution for our global energy requirements has been to produce more energy, but energy efficiency is becoming more important. Energy efficiency involves improving products or systems so that they do more work and waste less energy, thus conserving energy overall. General improvements in efficiency can be achieved by reducing energy use, improving the energy efficiency of processes, appliances, and vehicles, and increased use of public transport. Energy experts also advise that producing and using the most economical energy sources first, before moving on to more expensive forms, conserves both energy and resources.

Energy efficiency at home

Most of the energy used in domestic or commercial buildings is for heating, air conditioning, and lighting. Most buildings are highly energy inefficient, leaking energy as heat. New technologies and products enable the construction of energy efficient buildings (below) or superinsulated homes, saving the homeowner money and reducing carbon dioxide emissions. Superinsulated homes are often constructed from advanced insulation materials. They are designed to leak no heat, and gain heat from intrinsic heat sources (such as waste heat from appliances or the body heat of the occupants).

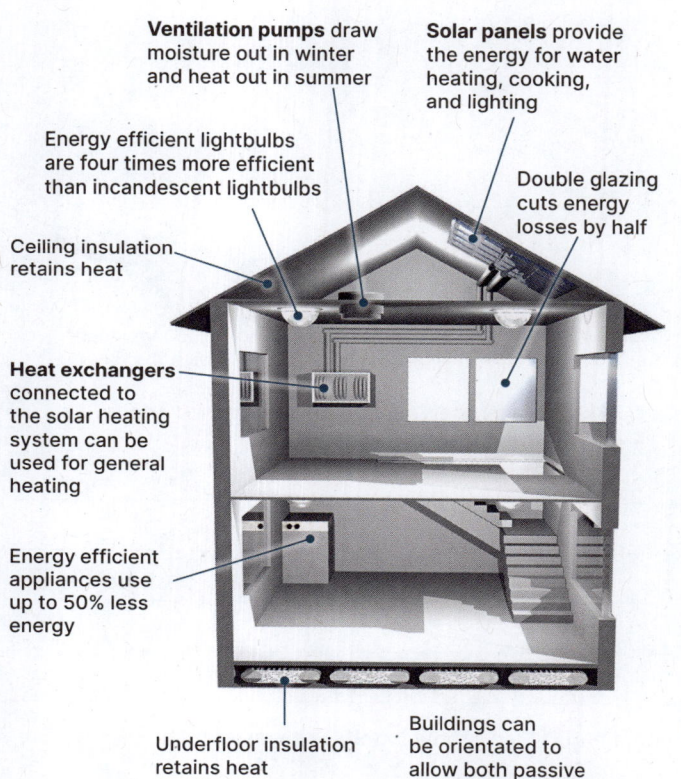

Ventilation pumps draw moisture out in winter and heat out in summer

Solar panels provide the energy for water heating, cooking, and lighting

Energy efficient lightbulbs are four times more efficient than incandescent lightbulbs

Double glazing cuts energy losses by half

Ceiling insulation retains heat

Heat exchangers connected to the solar heating system can be used for general heating

Energy efficient appliances use up to 50% less energy

Underfloor insulation retains heat

Buildings can be orientated to allow both passive warming and cooling

Energy efficiency in transportation

25% of the world's global energy is used for transportation, 80% of which is wasted because it can not be utilized by internal combustion engines. Many countries set fuel economy standards for their vehicle fleets. In the US, this is the Corporate Average Fuel Economy (CAFE). Fuel use targets in L/km decrease every year. To meet these standards, manufacturers must improve the fuel efficiency of their vehicles or retire vehicles that cannot be improved. The use of lighter, stronger materials in car manufacturing, coupled with improved aerodynamics and hybrid and battery vehicles, also aid fuel efficiency.

Hybrid vehicles (below) use two or more different power sources for propulsion, with the combination of combustion engine and electric batteries being the most common. Energy savings are gained by capturing the energy released during braking, storing energy in the batteries, using the electric motor during idling, and using both the gasoline and electric motors for peak power needs (which reduces fuel consumption).

The total **transport efficiency** of a vehicle depends on its fuel efficiency and the number of people it is transporting.

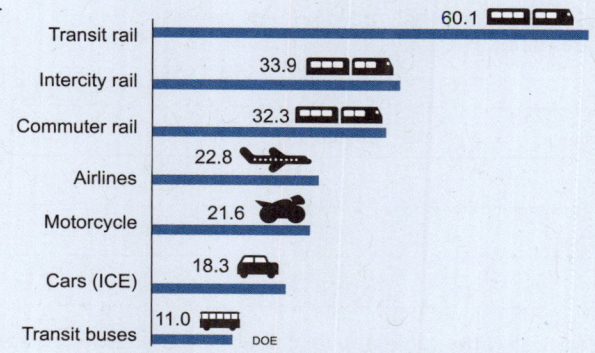

Transit rail	60.1
Intercity rail	33.9
Commuter rail	32.3
Airlines	22.8
Motorcycle	21.6
Cars (ICE)	18.3
Transit buses	11.0

Passenger kilometers per liter gasoline equivalent

1. Discuss methods of increasing energy efficiency in the home: _____

2. How is fuel efficiency different from transport efficiency? _____

LED lightbulbs, an example of efficiency

A move towards more energy efficient systems for industrial and domestic use has been driven by the demand for energy, increasing energy costs, and by a decreasing availability of natural energy resources. Improvements in energy use help to slow the use of resources but can also help the individual save on electrical and heating bills. LED (light emitting diode) light bulbs have replaced incandescent bulbs in everything from household lighting to torches. Reasons include: they are light-weight, use very little power, and can produce extremely bright lights that do not get hot when used. In many countries, incandescent light bulbs have been phased out.

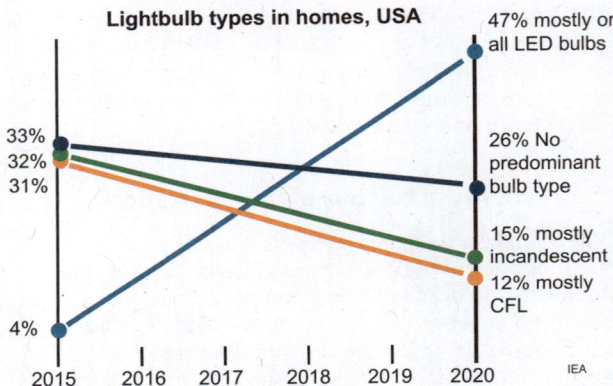

Lightbulb types in homes, USA
- 47% mostly or all LED bulbs
- 26% No predominant bulb type
- 15% mostly incandescent
- 12% mostly CFL

IEA

Energy efficient light bulbs are rapidly becoming the major part of the light bulb market as their price continues to drop. Incandescent bulbs, on the other hand are no longer a significant part of the market.

	Lumens				
	250	400	700	900	1100
Incandescent	25 W	40 W	60 W	75 W	100 W
Halogen	18 W	28 W	42 W	53 W	70 W
Compact fluorescent	6 W	9 W	12 W	15 W	20 W
LED	4 W	6 W	10 W	13 W	18 W

This table shows the light output (measured in lumens) of various light bulb types compared to their power use (**watt**s (W)). Incandescent bulbs last about 1000 hours compared to the 10,000 hours of a compact fluorescent light (CFL) and up to 50,000 hours of a LED light.

Saving money and lowering energy needs

LED bulbs are extremely efficient, converting around 80% of the electricity they use into light. Incandescent bulbs convert only about 5% of electricity in light. Early LEDs suffered from producing bright white light which was harsh on a person's eyes. Current bulbs can be either daylight white or warm yellow, similar to incandescent bulbs. LEDs can also be designed to produce almost any color of the rainbow on command by incorporating different colored LEDs into the bulb and integrating remote or wifi circuitry into the driver circuit. LED bulbs require integrated circuitry (drivers) to work, which increases their cost. These can sometimes fail before the LEDs themselves.

Incandescent bulb (tungsten filament)   CFL bulb (compact fluorescent light)   LED lamp (light emitting diode)

3. Explain why simply building new power stations is no longer seen as an acceptable solution to increased energy demands:

___

4. (a) Using the table above, calculate the energy required to run an LED bulb producing 900 lumens continuously for 2 hours (1 W = 1 joule per second (1 J/s)

___

(b) Compare this with an incandescent bulb producing the same number of lumens:

___

(c) If electricity costs $0.2 per kWh what is the cost of running each of the two types of bulbs?

___

135 Energy Security

Key Idea: Energy security is the ability to provide uninterrupted energy supplies at affordable prices.

The modern world relies on energy being available when needed. This includes electricity and fuel for transport but also access to the energy source, e.g. coal, or water. Energy security centers around being able to supply energy during peace, war, or natural disasters and being able to supply it at prices acceptable to the population. It is important, therefore, that countries are able to supply their own energy needs, or secure uninterruptible energy from other countries. Recent conflicts around the world have shown how vulnerable a country's energy supply can be.

What makes up energy security?

▶ It is easy to forget how reliant we are on easily accessible energy. Yet it only takes a short power cut or a spike in gasoline prices to remind us. The global demand for energy has increased every year, and this trend is likely to continue well into future decades.

▶ Issues in energy security include how to maintain energy supplies into the future as the world transitions to more renewable and low carbon energy supplies, how countries can retain sovereignty and control over their energy resources and supply, and how energy supplies can be maintained in the event of a natural disaster.

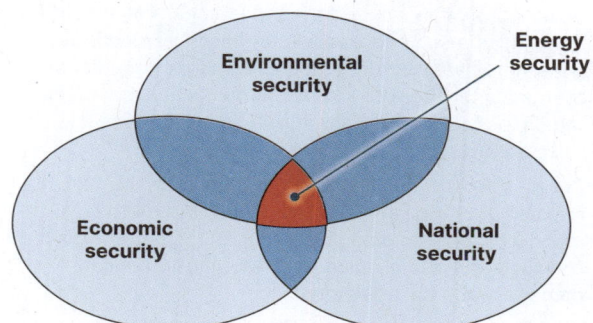

Mineral supply chains will remain important

As the world moves to **renewable energy** generation, it is important to understand and plan for the effects this will have on the energy supply and the need for resources. Although solid fuels such as coal may no longer be needed, minerals for building electric motors, solar panels, and computer chips must be secured. It is also important to understand the effect variable output renewables will have on the robustness of the energy supply chain.

Maintaining sovereignty over one's energy supply is important for countries to maintain their own independence. In 1973 OAPEC embargoed oil to countries supporting Israel, leading to the 1973 Oil Shock and fuel shortages across the Western Hemisphere. Following the 2022 Russian invasion of Ukraine, Russia cut gas supplies to Europe. European nations had to scramble to find new suppliers.

Natural disasters can affect energy supplies in many ways. The 2011 Tōhoku earthquake and following tsunami destroyed Japan's Fukushima nuclear power station, severely disrupting electricity supplies. Even minor natural events, such as storms, can bring down power lines and disrupt electricity transmission. Technological failures, such as the failure of hydroelectric dams or nuclear meltdowns, can both affect energy supply and the environment.

1. Why is energy security important? _____

_____

_____

2. Explain how each of the following may affect a country's energy supply:

   (a) Overseas conflict: _____

   _____

   _____

   (b) Internal natural disaster, e.g. large earthquake: _____

   _____

   _____

3. Why would having a diversity of energy supplies be beneficial for a country?

_____

_____

Space weather - the effect of the solar cycle

▸ Energy security is not only a matter of securing energy supplies. The transmission of electricity and the management of the infrastructure used in the transmission is also important. This includes not only the transmission lines but facilities such as substations where high voltage electricity is transformed to low voltage electricity for use in homes. This infrastructure can be vulnerable to electrical and magnetic disruptions.

▸ The Sun influences everything in the solar system, and that includes Earth-based technology. Charged particles streaming from the Sun can affect the Earth's magnetic field which, in turn, can affect electrical systems on Earth.

▸ The Sun goes through an 11 year sunspot cycle. During this time, sunspots, areas of magnetic disturbance on the Sun's surface, increase then decrease. Solar activity peaks with the sunspot peak.

▸ Solar storms may occur during these peaks in activity. They include solar flares and coronal mass ejections. Solar flares are outbursts of electromagnetic radiation that can interfere with radio communication and cause radio blackouts, especially in the high frequency (HF) radio band (3 to 30 MHz).

▸ During solar storms, charged particles from the Sun may reach the Earth and interact with the Earth's magnetic field, sending pulses of electric current down conducting material, such as power lines. This can damage electrical equipment.

▸ Earth has experienced a number of severe solar storms, the largest being in 1859 (the Carrington Event). Other large events have been recorded in 1972, 1989, 2003, and recently in May 2024.

▸ During these solar storms, power companies may take certain equipment such as power lines and substations off line to prevent possible damage.

▸ During the May 2024 solar storm, power companies in New Zealand took transmission lines out of service, power grid irregularities were reported in the United States, GPS signals were degraded, some satellites failed to transmit data, and auroras where seen at much lower latitudes than normal.

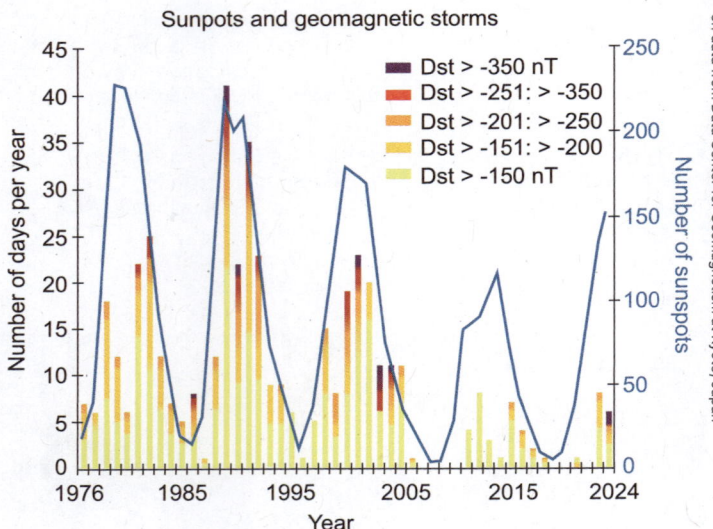

The disturbance storm time (Dst) index is a measure of the intensity of a solar storm and the effect it has on the Earth's magnetic field. The lower the number the weaker the magnetic field and thus the stronger the storm.

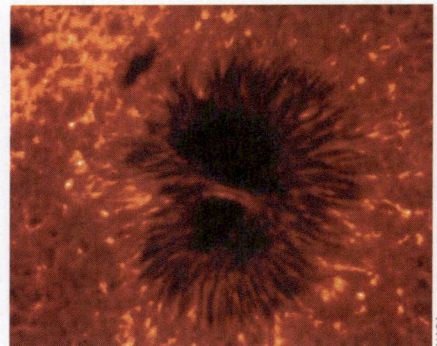

Sunspots come in pairs where the Sun's magnetic field punches through the surface, loops and plunges back through.

Solar flares are sudden releases of electromagnetic energy that can extend thousands of kilometers into space.

Magnetism and electricity interact (via electromagnetism). Disruptions in the Earth's magnetic field affect electrical equipment.

4. Explain how the three aspects of environmental, economic, and national security interact to produce energy security:

_____
_____
_____
_____

5. Why is the monitoring of the Sun's activity important for energy security? _____
_____
_____
_____

136 Energy Storage

Key Idea: Energy storage is an important part of energy security and reliability. It is important on both small and large scales.

We are familiar with energy storage on small scales, in the form of cells, e.g. AA (commonly but incorrectly called batteries) for portable devices such as torches or lead-acid batteries in cars. As **renewable energy** becomes more important, the storage of excess energy on a large scale also becomes important. This helps to add extra power to the grid times when power supply is not able to match increased demand, e.g. solar not working at night or a decrease in wind during certain times of day. Energy storage also provides energy security, protecting against the failure of a production facility and providing energy on demand.

Energy storage and use

The type of energy storage needs to be matched to its use. In most large scale cases, energy storage deals with storing excess potential energy that cannot be stored naturally, or recovering potentially wasted energy. Small scale storage is useful for portable devices and is a convenient way of storing energy for personal use.

Dlouhé stráne pumped storage plant, Czech Republic

Pumped storage ponds can be used to store potential energy. During times when there is excess water in a dam, and energy demand is low, the excess water can be pumped uphill into storage ponds. The water can later be released through turbines to produce extra power when needed.

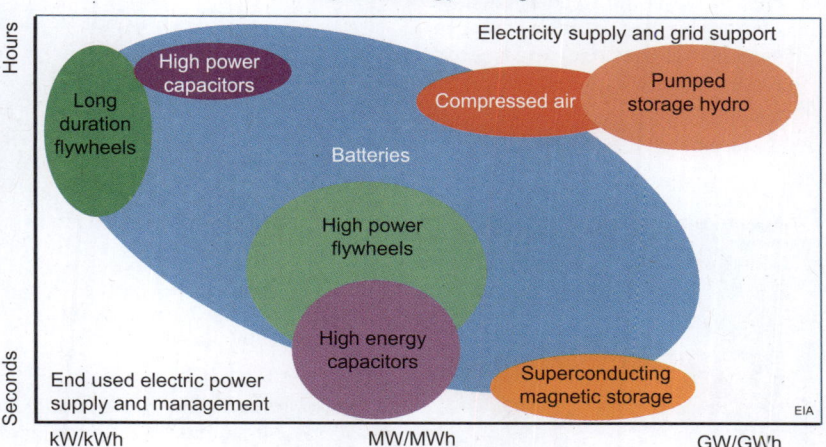

Batteries can be used in a domestic setting to store energy generated from home solar cells (or other generation devices). As domestic solar cells become more common there will be a greater desire to store excess power for use when the sun is not out, and so take greatest advantage of the solar cells.

Pressure air tank used to start diesel motor

Energy can also be stored in mechanical devices, including springs and flywheels. It can also in stored is pressurised vessels, such as air tanks. When the air pressure is released it can be used to power machinery. Compressed springs can release energy quickly to provide starting power for engines or spring loaded weapons (e.g. air rifles).

Underground natural gas storage Alkmaar, Netherlands.

Excess energy can be stored in geological formations. Geological storage methods involve storage of **natural gas** and hydrogen, compressed air, pumped storage, and thermal storage. These can be stored in depleted gas reservoirs, mines, or purpose drill boreholes. Geological storage has the potential to store many GWh of energy.

1. (a) Explain why there is a need for storing energy: _____

_____

_____

_____

(b) Explain why there is a need for several different ways of storing energy: _____

_____

_____

_____

Energy recovery systems

▸ Vehicle transportation produces about 10% of all greenhouse gases, mostly from the combustion of **fossil fuels**. Most of this is wasted as heat, especially during stopping and restarting. Recovering and using some of this wasted energy will effectively increase overall vehicle efficiency. It can therefore reduce transport's contribution to climate change and also reduce a country's overall energy demands.

▸ In vehicles, the kinetic energy recovery system (KERS) is becoming increasingly common. This comes in two forms, either a flywheel or, more commonly, an electric generator to charge a battery. Another method is the heat recovery system that uses hot exhaust gases to power an electric generator.

▸ A KERS can extend the range of the battery by up to 20%.

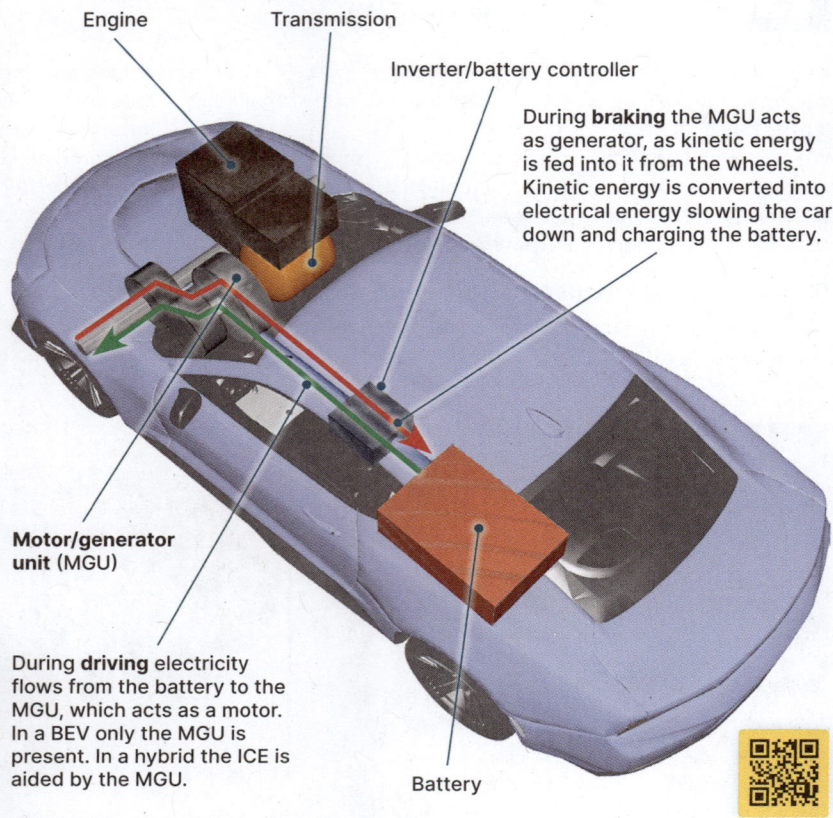

Engine • Transmission • Inverter/battery controller

During **braking** the MGU acts as generator, as kinetic energy is fed into it from the wheels. Kinetic energy is converted into electrical energy slowing the car down and charging the battery.

Motor/generator unit (MGU)

During **driving** electricity flows from the battery to the MGU, which acts as a motor. In a BEV only the MGU is present. In a hybrid the ICE is aided by the MGU.

Battery

In a flywheel system, the flywheel (left) is connected to the transmission. Kinetic energy from the wheels is transferred to the flywheel via the transmission. The flywheel can spin up to more than 60,000 RPM. When the car accelerates, kinetic energy from the flywheel to transferred back to the transmission. This system is entirely mechanical, and requires no heavy battery.

2. Why is regenerative braking an important development in vehicles?

_____
_____
_____

3. Explain how the different kinetic energy recovery systems work:

_____
_____
_____
_____
_____
_____

4. Why is a KERS never able to recover all the energy from a moving vehicle when engaged?

_____
_____

5. Identify a problem with storing energy from a primary source (i.e. a powerstation) instead of using it immediately:

_____
_____
_____
_____

137 Rechargeable Batteries and Energy Storage

Key Idea: Large capacity batteries are becoming important as storage for excess energy for electricity grids and as vehicle power sources.

Rechargeable batteries are important for storing excess generated energy or for powering electric vehicles. They are increasingly becoming a key part of storing energy for national grids from large scale power facilities. Battery technology currently centers around lithium-ion batteries. However, sodium-ion batteries are in development. These are cheaper as sodium is abundant and easy to obtain, and they are somewhat safer than lithium ion batteries. Sodium-ion batteries also charge faster than lithium-ion batteries. However, sodium-ion batteries have a much lower energy density (the amount of energy stored per mass (J/kg)).

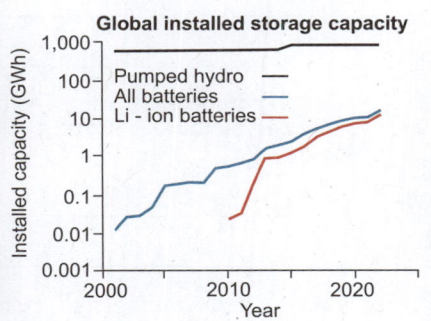

Large scale energy storage can be done using industrial scale battery storage power stations (above). These are useful because not only do they store large amounts of energy (around 1-2 GWh), but they can transition from standby to peak output (around 300MW) within seconds and maintain output for several hours. For these facilities, sodium-ion batteries are appropriate because space requirements and their lower energy density (compared to lithium-ion batteries) is not a major concern.

Cars have limited space for batteries and the lighter a car is, the more efficient it is. Lithium-ion batteries are therefore used in cars because of their high energy density. This reduces battery size and increases a car's range before recharging. However, it makes damaged batteries extremely dangerous. Some car manufacturers are now using sodium-ion batteries in their compact cars. This makes the cars much cheaper but reduces their range, although if used as town cars, this is generally not a problem.

Pumped hydro storage is still by far the major energy storage medium for electricity grids. It is, however, expensive to install and requires large areas e.g. lakes, that can be flooded and then drained often and rapidly. Battery storage has grown rapidly since the turn of the century. Lithium-ion batteries have led this increase since around 2010, although sodium-ion battery use is also beginning to increase. Note the y axis of the graph above is logarithmic.

Effect of battery storage on power grids.

Industrial scale battery storage power stations have become an important part of the electricity grid, especially as renewables such as wind and solar power continue to grow in use. Excess energy produced by solar power during the day can be used to charge batteries. When the sun goes down at the end of the day, the batteries' energy is added to the grid, reducing the need for **fossil fuels** to maintain the energy demands of users.

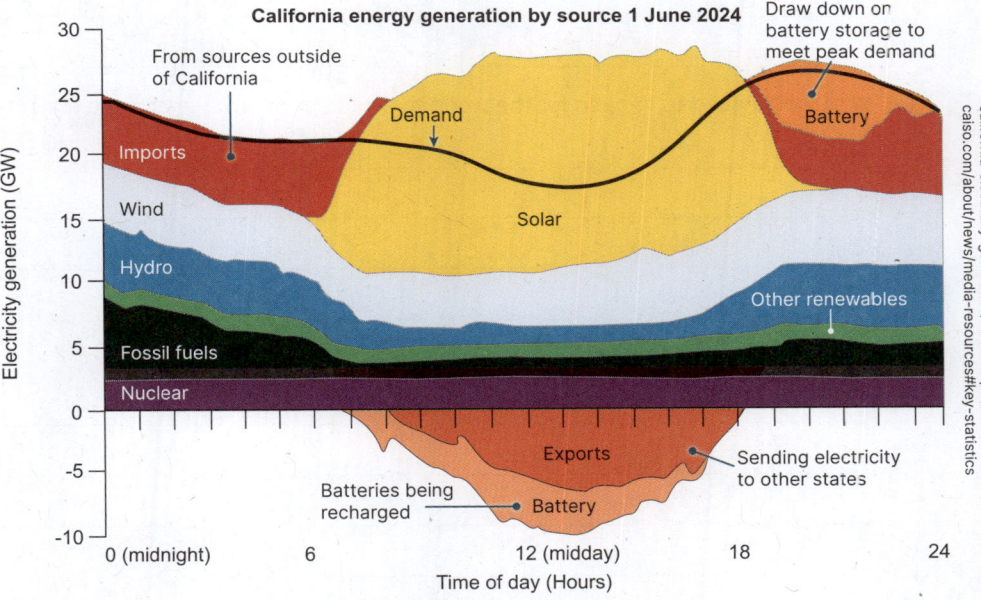

Data source

1. Explain the use of battery energy storage and production facilities in electrical grids: _____

_____

_____

2. Explain the importance of energy density: _____

_____

_____

138 Did You Get It?

1. Provide a general description of how a power station fueled by coal or oil produces electricity:

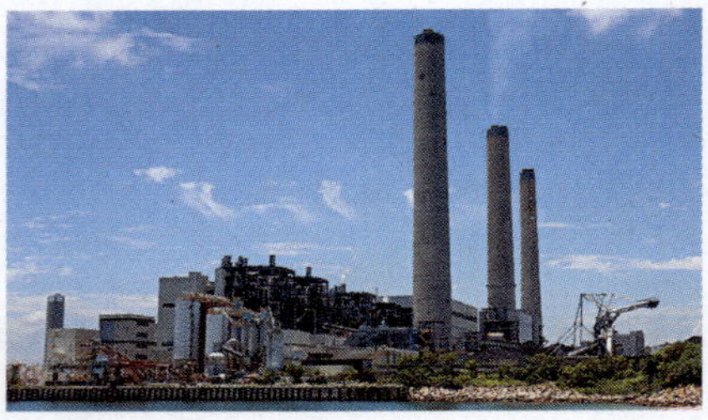

   _____
   _____
   _____
   _____

2. (a) From the graph on the right, compare the % of electricity produced by the US, China, and the world using fossil fuels:

   _____
   _____

   (b) From the graph on the right, compare the % of electricity produced by the US, China, and the world using renewable energy:

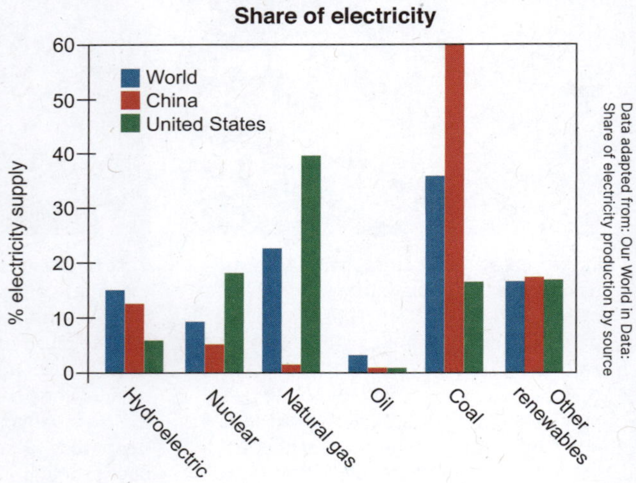

   _____
   _____

3. (a) Explain why fossil fuels are still the favoured fuel source for most of the world:

   _____
   _____
   _____
   _____
   _____

   (b) What would need to happen for the world to switch to renewables for the majority of their fuel?

   _____
   _____
   _____
   _____

4. Explain why energy efficiency is an important part of future energy planning:

   _____
   _____
   _____
   _____

5. Define each of the following:

   (a) Fission: _____
   _____

   (b) Renewable energy: _____
   _____

   (c) Efficiency: _____
   _____

Types of Pollution

Air pollution
- Sources of air pollution
- Formation and effects of smog
- Major air pollutants

Water pollution
- Sources of water pollution
- Effects on the environment
- Oil spills

Ozone depletion
- Formation of stratospheric ozone
- Depletion by CFCs
- Environmental effects of ozone depletion

Impacts and Treatments

Treating pollution
- Sewage treatment and disposal
- Waste management systems
- Monitoring water quality

Impacts of pollution
- Clean up costs
- Environmental effects
- Health effects

Disasters
- Environmental and industrial disasters
- Cyclones and hurricanes

Pollution has wide ranging and detrimental effects on aquatic and terrestrial environments.

Treating the environmental and health effects of pollution is costly and often difficult.

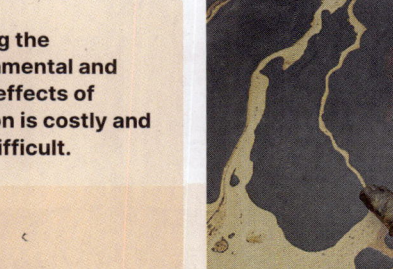

Global Change

Human activities have a global effect. Pollution and habitat destruction continue to damage vast areas of the Earth.

Strategies aimed at slowing these global changes are being implemented in places, but it could be many years before the benefits are seen

Human induced greenhouse gas emissions are leading to global climate change.

The harmful effects of human activity can be prevented or reduced through conservation.

Climate change
- Physical science
- Impacts of climate change
- Adaptation, mitigation, and legislation

Tipping points
- Polar and boreal regions
- Boreal and Amazon forest
- AMOC and subpolar gyres
- Coral reefs

Loss of biodiversity
- Habitat loss
- Exploitation
- Endangered species

Maintaining biodiversity
- Conservation efforts
- Nature reserves
- Laws and treaties

Climate Change

Conserving Biodiversity

SINGLE-USE SOCIETY

Rubbish left in the environment is a visible reminder that human activity has caused change in almost every part of the globe. Through remarkable industrial advancements, humans have created products to fulfil almost all needs and desires. However, humans readily dispose of objects that are considered easily replaceable. It is estimated that annual global waste production exceeds 2 billion tonnes, with nearly 40% of it being deposited in open landfills. Shockingly, 99% of products are discarded within 6 months of purchase. As our planet becomes increasingly burdened with refuse, it is imperative that we explore innovative solutions to minimize our environmental impact.

Anthropogenic evidence

Coots build floating nests consisting of twigs and other floating debris. They have been observed using plastic to construct nests because of the abundance of this waste product in our waterways,

A circular economy

More than the 3 Rs, the concept of the circular economy is all about keeping materials in circulation. This can be via methods such as maintenance, reuse, refurbishment, remanufacture, recycling, and composting. By adopting a circular way of using and reusing, there will be less need to keep exploiting Earth's resources for new products.

Out of sight, out of mind

The process of decomposition breaks down the chemical bonds that hold materials together. These materials are converted to simpler substances but some of these are greenhouse gases which are detrimental.

🔍 Take a Deeper Look

▶ Investigate any aspect of rubbish presented on this page and consider how it may be relevant to you, your school, or your community.
▶ Share your findings as a poster, digital media, or a form that sparks your creativity.

Chapter 7 Global Change
Pollution

Resource Hub
bit.ly/4czmNfk

Key Terms

- acid rain
- air pollution
- biomagnification
- endocrine disrupting chemicals (EDCs)
- environmental disasters
- environmental legislation
- environmental remediation
- eutrophication
- forever chemicals
- heavy metals
- human activity
- light pollution
- nanoplastic (microplastic)
- noise pollution
- nuclear accidents (pollution) persistent organic pollutants (POPs)
- oil spill
- ozone depletion
- photochemical smog
- plastic pollution
- pollutants
- pollution
- radioactive waste
- recycling
- sewage / sewerage
- stratospheric ozone
- waste management
- water pollution

Key Concepts

- Pollutants originate from various sources and activities.
- Pollution incurs environmental, health, and economic costs.
- Numerous pollutants have long-term impacts on both health and the environment.
- By minimizing the consumption of materials and waste, pollution can be significantly reduced.

Types of pollution and their effects

Learning Outcomes:

		Activity Number
☐ 1	Describe the main sources of environmental pollution. Explain how water becomes polluted.	139-140
☐ 2	Name the different forms of nitrogen that act as pollutants and the sources of these. Define 'eutrophication', 'Biological Oxygen Demand', and 'biomagnification.' Explain the process of eutrophication and how BOD is used to assess water quality. Explain how biomagnification causes the build up of toxins in organisms.	141-143
☐ 3	Describe the stages that waste goes through in a typical sewage treatment plant.	144
☐ 4	Discuss measures for reducing pollutants. Explain the meaning of the term 'circular economy' and describe how it can help reduce waste.	145
☐ 5	Explain why plastics are so widely used by humans and they are a major pollutant. Describe some technological approaches to reducing plastic pollution in the environment.	146
☐ 6	Describe why plastic pollution in the ocean is a danger to marine organisms. Describe the difference between micro- and nano-plastics. Explain why bioaccumulation and biomagnification are a problem with these plastics.	147-148
☐ 7	List the main sources of air pollution and distinguish between natural pollution and human caused pollution. Explain the causes and effects of acid rain. Describe steps that can be taken to reduce air pollution and actions taken by humans to mitigate ozone depletion.	149-153
☐ 8	Describe sources of noise pollution in both rural and urban areas. Name sources of pollution found in homes and some effects on health caused by them. Explain why light pollution is an issue for both humans and other organisms. Outline the effects different pollutants have on human health.	154-157

Environmental disasters and remediation

☐ 9	With reference to specific examples, outline why oil spills cause wide scale environmental pollution. Describe the various ways in which oil spills are cleaned up. Outline, in general terms, the effects of fossil fuels on human health throughout the lifecycle of extraction and use.	158-160
☐ 10	Describe the reasons for the Fukushima Daiichi nuclear disaster of 2011 and compare its effects to the Chernobyl disaster of 1986.	161
☐ 11	Describe the reasons for the disasters in Bhopal, Aberfan and Brumadinho dam and the effects on the local communities.	162-163
☐ 12	Explain what is mean by environmental remediation and describe some methods by which this can occur. Compare and contrast *ex-situ* and *in-situ* remediation.	164
☐ 13	Explain why the economic costs of pollution can be difficult to quantify, Compare direct and indirect costs. Describe examples of legislation that aim to protect humans and the environment from pollution.	165-166

139 Types of Pollution

Key Idea: Land, water, and air can become polluted by many human activities.

Pollutants added to the air, water, soil, or food that threatens the survival, health, or activities of organisms is called **pollution**. Pollutants can enter the environment naturally, e.g. from volcanic eruptions, or through **human activities**. Most pollution from human activity occurs in or around urban and industrial areas and regions of industrialized agriculture. Pollutants may come from single, identifiable point sources such as power plants, or they may enter the environment from diffuse or non-point sources, such as through land runoff. While pollutants often contaminate the areas where they are produced, they can also be carried by wind or water to other areas. The most commonly recognized forms of pollution are **air pollution**, including **ozone depletion** in the stratosphere, and **water pollution**. However, other less obvious forms of pollution, including **light pollution** and **noise pollution**, are also the result of concentrated regions of human activity.

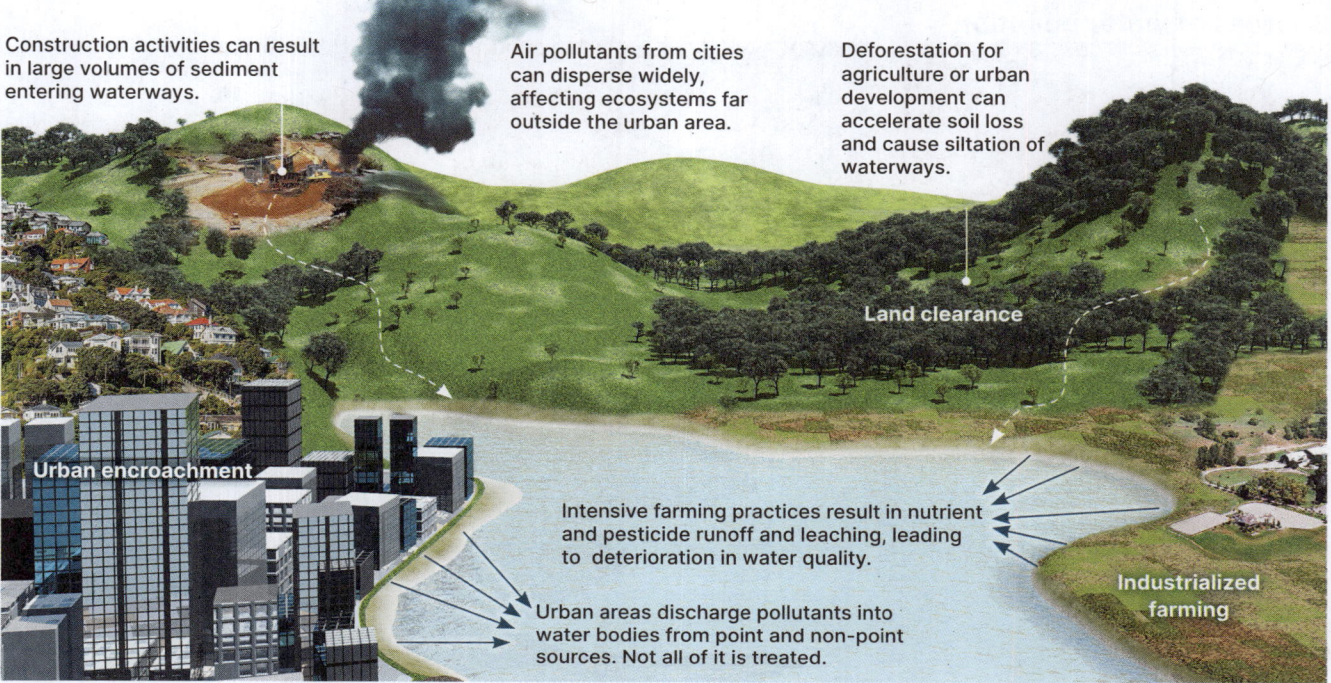

Construction activities can result in large volumes of sediment entering waterways.

Air pollutants from cities can disperse widely, affecting ecosystems far outside the urban area.

Deforestation for agriculture or urban development can accelerate soil loss and cause siltation of waterways.

Land clearance

Urban encroachment

Intensive farming practices result in nutrient and pesticide runoff and leaching, leading to deterioration in water quality.

Urban areas discharge pollutants into water bodies from point and non-point sources. Not all of it is treated.

Industrialized farming

Fertilizers, herbicides, and pesticides are major contaminants of soil and water in areas where agriculture is industrialized. Fertilizer runoff and leaching adds excess nitrogen and phosphorus to both ground and surface water, leading to **eutrophication**.

Soil contamination occurs via chemical spills, leaching, or leakage from underground storage. Runoff from mining and metal processing operations can carry **radioactive waste** and **heavy metals** such as mercury, cadmium, and arsenic.

Together with vehicle exhausts, power plants and industrial emissions are a major source of air pollution. SO_2 and NO_2 from these primary sources mix with water vapor in the atmosphere to form acids, deposited as **acid rain**, snow, or dry acid.

1. Identify the main sources of each of the following pollutants:

    (a) Pesticides and herbicides in waterways: _____

    (b) Sewage: _____

    (c) Oxides of sulfur and nitrogen: _____

    (d) Sedimentation and siltation of waterways: _____

2. Explain the impact of urbanization on the pollution load in a given region: _____

140 Water Pollution

Key Idea: Water pollution can come from various sources, including urban, industrial, and agricultural activities.

Pollutants may first enter the groundwater where they are difficult to detect and manage. Some enter surface waterways directly through runoff from the land, but most are deliberately discharged at single (point) sources. Some pollutants alter the physical state of a water body, i.e. its temperature, pH, or turbidity (a measure of cloudiness). Others involve the addition of potentially harmful substances. Even substances that are beneficial at a low concentration may cause problems when their concentration increases. One such form of **pollution** involves excessive nutrient loading of waterways by organic effluent, such as untreated sewage, containing human waste and detergents. Developing global and national initiatives to control **water pollution** is important because many forms of pollution cross legislative boundaries. Water conservation is required to enable more effective use of water, reduce the burden on wastewater systems, decrease pollution of surface and groundwater, and slow the depletion of aquifers.

Sources of water pollution

Sediment pollution: Erosion causes soil particles to be carried into waterways. In addition to erosion destroying topsoil, the increased sediment load may cause choking of waterways, build-up behind dams, and the destruction of aquatic habitats.

Sewage: Water contains human waste, and soaps and detergents from bathrooms, laundries, and kitchens. If this sewage is discharged into waterways without proper treatment, it can potentially cause disease outbreaks.

Thermal pollution: Many industrial processes, including electrical power generation (above), release heated water into river systems after burning coal or other fuel. The increase in water temperature reduces oxygen levels and may harm the survival of river species.

Inorganic plant nutrients: Fertilizer runoff from farmland adds nitrogen and phosphorus to nearby waterways. This nutrient enrichment accelerates the natural process of **eutrophication**, causing algal blooms and prolific, aquatic weed growth. This can lead to the suffocation of aquatic animals, and pollutes water for human use.

Detecting water pollution

Water pollution can be monitored in several ways. The nutrient loading can be assessed by measuring Biological Oxygen Demand (BOD). Electronic probes and chemical tests can identify the absolute levels of various inorganic pollutants, e.g. nitrates, phosphates, and **heavy metals**. The presence of indicator species (below) can give an indication of the pollution status for a waterway. This method relies on an understanding of the tolerance levels to pollution of different species that should be living in the waterway, e.g. worms, insect larvae, snails, and crustaceans.

Indicator species:

| Mayfly nymph | Dragonfly nymph | Freshwater shrimp | Hoglouse | Tubifex worm |

Low water pollution → High water pollution
Most sensitive → Least sensitive

- India is the world's largest user of water, mainly due to population size. Around 70% of the surface water is too polluted for human consumption.
- Turkmenistan is by far the largest per capita user of water, using 4x that of the US, mostly due to growing agricultural crops despite increasing desertification. The country faces significant environmental water issues because of irrigation runoff.

1. Describe three uses of water for each of the following areas of human activity:

   (a) Domestic and urban use: _____

   _____

   (b) Industrial use: _____

   _____

   (c) Agricultural use: _____

   _____

   _____

Water pollution events from accidents and disasters

Fukushima nuclear power plant, Japan.

Abandoned sulfide ore mine, Portugal

Flooded village in Northern Bangladesh

Nuclear accidents: **Radioactive waste** can be released from nuclear power stations, as happened in 2011, in Fukushima, Japan. Water came into contact with radioactive material. This contaminated water was discharged into the sea, accidentally, during the event and later, on purpose, to disperse it.

Mine runoff: Acid drainage from mines and **acid rain** can severely alter the pH of waterways. Runoff from open-cast mining operations can be loaded with poisonous **heavy metals** such as mercury, cadmium, and arsenic, or sulfides, which can cause severe nerve damage and other human health problems.

Flooding events: During flooding events near human populations, disease-causing microbes from infected animals and humans can enter waterways, posing a significant risk. This can lead to drinking water being contaminated with human waste, resulting in the spread of diseases like cholera.

Oil tanker spill cleanup, 2020, Mauritius

Organic compound spills: Synthetic, often toxic, compounds may be released into waterways from oil spills, and accidental spills of waste products of manufacturing and farming processes, e.g. dioxin, PCBs, phenols, and DDT. These can build up in the food chain through **biomagnification** and poison human consumers.

2. Why might countries allow industries to operate despite the risk of water pollution from their activities:

_____
_____
_____
_____
_____

3. Suggest what is likely to happen to the regulations controlling industry after a major pollution accident such as an oil spill or environmental disaster?

_____
_____
_____

4. Sewage effluent may be sprayed onto agricultural land to irrigate crops and it eventually drains into waterways.

   (a) Describe an advantage of utilizing sewage in this way: _____
   _____

   (b) Describe a major drawback of using sewage effluent in this way: _____
   _____

   (c) Suggest an alternative treatment or use of the effluent: _____
   _____

5. When studying aquatic ecosystems, the species composition of the community (its biodiversity) in different regions of a water body over time is often recorded. In general terms, suggest how a change in species composition of an aquatic community could be used to indicate water pollution:

_____
_____

6. What events are likely to compromise safe drinking water important for human populations, such as in India or Turkmenistan, and what are the consequences?

_____
_____
_____

141 Nitrogen Pollution

Key Idea: Excess nitrogen accidentally or purposefully released into the environment acts as a pollutant.

The effect of excess nitrogen compounds on the environment is varied. Depending on the compound formed, nitrogen can cause smog in cities or algal blooms in lakes and seas. Nitrogen gas makes up almost 80% of the atmosphere but is unreactive at normal pressure and temperature. At the high pressures and temperatures reached in factories and combustion engines, nitrogen gas forms nitric oxide along with other nitrogen oxides, most of which contribute to atmospheric **pollution**. Nitrates spread onto pastures as fertilizers are washed into groundwater by rain and slowly make their way to lakes and rivers and eventually out to sea. This process can take time to become noticeable as groundwater can take many decades to reach a waterway. In many places where nitrate effects are only just becoming apparent, such as in Germany, the immediate cessation of their use could take a long time to have any effect. It might take many years before the last of the ground water carrying the nitrates reaches a waterway, leading to **eutrophication**.

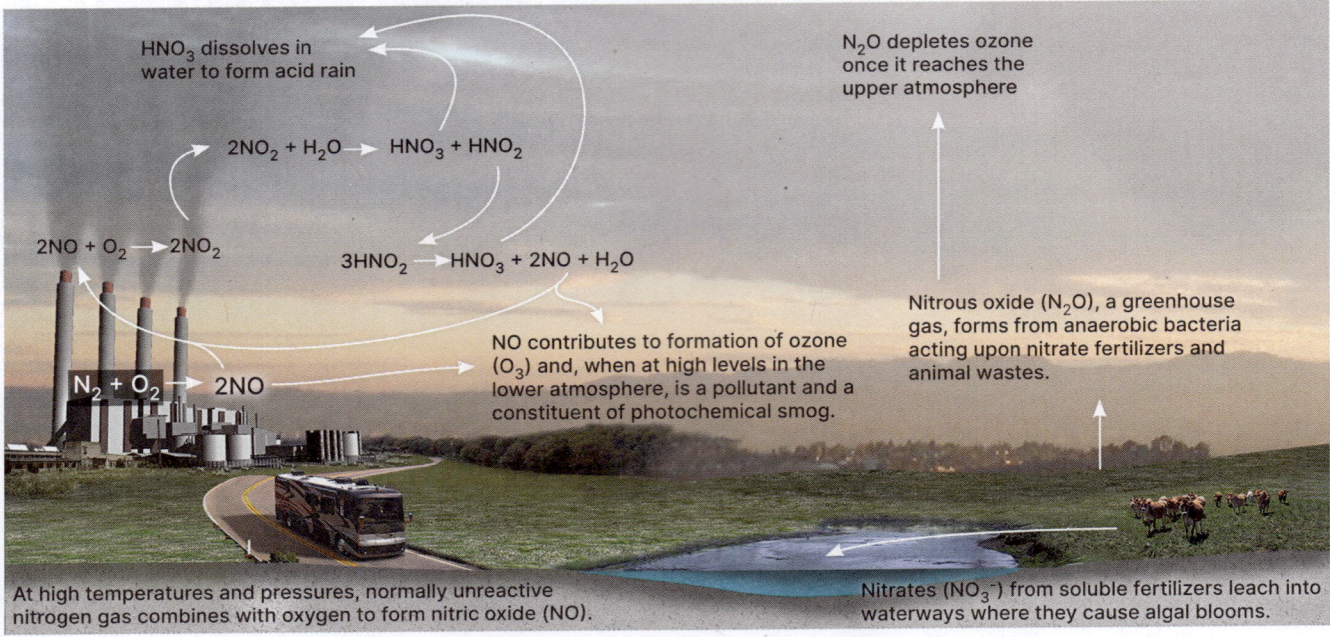

Hole in the pipe model of nitrogen inefficiency

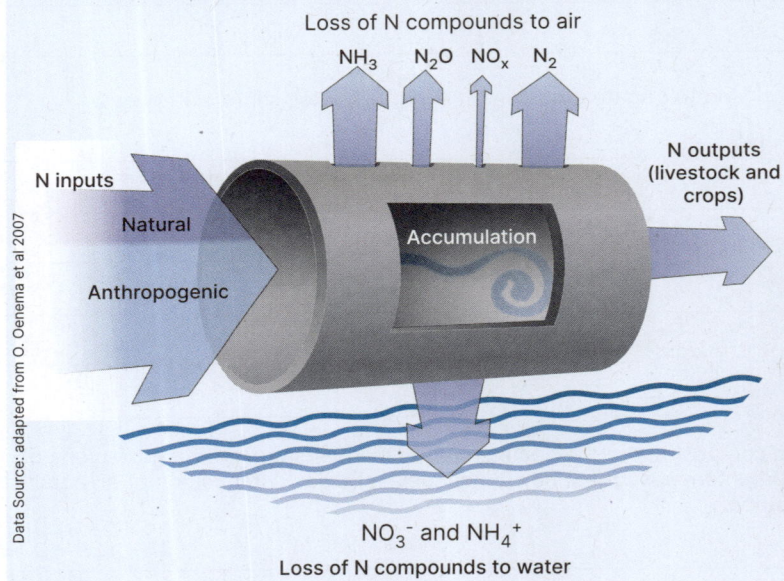

The 'hole in the pipe' analogy (left) represents the inefficiencies in nitrogen fertilizer use. It illustrates that, when nitrogen is added to the soil as fertilizer and is not immediately taken up by plants, it can be lost through two main pathways: leaching into waterways or being released into the air through bacterial action. This analogy highlights the potential environmental impacts of nitrogen pollution and emphasizes the need to minimize these losses through better fertilizer management practices.

Excessive nitrogen contributes to algal blooms in both coastal and inland waters, Florida

1. Describe the effect each of the following nitrogen compounds has on air and water quality:

   (a) NO: _____

   (b) N_2O: _____

   (c) NO_2: _____

   (d) NO_3^-: _____

The influence of the Haber process

- Early last century, the Haber process made nitrate fertilizers readily available for the first time. Since then, the use of nitrogen fertilizers has increased at an almost exponential rate.
- After 1940, the amount of food grown would not have been able to support the increasing human world population without use of synthetic fertilizer (see graph below). Significantly, this has led to an increase in the levels of nitrogen in land and water by up to 60 times those of 100 years ago.
- This extra nitrogen load is one of the causes of eutrophication of lakes and coastal waters.
- The rate at which nitrates are added to the water has increased faster than the rate at which nitrates are returned to the atmosphere as unreactive N_2 gas. This has led to the widespread accumulation of nitrogen.
- **Environmental legislation** measures introduced in the European Union in 1990 have helped to reduce use of nitrogen fertilizer.

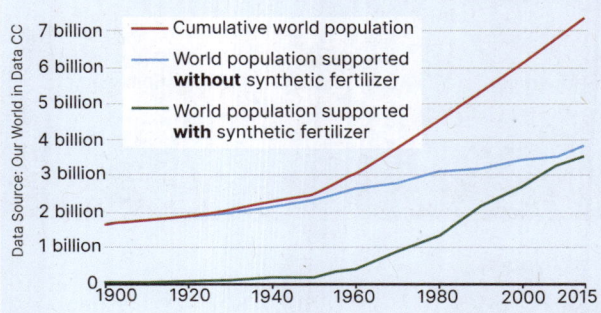

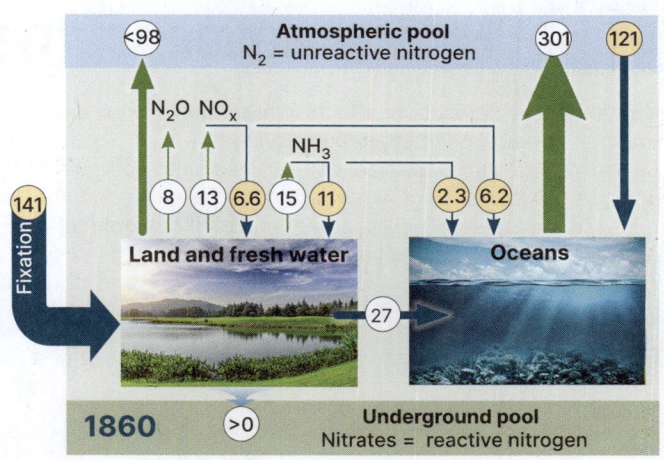

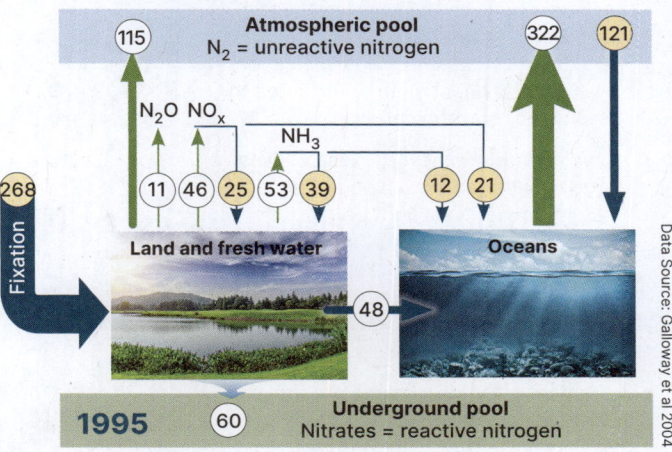

Changes in nitrogen inputs and outputs between 1860 and 1995 in million tonnes.

2. Explain why nitrogen fertilizer synthesized through the Haber process is necessary for humans but also detrimental to the environment:

___

3. (a) Calculate the increase in nitrogen deposition (input) to the oceans from 1860 to 1995 and compare this to the increase in release of nitrogen from the oceans:

___

(b) What effects are likely to be caused by nitrogen pollution in the oceans and waterways?

___

(c) What trend in nitrogen inputs and outputs would you expect continuing to and beyond 2024? Justify your answer:

___

142 Eutrophication and Water Quality

Key Idea: Eutrophication affects water quality and causes damage to aquatic and marine ecosystems.

Excess nitrogen from fertilizer use can leach into groundwater and run off into surface waters. This extra nitrogen load is one of the causes of increased enrichment (**eutrophication**) of lakes and coastal waters. An increase in algal growth increases decomposer activity, depleting oxygen and leading to the death of fish and other aquatic organisms. At the same time, eutrophic conditions allow undesirable species, tolerant of low oxygen, to increase in numbers. This can permanently disrupt ecosystem stability, removing food sources for many other animals. Many aquatic microorganisms also produce toxins. These can accumulate in the water, fish, and shellfish by **biomagnification** (see activity 143). The cycling of nitrates back into the atmosphere as N_2 gas has become unbalanced, and nitrogen is accumulating in the soil and water.

Eutrophication causes

▸ Eutrophication can occur naturally but is usually the result of **human activity**. Discharge or runoff of leached nitrate or phosphate-containing detergents, fertilizers, or **sewage** into a waterway are the main causes of eutrophication. Phosphorus enrichment contributes to freshwater eutrophication.

▸ The high nutrient levels cause excessive algal growth (an algal bloom). The algal bloom prevents sunlight penetrating far beneath the water's (turbid) surface and aquatic plants (macrophytes) begin to die because they cannot photosynthesize. Oxygen levels begin to fall at lower water levels.

▸ Eventually, the dead algae are broken down by microbes through the process of decomposition. During this process, oxygen is consumed, leading to a decrease in oxygen levels in the water, resulting in a condition known as hypoxia.

▸ The use of oxygen by decomposing microorganisms is known as biochemical oxygen demand (BOD). A high BOD can deprive other organisms of oxygen, causing the habitat to become inhospitable and leading to the death or migration of organisms.

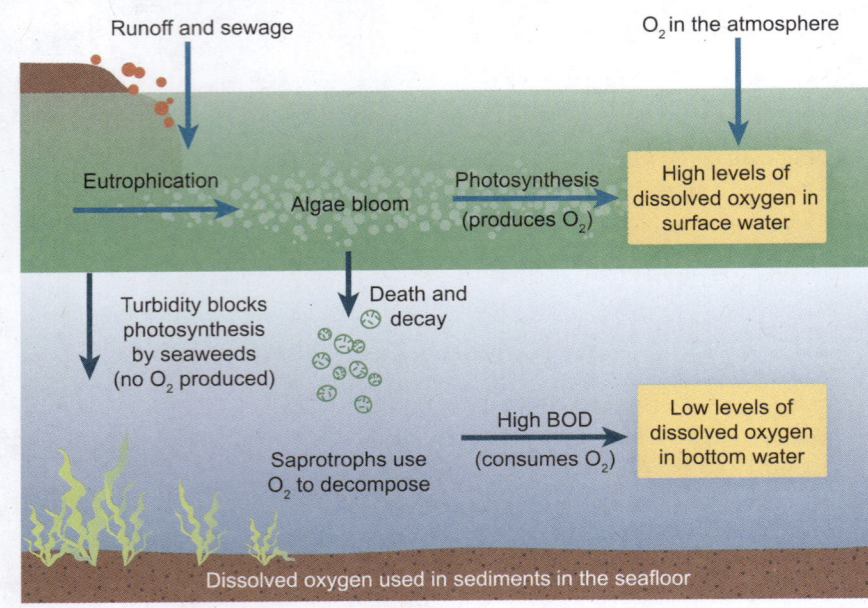

Oligotrophic waterway (nutrient poor, high oxygen, few organisms)

Hypereutrophic waterway (nutrient rich, low oxygen, high density of organisms)

Waterways with very low primary production due to correspondingly low nutrient levels are known as oligotrophic. They are often found near the top of watershed areas, such as in mountains, and are clear, often deep, with little evidence of aquatic organisms. Nutrient rich waterways have high primary production, are often murky or green with algae and, in an extreme state, known as hypereutrophic. They are in areas with run-off, usually agricultural. Waterways with rich biodiversity have nitrogen levels somewhere between these two states.

1. Describe the causes and consequences of eutrophication: _____

_____

_____

_____

_____

2. Describe the differences between an oligotrophic waterway and a eutrophic waterway: _____

_____

_____

_____

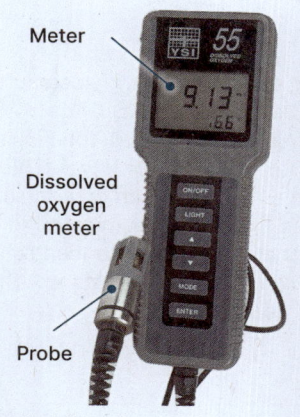

Biological Oxygen Demand

The amount of polluting organic material in a water body can be inferred by the biological oxygen demand (BOD). This is a measure of how much oxygen the organic material is using for its decomposition. BOD is measured as the weight (mg) of oxygen used by one liter of sample effluent stored in darkness at 20°C for 5 days. The more oxygen that is used, the greater the bacterial activity, and (therefore) the greater the pollution. A high BOD results in less oxygen being available for aquatic organisms such as fish and invertebrates, which will die, or shift away from the polluted area. It is a good measure of **water pollution** extent.

Dissolved oxygen (DO) varies with depth in a lake and from place to place in a stream or river, e.g. DO is lower above a waterfall than below it. At 20°C, the maximum DO is 9.07 mg L-1.

In order for a BOD to be significant, it needs to be compared to BODs taken from other parts of the stream or lake, and to a control. Two samples are taken at each site: one to be tested for DO immediately, the other to be tested after 5 days of storage.

If organic matter is discharged into a stream, the BOD increases markedly immediately after the point of discharge (right).

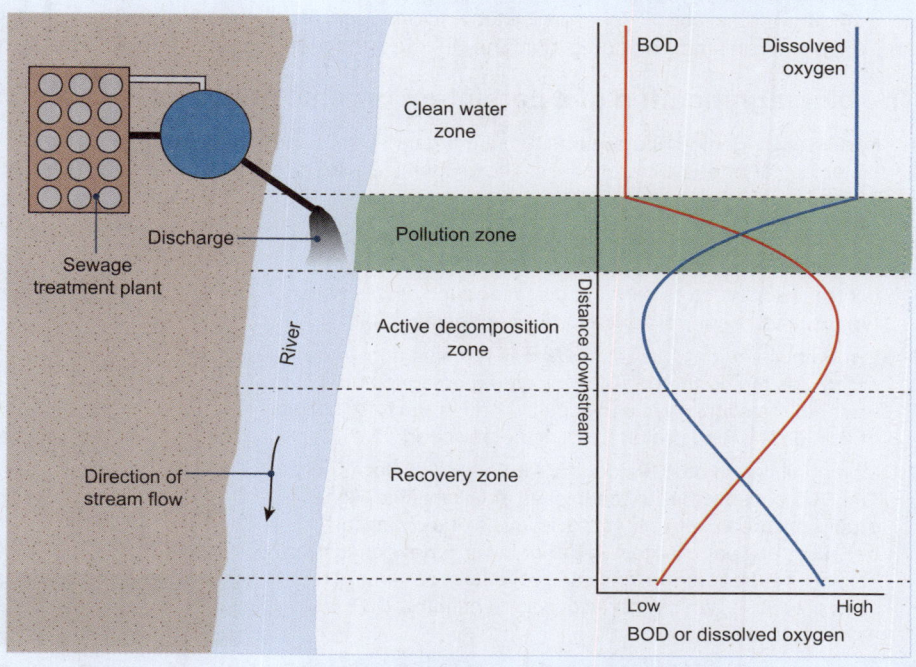

3. What is the link between BOD and dissolved oxygen in water: _____

4. Discuss the relationship between water quality and land use. Relate to use of nitrogen and phosphate fertilizers: _____

5. How could a BOD test identify the source of a pollutant in a water system when there is no visible sign of the source? _____

6. Discuss some of the ways in which nitrogen polluted waterways affect biodiversity? _____

143 Biomagnification

Key Idea: Persistent pollutants in ecosystems have the potential to cause harmful biomagnification.

Persistent toxins such as **heavy metals**, e.g. mercury, and organic pesticides and industrial chemicals, e.g. DDT, or **endocrine disrupting chemicals**, resist degradation and stay in the environment for a long time. They can leach into the surrounding waterways and soil due to unsafe practices and dumping of waste. These persistent toxins can be taken up by organisms, including humans, in their food or absorbed from the surrounding environment and accumulate in the tissues. This is called bioaccumulation. Once within an organism, toxins can be passed through a food chain, becoming more concentrated at each trophic level. This is called **biomagnification**. Some substances can be non-toxic at lower concentrations, but as they are accumulated in an organism, or through the food chain, the higher concentration then makes the substances toxic. Most **pollutants** will then result in health effects or even death.

The biomagnification of a persistent organic pollutant in a food chain

- **Persistent organic pollutants** (POPs) are organic compounds that are highly resistant to degradation by chemical, biological, or photolytic processes. These chemicals are often referred to as **'forever chemicals'** because they persist in the environment for a long time without breaking down.

- DDT is a man-made (synthetic) insecticide and was first produced in the 1940s. In the past, it was widely used to control insect vectors carrying diseases such as malaria and typhus, and to control agricultural insect pests.

- It soon became obvious that DDT was harming non-target organisms too. In the US, DDT sprayed onto agricultural crops washed into waterways and accumulated in the fatty tissues of fish. Fish-eating predators became poisoned.

- Biomagnification of POPs occurs within food chains (right). The DDT accumulates in the tissues of organisms. Higher order consumers may ingest toxic levels of a chemical because they eat a large number of lower order consumers. This concentration of toxins can prove fatal or harmful to apex predators, which can also include humans. DDT is now mostly banned.

- PCB (polychlorinated biphenyl), also a POP, was added to a range of products to stabilize them. After residues caused the deaths of 539 Japanese citizens who ate contaminated rice in 1968, this forever chemical was eventually banned.

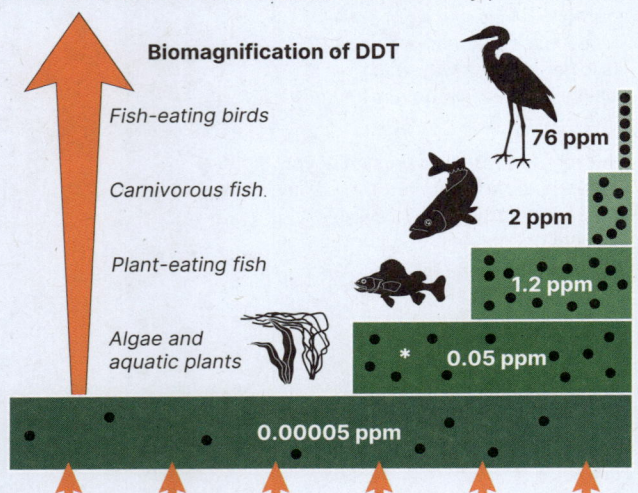

Biomagnification of DDT
- Fish-eating birds: 76 ppm
- Carnivorous fish: 2 ppm
- Plant-eating fish: 1.2 ppm
- Algae and aquatic plants: 0.05 ppm
- 0.00005 ppm

DDT enters the water as runoff from farmland sprayed with the insecticide. * ppm = parts per million

Carnivorous Weddell seal pup, Antarctica

Effect on apex predators: POPs tend to accumulate in top or apex predators, e.g. in the Antarctic, the Weddell seal; in the Arctic, the polar bear. The polar bear's diet is rich in fats, so the potential to ingest contaminants is very high. Legacy contamination of prohibited POPs still remains in some food sources, as they do not break down. POPs, which are also 'forever chemicals', are included in many human-made products.

Osprey eggs at risk of thinning from POPs

Eggshell thinning: A Swedish study in 2004 showed increasing levels of POPs (mainly DDT and PCBs) in many raptors, such as peregrine falcons and ospreys. DDT caused the birds' eggshells to be thinner than normal. Thinner shells make the eggs more fragile and many eggs broke before they could fully incubate, leading to the death of the chicks. This phenomenon has been linked to population declines in several carnivorous bird species as they accumulate toxins from their prey.

POPs accumulate in fatty tissue like salmon

Human health: Humans are exposed to POPs in food such as fish that has accumulated the toxins. POP exposure in humans has been linked to cancer, reproductive disorders, immune disorders, circulatory problems, birth defects, and endocrine disruption. The Stockholm Convention on POPs was established to protect human health and the environment through a number of pathways including reduced use and production of POPs.

1. What is the difference between bioaccumulation and biomagnification?

Bald eagles and PBDEs

▸ The biomagnification of DDT in bald eagles is a well-documented case study.

▸ Persistent chemicals from flame retardant chemicals, specifically PBDEs, found in many discarded fabrics and furniture, were found in bald eagles in Michigan. These chemicals do not dissolve in water and are taken up by other organisms, and eventually by the eagles, where it accumulates in their fat.

▸ Studies indicate these chemicals lead to behavioral and reproductive issues in other bird species but may still be subclincal (without visible effects) in bald eagles at early stages.

▸ Once the danger was recognized, the chemicals were mostly phased out. This corresponded to a drop in the concentration of PBDEs in the bald eagles.

▸ Bald eagles are suitable species to act as 'bioindicators', revealing whether persistent, forever chemicals are present in the environment.

Bald eagles feeding chicks meat

Mercury in food chains

▸ Just as POPs accumulate in tissues and increase in concentration in food chains, so too do heavy metals such as mercury. During the California gold rush of the 1800s, mercury was used to extract gold. Some mining sites still leach mercury into the Sacramento Delta and San Francisco Bay, highlighting just how persistent mercury is.

▸ Methylmercury is the most toxic form of mercury. Like POPs, methylmercury accumulates in the tissues of organisms. This occurs because it is taken in at a faster rate than it can be excreted (removed by metabolic processes) from the body.

▸ As methylmercury passes through successive trophic (feeding) levels in a food chain, it becomes more concentrated (biomagnification).

▸ When a predatory fish consumes smaller fish, it absorbs the mercury present in those smaller fish which, in turn, have acquired mercury from the organisms they consumed. As a result, mercury becomes more concentrated at higher trophic levels. This process also applies to humans, as consuming contaminated fish and shellfish is the primary source of methylmercury exposure for humans.

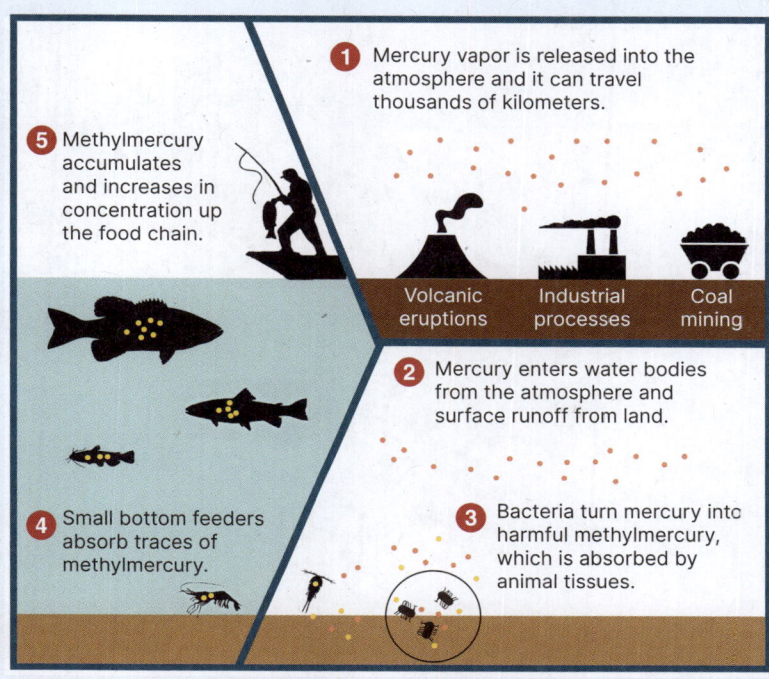

2. Suggest why bald eagles are suitable as a bioindicator species for monitoring persistent chemicals in the environment?

___

3. Why is biomagnification of methylmercury a particular concern for humans?

___

4. Heavy metals, including cadmium, manganese, copper, and chromium, sink into sediments in aquatic ecosystems. Marine nematodes (sediment eating worms) are an important food source for juvenile fish. Why would this relationship be significant for heavy metal biomagnification?

___

Endocrine disrupting chemicals and biomagnification

▸ Another group of forever chemicals that cause health issues in animals, including humans, are called endocrine disrupting chemicals (EDCs). EDCs are found in many commonly used products, such as industrial chemicals, pesticides, polymer/plastics, discarded medicines, including hormonal tablets, and personal grooming products. Most EDCs that enter the waterways accumulate in aquatic organisms as small particles. These can build up in ecosystems due to biomagnification.

▸ Endocrine disruptors are chemicals that interfere with the normal function of an animal's endocrine system (hormone-releasing glands), that controls reproduction, growth and development, and metabolism. They may mimic hormones or interfere with the endocrine system and disrupt normal processes.

▸ Some EDCs cause sex ratio imbalances (an unusually greater proportion of one sex than the other). This has been seen in fish populations near pulp and paper mills.

▸ Exposure to EDCs at critical development times has been shown to cause disruption to the development of ovaries, testes, or genitalia in humans as well. EDCs are also linked to birth defects, declining fertility, and lower numbers of males being born.

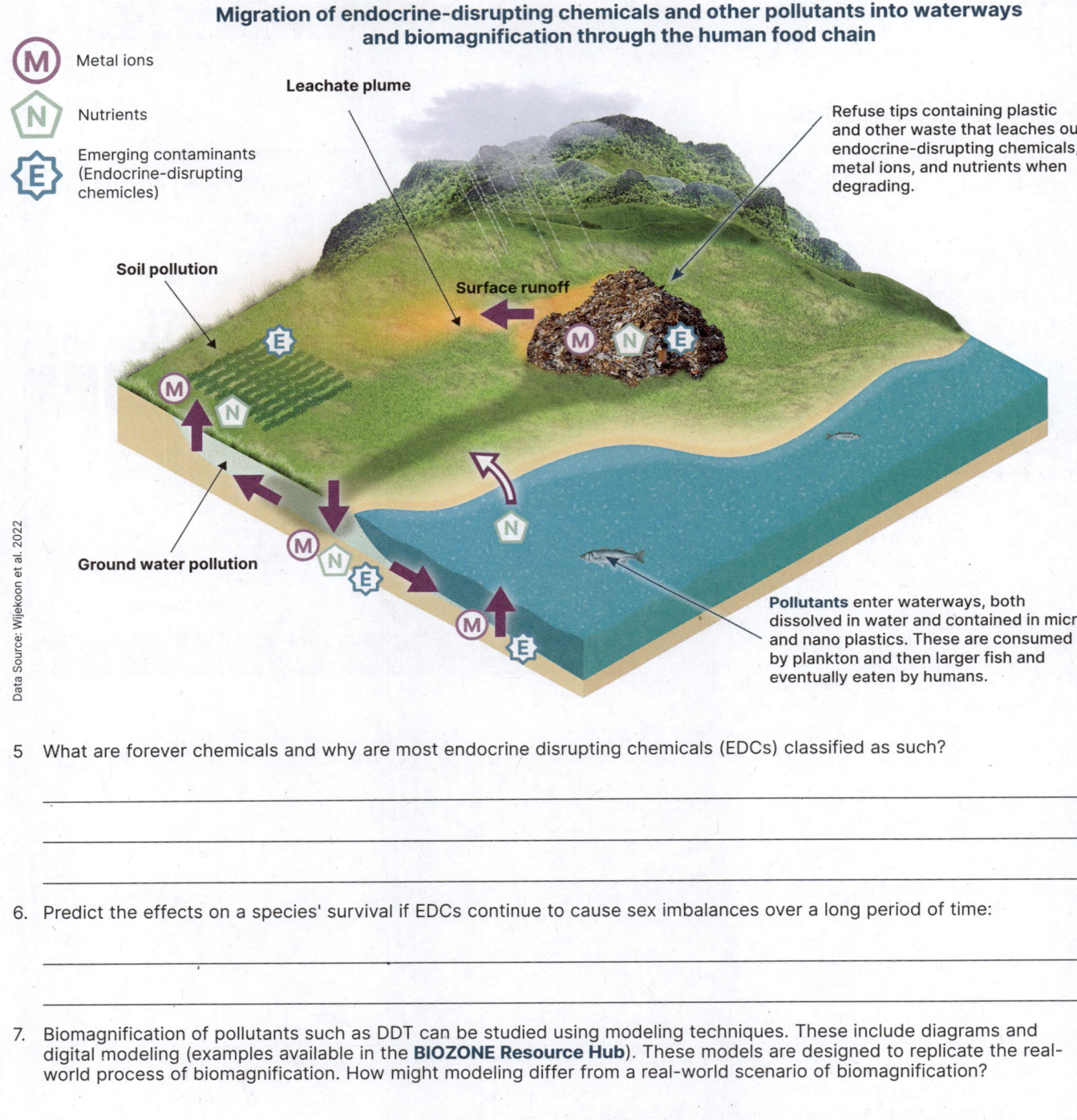

Migration of endocrine-disrupting chemicals and other pollutants into waterways and biomagnification through the human food chain

5. What are forever chemicals and why are most endocrine disrupting chemicals (EDCs) classified as such?

6. Predict the effects on a species' survival if EDCs continue to cause sex imbalances over a long period of time:

7. Biomagnification of pollutants such as DDT can be studied using modeling techniques. These include diagrams and digital modeling (examples available in the **BIOZONE Resource Hub**). These models are designed to replicate the real-world process of biomagnification. How might modeling differ from a real-world scenario of biomagnification?

144 Sewage Treatment

Key Idea: Sewage treatment can remove most pollution before water is released back into rivers, lakes, and oceans. Once water has been used by a household or industry, it becomes **sewage**. Sewage includes toilet wastes and all household water, but excludes stormwater, which is usually diverted directly into waterways. In some cities, the sewage and stormwater systems may be partly combined, and sewage can overflow into surface water during high rainfall. When sewage reaches a treatment plant, it can undergo up to three levels of processing (purification). Primary treatment is little more than a mechanical screening process, followed by settling of the solids into a sludge. Secondary sewage treatment is primarily a biological process in which aerobic and anaerobic microorganisms are used to remove the organic wastes. Advanced secondary treatment targets specific **pollutants**, particularly nitrates, phosphates, and **heavy metals**. Before water is discharged after treatment, it is always disinfected (usually by chlorination) to kill bacteria and other potential pathogens. Effective sewage treatment can be an efficient method of water reuse in dry regions.

Typical sewage treatment plant

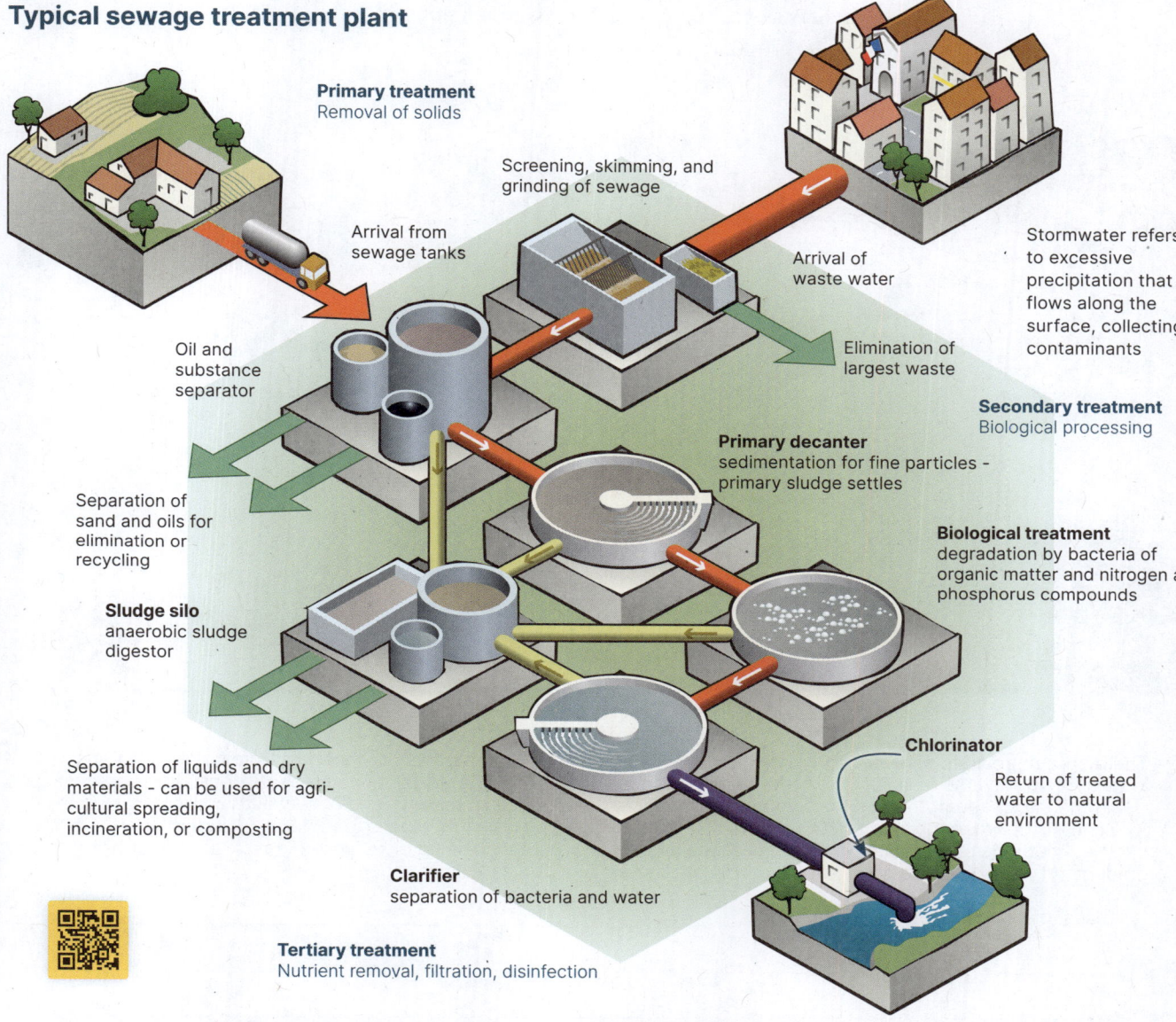

1. Using the diagram above for reference, investigate the sewage treatment process in your own town or city, identifying the specific techniques and problems of waste water management in your area. Make a note of the main points below:

   (a) Your urban area and treatment station: _____

   (b) The volume of sewage processed: _____

   (c) The degree of purification: _____

   (d) The treatment processes used (list): _____

   (e) The discharge point(s): _____

   (f) Problems of wastewater management: _____

   (g) Future options or plans: _____

145 Waste Management

Key Idea: Effective management of waste limits pollution and repurposes or recycles the by-products after treatment.

The disposal of solid and hazardous waste is one of the most urgent problems of today's industrialized societies. Traditionally, solid waste has been disposed of in open dumps and, more recently, in sanitary, scientifically designed landfills. Even with modern designs and better waste processing, landfills still have the potential to contaminate soil and groundwater. In addition, they occupy valuable land and their siting is often a matter of local controversy. More often, city councils and local authorities support initiatives for the reduction, reuse, and **recycling** of solid wastes. At the same time, they must develop strategies for the safe disposal of hazardous wastes, which pose an immediate or potential threat to environmental and human health. A programme of integrated **waste management** (below) combines features of traditional waste management with new techniques to reduce and incinerate wastes.

Components of integrated waste management

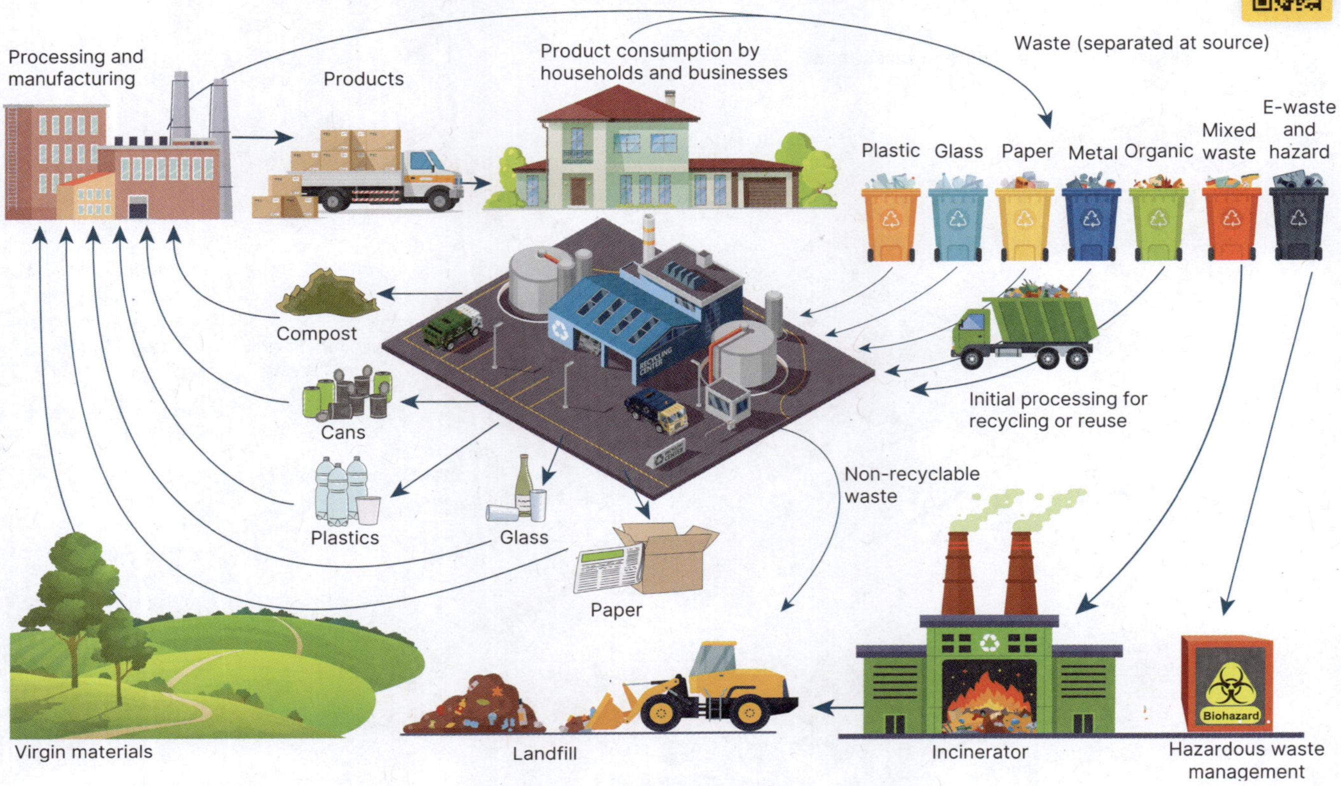

1. The diagram provides an overview of an idealized management system for waste materials from households and industries. It provides a starting point for comparing how different waste products could be disposed of or processed. Using the information provided for guidance, summarize the disposal, recycling, and post-waste processing options for each of the waste products listed below. List the important points in the spaces provided, including reference to disposal methods and particular problems associated with these, processing or recycling (if relevant), and useful end-products (if relevant). If required, develop this list, or part of it, as a separate report:

   (a) Glass, plastic and paper waste: _____

   _____

   _____

   (b) Metals and their alloys, e.g. aluminum, tin, and steel: _____

   _____

   (c) Organic waste: _____

   _____

   (d) Hazardous waste (including e-waste): _____

   _____

2. Identify a waste product that is not part of an integrated waste management program:

   _____

146 Reducing Waste

Key Idea: Reusing, recycling, or re-purposing end-of-life products can reduce the environmental impacts of waste. Much of the waste produced by industrialized countries contains valuable resources, which could be used again if properly processed. As resources become scarcer and competition for them grows, both individuals and companies are beginning to explore ways of reusing waste material. The reduction of solid waste presents a great challenge because its components can often be difficult to separate or break down, but this could be facilitated by the initial design of the product. The model of the circular economy is being adopted by many organizations to limit what is considered waste. All parts of the manufacturing chain can be designed to consider end-of-life products and materials as reusable components. Natural ecosystem processes can be utilized to release nutrients back into the biosphere.

Composting
Composting is a simple way in which biodegradable waste, such as food scraps and garden waste, can be broken down and nutrients returned to the earth. Composting is possible in almost every household, even those without a yard. A common trend today is to have small worm farms or cultured composts for kitchen scraps in small containers on or under kitchen benches. Outdoors, compost can be made in purpose-built containers or simply in a pile in the corner of the yard.

Reusable shopping bags
Reusing an object or material in an unchanged form is the simplest way of reducing waste. A growing number of retail outlets are offering their customers the choice of buying reusable shopping bags, while others now charge extra if customers use plastic bags. In New Zealand, single-use plastic shopping bags were banned countrywide in 2019. The second phase will be to ban any other PVC and polystyrene packaging in foods in 2025, as these types of materials are difficult to recycle.

Recycling
Not everything can be reused. **Recycling** broken or otherwise useless material stops waste building up in the home. Glass and metals can be recycled by melting them down and reshaping them, which saves the energy that would have been needed to extract and process new raw materials. Plastics may be chemically treated to break them into their monomer (small segments) components, or they can be heated and reformed.

The three Rs of waste reduction.
The priority of reduce, reuse, and recycle is based on reducing the volume of raw materials used, reusing products already manufactured (such as shopping bags and bottles), and recycling what cannot be reused. What cannot be recycled is then dealt with as true waste. However, in theory virtually everything can be recycled in some way to limit **pollution**.

1. Many countries have banned single-use plastics, including Kenya, Taiwan, Australia, and France. How will this ban contribute to reducing plastic waste?

2. Explain why recycling metals is particularly energy saving:

3. List three reasons why waste should be recycled:

4. Why is putting waste in a landfill the least preferred option of a waste management scheme?

5. What environmental issues are posed by using incineration in a management scheme?

Recycling, reusing, energy, and cost

▸ Recycling, like any resource extraction method, requires energy. This may be in the form of collecting and sorting material, transport to manufacturers, and the energy used in melting or remelting a material. For recycling to be beneficial, the energy required for the recycling process must be lower than the energy needed to produce a new item or material from the original resource.

▸ Recycling is not a complete solution because some materials, like wood, may result in lower-quality products during the recycling process (known as down-cycling). This often requires the addition of new substances. However, recycling still reduces the demand and prolongs the lifespan of products. This decreases the energy required for manufacturing new goods.

Glass recycling. Modern recycling plants can sort materials automatically using the physical properties of the materials.

▸ In general, metals can be 100% recycled. They are relatively easy to melt down and recover from scrap. Aluminum requires 96% less energy to make from recycled cans than it does to process from bauxite (ore). What matters the most when producing materials is the cost of extracting and refining the raw materials versus the cost of collecting, sorting, and reprocessing used materials recycled after use.

▸ New technology for recycling has helped to reduce the cost of collecting and sorting. For example, in San Francisco, no pre-separation of recyclable material is needed. Although some workers are still required, most of the separation work is carried out by the city's state-of-the-art recycling facility. Magnets remove steel, eddy currents separate aluminum, and vacuum tubes and separators remove plastics.

▸ Reusable objects, such as reusable shopping bags or take away coffee cups, have recently become fashionable. However, many people do not realize how long these items need to be reused before they become 'environmentally friendly'. For example, a single use paper cup has a cost of production and a cost of dumping, and these costs are fixed for every cup (see table below). A reusable cup has a much higher cost of production and a cost of cleaning. In this case, the production cost reduces over time while the cleaning cost remains the same.

Energy cost of reusable vs disposable cups

Cup type	Cup mass g/cup	Material specific energy MJ/kg	Energy per cup MJ/cup
Ceramic	292	48	14
Plastic	59	107	6.3
Glass	199	28	5.5
Paper	8.3	66	0.55
Foam	1.9	104	0.20

Data Source: Hocking (1994). Reusable and disposable cups: An energy-based evaluation.

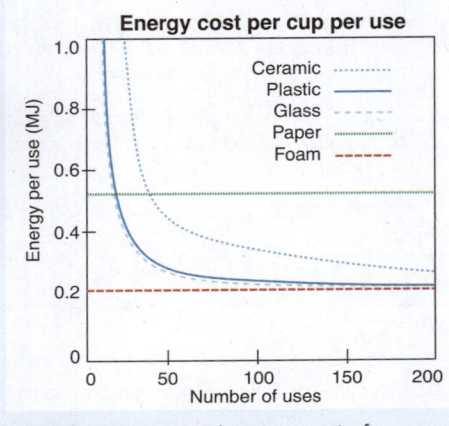

The manufacture of reusable items, such as ceramic cups, uses a large amount of energy. Energy is also used in heating water and manufacturing detergent to wash them. Disposable foam or paper cups require little energy to produce and none for cleaning. The graph above illustrates how many uses it takes before recyclable items (glass, plastic, or ceramic) reach the same energy per cup value as single use items (paper and foam). It does not account for the cost of disposal.

6. Estimate the energy used per use after 100 uses of:

(a) A ceramic cup: _____

(b) A plastic cup: _____

(c) A glass cup: _____

(d) A paper cup: _____

(e) A foam cup: _____

Energy savings using recycled material

Recycled material	Energy saving (%)
Aluminum	95
Plastic	88
Copper	75
Steel	60
Paper	60
Glass	34

7. Using information from your calculations above and the energy savings data, construct an argument on what you consider the most environmentally 'safe' material to use for drinks: _____

The circular economy and waste

▸ Traditional **waste management** and recycling aim to address waste-related issues and minimize environmental disruption. In contrast, the principles of the circular economy model focus on preventing a significant amount of products from becoming waste at the end of their lifespan.

▸ The circular economy principles also focus on reduction of greenhouse gas emissions as a result of waste reduction.

▸ Additionally, organic output from the consumer can be considered a resource, as an energy source or a supply of nutrients to feed livestock and plants. Instead of landfills, specially enhanced ecosystems would be used to release nutrients from the organic waste using natural processes.

▸ The gold standard of the circular economy is to implement changes that prevent the creation of the waste in the first instance. That may involve the redesign or reduction of products, including the packaging. Also, remanufacturing or reuse of products instead of building from new, preferably incorporating previously used components.

▸ If a product or substance was unable to be redesigned, reduced, or reused, then more traditional recycling would allow recovery of some material. Crucial to the success is the elimination of plastics and non-recyclable components.

▸ Traditional waste management may recover some energy by incineration of waste, but this process often adds **pollutants** to the atmosphere. In the circular economy model, the energy is recovered through natural fermentation processes as biogas.

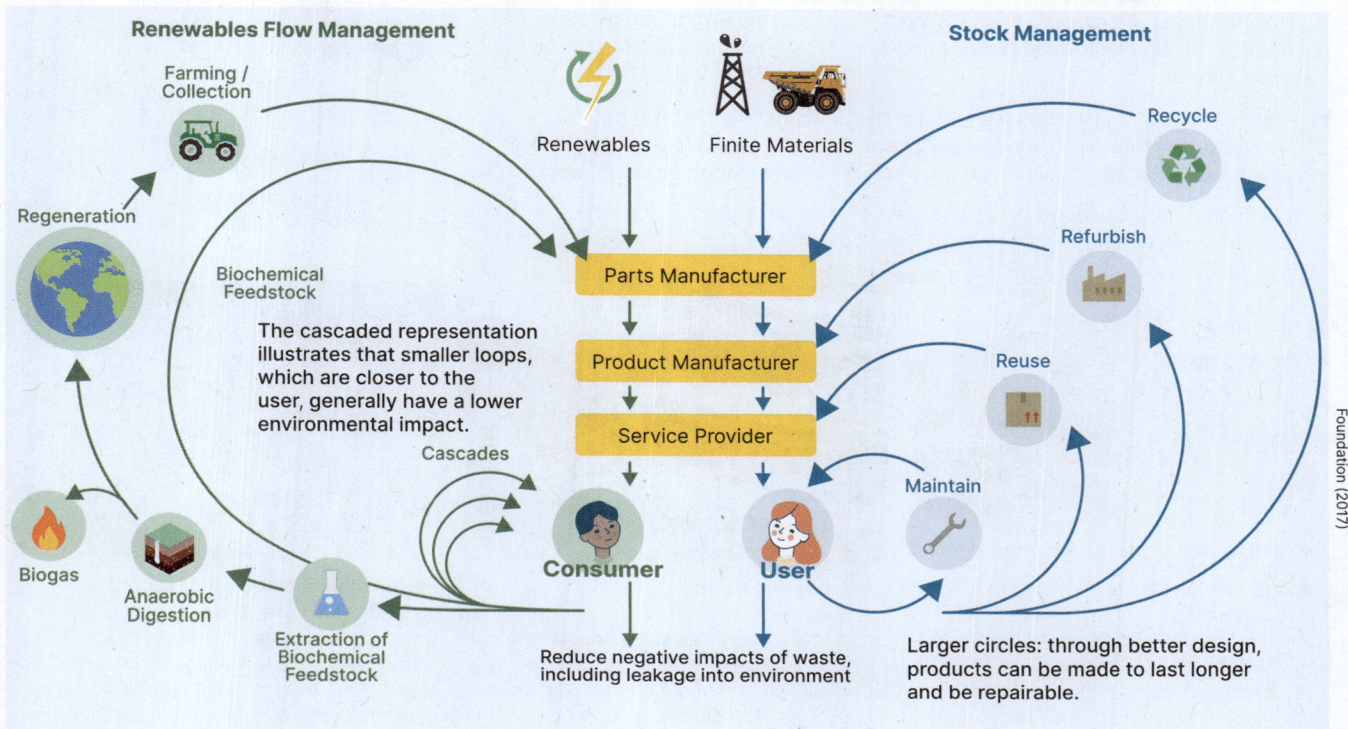

8. What is the difference between the circular economy model and traditional waste management?

___

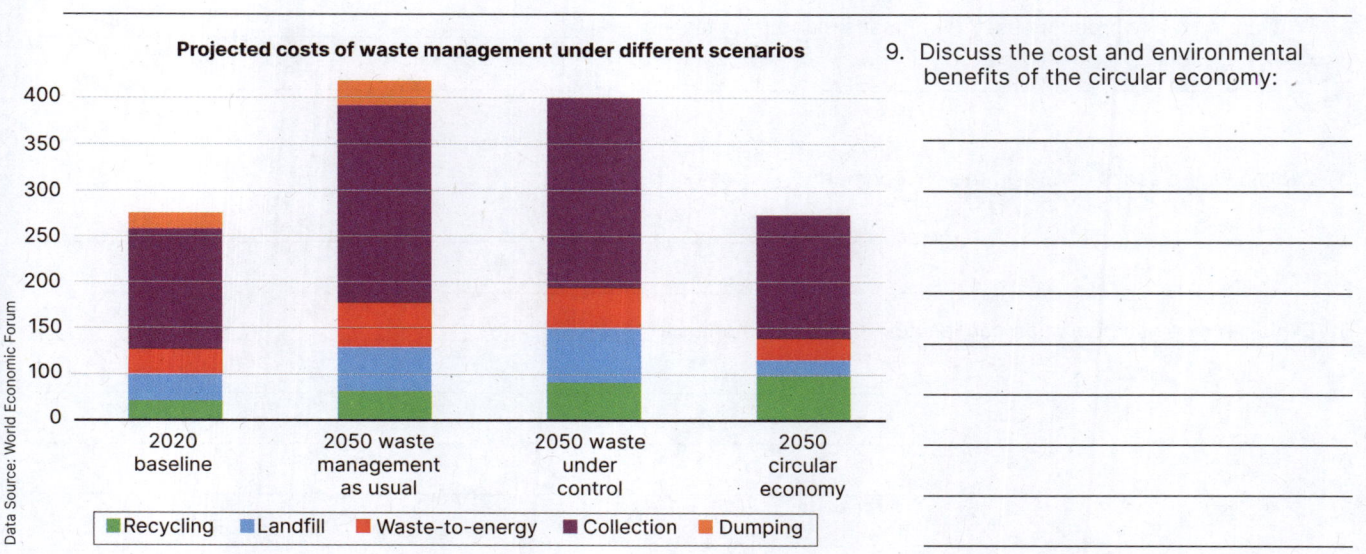

9. Discuss the cost and environmental benefits of the circular economy:

___

147 Plastics in the Environment

Key Idea: The widespread presence of plastic waste is causing significant harm to ecosystems and the organisms that inhabit them.

Almost every home and industry contains plastic: a synthetic substance that has only been introduced in the last 100 years. Plastic is widely used because it is convenient and easy to produce. However, its low cost of production also makes it easy to discard. Due to the ocean currents spreading the material, even the most remote parts of the world are now affected by **plastic pollution**. Currently, around 19-23 million tonnes of plastic are released into the environment each year as waste into lakes, rivers, and oceans. This problem will only be partially solved by plastic **recycling**. Instead, a shift to non-plastic products and the development of innovative breakdown processes that can tackle the issue of plastic permanence will be required.

Plastic permanence is a problem

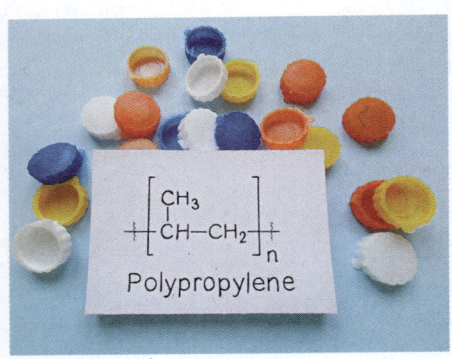

- The problem with plastic is its stability. In nature, organic material is broken down by enzymes and microbes that have evolved over billions of years to deal with the chemical bonds found in nature. Plastic degradation is limited due to the difference of chemical bonds in most plastics compared to those found in nature. This results in only a small number of organisms capable of breaking down plastic.

- Plastics have a long-lasting presence in the environment, persisting for hundreds of years. The excessive disposal of plastic products over the past 50 years has led to significant environmental and **waste management** challenges.

- Environmental groups are collecting plastic from polluted areas, but the problem then becomes how to get rid of it completely.

Colourful, cheap plastic is in most homes.

Plastic pellets can be molded into products.

Waterway plastic pollution, Tamil Nadu

Plastic is everywhere
The earliest forms of synthetic plastics, such as Bakelite (1909), cellophane (1912), PVC (1926), and teflon and nylon (1938), revolutionized many products. Mass use in medical supplies and shopping bags increased during the 1970s. Fleece clothing entered the market in 1993. In a relatively short time, plastic became an indispensable material: light, cheap to make, and waterproof. Shops offer a huge variety of plastic products, and nearly every food and drink is packaged in plastic.

Plastics and health
The production of plastic from fossil fuels and other chemicals can have detrimental effects on the environment, including the emission of greenhouse gases. Elevated numbers of cancer cases have occurred in Louisiana near plastic production plants. Certain additives such as BPA, found in plastic, including children's toys, have been found to negatively impact both the reproductive and immune systems. Additionally, when waste plastic is incinerated, harmful fumes are released.

Plastic pollution and water supply
Plastic waste in waterways that supply water for human use can make much of it unusable. India is one country with a significant plastic pollution problem, partly because of the concentrated population. In Tamil Nadu, above, the local government is attempting to solve their plastic pollution by importing several 'boom interceptors' from the Netherlands. However, without an effective recycling plan, this will just collect and move the plastic from one site to another.

1. Why are so many products used by humans made from plastic? _____

2. Explain why plastics persist in the environment: _____

3. Explain how plastic pollution can impact human communities: _____

Concentrating the problem

▶ **Human activity** has a global reach, affecting every part of the planet in some way. Even remote parts of the world are affected by activities thousands of kilometers away (also see page 248). The ocean currents have a tendency to gather and concentrate plastic waste in specific regions.

▶ The surface water of the oceans circulates in giant whirlpools called gyres. In the same way that you can concentrate debris in a small pool by swirling the water around, these gyres also concentrate floating debris. When this happens with floating plastic, giant areas of the ocean become plastic 'garbage patches'.

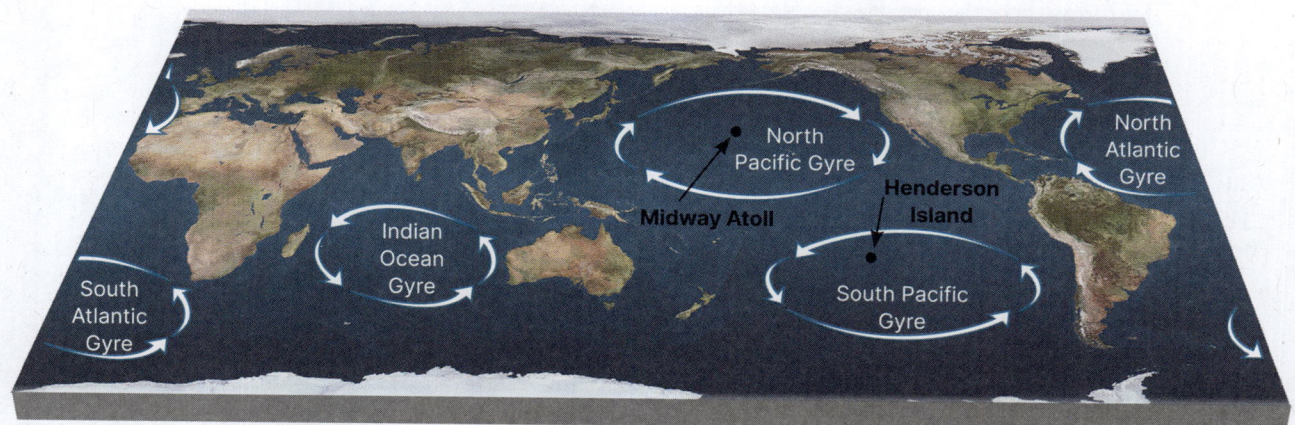

The **Great Pacific Garbage Patch** is an area within the North Pacific Gyre. Although given the name 'garbage patch,' most of the debris is not easily visible. The patches contain concentrations of waste and debris (mostly plastic) that have made their way into the ocean. The area covers around 1.2 million km², although definitions of the extent vary and the concentration of debris changes with the seasons. This makes an accurate estimate difficult.

Plastic and marine species

▶ Marine animals, such as dolphins, sea lions, seals, whales, and turtles, can become entangled in plastic waste in the oceans. The plastic, such as plastic bags and six-pack rings, can drown, starve, or suffocate the animals.

▶ The plastic net threads discarded from fishing boats can wrap around whales and remove tails or fins. The nets can also cover and smother coral reefs, preventing proper growth, and sometimes causing death.

▶ Consumption of plastic can cause starvation in animals, preventing ingestion and absorption of normal food, as well as leaching toxins into their bodies.

▶ Mammals that filter water, such as baleen whales, cannot be selective of food, and plastic can block their digestive systems.

Marine species, including turtles, can mistakenly consume plastic, often mistaking it for jellyfish, which can cause harm. Also, marine species can become entangled in plastic.

4. Explain how plastic waste becomes concentrated in certain areas of water:

5. What types of plastic are a particular problem for marine species?

6. Suggest what marine species would be most at risk from plastic pollution?

Henderson Island

▸ Henderson Island sits in the South Pacific Ocean. It is part of the Pitcairn group and is about 5000 km from the nearest significant land mass. The island is small, at only 37.3 km^2, and uninhabited.

▸ A study in 2017 measured the amount of plastic on the island's beaches. The research team measured the amount of plastic already on the beach, and then cleared a control area to measure the rate at which plastic accumulated.

Site	Mean density on beach (items per m^2)	Estimated total debris on beach (items)		Estimated island total including buried items and back beach (items)	
		Number	Mass (kg)	Number	Mass (kg)
North Beach	30.3	812,116	2985	7,634,052	4,744
East Beach	239.4	3,053,901	12,611	30,027,343	12,857
Total		3,866,017	15,597	37,661,395	17,601

Data Source: Lavers and Bond (2017).

Plastic debris on beach, Henderson Island. Plastic bottles often make up a large proportion of plastic rubbish. In water, eddies of the gyres often concentrate them into a single area. Henderson island is a UNESCO World Heritage site in recognition of its environmental value, but unfortunately it is also one of the most polluted islands.

Midway Atoll

▸ Midway Atoll (land area 6.2 km2) is in the North Pacific Ocean, 2,400 km west of Hawaii (the nearest significant land), and near the center of the North Pacific Gyre.

▸ As with Henderson Island, the circulation of the ocean washes vast quantities of plastic onto the beaches. The National Oceanic and Atmospheric Administration (NOAA) regularly removes plastic debris from the beaches. Since 1999, they have removed 125 tonnes of plastic.

▸ Marine species are particularly affected by plastic waste. On both these islands, and many others, plastic is mistaken for food by seabirds who eat it or feed it to their young. Every year on Midway Island, thousands of young albatrosses die from ingesting plastic products.

▸ The proportion of plastic deposited in the ocean is expected to increase at a rate of 4.8% each year until 2025, and at a rate of 3% from 2025 to 2050. At this rate, the ratio of plastic mass to fish mass could be roughly 1:1 by 2050.

▸ The wave action of the water against the sand breaks down plastic pollution into smaller pieces known as microplastics. These microplastics are small enough to enter the bloodstream of animals and can accumulate in their tissues. This can cause reproductive issues for animals such as seals.

Plastics don't break down in the environment, so if they are washed into waterways they can float for many thousands of kilometers. The rotation of the gyres concentrates any floating debris. This results in many tonnes of plastic waste washing up on the beaches of remote Pacific islands every year. The albatross chick (above) died on Midway Island after eating numerous plastic items, including a felt pen and a number of bottle caps.

US Fish and Wildlife Service

7. Explain how plastic thrown away in a city on the west coast of North America can end up in a Laysan albatross chick on Midway Atoll:

___

8. Work in small groups to discuss some ideas that you could take to reduce the plastic problem. Record ideas below:

___

9. Some scientists use digital models to predict plastic pollution movement in the oceans (see **BIOZONE Resource Hub** for an example). Consider how plastic might move from your own local destination around the ocean. Use information about the ocean current patterns from diagrams in this activity (and in activity 20), or the digital model simulation, to estimate the trajectory and possible final location of the plastic pollution. Describe below:

___

Solving the problem

▶ Increasingly, degradable plastics are being produced. These may be photodegradable (break down in sunlight) or made from a blend of sugars and other chemicals and can break down within 45 days. Although many degradable plastics break down into smaller pieces, those pieces persist in the environment for a long time.

▶ A subset of biodegradable plastic is compostable plastic. These products break down naturally into nutrients within a commercial compost heap, and no plastic residue remains.

▶ Both types of plastic still need to be disposed of correctly. If just placed in landfill, the breakdown will release methane, a potent greenhouse gas.

Plastic eating bacteria

▶ PET (Polyethylene terephthalate) plastic is widely used to make bottles. A few fungal species are known to digest PET, but until recently no bacteria were known to do so.

▶ However, in 2016 a Japanese research group collected 250 samples of sludge from a PET bottle recycling plant. They incubated these sludge samples with very thin PET film. After 15 days, they found that the PET film had vanished in some of the samples, indicating that something was breaking it down.

▶ Further analysis found that the bacterium *Ideonella sakaiensis* was responsible. This was the first bacterium shown to digest PET plastic. It does this by secreting the enzyme PETase, which breaks the PET molecule down into its single monomers. The enzyme MHETase then breaks the monomers down into compounds the bacteria can use.

▶ The technology is still in its early stages. If used commercially, it is predicted the bacteria could degrade more than 50 million tonnes of PET plastic annually. This is close to the global production of PET plastic. Currently, only 2.2 tonnes of the plastic is recycled annually.

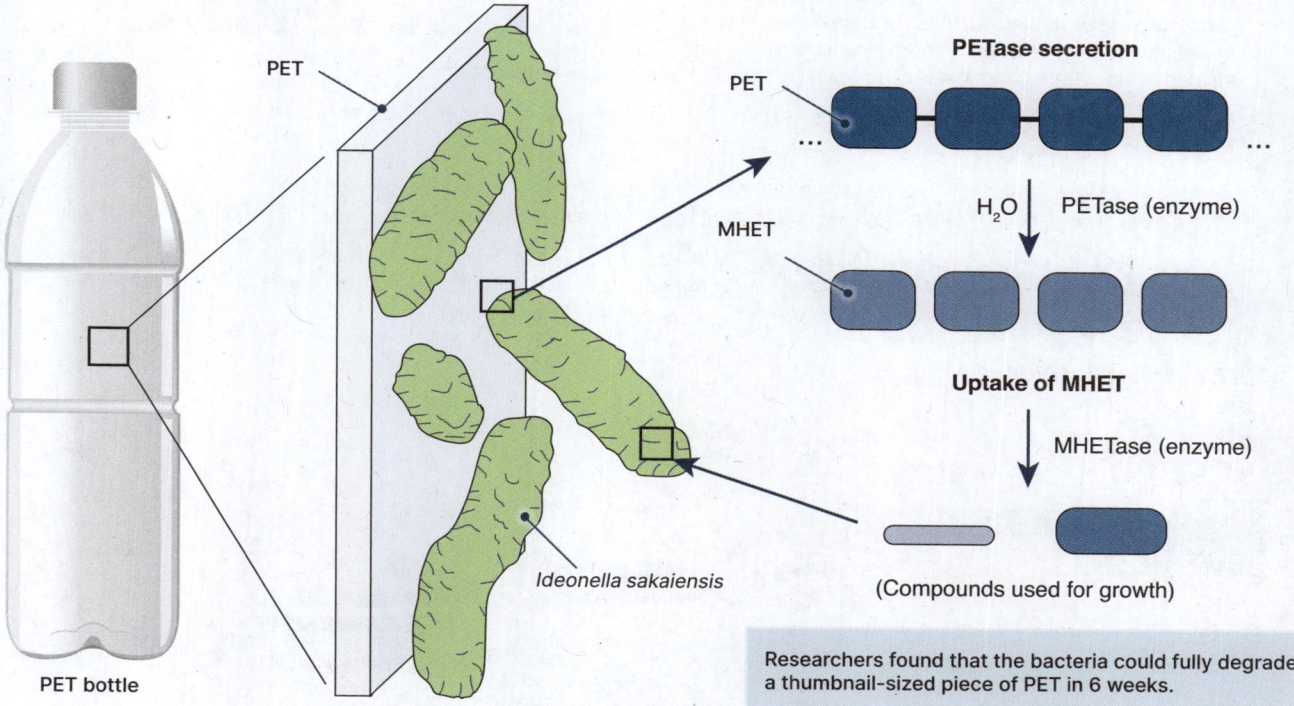

Researchers found that the bacteria could fully degrade a thumbnail-sized piece of PET in 6 weeks.

10. (a) Briefly describe how the bacteria *Ideonella sakaiensis* was found and isolated: _____

_____

_____

_____

(b) Evaluate the potential of this technology in reducing PET plastic waste: _____

_____

_____

_____

_____

_____

148 Microplastic and Nanoplastic Pollution

Key Idea: Microplastics and nanoplastics, which are tiny particles, are widely distributed and pose numerous issues.

Microplastics and **nanoplastics** are tiny pieces of plastic that enter the food chain and concentrate toxins. These can accumulate in fish and other marine animals, and then in humans that eat them. Nanoplastics (less than 1 μm) are even smaller than microplastics (less than 5mm) and they can enter cells and affect their functioning. Recently, scientists have found that nanoplastics are present in clouds and distributed worldwide. In addition to the persistent chemicals found in plastic, these tiny particles can also absorb other toxins and serve as a means of delivering them into the bodies of animals. Although large plastic materials can break down into smaller pieces or fibers, products like glitter and cosmetic microbeads in face scrubs are considerably worse. They are being banned from manufacture or sale in many countries, including in the USA where they are banned specifically through the Microbead-Free Waters Act of 2015.

Forever chemicals in plastics

▸ Many types of plastic contain **forever chemicals** which are especially dangerous as components of micro- and nanoplastic that can enter the body.

▸ PFAS (perfluoroalkyl substances) are found in many waterproof and greaseproof plastic products. By increasing the risk of developing cancer and causing hormone disruption, fetal disorders, and reproductive harm, they pose significant health risks. A recent study found 98% of participants had some evidence of PFAS in their bloodstream, breastmilk, or umbilical cord blood.

▸ Phthalates are commonly added to plastics, including cosmetic products, PVC, flooring, and tubing. They serve the main purpose of adding softness and flexibility to the plastic materials. Once microplastic and nanoplastic particles are inside the body, the chemical they contain leaches out and acts as an endocrine disruptor.

▸ Currently research is investigating the use of fungus, and technology such as pressure and plasma, to understand how forever chemicals may be degraded.

Many new plastic products, especially designed for children, promote being forever chemical free.

Movement of microplastics and nanoplastics through the food chain.

▸ Aquatic food chains are more likely to have higher concentrations of small particle plastics due to the distribution ability of water. Some animal groups take in micro- and nanoplastic directly AND through biomagnification.

▸ **Human activity** is the ONLY point of entry of the plastic into the food chain.

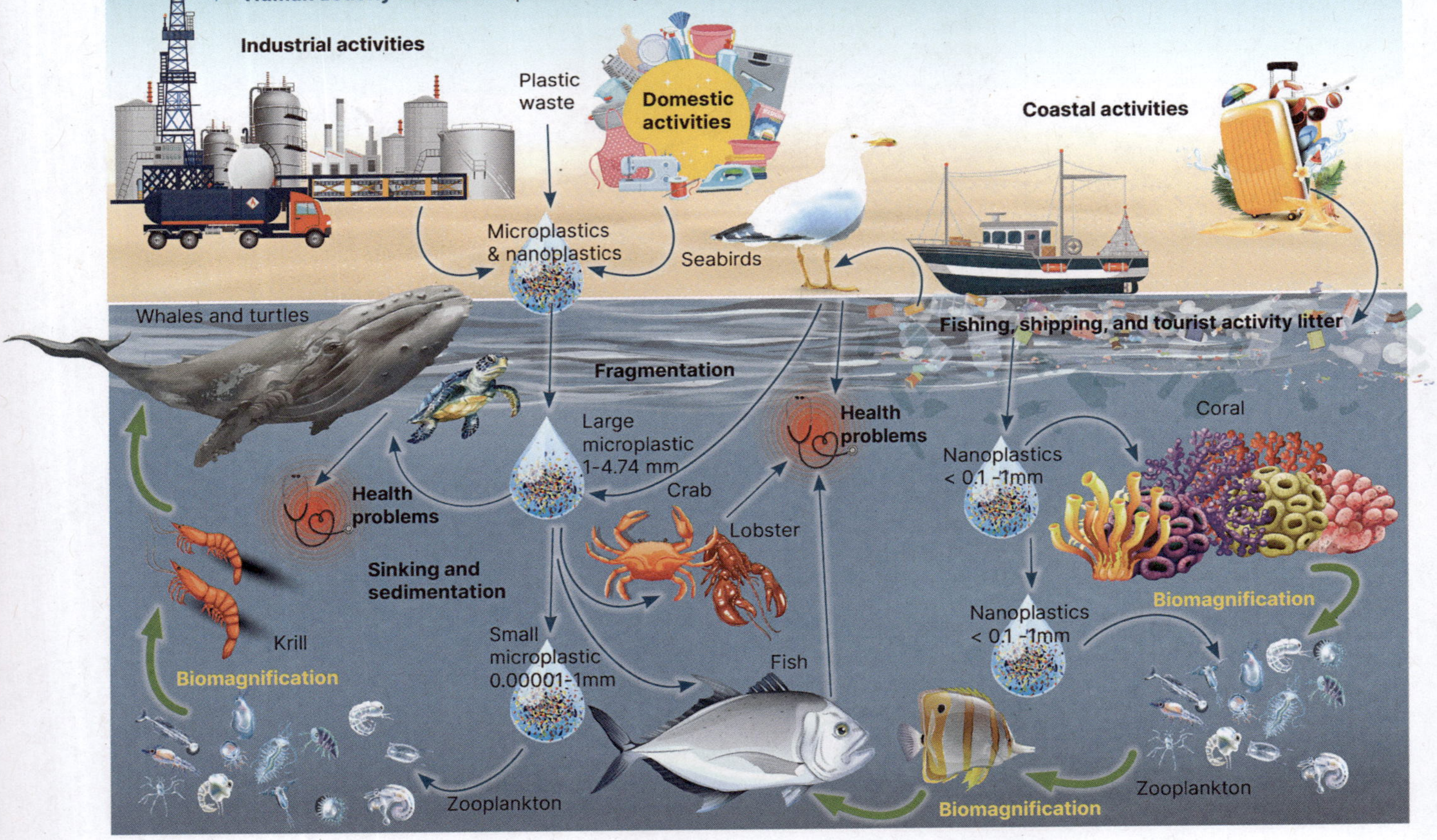

Plastic pollution and the media

- More than 62% of the global population has access to social media platforms. Various environmental agencies have turned towards social media as a means to spread awareness about **plastic pollution** and waste in the hopes of spurring action from citizens.

- Research has shown that, if claims are linked to a confirmed, valid study demonstrating that plastic pollution can be attributed to lifestyle choices, and if possible manageable solutions are provided, then social media posts can be an effective method to produce behavioral changes in people that reduce plastic use and encourage **recycling**.

- Social media posts have the advantage of delivering targeted messages, sharing the media between people and groups, and allowing a personal contribution with the chance to comment.

1. Scientists and conservation organizations are using social media to inform citizens about the problems created by plastic, microplastic, and nanoplastic pollution. What is your perspective on the effectiveness of these messages and how might that encourage you to take action against this problem?

    _____
    _____
    _____
    _____

2. Why is the size of microplastics and nanoplastics so significant to the effects they have on humans and other animals?

    _____
    _____
    _____

3. Movement of micro- and nanoplastics in aquatic ecosystems is an example of bioaccumulation and biomagnification (see diagram on previous page). Explain how these occur in this context:

    _____
    _____
    _____
    _____

4. Why are forever chemicals present in some plastics such a concern, especially in the form of micro- and nanoplastics?

    _____
    _____
    _____
    _____

5. Record any products that you may have used in the past or use currently that state whether they have removed forever chemicals or microbeads, or have information on safe disposal to prevent micro- or nanoplastics forming:

    _____
    _____

149 Air Pollution

Key Idea: Human activity can contribute to emission of pollutants into the atmosphere. This effect is more pronounced in urbanized regions.

Air pollution consists of gases, liquids, or solids present in the atmosphere at levels high enough to harm living things or cause damage to materials. **Human activities** make a major contribution to global **air pollution** although natural processes can also be responsible. Lightning causes forest fires, oxidizes nitrogen, and creates ozone, while erupting volcanoes give off toxic and corrosive gases. Primary air **pollutants** are emitted directly from the source, while secondary pollutants form when primary pollutants react in the atmosphere, e.g. **acid rain**. Air pollution tends to be concentrated around areas of high population density, particularly in Western industrial and post-industrial societies. In the last few decades, there has been a massive increase in air pollution in parts of the world that previously had little, such as Mexico city and some large Asian cities. Air pollution levels can be quantified using an Air Quality Index (AQI), linked to health risks. Air pollution does not just exist outdoors. The air enclosed in spaces such as cars, homes, schools, and offices may have significantly higher levels of harmful air pollutants than the air outdoors.

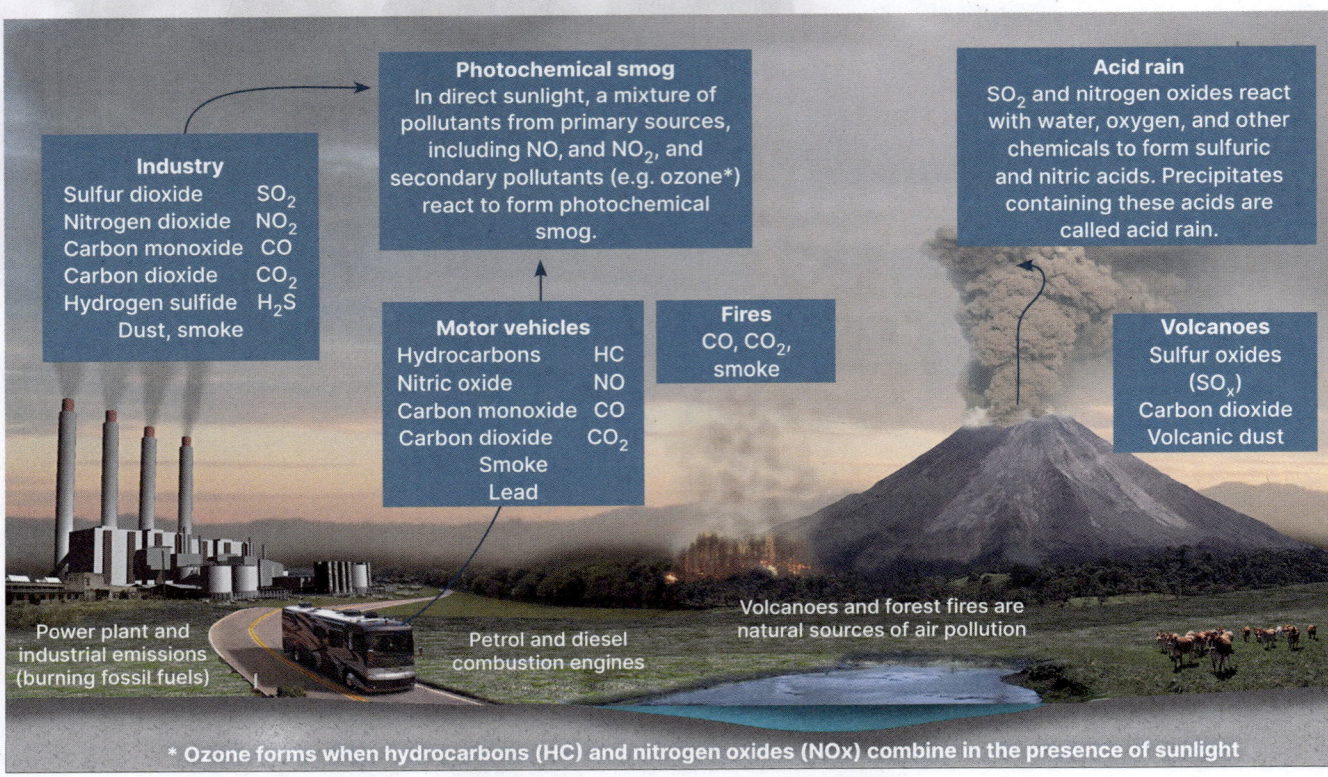

Industry
Sulfur dioxide SO_2
Nitrogen dioxide NO_2
Carbon monoxide CO
Carbon dioxide CO_2
Hydrogen sulfide H_2S
Dust, smoke

Photochemical smog
In direct sunlight, a mixture of pollutants from primary sources, including NO, and NO_2, and secondary pollutants (e.g. ozone*) react to form photochemical smog.

Motor vehicles
Hydrocarbons HC
Nitric oxide NO
Carbon monoxide CO
Carbon dioxide CO_2
Smoke
Lead

Fires
CO, CO_2, smoke

Acid rain
SO_2 and nitrogen oxides react with water, oxygen, and other chemicals to form sulfuric and nitric acids. Precipitates containing these acids are called acid rain.

Volcanoes
Sulfur oxides (SO_x)
Carbon dioxide
Volcanic dust

Power plant and industrial emissions (burning fossil fuels)

Petrol and diesel combustion engines

Volcanoes and forest fires are natural sources of air pollution

* Ozone forms when hydrocarbons (HC) and nitrogen oxides (NOx) combine in the presence of sunlight

1. Name a primary source of air pollution in large cities: _____

2. A major cause of air pollution is the burning of fossil fuels to supply energy for domestic or industrial purposes. Research four fossil fuels and their applications, and describe a negative environmental effect of their use:

   (a) _____

   (b) _____

   (c) _____

   (d) _____

3. Compare and contrast natural and human causes of air pollution sources, including the substances released:

   _____
   _____
   _____
   _____

4. What purpose would Air Quality Standards serve in different countries?

   _____
   _____

Natural sources of air pollutants

Natural processes and activities can cause air pollution. Natural sources are small contributors to total air pollution levels, but they can have significant effects in the short term, often within a localized region. Carbon dioxide and particulate matter are two natural pollutants of significance. Their natural sources are described below.

Wildfires and bushfires

Wildfires (bushfires) are large, uncontrolled fires. They can be started by lightning strikes, although many start as a result of human actions. As they burn, they release large amounts of CO_2, smoke, and particulates into the air. Regions close to the fires are most affected by the **pollutants**, but winds can carry smoke large distances. Smoke from the 2019/20 Australian bushfires affected air quality in New Zealand.

Dust (sand) storms

Dust storms (also called sandstorms) are common in areas with low precipitation levels, e.g. deserts. Strong winds drive loose particles of sand into the air, increasing particulate pollution. Some dust storms are huge, measuring over 1.6 km high. The Sahara Desert has always experienced dust storms, but they have increased 10 times as much since the 1950s. Dust from a large storm in the Sahara may carry as far as central Europe.

Volcanic eruptions

Volcanic eruptions release various pollutants such as sulfur dioxide, hydrogen chloride, carbon dioxide, and ash (particulate matter) into the atmosphere. These emissions can travel long distances. The materials released during an eruption can lead to respiratory problems, eye injuries, and even fatalities. To protect themselves, people often wear masks and use protective gear.

Monitoring air pollution

▸ The amount of air pollution in any one area around the globe can be measured by the Air Quality Index (AQI) (see **BIOZONE Resource Hub**). This index grades the quality from 0 to 500: the higher the value, the more air pollution is present.

▸ The AQI can be used to indicate the level of health risk to the nearby population, where levels over 200 indicate increased health risk for the population, and over 300 is considered to be a health alert emergency for all.

▸ The AQI can be increased due to atmospheric conditions, where polluted air gets trapped close to the ground, rush hour traffic releasing vehicle emissions, nearby volcanic activity, or forest fires.

▸ The Air Quality Index measures several aerosols, including $PM_{2.5}$, which refers to small particulate matter with a size of less than 2.5 micrometers. These fine particles pose the highest risk to human health.

Data source

Air Quality Index (AQI) scale

0 - 50	51 - 100	101 - 150	151 - 200	201 - 300	301 - 500
Good	Moderate	Unhealthy for sensitive groups	Unhealthy	Very unhealthy	Hazardous

5. Use an Air Quality Index measure (QR link above) to identify the current level of the local area closest to you. Record date and level below. You could also mark your AQI value on the scale to the left.

6. What sources of air pollution do you have in your local area? State whether it is rural or urban.

7. What initiatives could help your local town or area to reduce air pollution?

150 Cities and Air Pollution

Key Idea: The heat generated from cities can affect the microclimate and trap air pollution in the form of smog.

Cities contain vast amounts of radiative surfaces, and so tend to heat up very quickly during the day. This region of hot air modifies the local climate, resulting in quite different weather in and near the city than in the surrounding rural areas, especially in large inland cities. The heat island effect of cities produces a rising current of air above the city which results in heavier rainfall and higher air temperatures, while trapping **photochemical smog**.

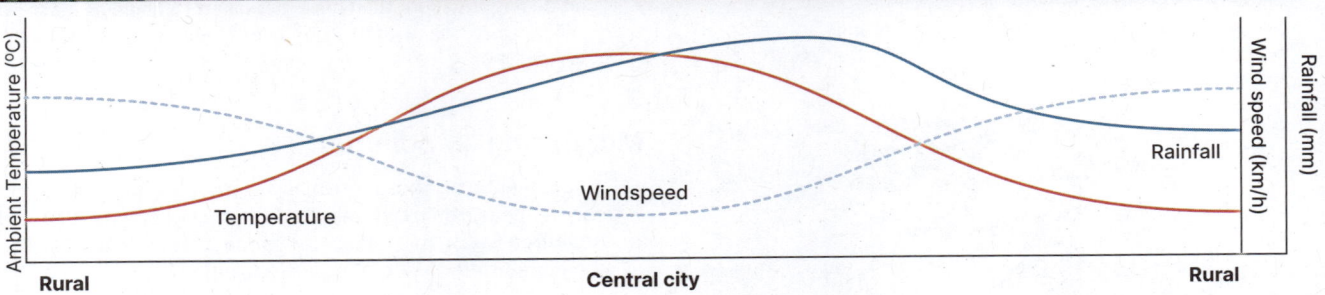

- The majority of urban areas are covered with concrete, asphalt, and iron so they tend to absorb energy and then heat up faster than surrounding rural areas and retain the heat longer. Average temperatures in a city can often be between 5-10° C higher than surrounding rural land.
- City environments also affect rainfall. Hot air rising above a city carries more moisture with it than cool air. Wind moves this air away from the city and, as it cools, the moisture is lost as rain.
- Runoff of water after rain is higher in the urban environment due to the large paved surface area. Stormwater systems catch much of this runoff (up to 70% of the rain that falls) and funnel it into rivers, which can cause them to rise and fall rapidly.
- Wind within a city is often hampered or channeled by buildings with some parts of a city acting like wind tunnels. However, overall wind speed may be 20-30% lower in the city than in rural areas.

Many areas around the world experience the weather phenomenon of temperature inversion. This occurs when a layer of warm air sits on top of a layer of cool air. The cool, dense air remains close to the ground. In urban areas, smoke and other air pollutants can become trapped and concentrated due to a phenomenon called temperature inversion. This prevents them from rising to higher altitudes. As a result, **pollutants** like photochemical smog can accumulate and linger in the city. This may create serious health issues.

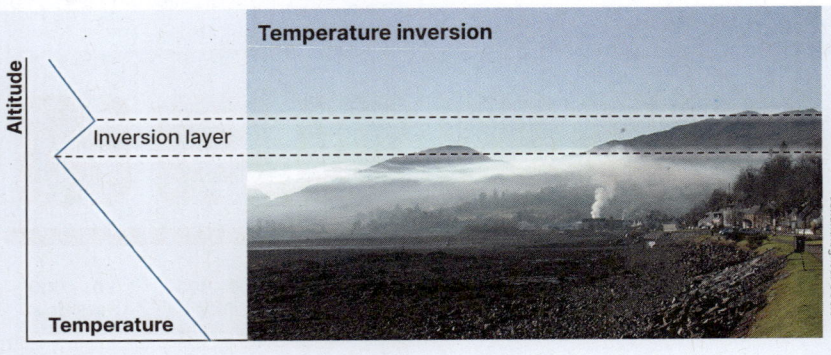

1. (a) Explain why the ambient temperature in a city is often higher than that of the surrounding rural land: _____

   _____

   (b) What is the effect of cities on rainfall? _____

   _____

2. Explain why an inversion layer above a city can have serious health effects: _____

   _____

   _____

Photochemical smog

- Smog (a joining of the words smoke and fog) is a type of **air pollution**.
- Summer smog is also called photochemical smog and forms mainly in the warmer months. It is characterized by high levels of ground level ozone.
- Nitrogen oxides (NOx) from vehicle emissions or **pollutants** from coal fired industrial processes react in the atmosphere in the presence of sunlight to form the secondary pollutants that make up photochemical smog.
- This is a serious problem in many large cities around the world and it can cause many health problems. It mainly affects the respiratory (breathing) system. Emphysema, asthma, chronic bronchitis, lung infections, and cancers are caused by, or made worse from, exposure to photochemical smog. Eye and nose irritation, birth defects and low birth weights are also linked to prolonged exposure to photochemical smog.

Smog day and sunny day (Fanhe, China)

How does photochemical smog form?

- The largest contributor to photochemical smog is vehicle emissions (exhaust fumes). Photochemical smog tends to form in the morning in large cities because vehicle emissions are high as a result of large numbers of people traveling to work.
- The nitrogen oxides produced by the car engines enter the atmosphere where, in the presence of sunlight, they react with volatile organic compounds (VOCs), specifically hydrocarbons, to form photochemical smog (diagram below). VOCs are found in the atmosphere as a result of **human activity** such as burning fossil fuels and also from naturally occurring processes.
- A feature of VOCs is that they have a high vapor pressure resulting from a low boiling point. This feature causes VOCs to evaporate (liquid → vapor) or sublimate (solid → vapor) at room temperature, i.e. they easily vaporize.
- Nitrogen oxides may also react with sunlight and then combine with molecular oxygen (O_2) to produce ozone (O_3). Whereas NO_2 levels peak in the morning, ozone production peaks in mid-late afternoon after the morning exhaust fumes have had time to react in the sunlight. The ozone in photochemical smog is called ground level ozone because it is located relatively close to the ground. This distinguishes it from **stratospheric ozone** which is formed naturally when sunlight energy splits molecular oxygen. Ground level ozone is regarded as a pollutant.

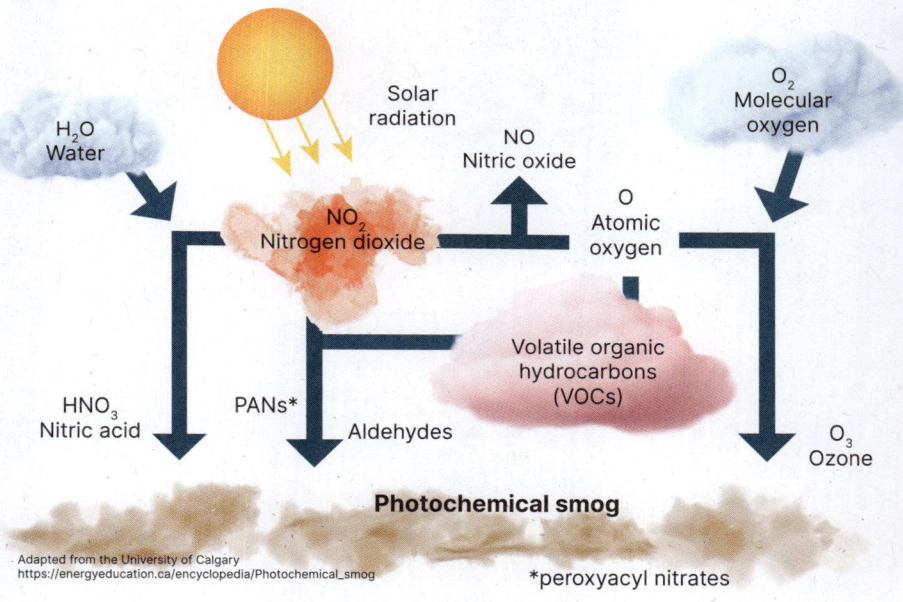

Photochemical smog over Mexico City, 2010

3. Suggest why most cities experience lower levels of smog at the weekend compared to weekdays:
_____
_____

4. Use the diagram above to outline how nitrogen oxides form the air pollutants in photochemical smog:
_____
_____
_____
_____

151 Acid Rain

Key Idea: Pollutants in the air can mix with water to fall as acid rain, damaging buildings and ecosystems.

Acid rain is not a new phenomenon. It was first noticed last century in regions where the industrial revolution began. In areas with heavy industrial activity, buildings and statues, especially marble, were being worn away by rain. Acid rain, more correctly termed acid deposition, can fall to the Earth as rain, snow, or sleet, as well as dry, sulfate-containing particles that settle out of the air. It is a problem that crosses international boundaries. Gases from coal-burning power stations in England fall as acid rain in Norway and Sweden, emissions from the United States produce acid deposition in Canada, while Japan receives acid rain from China. The effect of this fallout is to produce lakes that are so acidic that they cannot support fish, and forests with sickly, stunted tree growth. Acid rain also causes the release of **heavy metals**, e.g. cadmium and mercury, into the food chain by affecting the solubility of the metal salt or oxide (because solubility changes with pH). Acids also react with metals to form soluble salts. Changes in species composition of aquatic communities may be used as biological indicators measuring the severity of acid deposition. However, regulations implemented to reduce sulfate and nitrous emissions have significantly reduced the prevalence of acid rain worldwide.

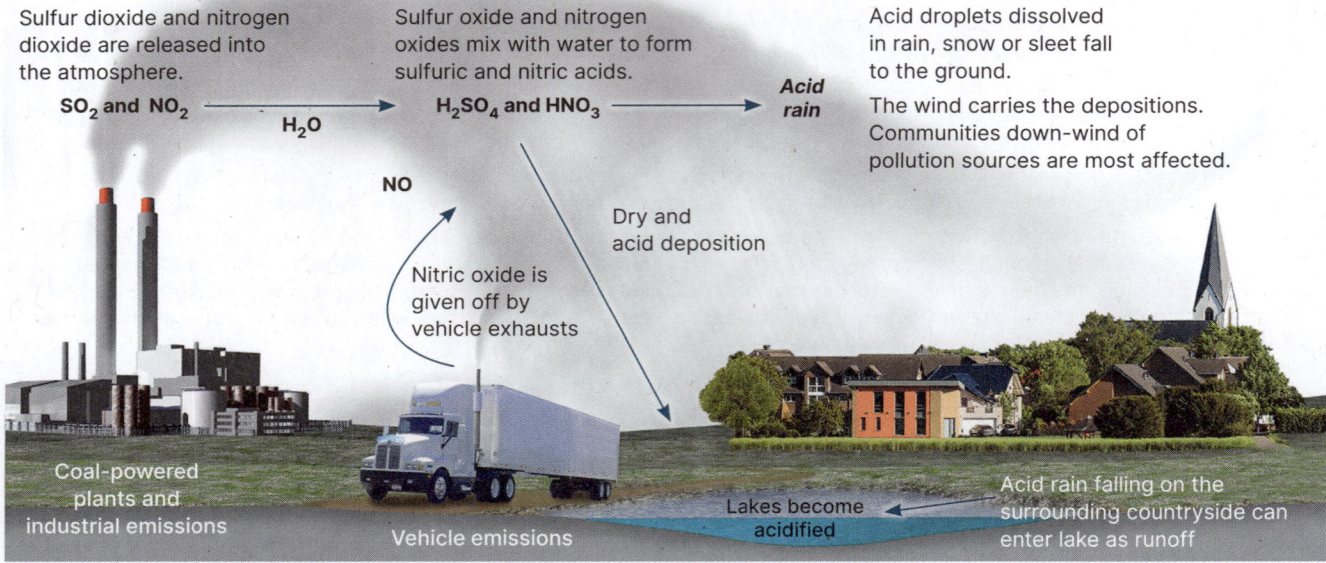

Cap-and-trade system to reduce acid rain

▸ In the 1980s, **environmental legislation** resulted in the adoption of many regulations in countries. These regulations aimed to limit the amount of sulfate and nitrous emissions released into the air by industries.

▸ In the US, the power plants were given 'allowances' for emissions, and if they released fewer, then they could financially benefit by selling unused allowances to other power plants. This system is known as 'cap-and-trade'.

▸ Lower future emissions targets resulted in lowering the cap incrementally, with correspondingly lowered **pollution**.

▸ Technology enabled the industries to respond with scrubbers to capture substances.

▸ The policies have almost eliminated acid rain across North America.

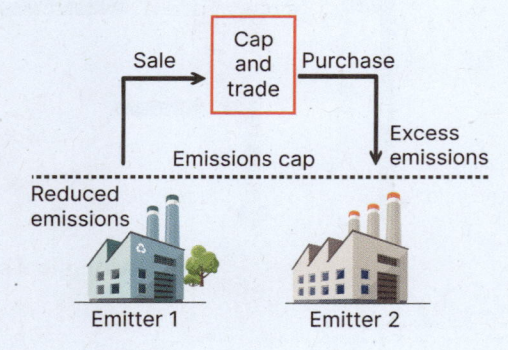

1. What is the link between air pollution and acid rain? _____
   _____
   _____

2. What brought about a reduction of acid rain, particularly in North America, after the 1980s?
   _____
   _____
   _____

152 Reducing Air Pollution

Key Idea: A shift in behavior, goods, and services choices can reduce air pollution at country, local, and individual level. At industry or government level, most countries have implemented **environmental legislation** to control and reduce **air pollution**. This legislation places limits on the emission of air **pollutants**. A transition from fossil fuel power stations to renewable energy sources for electricity generation is happening at both country and regional levels. However, households have the ultimate decision-making power when it comes to choosing their electricity providers, some of which offer renewable energy sources. Reducing air pollution can also be effective at a household or individual level. Town planning can facilitate a transition from fossil-fuelled vehicles to electric vehicles, public transport, safe-pathway facilitated cycling and walking. Some towns have regulations on backyard fires and open fires in houses. However, the use of these can be reduced by moving to low emitting heat pumps inside and compost bins outside. More companies are offering low emissions and low VOC household cleaners, paints, adhesives, and other products. Finally, fuelling vehicles and mowing lawns at times to avoid the heat of the day, can prevent most of the ozone formation.

Taking action to reduce air pollution in your own place

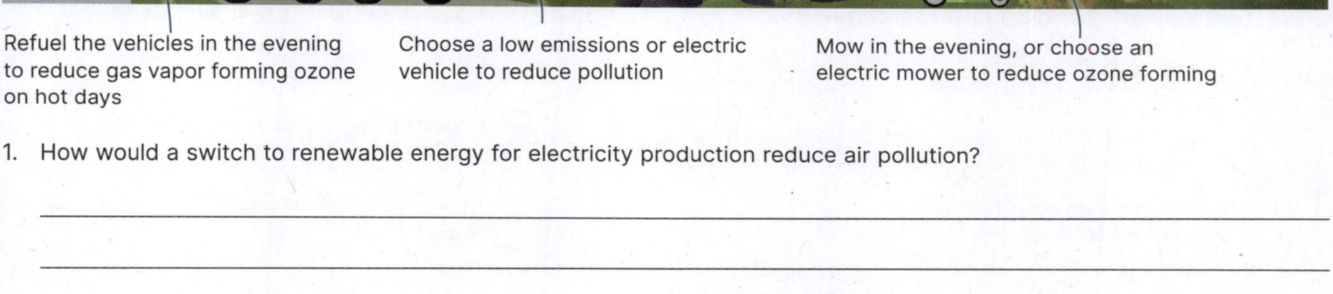

- Compost garden waste rather than burning
- Limit wood or coal burning house fires on high pollution cold days
- Choose low VOC household cleaners, paint, and other products
- Choose renewable energy for electricity without pollution
- Walk, bike or take the bus
- Refuel the vehicles in the evening to reduce gas vapor forming ozone on hot days
- Choose a low emissions or electric vehicle to reduce pollution
- Mow in the evening, or choose an electric mower to reduce ozone forming

1. How would a switch to renewable energy for electricity production reduce air pollution?
   _____
   _____
   _____

2. Why does a switch to low VOC products limit air pollution (hint: see activity on photochemical smog):
   _____
   _____
   _____

3. What measures have your local town, city, or region implemented to help reduce air pollution:
   _____
   _____
   _____

4. What personal actions could you or your family, or school, take to reduce their own air pollution emissions:
   _____
   _____
   _____

©2025 BIOZONE International
ISBN: 978-1-99-101409-2
Photocopying prohibited

153 Stratospheric Ozone Depletion

Key Idea: Ozone is an important molecule in the stratosphere, and its abundance is influenced by human activity.

In a band of the upper stratosphere, 17-26 km above the Earth's surface, exists a thin veil of renewable ozone (O_3). Ozone levels can fluctuate naturally through polar circulation and stratosphere temperatures. This ozone absorbs about 99% of the harmful incoming UV radiation from the sun and prevents it from reaching the Earth's surface. The problem of **ozone depletion** was first detected in 1984. Researchers discovered that ozone in the upper stratosphere over Antarctica was depleted during spring. Rather than a hole, it is more a thinning, where ozone levels typically decrease by 50% to 100%. Anthropogenic (human caused) ozone loss has also been observed over the Arctic. The main cause of ozone depletion is the widespread use of chemicals, specifically chlorofluorocarbons (CFCs). Since 1987, after the global support of the Montreal Protocol, nations have cut their consumption of ozone-depleting substances by 70%, although the phaseout is not complete and there is a small but significant black market in CFCs. Ozone loss is projected to diminish gradually until around 2070 when the polar ozone holes will return to 1975 levels.

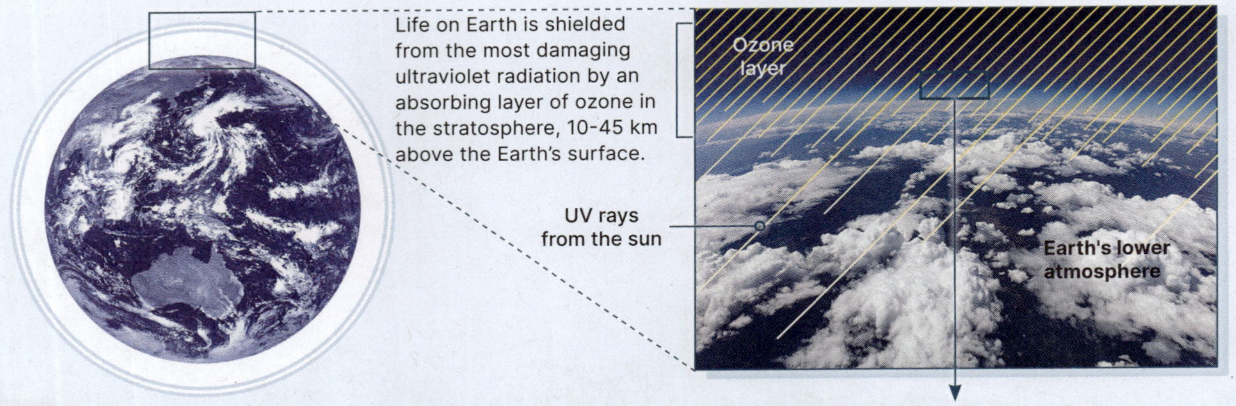

Life on Earth is shielded from the most damaging ultraviolet radiation by an absorbing layer of ozone in the stratosphere, 10-45 km above the Earth's surface.

UV rays from the sun

Ozone layer

Earth's lower atmosphere

In addition to health problems such as severe sunburn, skin cancers, and cataracts of the eye in both humans and other animals, high levels of UV-B radiation can also lead to immune system suppression in animals. It can also cause lower crop yields, a decline in the productivity of forests and surface-dwelling plankton, increased smog, and changes in the global climate.

Sources of ozone depleting chemicals

The chemicals below drift up to the stratosphere, where ultraviolet radiation causes release of free chlorine, a highly reactive chemical. Although remaining stable at lower altitudes, the CFC breaks apart and releases chlorine that acts as a reactive catalyst to break down ozone.

Chloro-fluoro-carbons (CFCs)
- Propellants for aerosol cans
- Coolants in air-conditioners
- Coolants (freon) in refrigerators
- Styrofoam insulation/packaging
- Medical sterilizers

Halons
- Used in many fire extinguishers
  Methyl bromide
- Used as a fumigant in agriculture
  Methyl chloroform
- Used to degrease metals
  Carbon tetrachloride
- Used in many industrial processes

UV light hits a CFC molecule and releases a chlorine atom

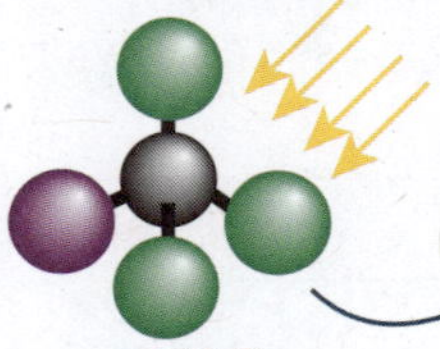

CCl_3F Chlorofluorocarbon (CFC)

The destruction of ozone by free chlorine

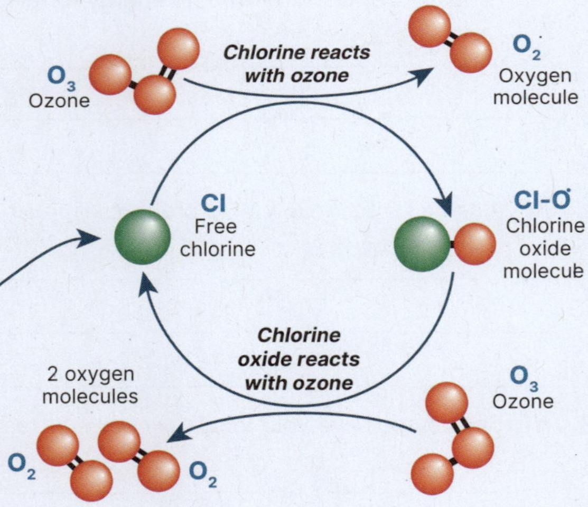

O_3 Ozone — Chlorine reacts with ozone → O_2 Oxygen molecule

Cl Free chlorine

Cl-O Chlorine oxide molecule

2 oxygen molecules — Chlorine oxide reacts with ozone — O_3 Ozone

O_2 O_2

1. Explain how human activity has influenced the growth of the ozone hole since 1980:

___

___

2. Why did the size of the ozone hole trend downwards after 1987? ___

___

©2025 BIOZONE International
ISBN: 978-1-99-101409-2
Photocopying prohibited

The ozone hole

▶ A large 'hole' in the **stratospheric ozone** layer develops over Antarctica each summer, dropping the ozone well below its normal level. The size and intensity of the hole grew from 1980 until 2000, as can be seen in the satellite data below. In recent years, a similar hole has developed over the Arctic. In 2006, the extent of the hole above Antarctica was the largest ever, but depletion levels were slightly less than 1999.

▶ The Dobson Unit (DU) is used to describe relative column ozone levels (the ozone between the Earth's surface and outer space). In the tropics, ozone levels are typically between 250 and 300 DU year-round. In temperate regions, seasonal variations can produce large swings in ozone levels. These variations occur even in the absence of ozone depletion. Ozone depletion refers to reductions in ozone below normal levels after accounting for seasonal cycles and other natural effects.

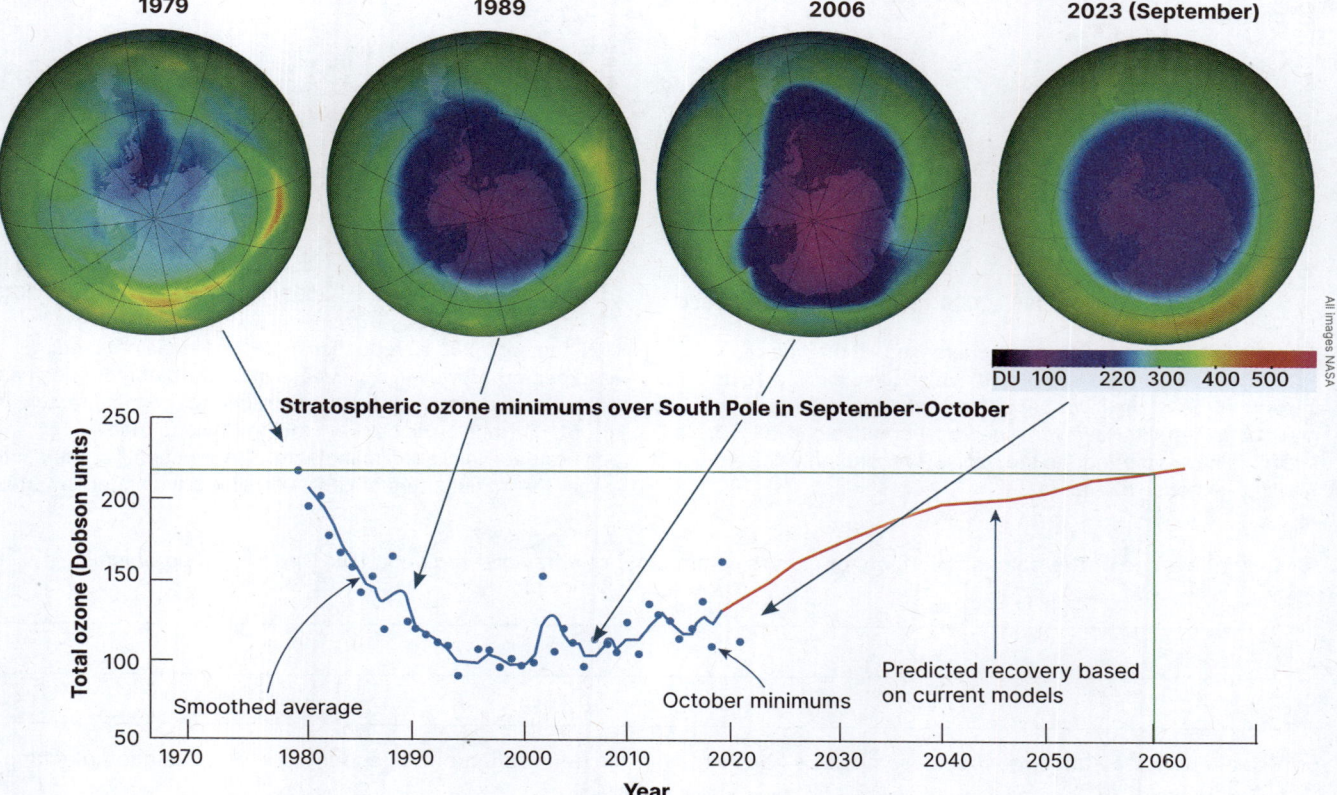

▶ Mapping of the ozone layer first began in 1979 and the hole over Antarctica was discovered shortly afterwards.

▶ The ozone layer surrounds the entire planet. Under certain conditions, this layer is thinned in places. The annual thinning is most well-known over Antarctica, however it also thins over the Arctic, although more rarely. In 2020, the Arctic experienced its largest ozone hole, nearly three times the size of Greenland. This lasted for over a month.

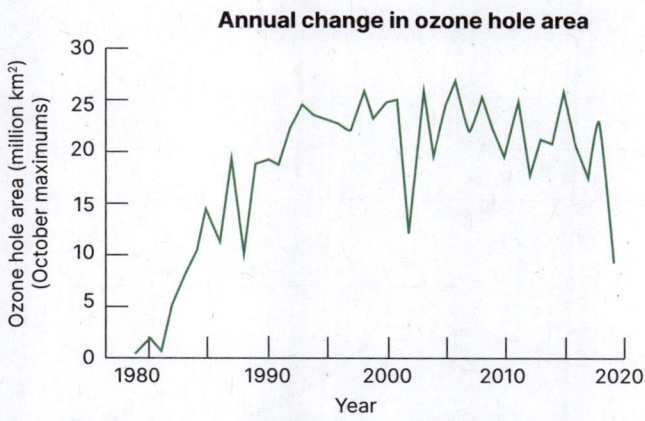
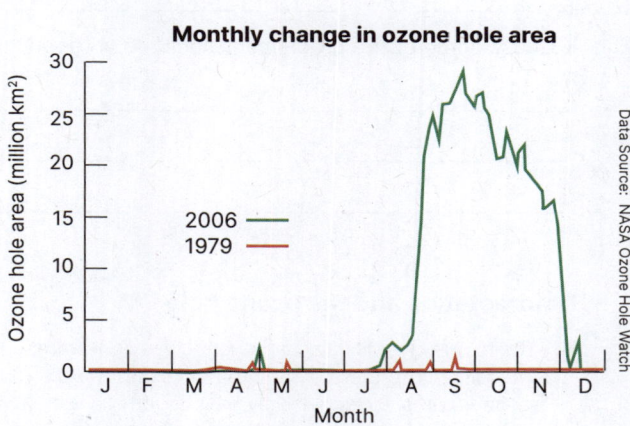

3. One Dobson unit is equivalent to a gas layer 10 μm thick (at 0°C and one unit of pressure). What does this tell us about the thickness of the ozone layer that protects organisms from harmful UV light?

4. Ozone is formed when UV light reacts with oxygen gas (O_2). What does this tell us about the early conditions for life on land before photosynthetic organisms evolved?

Replacing CFCs

▶ CFCs were developed in the 1920s and were in widespread use by the 1970s. However, the dramatic evidence of their effect on the ozone layer shown in the late 1970s was enough for countries to act collectively to ban their use.

▶ Since then, other chemicals to replace CFCs have been developed. However, these have their own sets of problems, the most notable being that most of them are very potent greenhouse gases.

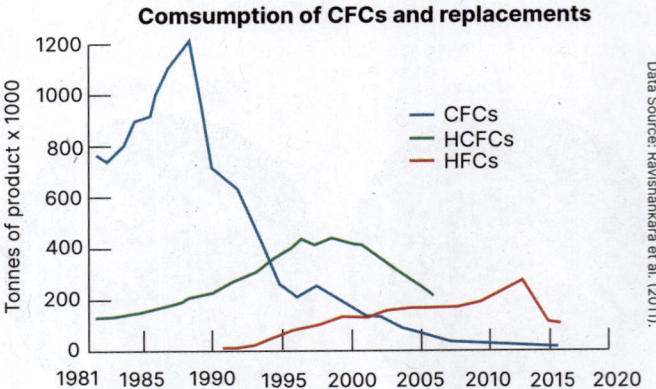

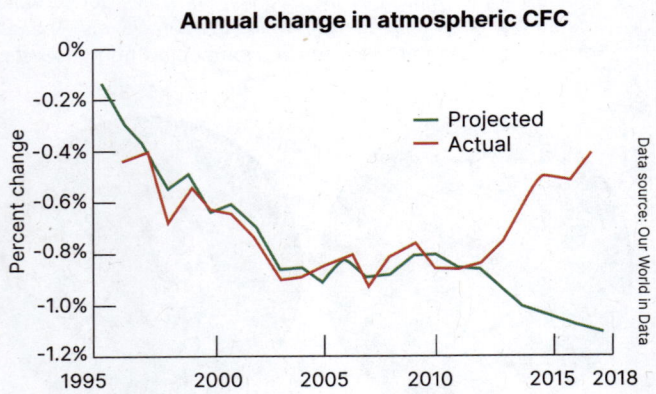

Since 1987, the use of CFCs has decreased. In their place, HCFCs and HFCs were developed, which have minimal to no impact on the ozone layer but unfortunately contribute to the greenhouse effect. However, a newer alternative called HFO seems to have no effect on the ozone layer and only has a weak greenhouse effect.

Until the early 2010s, reduction in CFC use was proceeding as expected. However, a decrease in the rate of reduction has been observed and this has been traced to the manufacture of CFC-11 (trichlorofluoromethane) in China for use in polyurethane insulation manufacture. International agreements meant the manufacture of CFC-11 should have ended in 2010.

5. Describe some of the damaging effects of excessive amounts of ultraviolet radiation (UV-B) on living organisms:

6. Explain how the atmospheric release of CFCs has increased the penetration of UV (especially UV-B) radiation reaching the Earth's surface:

7. Discuss some of the problems associated with reducing the use of ozone depleting chemicals:

Deforestation and the ozone hole

▶ The ozone layer is vital to prevent UV-B light from reaching the soil. Where there are thin patches and the UV-B radiation reaches forested land, it may have a detrimental effect on the soil rhizosphere (microorganisms), affecting nutrient availability for tree and other plant growth.

▶ Conversely, the loss of forest cover through deforestation reduces the amount of oxygen released via photosynthesis. Ozone needs a continuous supply of oxygen and breaks down after a few years.

▶ Ozone can also form close to ground level due to human caused pollution. This ozone is harmful to organisms. Trees have been found to be more effective at removing this polluting ozone than any mechanical devices.

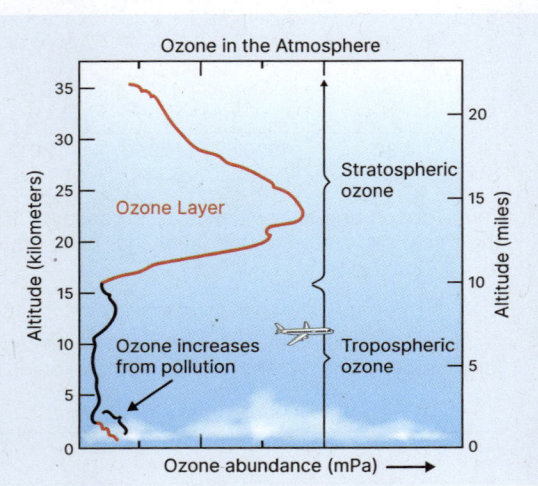

154 Noise Pollution

Key Idea: Noise pollution, often from human activity, can affect the wellbeing of humans and animals.

Pollution does not necessarily have to be chemical in its nature. With the development of industry, technology, and large human populations comes an increase in noise. Noise pollution is a growing issue in many areas, not just in cities, which are noisy places simply by their nature, but also in areas where noise or very loud, persistent sounds are uncommon. Aside from being irritating, excessive noise pollution can also affect breeding and feeding behavior of animals.

Cities are very noisy places, with the constant drone of machinery and traffic audible throughout day and night. This persistent noise can create excessive stresses on people and, in extreme cases, can lead to insomnia or psychosis.

Away from cities, noise can still be a problem. Farm machinery can produce constant, low frequency noise, while equipment such as bird scarers (above) produce very loud, irregular bangs which can be irritating to some.

Noise pollution from the engines and sonar of ships can travel long distances in the ocean, affecting the sonar and communication abilities of whales. As a result, whales may face challenges in navigating their surroundings. This could potentially contribute to strandings.

Acceptable noise levels for urban environments are set by local councils. Careful planning of suburban areas can reduce the amount of noise pollution. Measures such as separating industrial and residential areas decrease clashes between industrial and residential expectations of noise levels. Noise pollution can be reduced by better design and insulation of equipment and buildings.

Cause of sound	Loudness dB
Blue whale call	160
Space shuttle at launch	
Artillery fire	140
Small arms fire	
Jet aircraft at take-off	120
Rock concert	
Chainsaw	100
Lawn mower	
Alarm clock	80
Busy traffic	
Normal conversation	60
Moderate rainfall	
Quiet room	40
Isolated desert night	
Whisper	20
Normal breathing	0

Sound levels are measured in decibels (dB). The scale is logarithmic so that 30 dB has ten times more energy than 20dB and one hundred times more than 10 dB. Exposure to noise levels above 85 dB can cause permanent hearing loss, while pain is experienced at around 120 dB.

Noise pollution affecting birds

Research has shown that noise pollution from traffic is stunting the growth of birds in cities. Zebra finches (above) exposed to noise inside the egg were less likely to hatch, and the birds were born smaller and lighter. The noise had genetic effects, causing shorter, more eroded telomeres: a component of DNA. When birds reached adulthood, they were producing around half the number of offspring of birds not exposed to traffic noise. Behavior in other bird species was also affected by noise.

1. Describe some of the effects of noise pollution:

   _____

   _____

2. Describe some ways in which noise pollution can be reduced: _____

   _____

   _____

3. Identify some sources of noise pollution in your community: _____

   _____

   _____

155 Pollution in the Home

Key Idea: Chemicals that are part of a house or products inside the house can sometimes cause serious health issues. People are often surprised by the number of dangerous chemical **pollutants** in their homes, even when excluding chemicals such as pesticides and cleaning agents. Many of these chemicals, including '**forever chemicals**', are used as preservatives or are created during the use of an appliance. Others are part of the fabric of clothes or carpet, breaking apart to form micro- or **nano plastics**. Exposure to low levels of pollutants over an extended period of time can lead to significant health issues that are both expensive to treat and potentially dangerous. However, countries are now starting to implement stricter regulations to closely monitor and control these harmful chemicals.

Sources of indoor air pollution

▸ Often, when we think of air pollution, we think of problems we see outside, such as smog in cities. However, **air pollution** can also occur indoors when pollutants accumulate inside homes, schools, workplaces, and vehicles.

▸ Pollutants can concentrate quickly inside so that the air indoors can contain higher concentrations of pollutants than outside. There are many sources of indoor air pollution (below) and many have harmful health effects if present in high enough concentrations over time.

▸ Indoor pollutants can be human-made or natural, and the type of air pollutants present differs depending on a building's use, e.g. commercial or domestic.

▸ Indoor **pollution** varies between countries depending on income level and technological development. Indoor pollution is generally a bigger problem in low-income countries where an open fire is used to cook indoors.

Mold and mildew can grow in poorly ventilated spaces. Inhalation of spores from these organisms can cause breathing problems, including asthma.

Tobacco smoke contains particulate matter and dozens of toxins including formaldehyde, carbon monoxide, hydrogen cyanide, and nitrogen oxides. Many countries have banned smoking indoors in public places to stop non-smokers being exposed to second-hand smoke.

Open fires are commonly used for cooking and heating in developing countries. Homes are often poorly ventilated, so smoke, particulates, and CO can reach very high levels. Even in developed countries, **gas cookers and heaters** release CO, NO_2, and particulates.

Radon-222 (a radioactive gas).

Particulate material refers to substances such as asbestos, dust, and smoke. Asbestos, a naturally occurring silicate mineral, was commonly utilized in the production of various household and industrial items, including construction materials. Inhaling **asbestos** dust can lead to serious lung diseases, including cancer. While it is still permitted for use in the United States, many other countries have banned its use due to its harmful effects.

VOCs (volatile organic compounds) are released from many products found in our homes and workplaces. Common sources include furniture, carpets, cigarettes, solvents, paint, glue, cleaning products, air fresheners, and photocopiers.

Prior to 1978, most homes were painted with **lead-based paint**.

Sulfur dioxide and carbon monoxide (CO) are present in car emissions. Breathing in high concentrations of CO reduces oxygen availability to cells. CO is an asphyxiant (deprives someone of oxygen).

Radon-222

Radon-222 is a naturally occurring, radioactive gas. It is produced as an intermediate product when uranium decays to lead. It is present in soils and rocks, and mostly enters buildings from the soil through small cracks and spaces in a building's walls or foundations. It may be present in groundwater too. All buildings contain some radon, but levels are highest in lower levels, e.g. basements. In normal circumstances, radon is the single largest contributor to a person's background radiation dose.

Radon is a colorless, odorless, tasteless gas. It is easily inhaled. Prolonged exposure increases a person's chances of developing lung cancer. Radon is the second largest cause of lung cancer after smoking in the USA.

1. A list of indoor pollutants is provided below. Categorize their source as human-made, natural, or products of combustion.
   Radon, carbon monoxide, dust, mold, asbestos, smoke, VOCs, nitrogen oxides, formaldehyde, mildew.

Human-made	Natural	Combustion

Persistent organic pollutants, metal ions, and forever chemicals in the house

▸ Dangerous chemicals can be used in the manufacture of household appliances and furniture, and in substances such as paint to decorate the house or insulation.

▸ When a product breaks down over the life-time of use, some of these chemicals may leach out and be inhaled or ingested.

▸ **Environmental legislation** controls what chemicals can be incorporated into new products, and have reduced toxins, but legacy items that may remain in the house for many decades still persist.

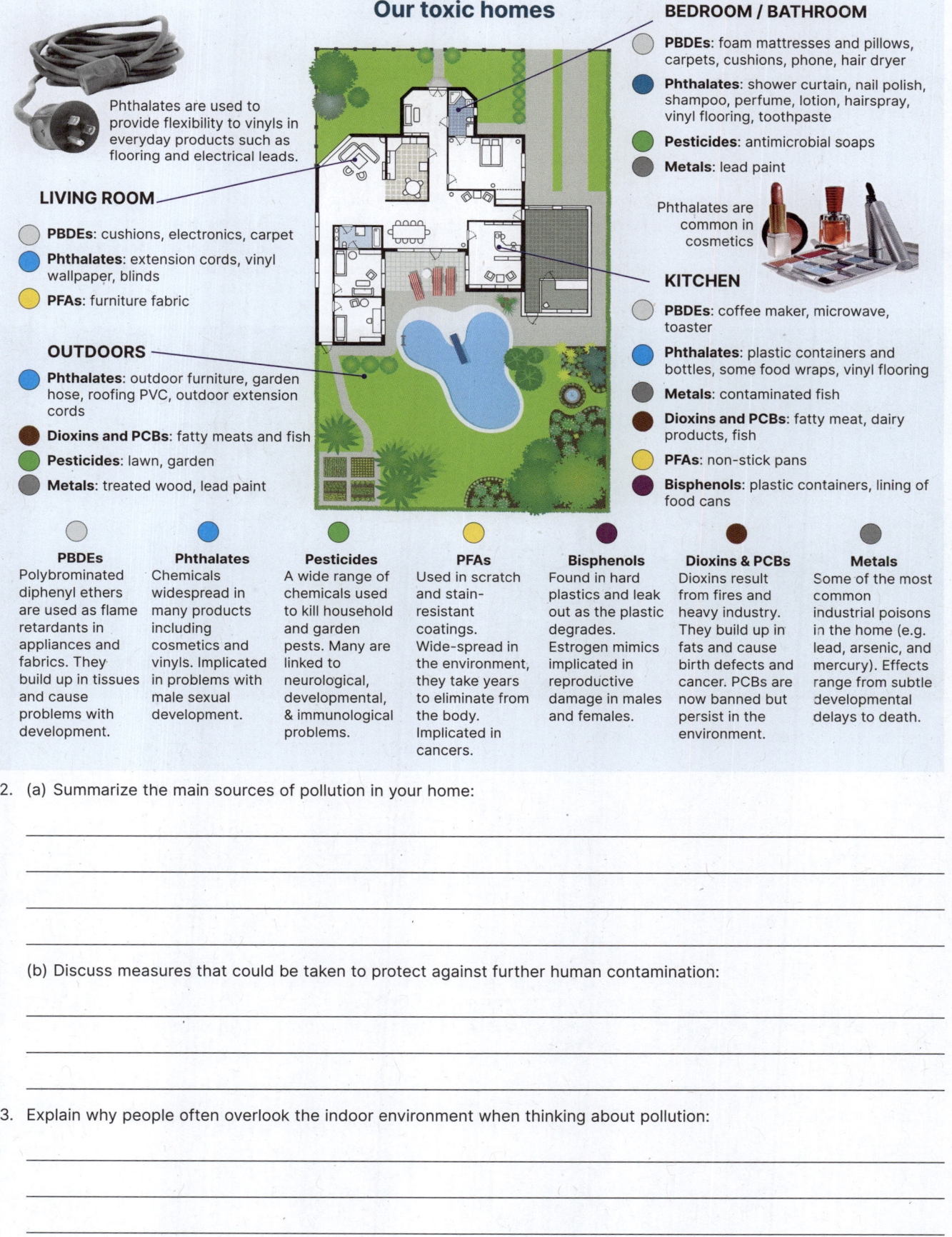

2. (a) Summarize the main sources of pollution in your home:

_____
_____
_____
_____

(b) Discuss measures that could be taken to protect against further human contamination:

_____
_____
_____

3. Explain why people often overlook the indoor environment when thinking about pollution:

_____
_____
_____

156 Light Pollution

Key Idea: Human-caused light pollution in populated areas is creating issues for both people and other organisms.

Almost all populated areas experience **light pollution** and people are unable to view a natural night sky. Cheaper electrical lighting and the advent of brighter LED bulbs have meant that light pollution continues to increase, causing a loss of night sky and star visibility in many places around the world. Light pollution inhibits astronomy, both amateur and scientific, diminishes aesthetics, wastes energy in powering artificial light, and can interfere with indigenous practices involving timing activities to star and lunar observations. Ecosystems can also be affected, creating mistiming for phenology (cyclic or seasonal behavior), such as migration, breeding, hunting, and pollination.

Where have all the stars gone?

- Recent data shows that Earth has experienced an almost a 10% loss of visible stars in the decade from 2011 to 2022 due to continued light pollution.
- 80% of the world lives under a night sky with reduced visibility due to light pollution. This increases up to 99% for most of Europe and the United States.
- In the USA alone, there are around 160 million outdoor lights and up to 30% of them are poorly shielded or aimed.
- Each day in the USA, wasted outdoor lighting contributes as much CO_2, a greenhouse gas, as does three million cars.
- Reducing light pollution can be achieved in your own homes or businesses by eliminating unnecessary lighting and lowering the lamp bulb wattage, shining light only where necessary and aiming downward with shades where possible, using timers to turn off lighting when not needed, and using longer wavelength, i.e. more yellow and red, light emitting bulbs.

Harbor view at night over Amsterdam port

Night sky showing Milky Way at Tekapo, New Zealand

1. (a) Summarize the main issues of light pollution: _____

   (b) What specific changes can be made in the usage of artificial light to effectively reduce light pollution?
   _____

2. (a) Describe the relative amount of light pollution in your area (low, medium, or high). (Links to light pollution data can also be found in the **BIOZONE Resource Hub**).

   Location: _____ Level of light pollution: _____

   (b) In a small group, discuss how light pollution may affect you personally. Record your main points below:
   _____

157 Health Effects of Pollution

Key Idea: All types of pollution have direct effects on the health of humans because of unintentional exposure.

The effects of **pollution** on health depend on concentration and type of **pollutant**, and extent of exposure. People living in cities are usually more exposed to pollutants than those in rural areas, although those living in or near intensively farmed areas using fertilizers and pesticides can also be exposed to high levels of toxins. While it is difficult to avoid some pollutants, such as carbon monoxide from busy traffic, others, such as cigarette smoke, can be more easily avoided. Some urban areas use an Air Quality Index (AQI) to alert residents to avoid outdoor air when it becomes too bad.

Air pollution

Air pollutants, such as lead, severely affect nerve function. CO reduces the blood's ability to carry oxygen and results in headaches and impairs thinking and reflexes. SO_2, NO_x and O_3 detrimentally affect respiratory function.

Sources: Fine particles, NO_x, SO_2, PCBs, O_3, Lead, CO

- Headache
- Nerve damage
- Respiratory illness
- Cardiovascular disease
- Vomiting and nausea
- Gastrointestinal illness
- Cancer
- Skin irritation

Soil contamination

Hazardous chemicals, Fertilizers, Pesticides

Pesticides based on organophosphates are extremely toxic to humans and other mammals. Fertilizers can cause types of cancer and respiratory illness.

Radiation

UV radiation, Radioactive waste

UV radiation from the sun causes thousands of cases of melanoma skin cancer every year, while radioactive waste can cause cancer and genetic fetal defects.

Water pollution

Mercury, Lead, Bacteria, Parasites

Heavy metals such as mercury and lead can cause nerve damage, while bacteria such as cholera or parasites such as giardia, cause intestinal illness.

Toxic gases and ultra-fine particles produced by industry cause lung diseases. Globally, fine particulate and ozone pollution are responsible for around 8.3 million excess deaths each year.

Water polluted by fertilizers, pesticides, heavy metals, and untreated **sewage** can cause serious health issues. In some countries, including many in Africa, barely 5% of sewage is treated.

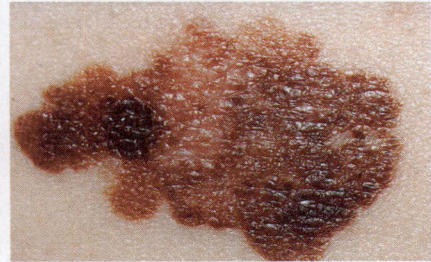

Higher levels of UV radiation as a result of **stratospheric ozone** deletion could cause an extra 300,000 cases of skin cancer and 1.5 million extra cases of cataracts per year.

1. Select one type of health-impacting pollution from above that you might be familiar with and discuss how it might affect people in your community:

___
___
___
___

158 Effect of Oil Spills

Key Idea: Accidental oil spills have occurred during the history of oil extraction, causing widespread and long-lasting damage to the aquatic ecosystems and organisms nearby.

Oil is arguably one of the most important chemicals in human economics. It provides power for transport and electricity, and the raw materials for many consumer products, including plastics. Billions of dollars a year are spent on removing it from the ground and billions more made in revenue from its sale. However, crude oil is a very toxic substance and removing it from reservoirs is fraught with difficulty and danger. Some of the biggest, man-made **environmental disasters** have occurred because of the search for and transport of oil. Oil tankers carry huge volumes of crude oil over the seas and are some of the largest ships afloat. As a result, there is enormous potential for disaster if one is grounded. The grounding of the Exxon Valdez is one of the most infamous examples.

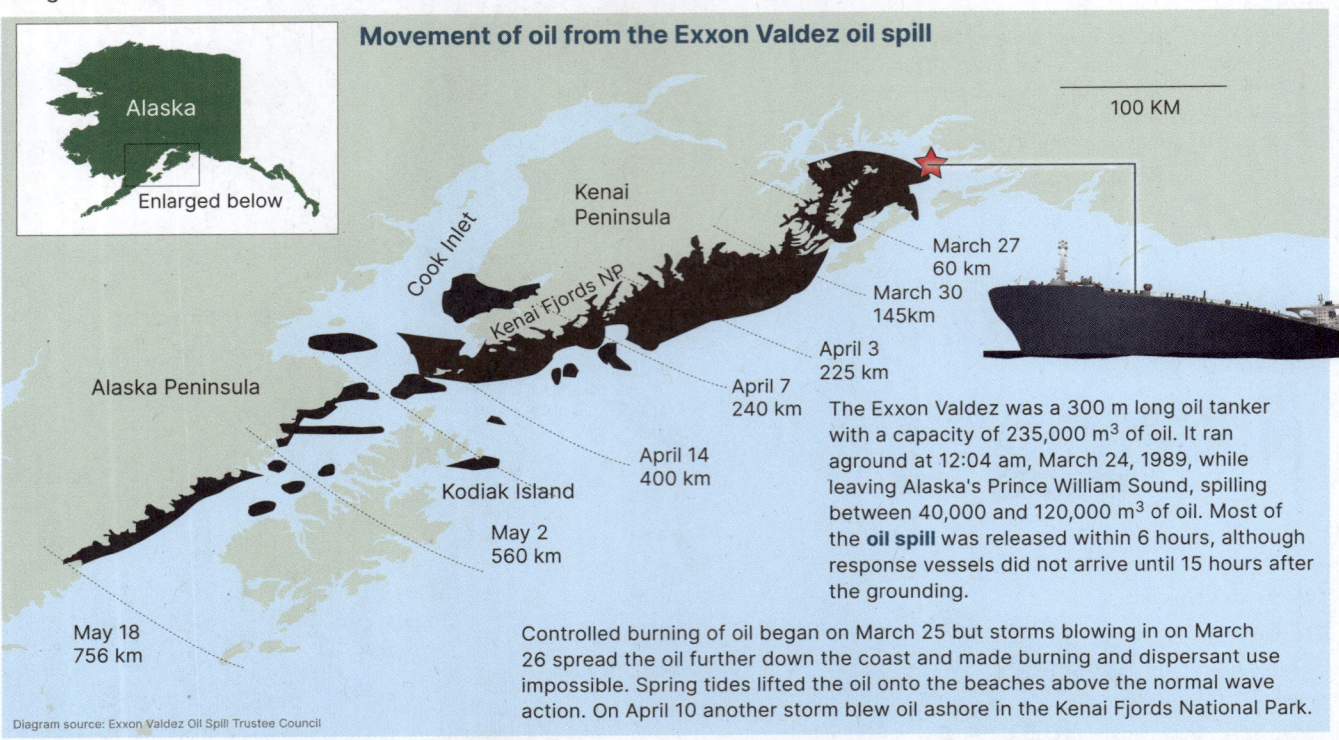

Movement of oil from the Exxon Valdez oil spill

The Exxon Valdez was a 300 m long oil tanker with a capacity of 235,000 m^3 of oil. It ran aground at 12:04 am, March 24, 1989, while leaving Alaska's Prince William Sound, spilling between 40,000 and 120,000 m^3 of oil. Most of the **oil spill** was released within 6 hours, although response vessels did not arrive until 15 hours after the grounding.

Controlled burning of oil began on March 25 but storms blowing in on March 26 spread the oil further down the coast and made burning and dispersant use impossible. Spring tides lifted the oil onto the beaches above the normal wave action. On April 10 another storm blew oil ashore in the Kenai Fjords National Park.

Diagram source: Exxon Valdez Oil Spill Trustee Council

There were several causes of the disaster. The crew of the Exxon Valdez had not had their mandatory rest period and were fatigued. They had also failed to maneuver the ship correctly (probably due to fatigue), and the radar system that could have informed the crew of a collision had not been repaired and was not operating.

The clean-up operation was made more difficult by the remote location of the oil spill. Food, equipment, and shelter had to be brought in for up to 11,000 workers, along with fuel and dispersant equipment and vehicles. The clean-up stopped in September due to the approach of the Alaskan winter, but was restarted in April 1990.

357 sea otters were treated after the spill, at an estimated cost of US$51,000 per otter. Fisheries in the area were closed, including black cod and Pacific herring. It is estimated that around 87% of the herrings' spawning grounds were oiled. Mollusks were found to contain higher than normal levels of aromatic chemicals after the spill.

Approximately a quarter of a million seabirds, 2800 sea otters, 300 harbor seals, 250 bald eagles, 22 killer whales, and countless fish were killed in the first weeks of the spill. Oil was still found 20 years later, not far beneath the surface of many of the affected beaches, despite one of the biggest clean-up operations in US history.

1. (a) Explain how the Exxon Valdez spill could have been avoided: _____

   _____

   (b) Describe some of the effects of the spill: _____

   _____

   (c) Explain why the clean up of this spill was particularly difficult: _____

   _____

©2025 **BIOZONE** International
ISBN: 978-1-99-101409-2
Photocopying prohibited

The Deepwater Horizon oil spill (the Gulf spill) and the ecosystem

▸ The coastlines of Louisiana, Mississippi, and Florida are important and delicate ecosystems. Their estuaries, marshes, and wetlands provide habitats for a range of species including oysters, mudcrabs, and many seabirds, and they act as shelter for fish fry before they return to the ocean. The Gulf of Mexico itself is home to many important commercial fish species and to 70% of the US shrimping industry.

▸ The oil that leaked during the Deepwater Horizon oil rig from the Macondo Prospect in 2010 placed all of these habitats at risk and many may remain so for many years to come.

▸ BP (British Petroleum) was the main oil company responsible for the Deepwater Horizon oil spill accident. After the event, they financially supported research into technology to remediate oil spills and education information in order to help improve their public image.

▸ After the Deepwater Horizon oil spill, a moratorium (temporary ban) was imposed on deep sea oil extraction and mining permits. This ban remained in effect for several years before it was eventually lifted.

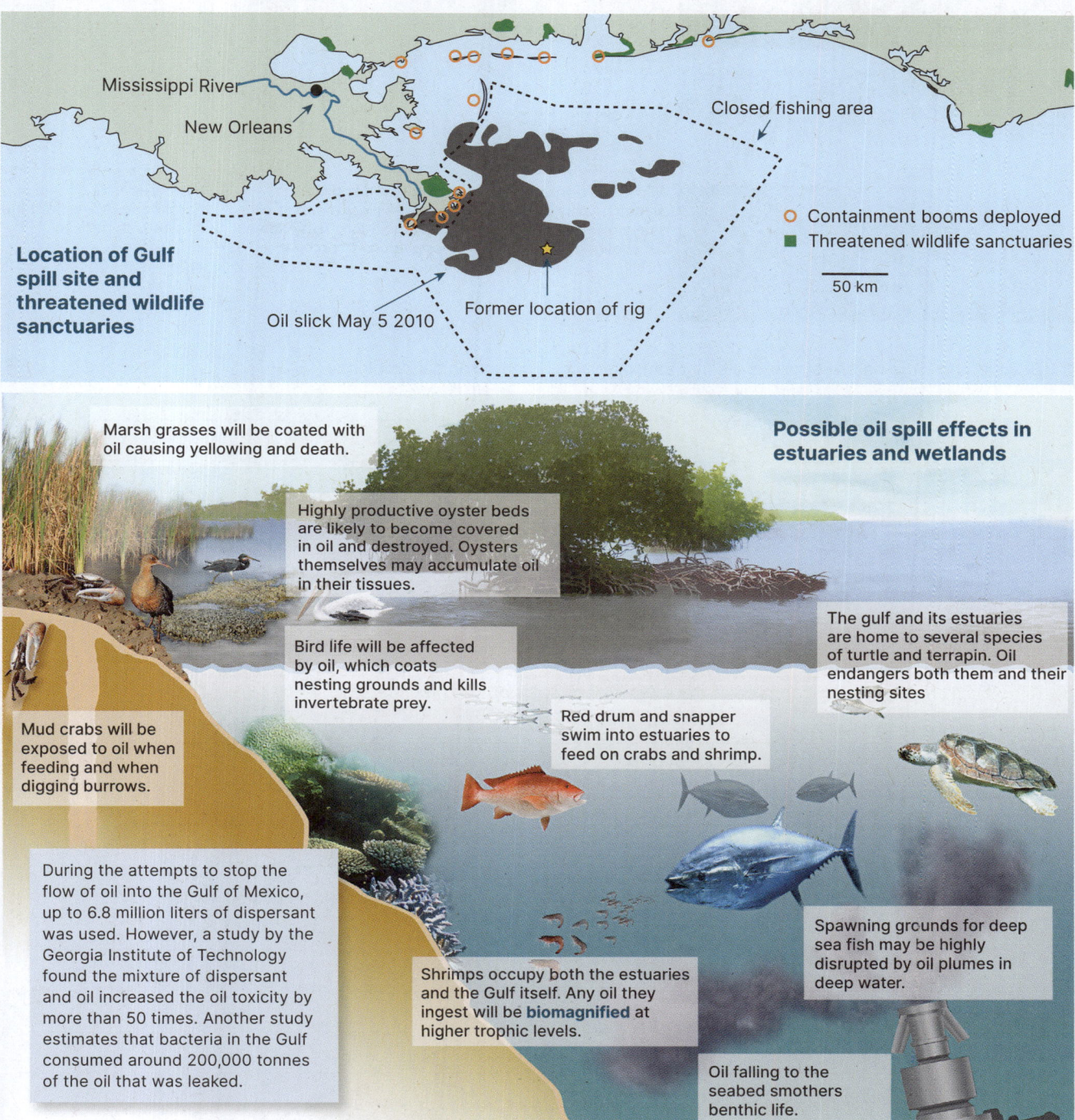

2. What factors influence the region and amount of damage that is experienced by aquatic ecosystems after an oil spill?

_____

_____

_____

Ecosystem recovery time

The rate of ecosystem recovery after an oil spill depends on the habitat involved, the local climate and conditions, and the type of oil spilled.

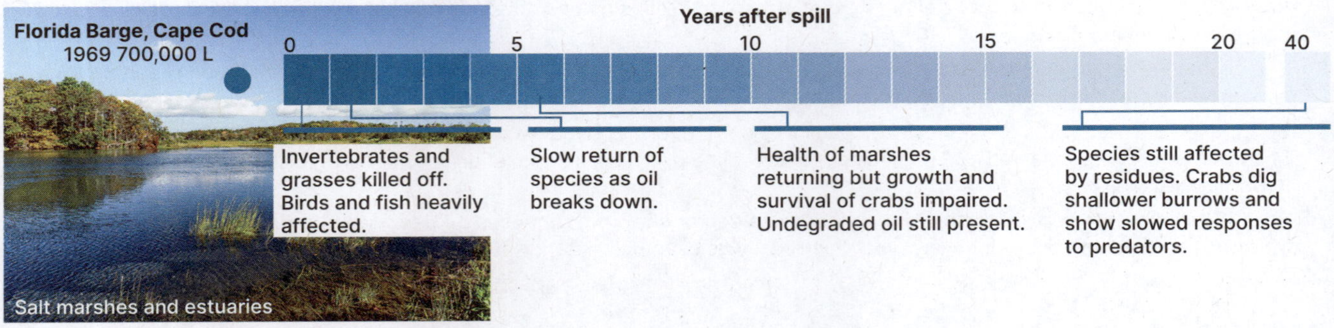

Florida Barge, Cape Cod
1969 700,000 L

Years after spill: 0, 5, 10, 15, 20, 40

Salt marshes and estuaries

- Invertebrates and grasses killed off. Birds and fish heavily affected.
- Slow return of species as oil breaks down.
- Health of marshes returning but growth and survival of crabs impaired. Undegraded oil still present.
- Species still affected by residues. Crabs dig shallower burrows and show slowed responses to predators.

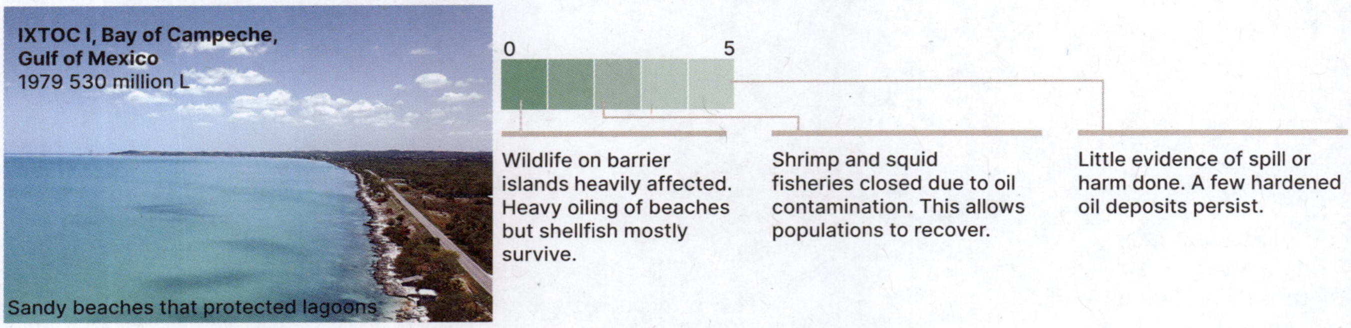

IXTOC I, Bay of Campeche, Gulf of Mexico
1979 530 million L

Sandy beaches that protected lagoons

- Wildlife on barrier islands heavily affected. Heavy oiling of beaches but shellfish mostly survive.
- Shrimp and squid fisheries closed due to oil contamination. This allows populations to recover.
- Little evidence of spill or harm done. A few hardened oil deposits persist.

3. Describe how the nature of oil results in it causing such large ecological problems:

_____
_____
_____

4. Describe some immediate effects of oil on the wildlife it comes in contact with:

_____
_____
_____

5. (a) Explain why it is difficult to predict the effects of oil spills on ecosystems:

_____
_____
_____
_____

(b) Explain why the size of the oil spill is not necessarily linked to the magnitude of the ecological effect:

_____
_____
_____
_____

6. Explain why oil spills are rarely just ecological disasters but economic ones too:

_____
_____
_____
_____

159 Cleaning Up Oil Spills

Key Idea: A range of methods are used to clean oil spills.

Oil spills can be extremely difficult to clean up due to the sticky nature of the oil. Heavy crude is one of the most difficult substances to clean as it forms thick, viscous slicks that are difficult to disperse. The environment also plays a major part in the clean-up: warm temperatures contribute to the evaporation of large quantities of volatile compounds from the oil and allow bacteria to quickly break it down. Wave action also helps to oxygenate and degrade the oil. Human efforts include confining and burning large slicks, using chemical dispersants, and cleaning beaches with steam or digging up sand and removing it for **remediation**.

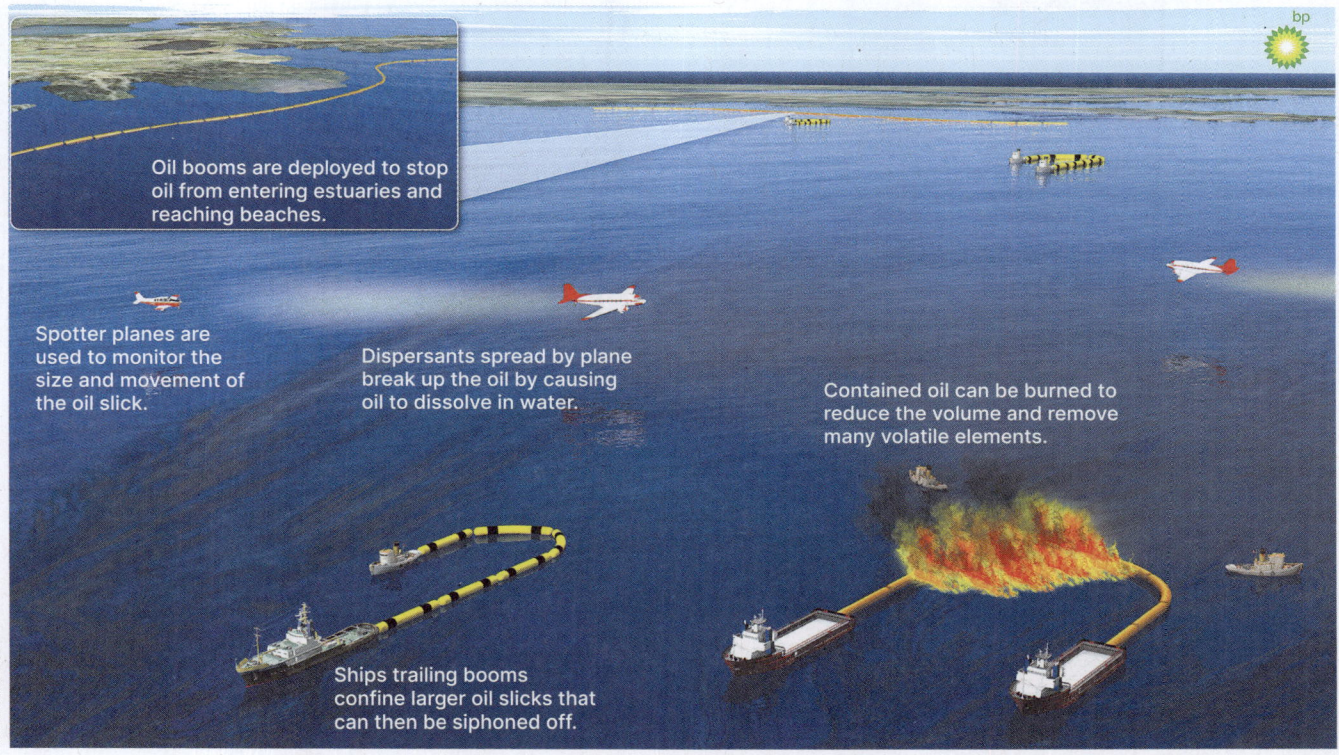

Oil booms are deployed to stop oil from entering estuaries and reaching beaches.

Spotter planes are used to monitor the size and movement of the oil slick.

Dispersants spread by plane break up the oil by causing oil to dissolve in water.

Contained oil can be burned to reduce the volume and remove many volatile elements.

Ships trailing booms confine larger oil slicks that can then be siphoned off.

Floating booms designed to stop oil movement are used to protect sensitive areas. More than 550 km of booms were deployed in the Gulf.

Oil captured by skimmer boats is set alight to reduce its volume, although with dense smoke. Around 38,000 m^3 of oil was burned in the Gulf.

Beaches may need to be dug up/removed for remediation. On accessible beaches, large graders can be used to scrape up large amounts of sand and oil and remove it from the beach. New sand may be brought in to maintain the beach.

Chemical dispersants are dropped from the air to clear surface slicks. These were also injected into the oil at the Mocondo wellhead to try to disperse oil before it reached the surface. Long term effects of using dispersant in this way is unknown.

Hot water and steam are used to clean rocks and beaches. This method was used during the Exxon Valdez clean-up but not during the Gulf clean-up as it has the potential to harm rock dwelling organisms such as barnacles.

Water quality tests are carried out on a regular basis to track the progress of the oil. Bacteria provide natural bioremediation and began to quickly break down the oil. A large plume of oil in deep water can be dispersed by bacterial action.

Skimmer boats collect oil floating on the surface. This can be burned or siphoned off and processed using centrifuges to separate the oil.

Bioremediation to clean up oil spills

- Bacteria can be used to metabolize hydrocarbons that form the oil in spills.
- Bioremediation, using organisms such as oil-eating bacteria, was used in the Deepwater Horizon oil spill in the Gulf of Mexico in 2010, to complement more traditional methods.
- In 2008, Oilzapper, a mixture of five bacteria that can metabolize hydrocarbons, was used to address an oil spill caused by a crude oil trunk line rupture near Gujarat in Western India. Over a period of five months, Oilzapper successfully achieved a biodegradation rate of 90-96% for the spilled oil. This was much more cost effective than allowing natural breakdown to occur in specially constructed sludge ponds, which are expensive to build and occupy land that cannot be used for any other purpose for a long time.

Oil contains hydrocarbons (compounds made up of hydrogen and carbon). CH_4

Hydrocarbon digesting bacteria (e.g. *A. borkumensis*) are introduced into the contaminated area.

The microbes metabolize the hydrocarbons. Hydrogen and carbon from the oil are added to oxygen to form water and carbon dioxide. H_2O CO_2

Chemical dispersants are added to an oil spill to break the oil up into smaller droplets. Nutrients are also added in the dispersant to encourage microbial growth.

The smaller oil drops provide more surface area for the bacteria to work on. As a result, the breakdown of the oil is much faster.

Not all the oil can be broken down by the microbes but, because there is less, it is more easily dispersed by ocean currents and the wind.

1. Describe the different uses of the floating booms: _____

2. Explain why collected oil is burned at the surface, and the effects of this: _____

3. Explain how dispersants work and issues involved in using them: _____

4. What are the benefits of using hydrocarbon-metabolizing bacteria for bioremediation? _____

5. Why might petrochemical companies be invested in developing, and promoting, new oil spill technology? _____

160 Fossil Fuels and Health

Key Idea: Every step in fossil fuel extraction, storage, and use can impact human health negatively.

Fossil fuels still constitute the majority of human energy needs, however the extraction, storage, manufacture into plastics, and combustion present a wide range of health risks to humans and other organisms. Coal mining, oil drilling, and natural gas fracking have significant negative impacts on the environment, including disruption and pollution. These activities also pose health risks to populations living near extraction sites, resulting in respiratory, cardiovascular, neurological, and liver diseases. Production of secondary products, such as plastics and fertilizers, can impact humans due to the leakage of toxic fumes and entry into waterways. Recent research (2021) suggests that the burning of fossil fuels that cause **air pollution** are estimated to be responsible for around 8.7 million deaths around the world per year.

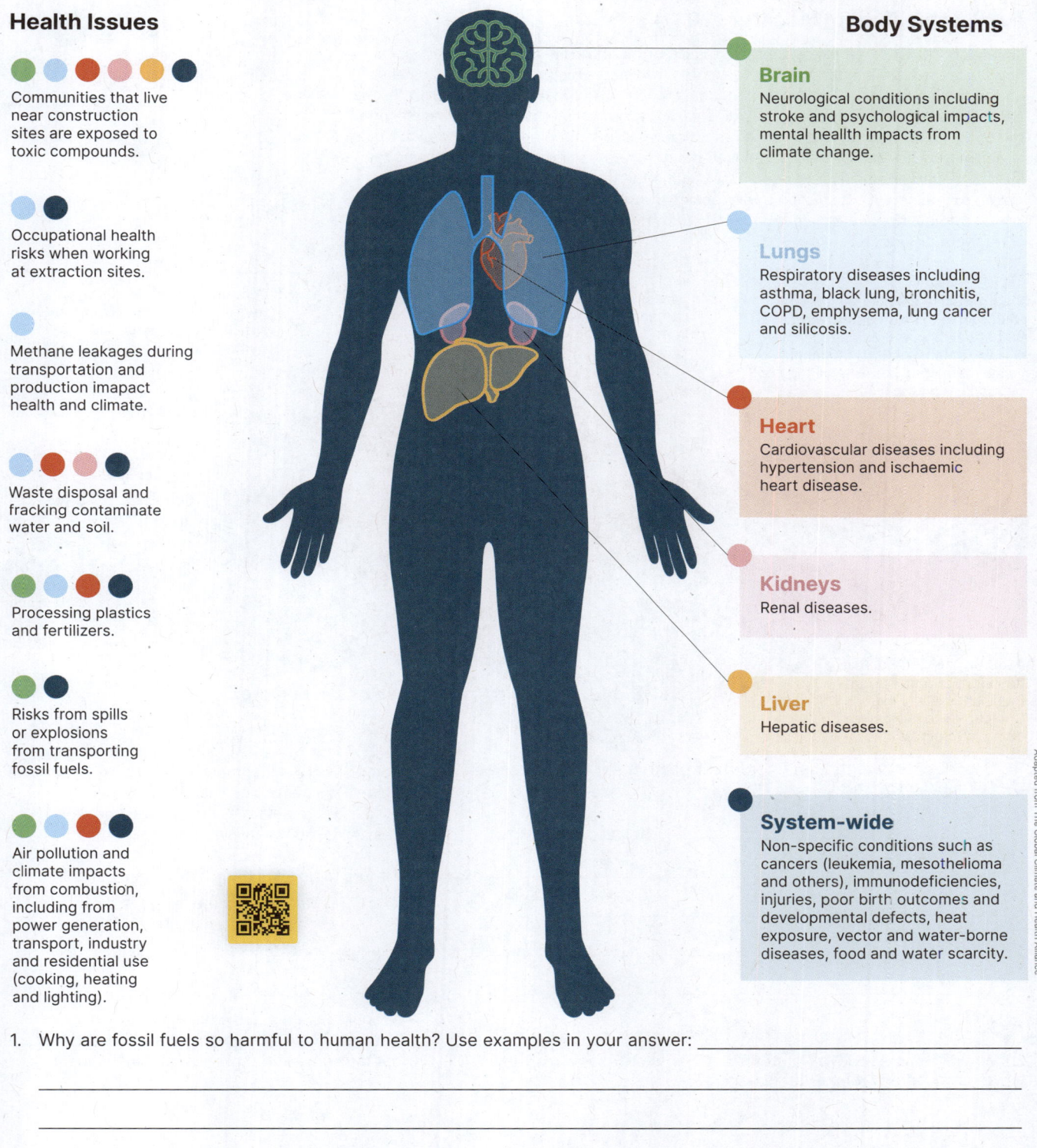

Health Issues

- Communities that live near construction sites are exposed to toxic compounds.
- Occupational health risks when working at extraction sites.
- Methane leakages during transportation and production impact health and climate.
- Waste disposal and fracking contaminate water and soil.
- Processing plastics and fertilizers.
- Risks from spills or explosions from transporting fossil fuels.
- Air pollution and climate impacts from combustion, including from power generation, transport, industry and residential use (cooking, heating and lighting).

Body Systems

Brain — Neurological conditions including stroke and psychological impacts, mental health impacts from climate change.

Lungs — Respiratory diseases including asthma, black lung, bronchitis, COPD, emphysema, lung cancer and silicosis.

Heart — Cardiovascular diseases including hypertension and ischaemic heart disease.

Kidneys — Renal diseases.

Liver — Hepatic diseases.

System-wide — Non-specific conditions such as cancers (leukemia, mesothelioma and others), immunodeficiencies, injuries, poor birth outcomes and developmental defects, heat exposure, vector and water-borne diseases, food and water scarcity.

Adapted from The Global Climate and Health Alliance

1. Why are fossil fuels so harmful to human health? Use examples in your answer: _____

161 The Effects of Nuclear Accidents

Key Idea: Nuclear energy production comes with large environmental risks if accidents cause the release of radiation. Nuclear power was once considered a solution to global energy demand. When nuclear power stations were first established in the 1950s, there were predictions that by the year 2000, these stations would be capable of producing 4.5 million megawatts per year. This amount of power generation would have been almost double the world's total electricity generation capacity at that time. However, spiraling costs, problems with storage of spent fuel, and safety concerns have shown these predictions to be wrong. The world has now had two level 7 nuclear events (an event that has widespread implications): the explosion of the number 4 reactor at Chernobyl, Ukraine, in 1986; and the disaster at Fukushima Daiichi, Japan, in 2011. Both of these events caused a range of health impacts to humans and the ecosystems.

Fukushima Daiichi disaster (2011)

On March 11, 2011, multiple reactor failures occurred at the Fukushima Daiichi nuclear power station, 220 km north of Tokyo, Japan, following a magnitude 9.0 earthquake and 10 m high tsunami. The three functioning reactors shut down immediately after the earthquake. Diesel generators provided electricity to circulate coolant but were flooded by the tsunami. Without coolant, the reactor cores overheated and melted down. Heat generated by fissionable material caused reactor water to boil to steam, exposing the fuel rods to air and producing hydrogen gas. Venting the gas from each of the three separate reactor containment vessels led to explosions that destroyed the outer reactor buildings. It was not until the end of September 2011 that the temperatures of the three reactor cores were brought to below 100°C.

Reactor 3: Partial core meltdown. Hydrogen explosion destroyed top half of building and damaged water supply for reactor 2. Containment vessel damaged, causing leakage of radioactive material.

Reactor 2: Full core meltdown. Hydrogen explosion from leaking gas destroyed upper part of outer building. Containment vessel damaged, causing leakage of thousands of liters of radioactive coolant water.

Reactor 1: Full core meltdown. Hydrogen explosion from vented gas destroyed outer building. Containment vessel undamaged.

Reactor 4: Fire in nuclear fuel storage pond. Spent fuel was partially exposed. Coolant leaked after ice damage to pipes.

Schematic of Fukushima Daiichi

Cause of the disaster

Although it was initiated by the earthquake and subsequent tsunami that flooded the backup generators, the full extent of the disaster at Fukushima Daiichi has been attributed to human error and a lack of safety systems. Investigations into the disaster found the following key points:

- A failure to develop basic safety requirements.
- Lack of preparation and the mindset to respond efficiently to a **nuclear accident**.
- Failure by regulators to properly monitor nuclear safety.
- TEPCO (the company running the station) failed to report changes to the emergency coolant systems, and failed to properly act on warnings that the generators were susceptible to flooding.

1. Describe the events that led to the three reactor meltdowns at Fukushima Daiichi:

2. Why are nuclear power stations accidents or reactor meltdowns considered environmental disasters?

Fukushima vs Chernobyl

Although both the Fukushima disaster and the Chernobyl accident are rated as level 7 nuclear events, the explosion at the Chernobyl nuclear power plant in 1986 released far greater amounts of radioactive material and affected a much greater area. The main difference is that the explosion at Chernobyl breached the containment building, allowing burning nuclear material to escape directly into the atmosphere and surrounding countryside.

Category	Fukushima Daiichi	Chernobyl
Reactors	3	1
Radiation released	370,000 tBq*	5.2 million tBq
Area affected	60 km radius	500 km radius
People evacuated	Tens of thousands	~300,000
Deaths	No direct deaths to date	64 due to direct exposure to radiation

*The Becquerel is the SI unit of radiation equal to one decay activity per second

The abandoned city of Pripyat, closest to the Chernobyl nuclear plant. The city was eventually overrun with forest and animals, e.g. feral dogs, bears, moose, foxes, and bison.

Fukushima Daiichi disaster health effects

The long-term effects of the disaster at Fukushima Daiichi will not be fully known for years, but studies already show high levels of radioactive contamination in many areas. The clean-up process is expected to take up to 40 years. Contamination includes:

- Radioactive particles leaked into the atmosphere and water have been detected, but continuing monitoring indicates the levels have mostly dropped below safe levels.
- Between March 21 and mid-July about 2.7×10^{16} Bq of caesium-137 was leaked into the sea, representing the largest ever individual leak of radioactive material.
- Radioactive material has been detected in the food supply including fish, beef, soybeans, and rice. Contaminated bluefin tuna have been caught off the coast of southern California.
- A study of populations of the pale grass blue butterfly from various places around Japan found a number of mutations in the populations closest to Fukushima including abnormal legs and wing shapes, dented eyes, and changes to color and spot patterns on the wings.

Chernobyl accident health effects

- Radioactive particles released by the explosion fell across a large area surrounding the plant, killing much of the nearby forest. In particular, an area of pine forest received an extremely large dose of radioactive particles. The pine forest was bulldozed and buried after the pines died.
- Winds blew radioactive material over a large part of Europe. Radioactivity in surrounding waterways fell rapidly due to dilution and deposition, but wildlife around the area have suffered higher than normal rates of mutation, with many animals dying from thyroid disease. Much of the radioactivity is now concentrated in the soil.
- The exclusion zone around the power plant has essentially become a wildlife refuge due to the evacuation of residents immediately following the disaster. Ecological succession has returned much of the land to forest or open meadow.
- Because of the long half-lives of many of the radioactive particles released, the area around Chernobyl will be contaminated for many generations to come.

3. What health risks do humans experience directly after a nuclear accident? Compare them to longer term health risks:

_____
_____
_____
_____

4. Small numbers of large mammals occupied the area now known as the Chernobyl exclusion zone before and 6 months after the nuclear accident. Mammal populations grew rapidly in number after the region was evacuated permanently of people. Wolf numbers increased substantially after 1993, and mainly preyed on boar, who also suffered an outbreak of swine fever at the same time. There appeared to be no correlation between radiation concentration and animal density. Discuss the environmental impact of the Chernobyl accident in relation to the ecosystem, including the role of humans:

_____
_____
_____
_____
_____
_____

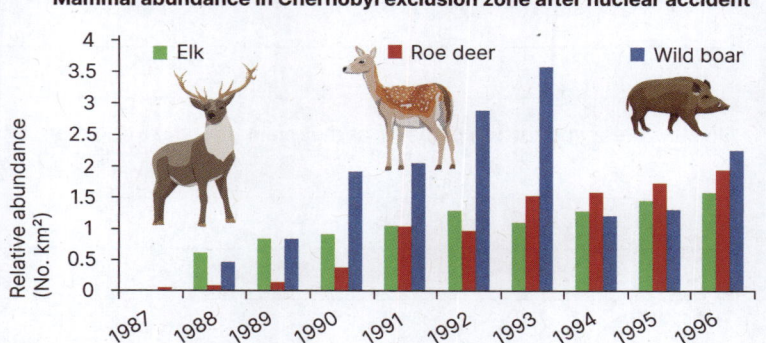

Mammal abundance in Chernobyl exclusion zone after nuclear accident

Data Source: Deryabina et al. (2015) Current Biology

162 Bhopal Disaster

Key Idea: The Bhopal chemical plant disaster was responsible for significant harm to the surrounding human population. Many industrial disasters can be traced back to the need to increase revenue and regain the often-vast sums of money involved. Industrial plants often operate twenty-four hours a day to increase production and plant efficiency. Cost cutting measures may reduce safety margins. Factors such as these were instrumental in the disaster at the Union Carbide factory in Bhopal, India. On the night of December 2-3 1984, harmful chemicals leaked from the plant, killing and injuring multitudes of people living around the area. Poor maintenance and a lack of safety were primarily to blame. Regulations were implemented after the event with the aim of preventing a repeat of this polluting disaster.

Bhopal disaster (1984)

▸ The Union Carbide plant was built for the production of the pesticide carbaryl, a relatively safe chemical. However, several of the chemicals used to make carbaryl are highly toxic and must be handled and stored with extreme care.

▸ The disaster occurred when water entered a tank containing the intermediate chemical, methyl isocyanate. Pressure from heating inside the tank caused toxic gases to be vented, covering parts of the city.

Cause of the disaster

▸ Hazardous intermediate chemicals were stored rather than used immediately.
▸ Poor plant maintenance.
▸ Failure of safety systems due to poor condition; gas scrubbers, flare towers, and water curtains under repair or insufficient.
▸ Safety systems, including refrigeration, were turned off.
▸ Plant location near to densely populated residential areas.

At the time, Union Carbide maintained that the disaster was the result of sabotage. However, in June 2010, eight officials from the industrial operation were convicted of negligence.

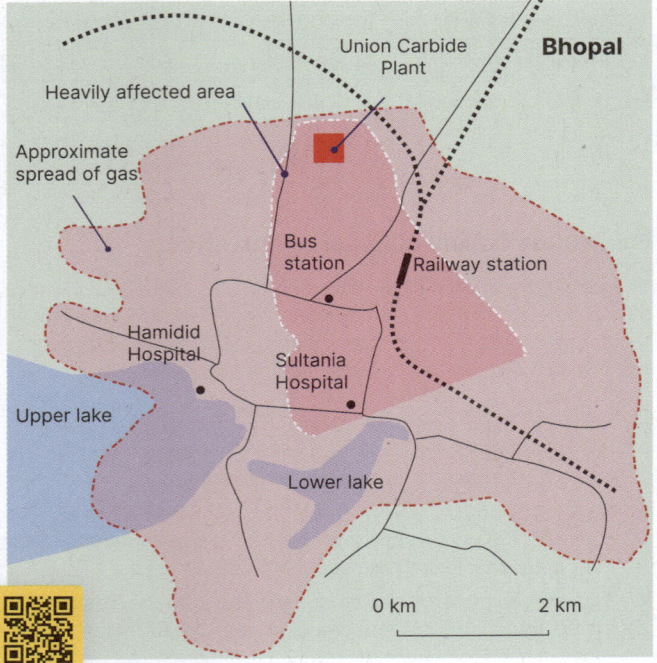

Effects of the disaster

▸ The gas cloud caused the immediate deaths of over 4000 people. Another 4000 died within a week and 8000 died in the following years. Between 200,000 and 500,000 people received injuries.
▸ Some research suggests animal and plant life around the plant were also severely affected by the toxic gas.
▸ After the disaster, the plant was shut down and the site cleaned up. However, toxic chemicals including mercury, lead, chloroform, and dichloromethane are still stored at the site and continue to leak into soil and water, resulting in dangerous chemical levels in the environment.
▸ The Indian government estimated the economic cost of the disaster at almost $4 billion. In 1989, Union Carbide agreed to an out of court settlement of $470 million, and to fund a hospital dedicated to treating disaster victims. The maximum compensation for individuals was ~US$3000.

On the site of the Bhopal disaster

1. Explain how the plant originally provided a beneficial service to Bhopal and India:

2. Give two reasons for the disaster and explain how they could have been avoided:

163 Mining Disasters

Key Idea: Environmental disasters in the mining industry are often due to human error and incorrect practices.

In the past, many mining sites used a dam or tip to contain tailings, or sludge waste, from metal or coal mining operations. The integrity of these structures can be affected by unsafe practices, engineering defects in construction, or overcapacity of sludge waste. When the tips or dams collapse, they can send a deadly mudslide of waste into nearby communities or waterways. The largest mining disaster in the UK occurred in 1966, in Aberfan, South Wales. Coal tailings flowed rapidly into the local village, killing 144, most of them children. More recently, a failure of a mining dam in Brazil resulted in over 270 deaths. Both **environmental disasters** were caused by human failures, which instigated extensive investigations to charge those responsible. The findings resulted in **environmental legislation** changes to prevent similar events occurring again.

Aberfan coal spoil disaster (1966)

The spoils from the number 7 coal tip covering the houses in the Aberfan village, with the Pantglas Junior School in the center.

▸ In the hills above the southern Welsh village of Aberfan, coal mining had been operating since 1875, becoming one of the largest coal pits in the area. Waste from the mining activities had been dumped into large spoil tips for at least 50 years prior to the disaster. Of the 7 tips, only one was in active use as of 1966. The National Coal Board (NCB) employed around 8000 miners, most of them from the local village.

▸ In previous years, the villagers had held meetings, concerned about small slips on the number 7 coal tip as well as flooding. Although the NCB had agreed to start some **remediation** work on drainage, this had not been started prior to the disaster.

▸ On the morning of the 21st of October, 1966, the coal tip began to slip, saturated after 3 weeks of heavy rain and having active springs running under the tip.

▸ Over 110,000 m³ of coal spoil slid down into the village, with the Pantglas Junior School directly in the line of fire. 109 young children and 5 teachers were buried alive, and the sludge rapidly solidified. Only 26 survivors were rescued from under the sludge at the school, and no more were found alive two hours after the disaster. In total, 144 people lost their lives due to this disaster.

▸ Although no-one at NCB was charged with negligence for the disaster, and only a £500 compensation was given to each family, petitioned from an original £50 offered by NCB, a new law, the Mines and Quarries (Tips) Act 1969, was written to prevent a similar disaster occurring again.

Brumadinho Dam disaster (2019)

▸ On the 25th of January, 2019, the tailings dam from a Vale iron ore mining site close to Brumadinho in Minas Gerais, Brazil, catastrophically failed, releasing a huge torrent of water and mud.

▸ The mudflow covered the administration buildings of the mining site and nearby houses and farms, killing over 272 people.

▸ Engineers and scientists discovered a slow creep of sediments behind the dam, which became unstable due to being built on top of waste from the mine rather than a solid and properly prepared ground surface.

▸ The Brumadinho dam disaster initiated legislation that banned similar tailings dams across many countries.

1. Why were the mining industries likely to have continued using unsafe tailing tips and dams?

2. What were the main outcomes that both mining disasters above had in common?

164 Environmental Remediation

Key Idea: Many types of pollution can be remediated, with different methods and technologies used to restore the land. As land becomes increasingly scarce, there is a greater incentive and requirement to redevelop land previously used for other purposes. For example, farmland near cities is often developed for residential use and abandoned industrial sites and landfills are often redeveloped into housing, parks, or commercial areas. Chemicals once used on these lands could present a risk to humans through contamination of soil or groundwater, or release of noxious gases. Before land can be redeveloped, it must undergo **environmental remediation**. This is the removal of contaminants in order to make the area safe for human health. The remediation method used depends on the extent and type of contamination. For example, polluted topsoil can be removed and treated off-site, or plants and bacteria may be placed *in situ* to absorb and break down the contaminants. A treated area is monitored over many years to ensure that no further leaching of contaminants occurs. Many new technologies have been developed to improve the effectiveness of remediation.

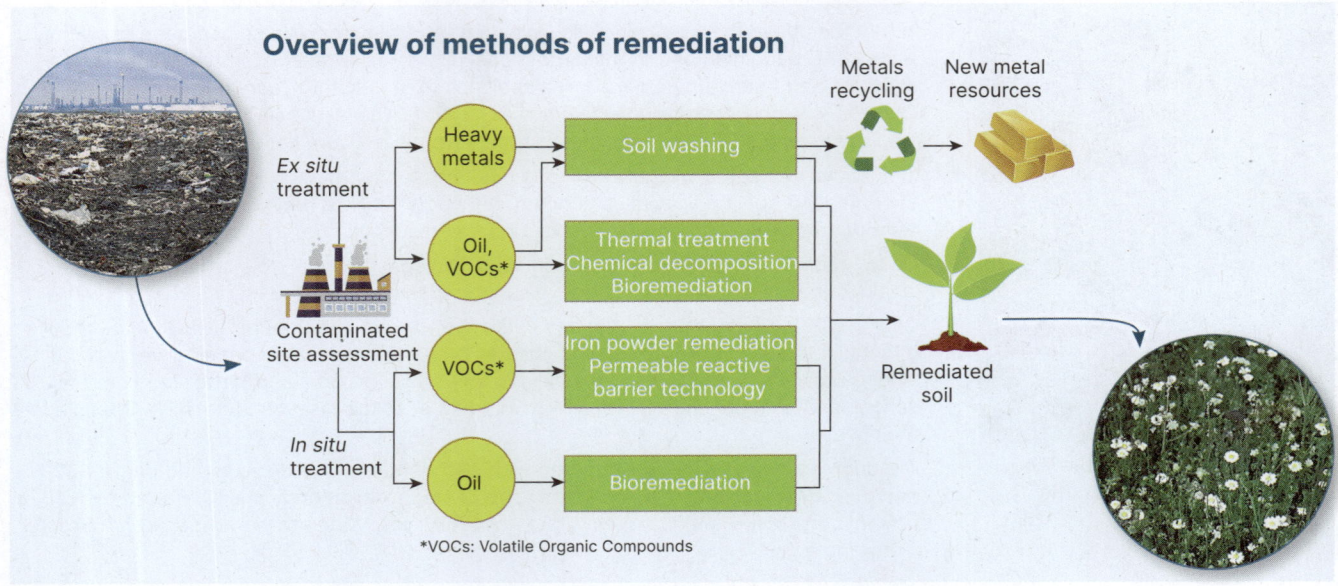

Contaminated sites in the US are classified as Brownfields or Superfund sites. Brownfields are industrial, urban, or commercial parcels of land that can be used again after remediation. Superfund sites are highly toxic, abandoned sites that the Environmental Protection Agency (EPA) has identified for extensive remediation, e.g. Love Canal in Niagara Falls.

The EPA offers a number of grants for environmental assessment, cleanup, and job training activities related to contaminated sites. The level of environmental remediation achieved depends upon the extent and nature of the contamination. Even after many years of remediation, some sites can never be made safe enough for humans to occupy. Typically, legislation requires the responsible parties to perform cleanups or reimburse the government for EPA-led cleanups.

1. Explain the purpose of environmental remediation: _____
   _____
   _____
   _____

2. Describe some benefits of remediation: _____
   _____
   _____
   _____

Technological solutions for environmental remediation

Land contamination remediation

Bioremediation and phytoremediation are examples of technologies that are used to restore land to its original state in place (*in situ*). Bioremediation utilizes bacteria, fungi, or plant processes to breakdown specific contaminants into safe byproducts, while phytoremediation specifically uses plants to uptake, stabilize, or degrade contaminants.

Ex situ (away from site) remediation

Some remediation technologies are used at locations away from the original contamination. Thermal treatment vaporizes and separates the contaminant at high temperatures, while land farming contains and treats the contaminants by a variety of methods in lined, leach-proof bins. *Ex situ* remediation tends to be expensive, but more thorough than in situ remediation.

Oil spill remediation

Limiting the extent of environmental damage is driving the innovation of new technology to effectively clean up and remove harmful **oil spills**. Sponges have been developed to absorb more than 30 times their weight in oil, and properties of cleaning magnetic soaps can be switched off and on to control the remediation process.

Water contamination remediation

Contaminants, such as **heavy metals** and industrial chemicals, can make water harmful to human and animal health, so prompt removal is vital. Membrane filtration can remove contaminant molecules that have different sizes and characteristics from water, while electrochemical technology uses electricity to attract and draw out charged contaminant particles.

3. A wide range of new technology has been developed to minimize or reverse environmental damage caused by polluting contaminants. The effectiveness of their application is dependent on many factors. Select one method of environmental remediation, and discuss in detail on where, when, and how it was used:

_____

_____

_____

_____

_____

165 The Economic Impact of Pollution

Key Idea: The extent of control and clean up of environmental pollution are often influenced by economic costs.

Although there is an economic cost to **pollution**, placing a monetary value on it is difficult and controversial. The various sectors involved (health officials, economists, and industry) often disagree about how to estimate the cost of pollution, as it is difficult to assign monetary values to the environment, health, and human life. It can also be difficult to determine the economic impact of pollution because, while a region as a whole may benefit from the economic activities of a polluter, specific groups within the region may suffer. Some economic costs associated with pollution are more easily determined than others. Direct costs, e.g. cleaning up an **oil spill**, are easily calculated, but indirect costs, e.g. estimating revenue losses, or repercussion costs, e.g. loss of public confidence, can be harder to quantify. Cost-benefit analysis is used to assess the cost of controlling pollution. The short-term and long-term costs and benefits for a variety of pollution control measures are compared and used to determine whether a control or regulation should be put in place. Environmental regulations, taxes, and pollution quotas are commonly used to control levels of pollution and to promote sustainable use.

Break-even costs of pollution clean up

▸ The ideal scenario after an environmental disaster or polluting event is for 100% **remediation** to return the area to its prior state. Many times, a pragmatic approach is taken by companies which is determined by cost and legal requirements.

▸ The cost of removing **pollutants** rises sharply as more pollutants are removed from an area (top graph, below left), until the clean-up costs exceed the harmful costs of the pollution.

▸ The point at which the costs of the pollution and the costs of clean-up are equal marks the break-even point. This point is determined by separately plotting the clean-up cost and the cost of the pollution to society. The two curves are then added together to reveal the total costs (below left).

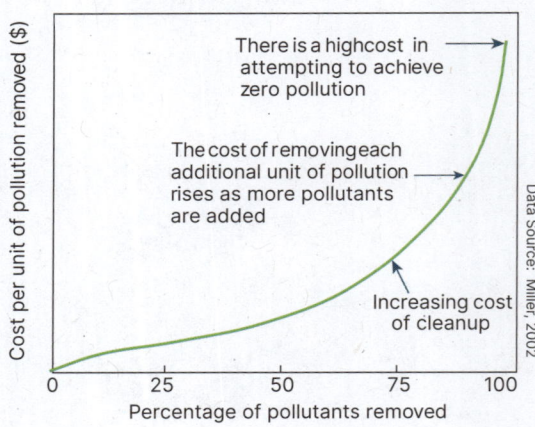

Direct and indirect costs of pollution events

Some costs of pollution can be assessed directly, such as the cost of clean-up, the cost to salvage equipment, such as oil tankers, the cost of relocation or rehabilitation of vulnerable species impacted by the event, and the cost of impact to local businesses. Other circumstances caused by the pollution event have more indirect costs, also estimated using non-market valuation. These include the 'value' of an unpolluted beach or environment, or the loss of aesthetic values, the loss of an endangered species, which is more than the market value of the meat or 'pet store' value, and the loss of recreational facilities. Some US government departments use the 'Habitat equivalency analysis' method to estimate the cost to restore a polluted ecosystem to its original state after a pollution event, or to gauge compensation.

Red Hill water crisis

In 2021, fuel from underground military tanks leaked into the O'ahu, Hawai'i water supply. Litigation cases to cover costs to the local community, such as human health, ecosystem damage, increased water testing and monitoring wells, provision of alternate water, stress, and local business loss, began in April 2024.

1. Identify each of the costs in the Red Hill water crisis as direct or indirect, distinguishing between the two cost types:
   _____
   _____

2. With reference to the break-even point, explain why total pollutant removal is often not cost effective:
   _____
   _____

166 The Role of Environmental Legislation

Key Idea: Environmental legislation aims to prevent pollution initially, but also ensure remediation and liability occurs.

Environmental legislation has a vital role in the protection of the environment, reduction of **pollution**, and public safety. Before the 1960s, there was little environmental consideration during the development of towns, factories, and many other human projects. Today, a number of laws and regulations are in place to make sure the environment is respected while developing a site. Legislation provides incentives and disincentives to ensure compliance. For example, a factory that builds a treatment pond may be given grants to cover the costs but a factory found to be openly dumping waste may be fined thousands of dollars. In the US, a number of environmental laws have been passed to regulate polluting industries and fine those who ignore them. These include the Clean Water Act (to prevent **water pollution**), the Clean Air Act (to avoid **air pollution**) and the Comprehensive Environmental Response, Compensation, and Liability Act, or Superfund (to enable the cleanup of contaminated sites). The Environmental Protection Agency (EPA) is the main governmental department that oversees most of the responsibility for compliance to these laws.

Environmental impact assessment and enforcement

Environmental impact assessment
In many countries, a developer must file a report on the possible impact of the development on the environment and the risks of pollution from the activity. The new developments must cater for comment and consultation, and scientific evaluation of the possible environmental impacts, including pollution. Whatever the country or name of the statement, environmental impact assessments are used to determine the positive and negative effects of altering the environment so that decisions on the extent of the development can be made, otherwise the development is prevented or penalized.

Responsibility for enforcement
The enforcement of environmental laws are often carried out by numerous different departments or agencies. Each department has its own area of responsibility regarding environmental issues. For example, in the United States the Department of Agriculture enforces the Soil and Water Conservation Act, while the Environmental Protection Agency enforces the Clean Water Act, and the Department of Interior enforces the Wild and Scenic Rivers Act. Individual states and local councils often have their own bylaws which are enforced within their area of jurisdiction.

1. Use information in this book and online to define or describe each of the following environmental agencies or laws:

(a) EPA: _____

(b) Superfund / CERCLA: _____

(c) European Green Deal: _____

(d) The Clean Air Act: _____

(e) UNEP: _____

(f) National Environmental Policy Act (NEPA): _____

Developing new environmental pollution legislation in the United States

▸ In 2021, a new US federal bill, to be voted on in Congress, was sponsored, called the **Break Free From Plastic Pollution Act**. The aim of the bill was to address **plastic pollution** and waste issues. The bill was reintroduced in 2023 and, as of mid-2024, the legislation was still in the process of potentially being passed by congress and becoming law.

▸ The Bill covers areas that are not addressed sufficiently in other current laws, including regulations to reduce products being made from plastic, decreasing the amount and type of plastic pollution, and protecting communities from plastic pollution, including health, economy, and climate protection.

▸ Bills need to go through a series of steps before they can be voted on by the federal government and signed into law by the President. Numerous lobby groups, both for and against the proposed laws, will influence final voting, as will the political party that supports the bill. Bipartisan support (support across both major parties) can often mean environmental legislation is likely to be passed and eventually become law.

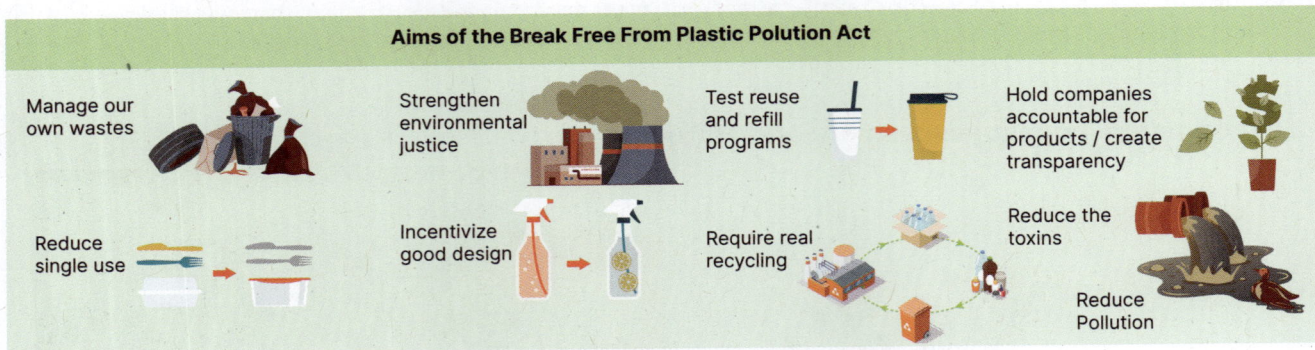

Aims of the Break Free From Plastic Pollution Act: Manage our own wastes; Strengthen environmental justice; Test reuse and refill programs; Hold companies accountable for products / create transparency; Reduce single use; Incentivize good design; Require real recycling; Reduce the toxins; Reduce Pollution.

2. Why is environmental legislation so important to limit and remediate polluting events?

3. Why do many of the large environmental disasters, like the Deepwater Horizon oil spill, result in new environmental legislation being passed into law?

4. Suggest what industries and groups might want to lobby (influence lawmakers) against the 'Break Free From Plastic Pollution Act' and why?

5. In small groups, decide on a new environmental law you would like to propose/sponsor to combat pollution. Assign a name, list the key aims of your new legislation, and give the reasons your new law is required. Record details below:

167 Did You Get It?

1. The 2010 Deepwater Horizon oil rig explosion and subsequent well gush in the Gulf of Mexico was responsible for the largest polluting marine oil spill to-date.

   (a) This event caused environmental pollution. Define this term:
   _____
   _____
   _____

   (b) Why does a large oil spill in water cause so many diverse problems in the environment:
   _____
   _____
   _____
   _____

   (c) How did legislation discourage oil companies from allowing a similar environmental disaster occurring in the future?
   _____
   _____
   _____

2. (a) Why does microplastic and nanoplastic pollution pose such an issue for human health?
   _____
   _____

   (b) How does biomagnification increase the risk of harm for forever chemicals being transferred to humans?
   _____
   _____

   (c) How can effective waste management reduce plastic waste? Provide examples in your answer:
   _____
   _____
   _____

3. (a) Explain the importance of the stratosphere ozone layer to humans and other organisms:
   _____
   _____

   (b) How did human activities contribute to a depletion of ozone, and then eventual repair after 1987?
   _____
   _____
   _____

   (c) Why is troposphere ozone considered a pollutant?
   _____
   _____
   _____

Chapter 8 — Global Change
Conservation

Resource Hub
bit.ly/4ctYVK0

Key Terms
- biodiversity
- biodiversity hotspots
- conservation
- deforestation
- ecosystem
- ecotourism
- endangered species
- *ex situ* (conservation)
- extinction
- habitat fragmentation
- human activity
- *in situ* (conservation)
- introduced (invasive) species
- IUCN (International Union for Conservation of Nature)
- legislation (conservation)
- population
- rewilding
- Sixth Mass Extinction
- species
- sustainability
- WWF (Worldwide Fund for Nature Organization)

Key Concepts
- Human activity is the primary driver of global biodiversity loss across various ecosystems.
- Anthropogenic activities have led to the classification of certain species as endangered or at risk of extinction.
- Sustainable management of environmental resources is crucial to prevent damage to ecosystems.
- Conservation practices play a vital role in reducing biodiversity loss and restoring ecosystems.

Managing Biodiversity

Learning Outcomes:

	#	Outcome	Activity Number
☐	1	Explain the meaning of the term 'biodiversity hotspot'. List and describe factors that affect biodiversity.	168-169
☐	2	Describe the importance of insects in ecosystems and to human food security. Explain what factors negatively affect insect populations.	170
☐	3	List the main causes of tropical deforestation. Explain the importance of tropical forest ecosystems.	171
☐	4	Explain what is meant by fragmentation of habitats and why this is a threat to biodiversity.	172
☐	5	Explain, giving examples, the effects of introducing species into areas of the world other than their native habitat and why there may be a need for their control in such areas.	173-174
☐	6	Explain, giving examples, how diseases can be transported around the word and affect organisms, including humans.	175
☐	7	Describe the main factors that cause the decline of species numbers. Describe the roles of the IUCN and CITES.	176
☐	8	Explain, giving examples, what is meant by the 'sixth mass extinction'.	177

Managing Resources

	#	Outcome	Activity Number
☐	9	Describe how human populations and their needs can come into conflict with wildlife. Using the example of the Maasai Mara National Reserve, describe how this conflict has been managed.	178
☐	10	Describe the role of ecotourism in areas such as the Galápagos Islands. Explain why tourist numbers must be managed carefully.	179
☐	11	Compare, using examples, *in situ* and *ex situ* conservation.	180-181
☐	12	Describe the role of rewilding as a means of ecosystem restoration.	182
☐	13	Describe the roles of both local and global legislation in conservation.	183
☐	14	Describe the relationship between conservation and sustainability.	184

168 Biodiversity Hotspots

Key Idea: Biodiversity hotspots are significant ecological regions, often with many endemic species, and under threat. The **species** is the basic unit by which we measure biological diversity or **biodiversity**. Biodiversity is not distributed evenly on Earth, but clustered in certain parts of the world, called **biodiversity hotspots**. These regions are biologically diverse and ecologically distinct regions under the greatest threat of destruction from **human activity**. They are identified based on the number of species present, the amount of endemism (species unique to a specific geographic location), and the extent to which the species are threatened. Unfortunately, biodiversity hotspots often occur near areas of dense human habitation and rapid human **population** growth. Habitat destruction and human-induced climate change are major threats to biodiversity hotspots. Biodiversity hotspots make up less than 2% of Earth's land surface but support nearly 60% of the world's plant and vertebrate species. Most are located in the tropics. Their **conservation** is considered central to securing global biodiversity.

Key biodiversity hotspots

1. What are biodiversity hotspots? _____
   _____
   _____

2. Looking at the map, where are most of the hotspots concentrated? _____
   _____

3. Many of the biodiversity hotspots coincide with regions of very high human population density. How does high population density create greater risk of biodiversity loss in these regions?
   _____
   _____
   _____
   _____

169 Loss of Biodiversity

Key Idea: Biodiversity loss around the world is accelerating. In many cases, this is due to human activity.

The world is facing a **biodiversity** crisis that will require committed efforts to reverse. Biodiversity loss is being driven by human **population** growth. This is leading to many environmental issues, as demand on **ecosystems** is affected by human competition for resources and land. According to the United Nations, the human population reached 8.0 billion in late 2022 and could peak at almost 10.4 billion by the mid 2080s. The current population is using up Earth's resources faster than they can be renewed by natural processes, depleting ecosystems that support biodiversity. Vast areas of land (and sea) are exploited for their resources and land is needed to grow food. The many factors that are causing the global decline in biodiversity and increasing number of species **extinctions** can be summarized as HIPPCO (Habitat destruction, **Invasive species**, (human) Population growth, **Pollution**, Climate change, and Over-exploitation). Additionally, genetic information is lost, reducing the capacity of the **species** to adapt and increasing the risk of inbreeding.

Commercial fishing

Suburban sprawl

Native forest tree cutting

Fishing techniques have become so sophisticated, and large scale, that hundreds of tonnes of fish can be caught by one vessel each trip. As a result, some fish stocks have plummeted as the human population has increased and fish stocks have been overexploited. Many fish species are on the brink of collapse. The same overharvesting trends can be seen in terrestrial populations, as wild animals are hunted for food, trophies, and pet supply.

Urbanization increases the density of the human population in an area. This makes demands on the environment to supply housing, industry, transportation, and waste management. Urban sprawl is the development of large areas of land for housing. Population growth leads to more houses and infrastructure being needed. These often use undeveloped or previously agricultural land, reducing habitats for other organisms.

Humans changed to a sedentary, agricultural lifestyle, from a hunter-gatherer mode 10,000 years ago. This meant that land needed to be cleared and protected from wildlife for the exclusive use of humans and their livestock and crops. 44% of the world's land area is now dedicated exclusively for agriculture, while other land is mined for resources. **Deforestation** is driven by demands for timber, land for fast growing plantations of trees, and urbanization.

Penguin entangled in plastic pollution

Rapid and extensive global travel

Possum pest in New Zealand

Humans produce a huge amount of waste. Not all is, or can be, recycled and so it is dumped in landfills. Pollution is a major problem in parts of the oceans and on land. Activities causing pollution include deliberate dumping of rubbish, runoff from the land and contaminated discharge, and industry waste. 12.7 million tonnes of plastic finds its way into the sea each year, harming or killing wildlife. Microplastics can also have severe detrimental effects and are now found in all parts of the world.

Disease spread is facilitated by global transport of plants, animals, and humans; often inadvertently. Food and plants carry diseases across international borders. These are then spread to species that originally were protected by geographical barriers. Some countries, such as New Zealand, have strict biosecurity regulations to prevent spread of diseases such as rabies. Foot and Mouth disease, carried in unpermitted meat, would decimate the meat and dairy industry.

Introduced species are those that have evolved at one place in the world and have been transported by humans, either intentionally or inadvertently, to another region. Some of these introductions are beneficial, but invasive species have a detrimental effect on the ecosystems into which they have been imported. They can include plants or pests such as insects which damage plants. Additionally, animals, like the brush-tailed possum in New Zealand, can out-compete or kill off existing species.

1. Identify some factors that threaten the biodiversity of an ecosystem local to where you live and explain the threat:

Loss of mixed dipterocarp forest in Southeast Asia

Forest ecosystems in Southeast Asia are home to some of the most biodiverse habitats in the world. Unfortunately, these areas also have some of the highest rates of deforestation and ecosystem loss.

Mixed dipterocarp forests are made up of a family of plants of around 700 species that provide a wide range of fruits, nectar, pollen, and habitats for countless other species, such as orangutan. Some dipterocarp trees are over 80m tall, and their valuable wood resource has made them a prime target for forestry, for building and as a source of income.

Certain regions in Southeast Asia have experienced significant loss of their original forest cover, with more than 50% of the land affected. This loss is primarily attributed to deforestation for timber and the conversion of land into monoculture plantations. The demand for commodities such as palm oil, rubber, and wood pulp from Eucalyptus and Acacia trees has driven the establishment of these plantations.

Mining and hydroelectric schemes also contribute to forest loss and degradation, and pollution of waterways. However, these industries provide employment. Encroaching urbanization from human population growth is spreading into forested regions.

Aside from habitat loss, the biodiversity in these forests has been impacted by hunting for food and trade. This often involves illegal poaching of protected species to profit from bushmeat sales and supplying the pet-trade.

Southeast Asian forest

2. Explain why the current rate of global biodiversity loss is so significant compared to loss from natural causes:

3. Mixed dipterocarp forests are economically and socially important to Southeast Asia. How does this complicate efforts to reduce or prevent the loss biodiversity of these ecosystems?

4. Demand for food increases as the population grows. Modern farming techniques favor monocultures to maximize yield and profit. However, monocultures, in which a single crop type is grown year after year, are low biodiversity systems and food supplies are vulnerable if the crop fails. The UN estimates that 12 plant species provide 75% of our total food supply. Suggest some consequences of creating a low biodiversity system:

5. Considering the HIPPCO factors, select one human activity leading to extinction or loss of biodiversity and suggest some actions that could reduce or reverse it. You might choose to provide some animal examples in your discussion:

170 Where Have All the Insects Gone?

Key Idea: Insect biodiversity loss has far reaching implications for ecosystem stability and human food systems.

The **biodiversity** of insects is continuing to decrease, with some research suggesting that around 40% of insect **species' populations** are decreasing. Additionally, some are at risk of **extinction**. This loss impacts both **ecosystems** and humans; insects are key pollinators, and most human food crops rely on them for fruit, vegetable, and cereal production. A reduction of pollinators would lead to food insecurity. Insects are near the base of most food webs, providing an essential food source for many other animals. They are also involved in recycling nutrients back into the soil. **Human activity** has been the primary cause of insect loss and the subsequent loss of habitats. However, it is possible to prevent or reverse this decline in insect biodiversity by making changes in human behavior and activity. This can start from government and industry levels and extend to individual and household actions in daily life.

Threats to insect biodiversity

Deforestation
Lost of habitat and plants, some of which the insects have a mutualistic relationship with.

Increased wildfires
Loss of habitat and insect death.

Increased nitrogen use
Affect abiotic conditions of insects habitats, change **biodiversity** of plants.

Introduced pests
Outcompeting, predating, and spreading disease to endemic insects.

Pollution
All types of **pollution**, including air, water, and soil are affecting insects globally.

Droughts
Loss of habitat, dehydration of insects, loss of food sources.

Landscaping
Native plants and habitats replaced by introduced grass and shrubs.

Urbanization
Habitats and plants for insects are lost, and more insecticides are used.

Intensive Agriculture
Monoculture, reducing biodiversity of plants and food source, and greater use of pesticides.

Climate change
Increased heatwaves that kill insects and insect food supply. Shift of habitat range, which impacts migrating insects. Sea level rise.

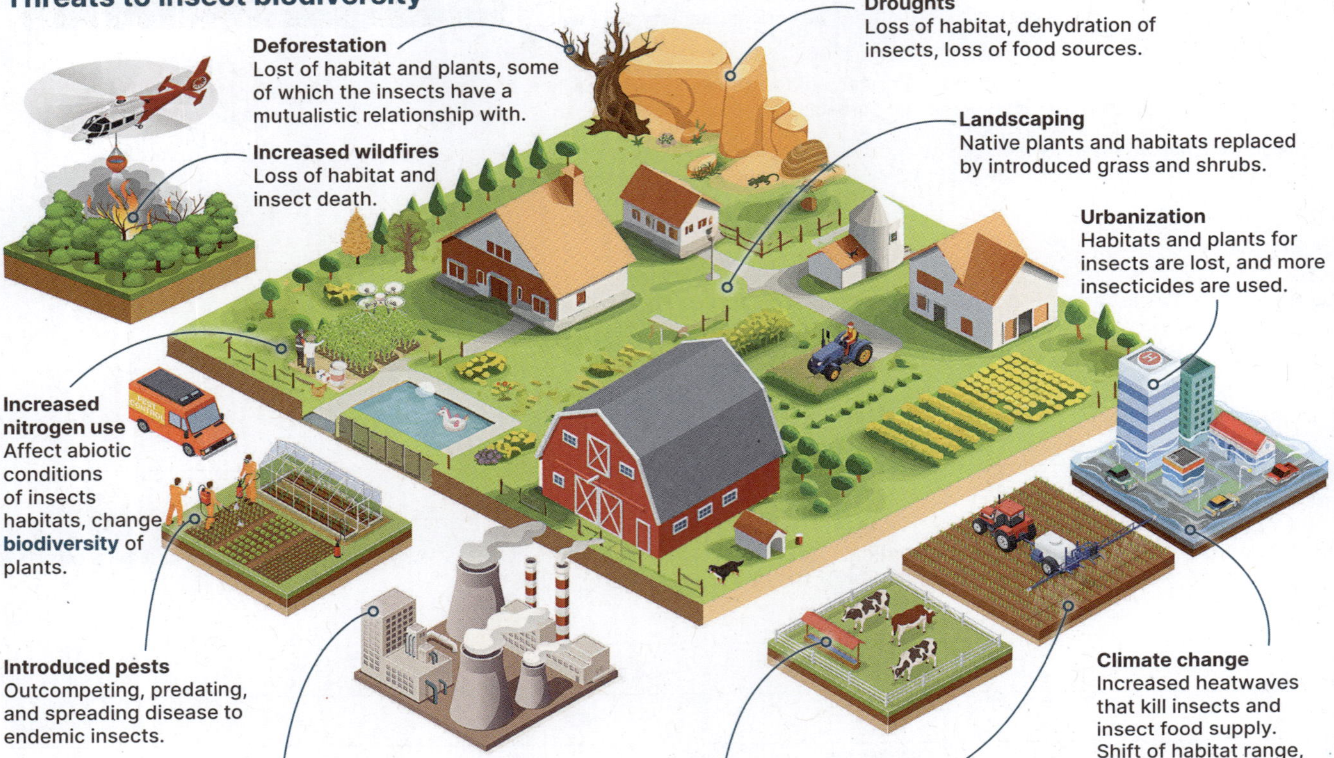

1. Considering the threats to insect biodiversity above, discuss which of these are occurring in your local region and why would they result in insect biodiversity loss:

_____
_____
_____
_____
_____

Why do we need insects?

▸ After the death of plants and animals, it is crucial to recycle the nutrients they contain back into the soil through a process known as decomposition. This allows plants to have better access to nutrients and helps maintain the nutrient levels in the soil, preventing depletion.

▸ Insect detrivores, such as beetles, flies, termites, and cockroaches, play a crucial role in nutrient recycling in **ecosystems** by breaking down organic matter. However, some of these insects are often seen as pests by humans and are targeted with pesticides, leading to their large-scale killing. Similarly, in freshwater aquatic ecosystems, fly larvae like mayflies, caddisflies, and craneflies also contribute to nutrient recycling, but they are negatively affected by pollution in waterways.

▸ Insects play various other important roles in ecosystems. They contribute to pollination, help control pests that can damage plants, and serve as a significant food source for animals. Additionally, research has estimated that the ecosystem services provided by insects have a global value of approximately US$57 billion per year.

▸ The multitude of tasks that insects perform would be all but impossible to replicate by any other means.

The importance of pollination

▸ Many flowering plants rely on insect pollination, and this includes 75% of all agricultural crops and 85% of wild plants.

▸ Many of the agricultural cropping practices have been modernized by using synthetic fertilizer and planting new genetically modified plants. However, without adequate numbers of pollinators, human populations would soon fall into food insecurity.

▸ Some plants are extremely dependent on a particular species of pollinator. For example, the Madagascan orchid requires a unique pollinating moth species, the local hawkmoth, to have a 22cm long proboscis to reach the nectar. With loss of this moth, the plant species would become extinct as well.

2. How would the loss of insects, as pollinators, directly impact human populations?

_____
_____
_____

3. How do insects act as detrivores? _____

_____
_____
_____

4. What actions could you, your family, or your school take to increase insect biodiversity in your local area?

_____
_____
_____
_____
_____
_____

171 Tropical Deforestation

Key Idea: Deforestation continues to reduce the world's tropical rainforests, with large areas still at significant risk.

Tropical rainforests prevail in places where the climate is very moist and warm throughout the year. They account for 45% of the world's forest type. Almost half of the world's rainforests are in just three countries: Indonesia in Southeast Asia, Brazil in South America, and Zaire in Africa. Much of the world's **biodiversity** resides in rainforests, particularly in **biodiversity hotspots**. **Deforestation** will contribute to climate change by significantly reducing the amount of carbon capture and storage through photosynthesis. In the Amazon, 75% of deforestation has occurred within 50 km of Brazil's roads. Many potential drugs could still be discovered in rainforest plants, and loss of **species** through deforestation may mean they will never be found. Rainforests can provide economically sustainable crops for local people.

Deforestation

▶ At the end of the last glacial period, about 10,000 years ago, forests covered around 45% of the Earth's land surface. Forests now cover about 31% of Earth's surface. These include the cooler temperate forests of North and South America, Europe, China, and Australasia, and the tropical forests of equatorial regions.

▶ Over the past 5000 years, approximately 1.8 billion hectares of forest cover have been lost. In the past decade alone, there has been a net loss of around 4.7 million hectares of forest each year.

▶ The temperate regions, such as Europe, where human civilizations have existed for a long time, have experienced significant damage due to deforestation. However, currently, the majority of deforestation is happening in tropical regions.

▶ The intensive clearing of forests during the settlement of newly discovered lands has resulted in significant landscape alterations and permanent changes, leading to a decrease in biodiversity.

Causes of deforestation

▶ Deforestation is the end result of many interrelated factors which often center around socioeconomic drivers. In many tropical regions, most deforestation is the result of small-scale family farming: poverty can be partially solved by clearing small areas of forest and making family plots (subsistence farming).

▶ However, huge areas of forests have been cleared for agriculture, including ranching and palm oil plantations. These produce revenue for governments through taxes and permits, creating an incentive to clear more forest.

▶ Just 14% of deforestation is attributable to commercial logging although, combined with illegal logging, it may be much higher.

Causes of deforestation: Subsistence farming 48%, Commercial agriculture 32%, Logging 14%, Fuel 5%

Tropical forest deforestation risk areas - projected 2010-2030

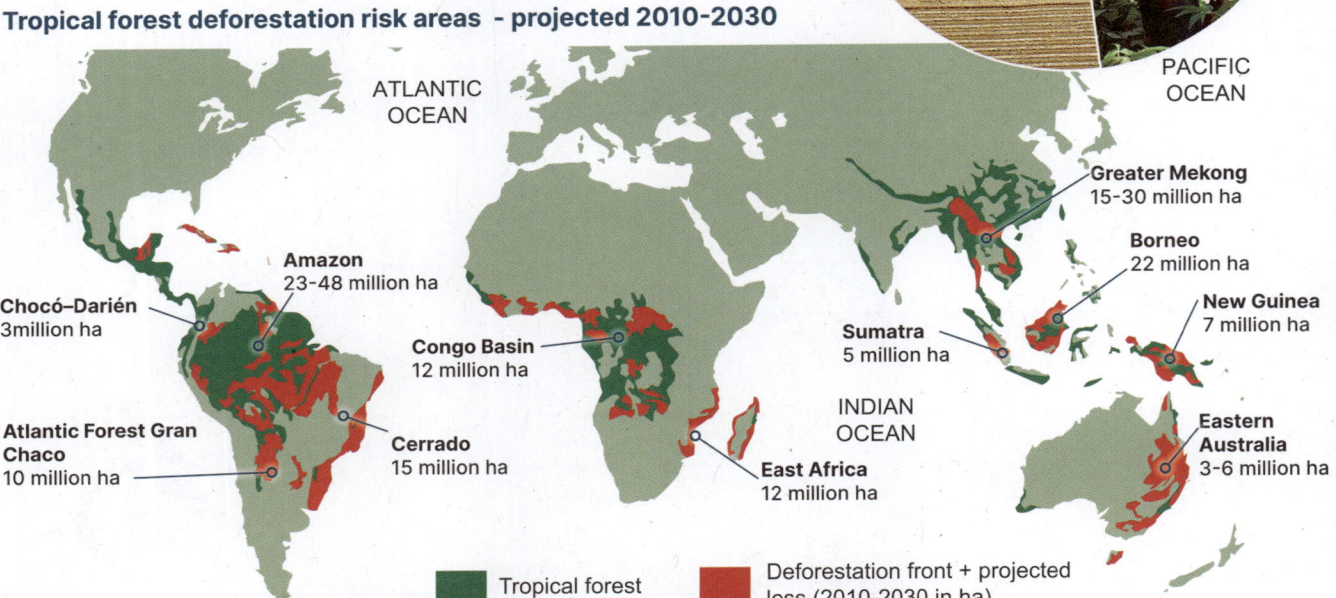

- Chocó–Darién 3 million ha
- Amazon 23-48 million ha
- Atlantic Forest Gran Chaco 10 million ha
- Cerrado 15 million ha
- Congo Basin 12 million ha
- East Africa 12 million ha
- Greater Mekong 15-30 million ha
- Borneo 22 million ha
- Sumatra 5 million ha
- New Guinea 7 million ha
- Eastern Australia 3-6 million ha

Tropical forest | Deforestation front + projected loss (2010-2030 in ha)

It is important to distinguish between deforestation involving primary (old growth) forest and deforestation in plantation forests. Plantations are regularly cut down and replaced, which can artificially inflate a country's apparent forest cover or rate of deforestation. The loss of primary forests is far more important as these are refuges of high biodiversity, including rare species, many of which are not found anywhere else in the world.

New road to Ilheus, Bahia, Amazon

The building of new road networks into regions with tropical rainforests causes considerable environmental damage. In areas with very high rainfall there is an increased risk of erosion and loss of topsoil.

Amazon leaf mantis

Up to 80% of Earth's terrestrial species are found in tropical rainforests. Many tropical species are endemic to specific parts of the forest or tree species. Loss of these areas could exterminate hundreds of species.

Palm oil plantations in rainforest

The soil in tropical rainforests in typically poor and the layer thin. Once the trees are removed and crops planted the soil quickly loses its nutrients and productivity decreases within a few seasons.

1. Identify the three main human activities that cause tropical deforestation and briefly describe their detrimental effects:

    (i) : _____

    (ii) : _____

    (iii) : _____

2. Describe the trend in tropical deforestation compared to temperate deforestation over the last 300 years:
   _____
   _____

3. Deforestation in tropical regions can be hard to reverse, even after plantations and crops have been removed and new trees are planted. Suggest why this might be so:
   _____
   _____
   _____

4. Why might building roads through primary forests have long-term negative effects on the ecosystem beyond the immediate damage caused by road construction?
   _____
   _____
   _____

5. Suggest some advantages of stopping deforestation in Borneo for both animals and humans:
   _____
   _____
   _____
   _____
   _____
   _____
   _____

Borneo orang-utan

Deforestation is causing habitat loss for Borneo orang-utan (*Pongo pygmaeus*) which is projected to affect 26,000 apes by 2032: nearly one quarter of the total population. The orang-utan is called an 'umbrella species'. It needs large tracts of habitat to survive and its **conservation** will also indirectly protect many co-habiting species.

172 Habitat Fragmentation

Key Idea: Habitat fragmentation is detrimental to ecosystems as it reduces species diversity, disrupts gene flow, and increases the likelihood of species extinction.

Habitat fragmentation is a major concern for global **biodiversity**. **Human activities** such as urbanization and road construction are encroaching upon natural areas, resulting in a loss of **species** diversity. This loss not only affects the stability and resilience of **ecosystems** but also hinders their ability to adapt to environmental changes.

Habitat fragmentation, whether caused by natural processes or human activities, leads to the separation of species **populations**. This division prevents them from interacting with each other, which can ultimately result in their local **extinction**. This occurs because a population becomes isolated and too small to effectively breed or experiences inbreeding, and the flow of genes between fragmented areas stops. If this pattern persists across fragmented habitats, it can ultimately lead to the complete extinction of the species.

Habitat fragmentation and biodiversity

▸ Habitat fragmentation is the process by which large habitats become divided up into smaller ones, usually with areas of completely changed (and often uncrossable) land between them. This can happen naturally (e.g. lava flows dividing areas of forest) but more often it occurs as a result of human activities.

▸ Habitat fragmentation can be a driver of evolution, creating greater biodiversity by separating species' populations. However, this is usually a response of smaller organisms, such as insects and small lizards in island ecosystems.

▸ Usually habitat fragmentation causes a loss of biodiversity, especially in larger animals that are territorial or require large areas of land to find food. Habitat fragmentation reduces population sizes and can reduce gene flow because individuals are unable to move easily between habitat fragments. This can lead to inbreeding because access to mates is limited.

▸ **Invasive** plant species are more able to invade fragments due to more open edges, which often provide disturbed land where they can easily become established.

▸ The degree of fragmentation of a species' habitat is a significant predictor of the likelihood of a species going extinct. The **IUCN** (International Union for Conservation of Nature) lists species from least concern to critically endangered (see activity 176). When the species in these categories are matched against the degree of their habitat's fragmentation a clear pattern emerges (right).

Forest fragmentation, Brazil

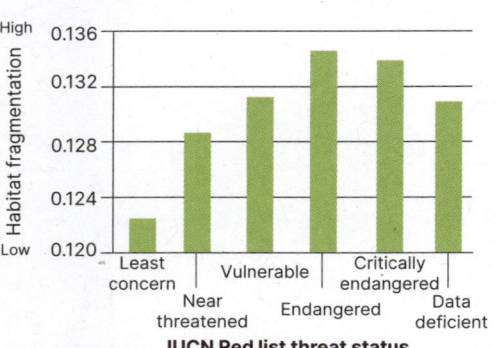
Fragmentation vs IUCN status

Data source: Crooksa (2017). Quantification of habitat fragmentation reveals extinction risk in terrestrial mammals

1. The mountain gorilla lives in just two separated groups in the hills of Rwanda, Uganda, and DRC. Fragmentation of surrounding land due to farming, deforestation for firewood, and demand for living space are creating challenges for the remaining apes. Explain how fragmentation increases the risk of mountain gorilla extinction:

   _____
   _____
   _____
   _____

2. The Texas ocelot, *Leopardus pardalis*, is an endangered wild cat that was once found across Texas and is important to the ecosystem. Two small populations are isolated from each other due to surrounding farm and human occupied land. Giving reasons, suggest what effect human activity might have on the biodiversity of the Texas ocelot ecosystems?

   _____
   _____
   _____
   _____

©2025 BIOZONE International
ISBN: 978-1-99-101409-2
Photocopying prohibited

173 The Impact of Introduced Species

Key Idea: When introduced species compete with existing species for essential resources, they frequently disrupt the stability of the ecosystem they are brought into.

Introduced species are those that have evolved at one place in the world and have been transported by humans, either intentionally or inadvertently, to another region. Some of these introductions are beneficial, e.g. introduced agricultural plants and animals, and Japanese clams and oysters (the mainstays of global shellfish industries). Invasive **species** are introduced species that have a detrimental effect on the **ecosystems** into which they have been imported. They number in their hundreds with varying degrees of undesirability to humans. Humans have brought many exotic species into new environments for use as pets, food, ornamental specimens, or decoration, while others have hitched a ride with cargo shipments or in the ballast water of ships. Some have been deliberately introduced to control another pest species and have themselves become a problem. Some of the most destructive of all invasive species are aggressive plants. These include the mile-a-minute weed, a perennial vine from Central and South America, Miconia, a South American tree invading Hawaii and Tahiti, and Caulerpa seaweed, an aquarium 'escapee', now found in the Mediterranean Sea.

Kudzu: a deliberate introduction

▸ Kudzu (*Pueraria lobata*) is a climbing vine native to south-east Asia. It was first introduced to the United States in 1876 as an ornamental plant.

▸ Kudzu was widely planted during the Dust Bowl Era to try to conserve soil and, in 1940, the government paid farmers to plant it. However, by 1953 the payments stopped as kudzu escaped farms and invaded woodlands, outcompeting native plants for light and space. By 1970, it was declared a weed and in 1997 was placed on the noxious weeds list.

▸ Huge amounts of money are spent every year to try to control it. Today, investigations indicate that kudzu is estimated to cover 3 million hectares of land in the south-eastern US.

Red fire ant: an accidental introduction

▸ Red fire ants (*Solenopsis invicta*) were accidentally introduced into the south-eastern states of the United States from South America in the 1920s and have spread north each year.

▸ Red fire ants competitively displace **populations** of native insects and ground-nesting wildlife. They also damage crops and are very aggressive, inflicting a nasty sting.

▸ Investigations from the USDA estimates damage and control costs for red fire ants at more than $6 billion a year.

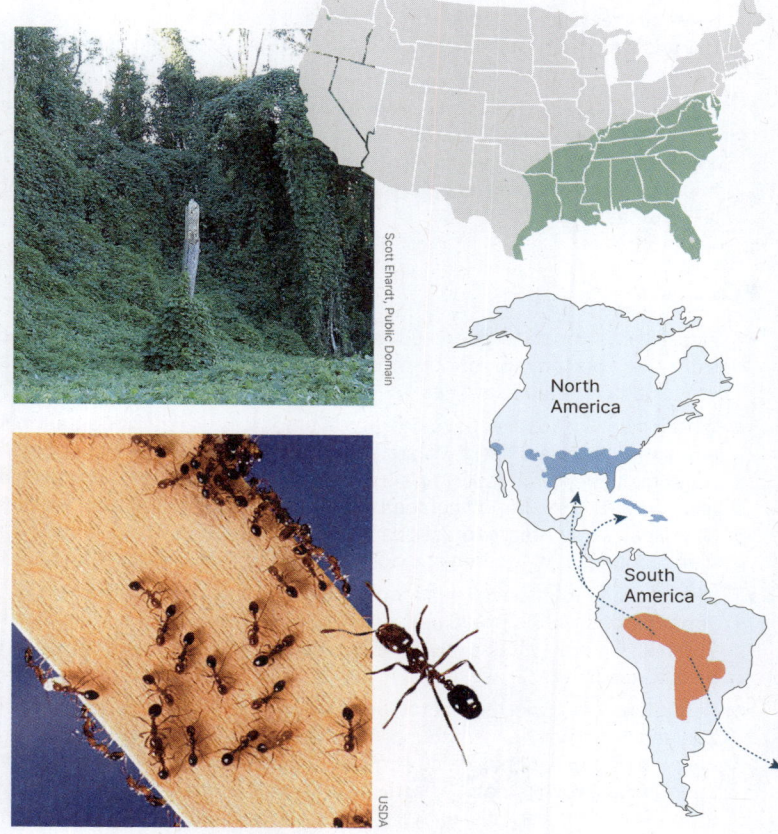

1. Explain why many introduced species become invasive when brought into a new area:

   _____
   _____
   _____

2. European whalers first introduced the domestic rabbit to New Zealand in the 1800s, in part for hunting sport but also as a food source. The rabbits had no natural predators, as New Zealand had no native land mammals, and they soon spread. After the rabbit populations spiralled out of control, ferrets were imported to control them. Ferrets found the many species of flightless New Zealand birds a much easier prey. Suggest some impacts both introduced species may have had on New Zealand ecosystems:

   _____
   _____
   _____
   _____
   _____

174 Control of Introduced Species

Key Idea: Introduced species can often become pests, causing economic and environmental damage, and requiring control measures to be put in place.

The movement of non-native **species** across borders by humans is now heavily regulated in many countries. However, in the past, there were minimal controls, and animals were transported between countries without considering the impact they would have on the environment upon release. During the colonization of the Americas, Australia, and New Zealand, numerous animals and plants were intentionally introduced from Europe, along with the parasites they carried. In the absence of their natural controls and predators, these have become pests, causing wide ranging damage. The cost of controlling pest species can involve large amounts of both funding and voluntary labor. 'Escaping' imported non-native ornamental plants have also extracted a huge cost to control. The most common 72 invasive tree species cost over one billion $US for global management alone, since 1960.

Why control introduced species?

Economics

▸ Invasive **introduced species** carry an economic cost, both as a result of the damage they do and because of the cost of controlling or removing them. In the UK, it is estimated that invasive species have an economic cost of up to £2 billion a year.

▸ In New Zealand, the economic cost of invasive species is estimated at over NZ$12.4 billion from 1968-2020. New Zealand aims to become 'predator free' (removal of all introduced mammal pests) by 2050. This will require government, businesses, **conservation** groups, and individuals to help financially and as volunteers.

Ecology

▸ Invasive species damage the **ecosystem** into which they are introduced. Generally, they prey on or out-compete native species, reducing numbers of both prey and competitor species.

▸ Where invasive species have no natural predators or other **population** controls, population numbers can escalate. For example, a small number of brush-tailed possums were introduced to New Zealand from Australia in 1837 for the fur trade. In 1980, the population peaked at 70 million. Possums are considered New Zealand's primary forest pest, consuming around 21,000t (metric tons) of vegetation a night and consuming eggs, chicks, invertebrates, and lizards.

▸ In Australia, rabbits were declared pests just 30 years after their successful introduction in 1859. In 1901, the No. 1 rabbit fence was constructed (at nearly 2000 km long) to try to exclude rabbits from the western agricultural areas. It was a complete failure.

▸ Predation is an important factor in the need to control introduced species. New Zealand in particular has had major problems with introduced predators affecting native bird populations, which evolved in the absence of mammalian predators. Various mustelids (stoats, weasels, and ferrets) were introduced to control rabbits in an effort that went disastrously wrong as they quickly switched to birds as prey.

▸ Rats arrived with trading ships. They have caused enormous damage to populations of native birds. In the UK, the American mink, which likely escaped captivity in fur farms, is now severely reducing the numbers of the native European water vole.

▸ Disease can be carried by introduced species. The gray squirrel was introduced to the UK from America. It is larger and more aggressive than the native red squirrel and easily out competes it in certain habitats. The gray squirrel also carries the squirrelpox virus, which is fatal to red squirrels.

▸ In New Zealand, the possum can carry bovine TB (tuberculosis) which can be spread to beef and dairy herds.

In New Zealand, brush-tail possums are considered pests and are controlled through methods like poison, traps, and hunting due to their negative impact on native birds, animals, and plants. Interestingly, although they are harmful pests in New Zealand, they are protected species in their native country, Australia.

Gray squirrel, not a native to UK, has now spread up into mid-Scotland. The red squirrel, once widespread through the UK, is now found in only a few limited areas.

1. Using examples, explain why it is necessary to control invasive species, despite the cost:

_____
_____
_____
_____

175 The Impact of New Diseases

Key Idea: Novel diseases can have a rapid and devastating impact on species with limited immune protection from them.

When **species** of plants and animals, including humans, are introduced into new regions, whether intentionally or accidentally, the diseases they carry can also spread to new hosts. During the 'Age of Discovery' from the mid-15th to mid-16th century, Europeans brought diseases like smallpox and measles to the New World. This resulted in devastating impacts on the indigenous **populations** of continents, such as South America. Relocation of exotic animals and plants also carried their diseases as well, some of which jumped to similar native species. Mobility in other species, such as migrating birds, allow diseases to spread fast across nearly all areas of the world. Furthermore, climate change is creating more favorable conditions in previously inhospitable areas, enabling certain **invasive species** to expand their habitat. This facilitates the spread of vector-borne diseases, such as Dengue fever, which is transmitted by *Aedes* mosquitoes that prefer warmer climates. As a result, these diseases can now also extend beyond their original range.

Dutch elm disease

Dutch elm disease (DED) is caused by the fungus *Ophiostoma ulmi* and affects trees in the elm family (*Ulmus* and *Zelkova*) causing wilt. The disease is spread by the elm bark beetle. It originated in Asia and was accidentally introduced to Europe, North America, and much later, New Zealand. Where the disease has been introduced it has quickly caused the loss of large numbers of elm trees by clogging their vascular tissue.

DED in the United Kingdom

DED appeared in Europe in 1910 and a much deadlier strain appeared in 1967. In the UK, these strains have killed 60 million trees. France has lost nearly 90% of its elm trees.

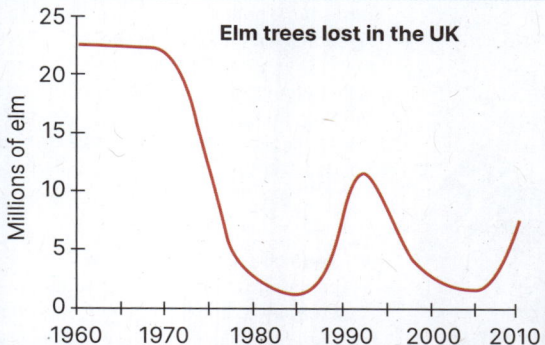

Tree numbers drop rapidly as management removes infected trees. New trees grow, but new infections, outbreaks, and repeated management reduces numbers.

Elm afflicted by DED

H5N1 bird flu

▸ The H5N1 bird flu, or avian influenza A, is a viral disease that was first detected in 1996, amongst a flock of geese in China. It has spread around most of the world since, and across different species.

▸ The risk of human infection is low, with only one recent case of cow to human transmission reported in Texas in 2024. However, there is a significant concern regarding the risk to birds and other species, both in the wild and in domesticated settings.

▸ Some viral strains are particularly deadly (high virulence). Combined with the high infectiousness (ability to spread), wild bird populations are at risk when migrating individuals return from their over-wintering bases.

▸ In 2024, scientists have found over 500 dead Adelie penguins in Antarctica and suspect the H5N1 virus may be the culprit. Traces of the virus were found in nearby skua seabird colonies, which predate on Adelie chicks. Scientists are also concerned for other more vulnerable Antarctic species, such as the emperor penguin.

▸ Remote bird species that are already impacted by climate change and habitat destruction, may now have to contend with another factor causing **biodiversity** loss.

One of a number of dead gannets found on an isolated beach in 2022, suspected of being infected with H5N1.

1. What factors influence the spread of disease into new areas? _____
_____
_____

2. What are the risks associated with H5N1 bird flu? _____
_____
_____
_____

Zika virus: An example of global disease spread and its containment

▸ The Zika virus was initially discovered in 1947 in the Zika Forest in Uganda. Mosquitoes played a crucial role in transmitting the virus to non-human primates by transferring infected blood between individuals through skin-piercing bites. Soon after, cases of Zika infections in humans were identified, mainly in tropical areas where the mosquitoes naturally thrive.

▸ Since then, Zika has spread slowly across the globe, with outbreaks in the Americas in 2015 and 2016. Due to the disease spreading only when *Aedes aegypti* mosquitoes are present, Zika has been limited to the same regions, albeit widely spread.

▸ Research has projected that by 2050 viral diseases spread by tropical mosquitoes will increase 20% due to climate change. They include Zika, dengue, and chikungunya.

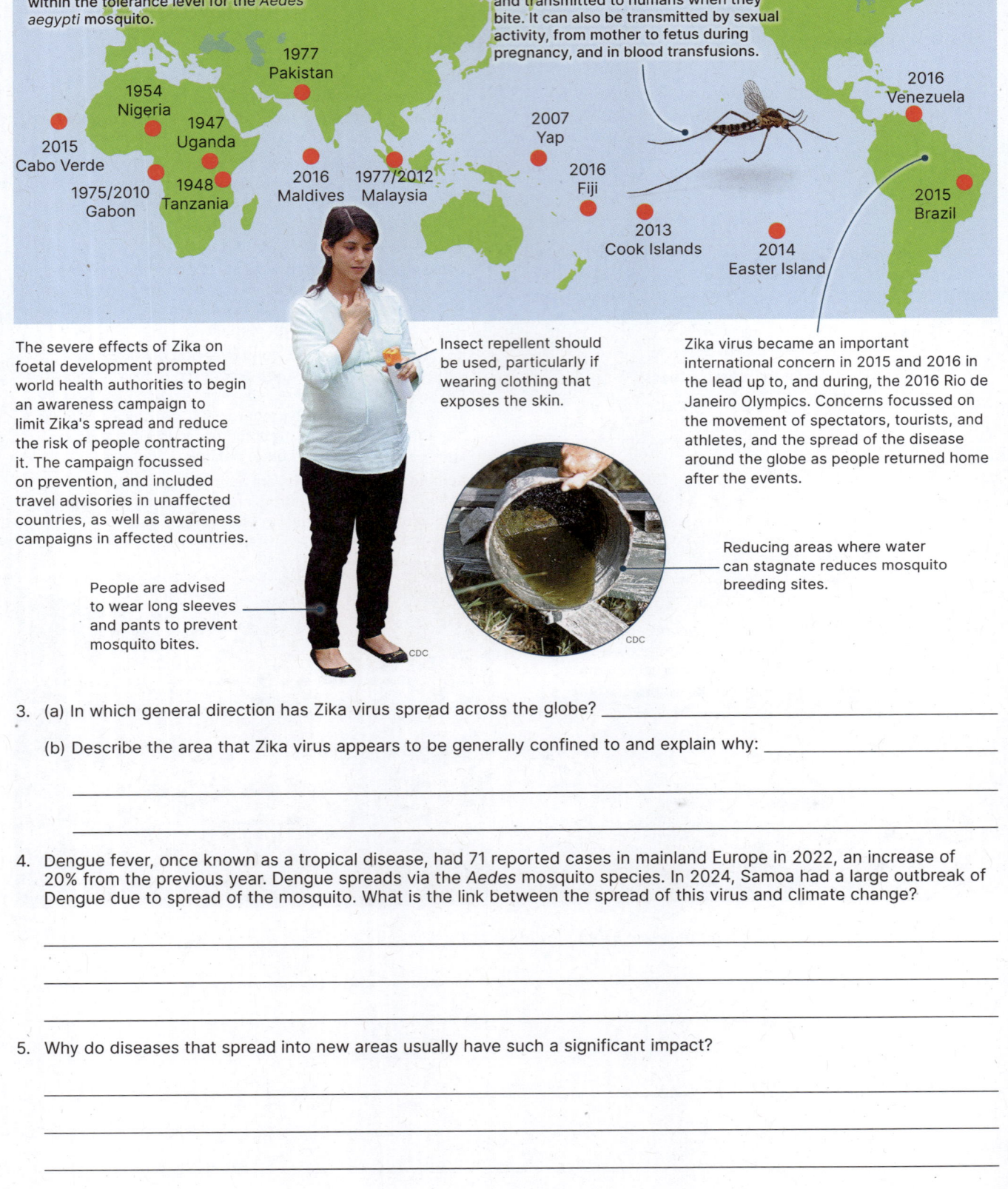

3. (a) In which general direction has Zika virus spread across the globe? _____

   (b) Describe the area that Zika virus appears to be generally confined to and explain why: _____
   _____
   _____

4. Dengue fever, once known as a tropical disease, had 71 reported cases in mainland Europe in 2022, an increase of 20% from the previous year. Dengue spreads via the *Aedes* mosquito species. In 2024, Samoa had a large outbreak of Dengue due to spread of the mosquito. What is the link between the spread of this virus and climate change?
   _____
   _____
   _____

5. Why do diseases that spread into new areas usually have such a significant impact?
   _____
   _____
   _____

176 Endangered Species

Key Idea: Human activity is endangering some species. **Species** under threat of severe **population** loss or **extinction** are classified as either endangered or threatened. An **endangered species** is one with so few individuals that it is at high risk of local extinction, while a threatened (or vulnerable) species is likely to become endangered in the near future. While extinctions are a natural phenomenon, the rapid increase in the rates of species extinction in recent decades is of major concern. It is estimated that every day up to 200 species become extinct as a result of human activity. Even if a species is preserved from extinction, remaining populations may be too small to be genetically viable. Human population growth, rising non-sustainable resource use, poverty, and lack of environmental accountability are the underlying causes of premature extinction of organisms. The biggest direct causes are **habitat fragmentation** and loss, or degradation and the accidental or deliberate entry of non-native **introduced species** into **ecosystems**.

Human causes of species decline

Rhinoceros horn cut due to poaching

Urban development destroys habitats

Polluted discharge into waterway

Illegal poaching and trafficking
Species may be hunted or collected illegally for commercial gain, even when they are in a reservation area. Illegal trade and poaching threatens the population viability of some species. Some species are hunted because they interfere with human use of an area or for food.

Habitat destruction
Natural habitat can be lost through clearance for agriculture, urban development and land reclamation, or trampling and vegetation destruction by introduced pest plants and animals. Habitats potentially suitable for a threatened species may be too small and isolated to support a viable population.

Pollution
Plastics and toxic substance **pollution** created and released by **human activity** into the environment, e.g. from industry, cause harm directly or accumulate in food chains. Estuaries, wetlands, river systems, and coastal ecosystems near urban areas are particularly vulnerable.

Weasel with stolen egg

Emperor penguin on melting ice shelf

Valuable bluefin tuna is overharvested

Introduced exotic species
Introduced predators (e.g. rats, mustelids, and cats) prey on endangered birds and invertebrates. Introduced grazing and browsing animals (e.g. deer, goats) damage sensitive plants and trample vegetation. Weeds may out-compete endemic species.

Climate change
Entire habitats may be changing and cause breeding and feeding locations to be no longer suitable. Extreme weather events may cause irreparable damage to habitats. Species in polar regions are especially susceptible to population decrease due to global warming, leading to ice melt.

Overharvesting
Commercial harvest of fish, shellfish, and wild game can capture numbers beyond which a healthy population can sustain or regenerate. Some indigenous peoples also rely on capturing wild foods to eat. Quota limits are imposed to allow numbers to regenerate.

1. Endangerment of individual species and biodiversity loss can often be caused by similar human activities. Discuss some similarities and differences between these two phenomena:

   _____
   _____
   _____
   _____

2. Why do small populations of a species make them more likely to be endangered?

   _____
   _____

The state of the endangered species

▸ Endangered species are found in nearly every class of living organisms. They include gut bacteria in humans, rare plants and fungi found in vulnerable habitats, such as alpine regions, and many animals, some only recently discovered. The protists, single-celled organisms, appear to be the only group not at risk.

▸ It is natural for species numbers to fluctuate over time in response to changing environments and ecosystems. However, the alarming increase in the number of endangered species is largely attributed to human activity.

The IUCN Red List

▸ The **IUCN** (International Union for the Conservation of Nature) is an organization involved in **conservation** and **sustainability** projects.

▸ The organization publishes updated data on the status of endangered species, called the Red List (link right).

Data source

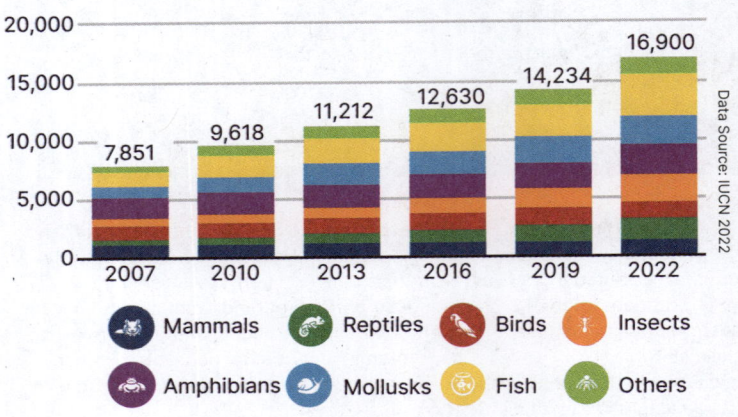

Number of threatened animal species over time (IUCN)

- 2007: 7,851
- 2010: 9,618
- 2013: 11,212
- 2016: 12,630
- 2019: 14,234
- 2022: 16,900

Data Source: IUCN 2022

Mammals, Reptiles, Birds, Insects, Amphibians, Mollusks, Fish, Others

Amur Leopard

Status: CR — Critically Endangered
Population: Around 100 individuals
Weight: 25 – 43 kg
Scientific Name: *Panthera pardus orientalis*
Habitats: Mixed Korean pine and deciduous forest

Steller's sea eagle: A vulnerable species

▸ Steller's sea eagle (*Haliaeetus pelagicus*), is endemic to coastal Asia and is one of the heaviest eagles in the world.

▸ The population of Steller's sea eagle is currently estimated to be around 5000 birds and this population is declining as a result of various human activities.

▸ Climate change-induced extreme weather events are flooding eagle nest areas on rivers.

▸ The eagles' habitats are being disrupted by demand for space and resources by humans, such as logging.

▸ Pollution from industry is harming the eagles, as is lead shot from carcass left by hunters.

▸ Overfishing in the eagle habitat is reducing the amount of food available for the birds, such as salmon.

▸ Eagles are being killed by fur trappers as they compete for the same food source.

3. List some critically endangered species from the IUCN red list (QR link above), including location and numbers:

4. Research a local living species on the IUCN red list, providing its status, the range, population, and some causes leading to its endangerment:

Decline of the elephants

▶ Both African and Asian elephant species are listed as endangered, while the African forest elephant is critically endangered according to the International Union for the Conservation of Nature (IUCN).

▶ Where elephants live near to agricultural areas, they raid crops and come into conflict with humans. The ivory trade represents the greatest threat to the African elephant. Elephant tusks have been sought after for centuries as a material for jewelry and artworks. In Africa, elephant numbers declined from 1.3 million to 600,000 during the 1980s. At this time, as many as 2000 elephants were killed for their tusks every week. Despite the investment of large amounts of money in fighting poaching, elephant populations continued to fall in many countries.

▶ From 1975 to 1989 the ivory trade was regulated under CITES (Convention of International Trade in Endangered Species), and permits were required for international trading. In 1997 Botswana, Namibia, and Zimbabwe, together with South Africa in 2000, were allowed limited commercial trade in raw ivory.

Two species of African elephant are currently recognized: the savannah elephant (*Loxodonta africana*) and the less common forest elephant (*Loxodonta cyclotis*). Recent evidence from mitochondrial DNA separated the elephants into two distinct species.

In 1989 the Kenyan government publicly burned 12 tonnes of confiscated ivory. With the increased awareness, the United States and several European countries banned ivory imports. Game wardens weigh confiscated ivory tusks and rhinoceros horns for records.

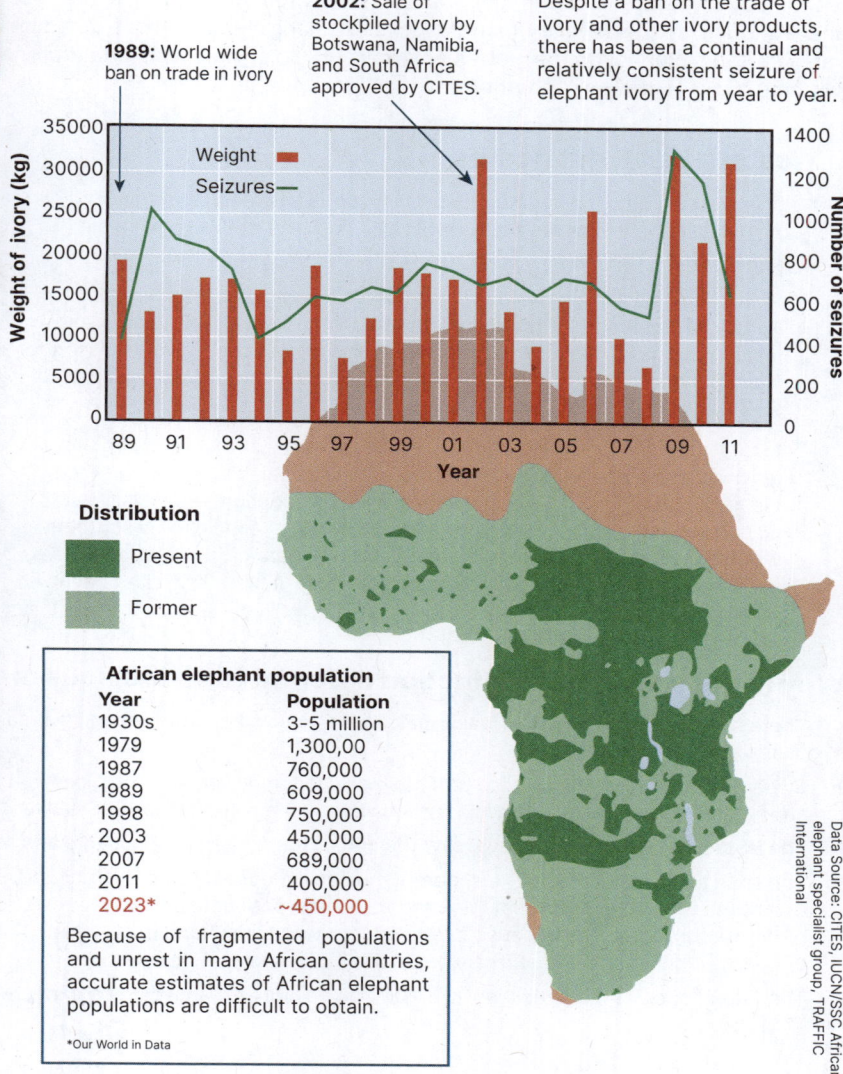

1989: World wide ban on trade in ivory

2002: Sale of stockpiled ivory by Botswana, Namibia, and South Africa approved by CITES.

Despite a ban on the trade of ivory and other ivory products, there has been a continual and relatively consistent seizure of elephant ivory from year to year.

African elephant population

Year	Population
1930s	3-5 million
1979	1,300,00
1987	760,000
1989	609,000
1998	750,000
2003	450,000
2007	689,000
2011	400,000
2023*	~450,000

Because of fragmented populations and unrest in many African countries, accurate estimates of African elephant populations are difficult to obtain.

*Our World in Data

Data Source: CITES, IUCN/SSC African elephant specialist group, TRAFFIC International

5. Outline the action taken in 1989 to try and stop the decline of the elephant populations in Africa:

_____

_____

_____

6. In early 1999, Zimbabwe, Botswana, and Namibia were granted a one-time, CITES-approved experimental sale of ivory to Japan. This sale involved 5446 tusks (50 tonnes) and generated approximately US$5 million for the governments involved. Suggest some advantages and disadvantages of that agreement:

_____

_____

_____

_____

177 The Sixth Mass Extinction

Key Idea: Evidence suggests that a current, sixth mass extinction event is due to anthropogenic (human) causes.

Human (anthropogenic) **activity** dominates Earth today. Humans can be found almost everywhere on the globe. As humans have spread across the planet, from Africa into Europe and Asia, and across into the Americas and beyond, and expanded in **population** size, they have changed the environment around them to suit their needs. How these changes have occurred has varied according to the technology available and the general social environment and attitudes at the time. Human-associated change has had a profound impact on the globe's physical and biological systems. Only in the past century have we begun to fully evaluate the impact of human activity on the Earth. Humans have intentionally or unintentionally caused the **extinction** of a significant number of species and have pushed many more **species** to the brink of extinction. The loss of a significant number of species, directly or indirectly caused by human activity, has led to the designation of this period in history as the **sixth mass extinction**. There is ongoing debate about the exact start of this extinction event, its magnitude, and the extent of human involvement. However, it is evident that numerous species are currently being lost, and many species that existed before humans encroached on their habitats are no longer present.

Western black rhinoceros

▸ The western black rhinoceros (*Diceros bicornis longipes*), a genetically unique subspecies of the black rhino, was declared officially extinct in 2011 by the **IUCN**. The species once was distributed across central West Africa.

▸ Rhino horn is still highly valued for traditional medicines, often in Asia. Poachers shoot and kill the rhino, cutting off the horn to sell illegally and leaving the dead rhino behind. Poaching was almost certainly the reason for the rapid decline of the rhino population, including the western black, with 96% of all black rhino disappearing between 1970 and 1990.

▸ In 2018, the last male western black rhino passed away while in captivity, leaving behind only two remaining females: Najin, his daughter, and Fatu, his granddaughter. Conservationists are now exploring the possibility of using preserved eggs in a closely related surrogate species, the southern white rhino, or utilizing genetic material from Sudan for future efforts. Unfortunately, the genetic diversity of the western black rhino has been irreversibly lost.

Naijin and Fatu, the last two western black rhinos

Why is the sixth mass extinction different from previous mass extinctions?

▸ There have been five previous major mass extinctions: short (by geological measurement) periods of time in which many distinct species die out.

▸ Evidence for the causes of these point to extreme temperature swings, ocean acidification, extensive volcanism, and asteroid strike, all occurring without any human influence (as humans hadn't yet evolved).

▸ The last mass extinction, 66 Mya, saw the end of the non-avian dinosaurs and most marine reptiles.

▸ The sixth mass extinction is the apparent, human-induced loss of much of Earth's **biodiversity**. Several human activities contribute to this: 40% of all land is used for food production; agriculture uses 70% of the world's available freshwater; and 90% of **deforestation** occurs because of a need for more agricultural land. In fact, human activities to produce enough food to feed the population is the biggest threat to Earth's biodiversity.

▸ The rate of species extinction is 100 - 1,000 times higher than would be expected as a background extinction rate.

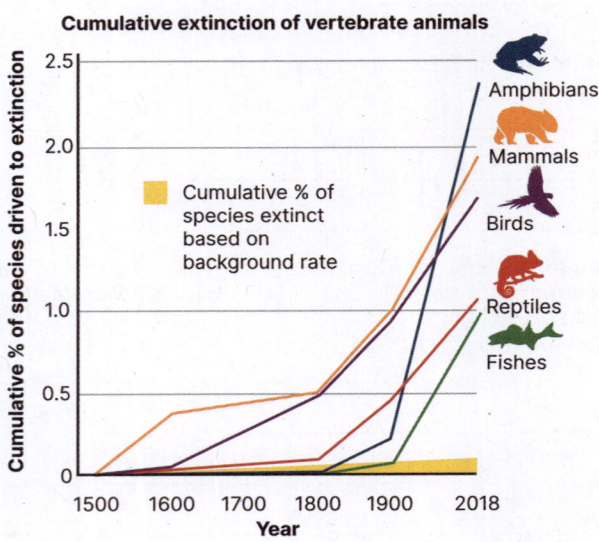

1. (a) Why would extinction still be occurring at a background rate in the graph (left)?

_____

_____

_____

(b) What events would have occurred from 1800 onwards to cause a shift in extinction rate?

_____

_____

_____

_____

_____

Estimating extinction rates

▶ Estimates of the rate of species loss can be made by using the background extinction rate as a reference. It is estimated that one species per million species per year becomes extinct. By totalling the number of extinctions known or suspected over a time period, we can compare the current rate of loss to the background extinction rate.

▶ Birds provide one of the best examples. There are about 10,000 living or recently extinct species of birds. In the last 500 years, an estimated 150 or more have become extinct. From the background extinction rate, we would expect one species to become extinct every 100 years (10,000/1,000,000 = 0.01 extinctions per year = 1 extinction per 100 years). It then becomes apparent that 150/500 = 0.3 extinctions per year or 30 extinctions per century, 30 times greater than the background rate.

▶ The same can be calculated for most other groups of animals and plants.

Organism*	Total number of species (approx)*	Known extinction (since ~1500 AD)*
Mammals	5487	87
Birds	9975	150
Reptiles	10,000	22
Amphibians	6700	39
Plants	300,000	124

*These numbers vastly underestimate the true numbers because so many species are undescribed.

Data Source: Our World in Data

2. Define the sixth mass extinction: _____
_____
_____

3. Use the data in the table to calculate (1) the rate of species extinction per century for each of the groups, and (2) how many times greater this is than the background extinction rate:

   (a) Mammals: _____

   (b) Reptiles: _____

   (c) Amphibians: _____

   (d) Plants: _____

Caribbean monk seal

▶ The Caribbean monk seal (*Neomonachus tropicalis*) is one of three species of monk seal adapted for warmer water. The seal clade evolved about 23 million years ago. Elephant seals, Antarctic seals, and monk seals separated from the other seal groups around 17 million years ago. Mitochondrial DNA sequencing has shown that the monk seals are the most distantly related to other seal species.

Caribbean monk seal reconstruction

▶ The Caribbean monk seal was originally widely distributed across the Caribbean and the populations were numbered in the hundreds of thousands. Hunting for the seals started to increase once Europeans arrived in the area and huge numbers were slaughtered in the 17th-19th centuries for oil and food. By the 1950s, the Caribbean monk seal was extinct.

▶ Two other species, the closely related Hawaiian monk seal and the Mediterranean monk seal, are both endangered and at risk of also becoming extinct. Monk seals are an important source of information for seal phylogeny. A recent fossil find of a possible fourth species of monk seal, now extinct, has been found on the west coast of New Zealand and sheds light on the origin of the group.

4. Research a species that has recently become extinct in your region and is listed on the IUCN Red List (link also in **BIOZONE Resource Hub**). Filter by your region and extinct category. Provide the scientific name and details of the species, as well as the timeframe and reasons for its extinction.

_____
_____
_____
_____
_____

178 Managing Environmental Resources

Key Idea: Consideration of both human and environmental needs for services and resources can be difficult and requires education, incentive, and compromise.

The growing human **population** requires land for food and resource production. This has resulted in increased pressure on untamed lands. However, in many countries, limited government funding is available for innovative methods to enhance the efficiency of food and resource production. Framers therefore often resort to a simple solution of clearing more land to increase food production. Rainforests in particular can provide money from harvested wood before the ground is plowed and sowed with crops. Ironically, the rise of **ecotourism** in natural areas has led to similar issues as businesses capitalize and construct tourist towns. The influx of people further encroaches into natural areas, creating a cycle of development and expansion. Striking a balance between human needs and preservation of the environment is challenging, as these needs often conflict with one another.

How does resource conflict between humans and wildlife occur?

▸ Human activity impinges on wildlife wellbeing e.g. hunting.
▸ Wildlife impinges on human wellbeing e.g. attacks by wildlife and/or destruction or raiding of crops.
▸ There is competition for space and resources between humans and wildlife e.g. need for grazing and shelter.
▸ Wildlife protection, **legislation**, or wildlife-based industry, e.g. tourism, impinges on the rights of humans and access to resources.
▸ The local human population can have negative attitudes to wildlife **conservation**.

Conflict between humans and wildlife occurs more often when the two are close together. Livestock are at risk of being caught by predators if they are not adequately protected, such as by secure hutches or compounds.

Wildlife needs to travel, just as humans do. Problems occur when humans and wildlife use the same routes. Many countries have installed wildlife crossings to avoid collisions between animals and vehicles.

Resources under the ground are difficult to reach without some disruption to the landscape. Legislation requiring mines to return the land to its original state after mining ends is becoming more common.

Reducing conflict between humans and wildlife

▸ **Education:** instilling an ethical belief in conservation rather than a material one (change from "what's in it for me?" to "how does the greater system benefit?"). This includes increasing public participation in conservation.
▸ **Producing jobs:** economic benefits of wildlife conservation.
▸ **Compensation:** for losses due to wildlife infringement on crops or livestock or loss of land use.
▸ **Land planning:** Ensuring that where land is to be used, wildlife is not adversely affected.

1. Suggest why logging rainforests might appeal to neighbouring farmers needing to expand their land: _____

2. Suggest some situations in which locals might have a negative attitude towards wildlife: _____

3. Explain why public participation is particularly important in resource management: _____

Case study: The Maasai Mara

▸ The Maasai Mara National Reserve is a region in southwestern Kenya covering 1500 km². It is part of the much larger Mara-Serengeti ecosystem, which covers around 25,000 km².

▸ Considerable change has occurred in the Maasai Mara region since the start of last century. Early in the 20th century, the region was much less populated, and the land was used mainly for nomadic agriculture (raising cattle). The arrival of European settlers forced many of the Maasai off their traditional lands. In 1945, more land was turned into reserves and the Maasai land placed in Trust. As a result of changes in governments and ideals, Trust land was redesignated as group ranches.

▸ This encouraged the Maasai to subdivide the land to acquire individual titles to secure legal rights to lands, rather than risk losing them outright. Privatization led to an increase in mechanized farming and a reduction in wildlife.

▸ The changes in land use in the Maasai Mara region have had a significant impact on the wildlife population. Specifically, the number of wildebeests in the area decreased from around 150,000 in 1977 to 4,000 in 2010. Similarly, the population of water buffalos dropped from nearly 40,000 to approximately 5,000. On the other hand, the number of livestock, particularly cattle, sheep, and goats, increased during this period.

▸ Over time, it was realized that modern farming methods limited the range of wildlife and also grazing options for livestock. In 2005, many landowners in the northern Maasai Mara began consolidating their land into conservancies, aiming to generate income through tourism. This included establishing partnerships with tourism operators. The success of this approach has seen a rapid expansion of conservancies.

Wildlife and livestock changes in the Maasai Mara

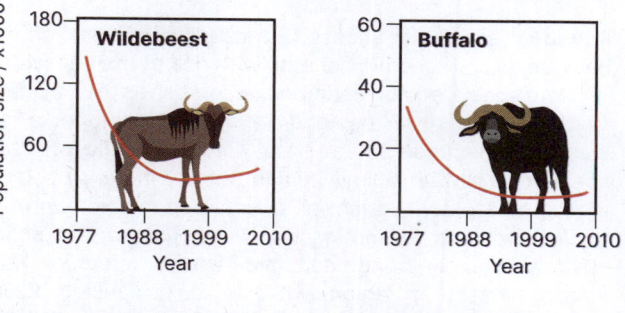
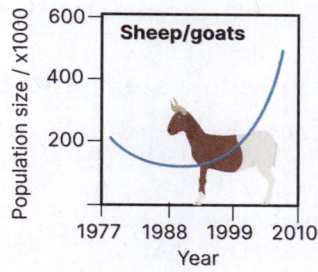
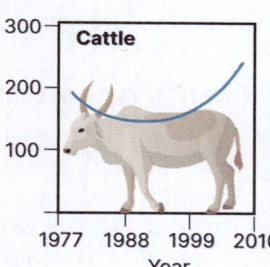

Data Source: Mundia and Murayama (2009)

The Maasai Mara National Reserve and surrounding conservancies

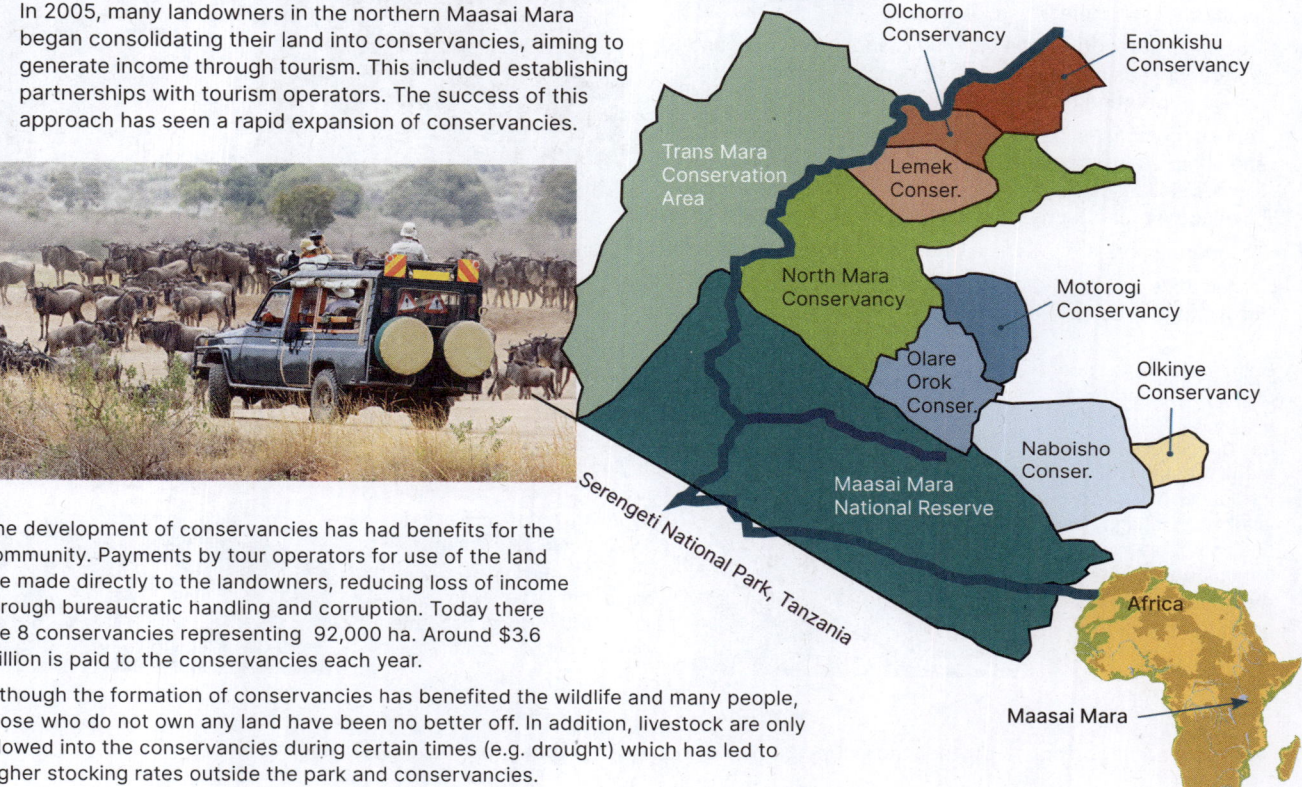

The development of conservancies has had benefits for the community. Payments by tour operators for use of the land are made directly to the landowners, reducing loss of income through bureaucratic handling and corruption. Today there are 8 conservancies representing 92,000 ha. Around $3.6 million is paid to the conservancies each year.

Although the formation of conservancies has benefited the wildlife and many people, those who do not own any land have been no better off. In addition, livestock are only allowed into the conservancies during certain times (e.g. drought) which has led to higher stocking rates outside the park and conservancies.

4. (a) What are the conservancies surrounding the Maasai Mara National Reserve and what is their role or purpose?

(b) In what way have these conservancies helped reduce conflict between humans and wildlife?

179 Ecotourism in the Galápagos Islands

Key Idea: Ecotourist access to the Galápagos Islands must balance local economic benefits with risk to the ecosystem. Areas of special ecological importance, such as the Galápagos Islands, are in great demand for visits from tourists, who want to experience and interact with potentially vulnerable species. One way to limit human activity in sensitive areas is to promote **ecotourism** as a viable economic alternative to development. Promoting tourism in unique and pristine environments can bring economic benefits and employment opportunities to a region, while typically causing minimal damage to the landscape. Controlling access to these areas through permits allows for better management and preservation of the environment. This ecotourism has become a major contributor to the economies of many countries, such as Ecuador who manage the Galápagos Islands. However, ecotourism can result in large groups travelling to sensitive ecological areas, albeit with a guide. The impact of catering to the needs of these tourists, and managing their resultant waste, can place heavy demands on pristine and fragile **ecosystems**. The process requires careful planning and adequate infrastructure to cater for increased numbers of visitors.

Balancing ecotourism and ecosystem protection

▸ The Galápagos Islands are 1000 km off the coast of Ecuador, the caretaker country. They are home to over 2000 endemic **species**, some of them classified as critically endangered.

▸ The Ecuadorian government must balance the desire for income from providing guided tours with the need to protect the valuable ecosystem.

▸ The experience of the ecotourism industry in the Galápagos Islands, 97% of which are a National Park, illustrates the need for careful management.

▸ More than 180,000 people visit the islands each year, and ecotourism has brought costs and benefits to the islands, whose **population** has more than tripled in the last 15 years.

▸ Tourism companies have developed a waste recycling station, and some are using hybrid fuel boats to limit the amount of fossil fuel leakage. Registered tourist guides are needed to have access, and tourists are required to stick to planned walking paths (right) and not interfere with the wildlife.

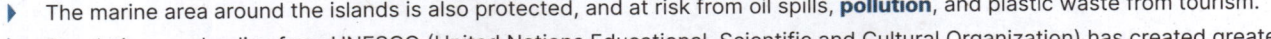

▸ The marine area around the islands is also protected, and at risk from oil spills, **pollution**, and plastic waste from tourism.

▸ Regulations and policy from UNESCO (United Nations Educational, Scientific and Cultural Organization) has created greater local and international awareness of the islands' ecology but the high tourist traffic also puts that ecology at risk of damage.

Despite the large impact of the tourist industry on the islands, there has been some progress in the reduction of introduced pests. The island of Santiago has had its population of feral pigs eradicated and the population of feral goats there has also been reduced.

Feral pig

The influx of humans to the Galápagos has had serious impacts on the islands outside the National Park. At least two ships bringing fuel for tourist boats have run aground, one spilling 3 million litres of fuel oil that killed 60% of the marine iguanas on the island of Santa Fe. Sea cucumber populations have also been devastated by local fishermen selling to overseas markets.

Marine iguana

1. Why is it so important to protect the ecosystem and the species that live on the Galápagos Islands?

   _____
   _____
   _____

2. With reference to the Galápagos Islands, discuss some of the benefits of ecotourism:

   _____
   _____
   _____
   _____
   _____

Benefits	Costs
Economic	**Economic**
• Greater development (upgrading of communications and infrastructure)	• Restrictions placed on local fishing industry because of National Park
• Larger population means more sustainable local economy	• Mass immigration of people looking for jobs
• More jobs (in both tourism and infrastructure)	• Most tourist dollars do not go to the Galápagos economy
Environmental	**Environmental**
• Greater local understanding of the importance of the islands	• Issues with rubbish disposal
• Development of local wildlife services (National Park Service)	• Disturbance of plants and animals by tourists
• Greater global awareness of environmental issues to do with the islands	• Increased travel increases the risks of new species introductions

3. Referring to the chart above, explain how ecotourism may negatively impact the Galápagos Islands ecosystems:

   (a) Increased tourist traffic: _____

   (b) Higher local population: _____

   (c) Support for National Park services: _____

   (d) Greater environmental awareness: _____

4. What general guidelines for ecotourism from Galápagos Islands can be applied to ecotourism in other countries?
   _____

5. If you were running an ecotourism operation in the Galápagos Islands, what rules would you have for your customers? Work in small groups to develop a list and record the rules below:
   _____

180 In Situ Conservation

Key Idea: *In situ* (on site) conservation methods manage ecosystems to protect biodiversity and species within the natural environment.

In situ conservation is implemented directly at the site where the **conservation** efforts are needed. It encompass various approaches including ecological restoration, reintroduction of **species** that have been lost, and the establishment of **legislation** to protect **ecosystems** that hold significant value. Nature reserves are areas designated as having wildlife, habitat, or natural formations needing protection, and often contain rare or nationally important species of plants or animals. Some species, such as corals, are difficult to conserve outside their natural habitats, so removal is a last option. **Rewilding** and reclamation of degraded ecosystems is a key component of in situ conservation. It can also involve removing **introduced species** and pests.

A case study in *in situ* conservation

Sanctuary Mountain Maungatautari is a 3400 hectare conservation 'mainland island', surrounded by 47 km of predator proof fence, situated in the North Island of New Zealand. Endemic New Zealand birds face a unique challenge, having evolved without mammalian predators, and many species are endangered due to introduced pests, such as possums, rats, and mustelids, as well as habitat loss. A project to recreate a pre-human temperate New Zealand ecosystem was established in 2001. The sanctuary has become an 'ark' to preserve birds, reptiles, insects, fungi, and plants.

How the project is proceeding

- Predator-proof fence built and maintained.
- Initial pest eradication and constant monitoring, both by rangers and a large volunteer group.
- Replanting of bare land into native species, vital for food sources and habitat for the New Zealand species.
- Reintroduction of **endangered species** originally extinct from the area, such as kākāpō, tuatara, hihi, kiwi, kākā, New Zealand robin, and takahē.
- Educate schools and visitors to promote conservation.

Advantages of *in situ* conservation

- Species left in the protected area have access to their natural resources and breeding sites.
- Species will continue to develop and evolve in their natural environment, thus conserving their natural behavior.
- *In situ* conservation:
  - is able to protect many species at once and allow them greater space than those in captivity.
  - protects larger breeding **populations**.
  - is less expensive and requires fewer specialized facilities than captive breeding.

Disadvantages of *in situ* conservation

- Controlling illegal exploitation and poaching of *in situ* populations can be difficult.
- Habitats that shelter *in situ* populations may need extensive restoration, including pest eradication, and ongoing control.
- Populations may continue to decline during restoration.

Maungatautari mountain surrounded by farmland and lakes

Tuatara have been introduced back into Maungatautari. This unique reptile is the only remnant of its order, present before dinosaurs. It was once widespread through New Zealand, but was eventually reduced to living on a few small islands.

1. Explain why *in situ* conservation commonly involves collaboration of many people and organizations to protect species:

___

2. Why is *in situ* conservation not suitable for every endangered species?

___

Saving the kākāpō

▸ The kākāpō is the only flightless, nocturnal parrot species on Earth. It is endemic to New Zealand and is found there on a few highly protected offshore islands. It has recently been introduced to Sanctuary Mountain Maungatautari.

▸ Kākāpō were once widespread across New Zealand but the first Polynesian settlers in the mid-1300s found them easy to catch as food. The birds had adapted to freeze in place when they felt threatened due to the absence of co-evolved predators.

▸ By the time of European arrival, 500 years later, kākāpō population numbers had already dwindled. Introduced pests decimated the remaining birds which were thought to be extinct until an isolated population was discovered.

▸ Intensive conservation measures have helped numbers recover, from 51 individuals in 1995 when the Kākāpō Recovery Plan began, to 248 as of 2023.

▸ Many conservation methods are *in situ* and include pest free reserves, vigilant monitoring using pest detector dogs, radio tracking, and supplementary feeding, especially around breeding season.

New Zealand kākāpō

Prioritizing conservation efforts

▸ Saving species and conserving ecosystems can be a huge financial task, often involving input from scientists, organizations, and volunteers. Lack of resources, including funding, means that some species are preferentially conserved over others.

▸ The EDGE of Existence programme promotes prioritized conservation efforts to save valuable species at high risk of extinction. They engage in research, support conservation efforts and the work of conservationists, and lobby policy makers to protect targeted species. They shine a spotlight on 'flagship' species such as the kākāpō, above, and aye aye, right, to lobby for their protection.

▸ Deciding which species to prioritize for conservation efforts raises many ethical, environmental, political, social, cultural, and economic questions. Different groups have vested interests in conserving different species.

Nocturnal aye aye of Madagascar

3. Kākāpō are a 'flagship species': an ambassador for conservation, much like the giant panda. Their popular status has raised millions of dollars from businesses, government, and the general public to fund their hugely expensive conservation programme. What might be some advantages and disadvantages of this type of funding model?

_____
_____
_____
_____

4. Using a debate format, identify an endangered species, preferably local, that has been prioritized for *in situ* conservation (link to EDGE site listing species in **BIOZONE Resource Hub**). Divide into two groups, and plan your debate points to cover the implications listed on this page. List your debate points below. Argue yay or nay for prioritization.

_____
_____
_____
_____
_____
_____

181 Ex Situ Conservation

Key Idea: *Ex situ* conservation methods operate away from the natural environment and are useful where species are critically endangered and cannot be protected in the wild.

***Ex situ* conservation** is the process of protecting an **endangered species** outside its natural habitat. It is used when a **species** has become critically low in numbers or *in situ* methods have been, or are likely to be, unsuccessful. Sometimes **ecosystems** are too degraded to be safe for the species. Zoos, aquaria, and botanical gardens are the most conventional facilities for *ex situ* conservation. They house and protect specimens for captive breeding programmes, with a focus on increasing genetic diversity, and often plan to reintroduce them into the wild to restore natural **populations**. The maintenance of seed banks by botanic gardens and breeding registers by zoos ensures that efforts to conserve species are not impaired by problems of inbreeding.

The important role of zoos and aquaria

- In the past, zoos were created as a form of entertainment for people who were unlikely to view the species in the wild. The animals were often exotic and kept in unsuitable small cages or enclosures.
- Modern zoos tend to concentrate on particular species. Their key goal has now become conservation and potential reintroduction of bred animals back into their natural habitats. Zoos are part of global programmes that work together to help retain genetic diversity in captive bred animals.
- In addition to their role in captive breeding programmes and as custodians of rare species, zoos have a major role in public education. They raise awareness of the threats facing species in their natural environments and engender public empathy for conservation work.
- Some species, such as the New Guinea singing dog, Przewalski's wild horse, and the European bison would most likely be extinct without the conservation efforts made by the zoos in which they have been bred.

The okapi is a species of rare forest antelope related to giraffes. Okapi are only found naturally in the Ituri Forest, in the northeastern Congo rainforests. An okapi calf was born to Bristol Zoo Gardens (UK) in 2009, one of only about 100 okapi in captivity.

Captive breeding and re-release

- Individuals are captured and bred under protected conditions. If breeding programs are successful and there is suitable habitat available, captive individuals may be re-released to the wild where they can once more establish natural populations. Zoos now have an active role in captive breeding.
- There are problems with captive breeding; individuals are inadvertently selected for fitness in a captive environment and their survival in the wild may be compromised. This is especially so for marine species. Additionally, survival behavior may not have been taught by parents. However, for some taxa, such as reptiles, birds, and small mammals, captive rearing is often very successful.

Above: The California condor is a success story of captive breeding. The condor population dwindled to just a few individuals by the time the remaining birds were taken into zoos in the 1980s. Since then, four captive breeding sites have increased numbers back to 561 birds, as of 2023. Release back into their natural habitat aims to rebuild a wild population once more.

Right: A puppet 'mother' shelters a takahe chick. Takahe, a rare rail species native to New Zealand, were brought back from the brink of extinction through a successful captive breeding program.

Right: The giant tortoise population of Española in the Galápagos islands was reduced to 15 individuals by the 1960s. A breeding programme on the larger Santa Cruz island has since restored numbers to around 2000 tortoises. One 130-year old male, Diego, (right), imported from San Diego Zoo, is thought to be the father of at least 800 of them - 40% of all new tortoise, and he has now been released back to the wild.

1. Describe the key features of *ex situ* captive breeding conservation methods:

2. Explain why some animal species are more well suited to captive breeding than others:

The role of botanic gardens

- Botanic gardens have years of collective expertise and resources and play a critical role in plant conservation. They maintain seed banks, nurture rare species, maintain a living collection of plants, and help to conserve indigenous plant knowledge. They also have an important role in both research and education.
- The Royal Botanic Gardens at Kew, London (above) contains an estimated 25,000 species, 2,700 of which are classified by the ICUN as rare, threatened, or endangered. Kew Gardens is involved in both national and international projects associated with the conservation of botanical diversity and is the primary advisor to CITES on threatened plant species.
- Kew's Millennium Seed Bank partnership is the largest *ex situ* plant conservation project in the world: networking with over 50 countries, they have banked 10% of the world's wild plant species.

Seed banks and gene banks

- Seed banks and gene banks around the world have a role in preserving the genetic diversity of species.
- A seed bank stores seeds as a source for future planting in case seed reserves elsewhere are lost. The seeds may be from rare species whose genetic diversity is at risk, or they may be the seeds of crop plants, in some cases of ancient varieties no longer used in commercial production.
- The Svalbard Global Seed Vault, Norway (above), has over 1.3 million seed samples, with capacity for millions more.

3. Describe three key roles of zoos and aquaria and explain the importance of each:

   (a) _____

   (b) _____

   (c) _____

4. Explain the importance of gene and seed banks, both to conservation and to agriculture:

5. Compare and contrast *in situ* and *ex situ* methods of conservation, including reference to the advantages and disadvantages of each approach:

182 Rewilding

Key Idea: Rewilding is a method of restoring ecosystems.
Rewilding is a crucial approach in **ecosystem** conservation, aiming to restore damaged ecosystems by reintroducing key biotic components such as apex predators, keystone **species**, and plant species, while removing pest species. Additionally, rewilding involves reconnecting fragmented ecosystems through the establishment of land bridges or restoration of the land between them. By reintroducing these components, rewilding allows for minimal human involvement, enabling natural processes to restore stability to the ecosystem. **Human activities** are the primary drivers of ecosystem damage. However, rewilding offers a promising approach that demonstrates the potential for even the most depleted areas to recover and return to their original state.

Reintroduction of blue wildebeest into the Serengeti

Blue wildebeest in the Serengeti, Tanzania, were decimated by viruses introduced by domestic cattle and were reduced to fewer than 300,000 individuals by the mid-20th century. Overgrown grasslands became a wildfire hazard, with up to 80% of them destroyed by fire each year. The small numbers of wildebeest could not support sufficient apex predators, and the food web collapsed. Rewilding of the ecosystem began with disease control in the wildebeest, restoring their numbers up to 1.5 million in less than a decade. Grasslands were once more utilized by the wildebeest, sometimes coined the 'lawnmowers' of the Serengeti for their ability to maintain the vegetation in a healthy condition, allowing other herbivore species to increase in number as well. With apex predators, such as lions, returning, as well as scavengers like hyenas, the Serengeti regained ecosystem stability.

Knepp Castle Estate rewilding project

▸ Knepp Castle Estate, a 951-hectare area in West Sussex, England, was previously intensively farmed but transitioned to a rewilding approach in 2001.

▸ As part of this shift, sheep and cattle were taken off the land and internal fencing was dismantled to enable the free movement of both native and introduced herbivores.

▸ The river and wetland habitats were rejuvenated using a cost-effective approach of eliminating artificial drainage systems. Over the last twenty years, natural processes have led to the development of unique and diverse habitats in the area.

▸ On the protected land, herbivores like pigs, ponies, deer, and cattle were introduced as substitutes for extinct species such as elk (moose) and bison. These herbivores play a crucial role in shaping and preserving habitats suitable for native species such as peregrine falcons, turtle doves, and purple emperor butterflies, which have suffered habitat loss in other areas. Additionally, the project aims to introduce pine martens and wild cats to further enhance the ecosystem.

▸ Networking to encourage surrounding land owners to contribute their efforts and conversion of some of their farms has allowed the project to expand, with the aim of developing a wild land corridor across a much bigger area.

Knepp castle estate, above, and the rare purple emperor butterfly, inset, now breeding on the estate

1. What are the key features of rewilding conservation? You may choose to select an example in your answer:

_____
_____
_____
_____

183 Conservation and Legislation

Key Idea: There are several global conservation agreements designed specifically to protect wildlife and habitats.

Legislation and agreements have been put in place at both local and global levels to address environmental and **conservation** concerns. These measures are specifically designed to ensure long-term protection and conservation of wildlife **species** and their habitats, as well as to regulate and prevent the illegal trade of animal parts. These laws and legislation grant different government departments the authority and responsibility to enforce compliance in the management of species and land, while also promoting awareness of conservation efforts. Non-compliance with the conditions outlined in the legislation can result in legal consequences and implications for trading activities. Other large organizations, such as **WWF** (World Wide Fund for Nature) and Greenpeace, can use their prominence to lobby governments to enact protection legislation for **endangered species** or vulnerable **biodiversity hotspots**.

International agreements

There are several international treaties and conservation agreements between governments designed to conserve **biodiversity**. Two such agreements are the Rio Convention on Biological Diversity and the Convention of International Trade in Endangered Species (CITES).

Rio Convention on Biological Diversity

The Convention on Biological Diversity became active in 1993. It aims to develop strategies for the conservation and sustainable use of resources while maintaining biodiversity. It has three main goals:
- Conservation of biodiversity.
- Sustainable use of its (biodiversity) components.
- Fair and equitable sharing of the benefits arising from genetic resources.

CITES

CITES aims to ensure that trade in animal and plant **species** does not threaten their survival in the wild. Trade on products are controlled or prohibited depending upon the level of threat to each species. More than 35,000 species are protected under the agreement.

CITES banned trading in ivory (right) or ivory products in 1989. However, poaching is still prevalent and large seizures of ivory by authorities still occur.

World Wide Fund for Nature

The World Wide Fund for Nature (WWF) is a non-governmental organization focussed on the conservation of biodiversity and reduction of humanity's ecological footprint. It is the world's largest conservation organization and operates in more than 100 countries.

One of the approaches the WWF and other NGOs use in conservation efforts is to convince governments to adopt and enforce environmentally friendly policies. One way to do this is to publicly release details when a government proposes a course of action with negative effects on the environment. For example, in the past few years WWF has introduced a number of bills to combat Amazon **deforestation**, protect coral reefs, and push for The North American Grasslands Conservation Act.

Domestic agreements

Most countries have country-specific legislation that focuses on the species and habitats within their borders. The US signed the Endangered Species Act (ESA) in 1973, and in partnership with CITES, the legislation has had many success stories, claiming that the Act prevented 99% of the targeted species from becoming **extinct**. Species saved include the bald eagle (right), gray wolves, manatees, and the American alligator.

1. Explain the importance of international conservation legislation in species and habitat protection:

   _____
   _____
   _____

2. Why is conservation legislation at country level also very important: _____
   _____
   _____

3. Describe how a organization such as WWF can influence conservation legislation: _____
   _____
   _____
   _____

184 Conservation and Sustainability

Key Idea: Conservation aims to safeguard and revive ecosystems and the species that inhabit them. Sustainability involves the responsible management of the resources found within these ecosystems.

Conservation involves the care and protection of both living systems and resources in the environment. The conservation of living systems focuses heavily on the management of **endangered species** so that their numbers remain stable or increase over time. Many living systems have no directly measurable economic value but are important for global **biodiversity**. Many people also support the moral value of conserving as many living systems as possible and that humans do not have the right to exterminate other organisms.

Conservation of resources focuses on the efficient use of resources so that remaining stocks are not wasted. Many of these resources are scarce or economically important so require prudent use. Others are damaging to the environment, and it is better to use less of them. In recent decades, there has been a growing acknowledgement that humans cannot afford to continue to waste natural resources. **Sustainability** is a subset of conservation, as it encompasses the goal of maintaining resources in a way that allows for their long-term viability. Sustainability encompasses environmental, economic, and societal aspects and, when combined with conservation, forms the foundation of the discipline of environmental science.

Conservation

- Living systems
  - Plant conservation
  - Animal conservation
  - Habitat conservation
- Resources
  - Energy conservation
  - Soil conservation
  - Water conservation

Sustainability

Socio-economic
- Ethical business practices
- Workers rights
- Fair trade

Economy
- Profit and cost saving
- Growth and development

Society
- Living standards
- Education and opportunity
- Equal rights

Environmental-economic
- Energy efficiency
- Incentives to use renewable resources

Environment
- Renewable resources
- Pollution prevention

Socio-environmental
- Environmental justice
- Care with resources
- Local and global considerations

▸ Sustainability can be represented conceptually as the intersection of the environment, society, and economics. Sustainable development must take into account all three of these concepts. Examples of cities that have used the concept of sustainability include Vancouver, San Francisco, Oslo, Curitiba, and Copenhagen.

▸ Curitiba, in Brazil, is a particularly good example of a city putting in sustainable plans that also enhance public well being. In the 1970s the Curitaba authorities redesigned the city to include new parks (producing 52 m² of green space per person), pedestrian only urban and business areas, strictly controlled urban planning, and a bus rapid transit system (buses that act like trains).

1. Describe the relationship between conservation and sustainability: _____

2. Discuss the impact of society and economy on the conservation of living systems and resources: _____

185 Did You Get It?

1. The California condor has once more begun to breed successfully in the wild. The species almost became extinct due to poaching, DDT and lead poisoning, and habitat destruction.

   (a) Both *in situ* and *ex situ* conservation methods were used at different stages for the condors. Explain the differences between those methods, using information from the graph (right) to illustrate:

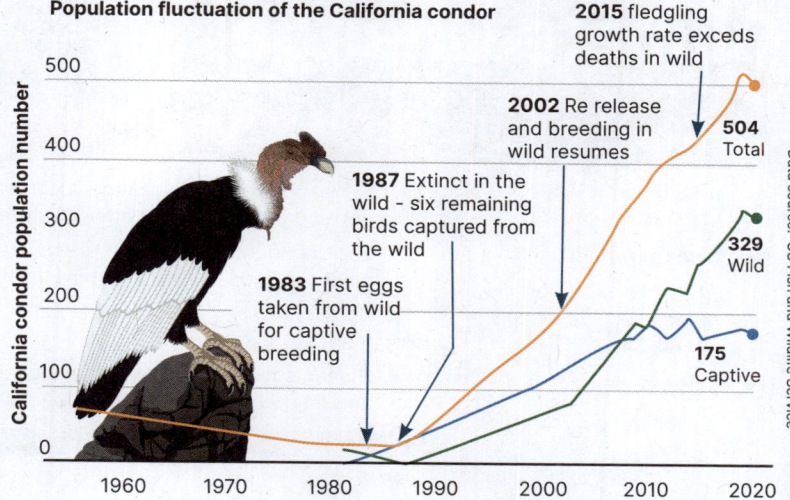

Population fluctuation of the California condor

1983 First eggs taken from wild for captive breeding

1987 Extinct in the wild - six remaining birds captured from the wild

2002 Re release and breeding in wild resumes

2015 fledgling growth rate exceds deaths in wild

504 Total
329 Wild
175 Captive

Data source: US Fish and Wildlife Service

   _____
   _____
   _____
   _____
   _____
   _____
   _____

   (b) What suggests that detrimental human activities were eventually reduced or eliminated in the condor habitat?

   _____
   _____
   _____
   _____

2. The California condor is classified as Critically Endangered, according to IUCN. What purpose do these labels provide?

   _____
   _____

3. The California condor was on the brink of being included among the species lost in the 'sixth mass extinction.' What specific events would have been responsible for this, and how would they differ from previous mass extinctions?

   _____
   _____
   _____

4. (a) Biodiversity loss is occurring in all groups of species, but decline in insect orders is especially significant (see graph below). Why does the loss of insects potentially impact biodiversity of ecosystems generally?

   _____
   _____

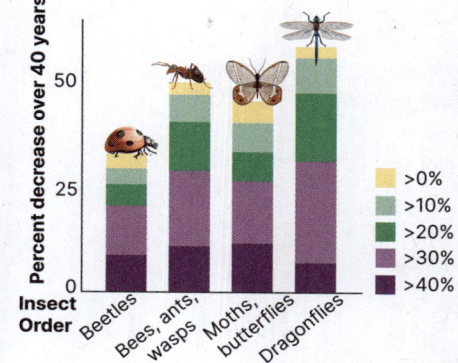

Data source: D.Wagner, 2020

   (b) Discuss what types of human activity are likely to be contributing to insect species endangerment and biodiversity loss, using insect examples (left):

   _____
   _____
   _____
   _____
   _____
   _____

©2025 BIOZONE International
ISBN: 978-1-99-101409-2
Photocopying prohibited

Chapter 9 Global Change
Climate Change

Resource Hub
bit.ly/4ct9m0c

Key Terms

- adaptation (climate change)
- albedo (-effect)
- AMOC (Atlantic meridional overturning circulation)
- anthropogenic
- bleaching (coral reef)
- boreal forest
- boreal permafrost
- climate change
- carbon capture and storage (CCS)
- carbon dioxide (CO_2)
- carbon footprint
- carbon sequestration
- carbon sink
- carbon trading
- climate justice
- climate legislation
- climate risk
- Conference of Parties (COP)
- emissions
- extreme weather event
- global warming
- greenhouse effect
- greenhouse gas (GHG)
- ice sheet
- IPCC (Intergovernmental Panel on Climate Change)
- net zero carbon
- ocean acidification
- Paris Agreement
- permafrost
- polar
- positive feedback cycle
- sea level rise
- solar radiation modification (SRM)
- subpolar gyre (SPG)
- tipping point

Key Concepts

- Evidence for climate change, including climate models, is used by governmental bodies to develop climate legislation and future greenhouse gas emissions goals.
- The enhanced greenhouse effect has led to many different effects on Earth's physical and biological systems, resulting in climate change.
- Global warming is pushing the Earth's systems towards tipping points, driven by positive feedback cycles.
- Human society requires both mitigation and adaptation strategies to respond to current and future disruptions caused by climate change.

The science of climate change

Learning Outcomes:	Activity Number
1. Explain that changes in Earth's climate are normal but changes over the last 200 years can be attributed to anthropogenic activity.	186
2. Describe the techniques used to collect evidence of past and current climate in order for changes to be observed.	187
3. Explain the role of the IPCC and outline the aim of the Paris Agreement in the context of Climate Change legislation.	188
4. Describe the major predicted effects of climate change on the Earth by the following: the enhanced greenhouse effect; the pH and temperature of the oceans, sea level rise, extreme weather events, frequency and severity of wildfires and megadroughts, and changes in albedo.	189-196
5. Describe the effects of climate change on wildlife with reference to biodiversity and location changes to habitats. Define the term, range shift.	198-199

Tipping points

6. Explain how a positive feedback cycle works and give an example.	200
7. Define the term, 'tipping point,' in the context of climate change and name the six most imminent tipping points.	201
8. Explain the outcomes if each of these examples reached their tipping point and state whether each would have a local effect, global effect, or both: Greenland and West Antarctic ice sheets, Boreal permafrost and forests, the Amazon rainforest, coral reefs, AMOC and the subpolar gyres.	202-208

Climate change and society

9. Name three factors that determine climate risk.	209
10. Describe some effects that climate change will have on agricultural activity.	210
11. Explain the meaning of mitigation in the context of climate change and some adaptations that humans will have to make in response to changes.	211
12. Describe the Climate Action Movement and its aims.	212
13. Describe the purpose and methodologies behind carbon trading and both natural and technology-assisted carbon capture. Define the terms carbon footprint and net zero carbon.	213-217
14. Describe three methods by which humans could modify the amount of solar radiation received by Earth and their purpose.	218

186 What's the Concern for Climate Change?

Key Idea: Climate change is happening now, and our responses will determine future impacts.

Over the past two centuries, human activities and industrialization have led to a rise in **greenhouse gas** levels in the atmosphere, affecting the climate. In more recent times, the term 'Anthropocene' has been used to describe the current geological epoch. This term, although not officially recognized as a geologic designation, highlights the influence of human-induced changes on the climate, known as **anthropogenic** forcing. Notably, **global warming** has been a prominent consequence of these activities, with the Earth's average surface temperature increasing by at least 1.19°C in 2024 compared to the mid-19th century.

To determine the precise average global temperature, data is collected from (100,000 plus) weather stations worldwide, along with weather balloons, ships, buoys, radars, and satellites that record daily temperature variations. This data is used to calculate an **average global temperature**. The average temperature is then compared with pre-industrial temperature data, obtained before substantial industrial source greenhouse gas **emissions**. The recent rise in the global average means specific regions are encountering notably higher and more harmful temperature extremes. Unprecedented heatwaves in **polar** areas and regions already struggling from human habitation are driving certain Earth systems towards irreversible tipping points.

The world is warming - What's the big deal about 1.5°C anyway?

Many students may have heard about a **'1.5°C'** global warming 'line in the sand' not to be overtaken in order to prevent the worst impacts of climate change. Yet, despite the seemingly small '1.5°C' target, exceeding this threshold can activate **tipping points** in the climate system, causing irreversible changes due to **positive feedback cycles** that intensify the initial warming effects. The significance of global temperature rise lies in its potential to disrupt ecosystems, weather patterns, and cause **sea level rise**, highlighting the urgent need to address anthropogenic climate forcing to mitigate these impacts.

Data source

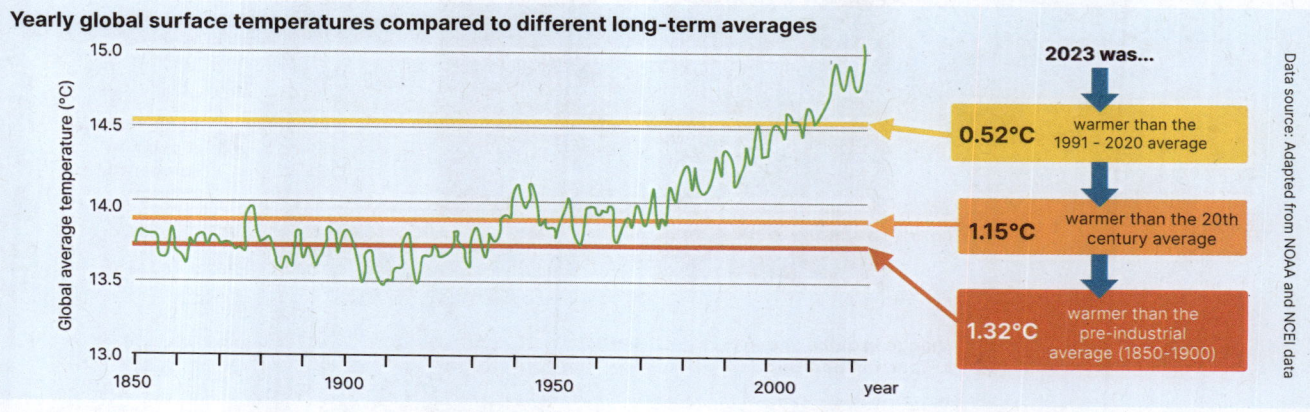

1. The image below shows a visual representative of the last 200 years of global temperature change, using the average from 1961-2010 as a measuring stick. Blue is colder, the darkest over 0.7°C colder, and the red is warmer; again the darker colors showing an extreme of over 0.7°C warmer. In a small group, discuss the implications of the data presented as a means to raise awareness of climate change. Record your thoughts below:

_____
_____
_____
_____

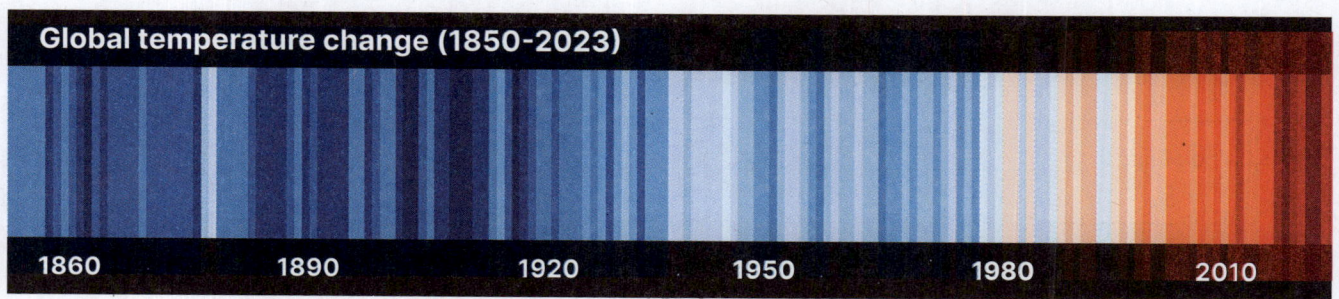

2. Two sets of data that change in proportion to each other, either negatively or positively, show correlation. However, this does not necessarily imply causation, where changes in one factor causes changes in another. Suggest how scientists might show causation between global temperature rise and anthropogenic-only climate forcing:

_____
_____
_____

What are the signs of climate change in the data?

The data collected on climate indicators is constantly evolving and can quickly become outdated. Scientific publications can only report on the data available at the time of publication, which may not reflect the most current information. However, these changes in data can still provide valuable insights into trends and patterns. The infographic provided below illustrates the changes in key climate indicators from 2021, when the **IPCC** Sixth Assessment Report was published (see activity 187), along with data from 2023. This two-year period has shown significant shifts in climate change data.

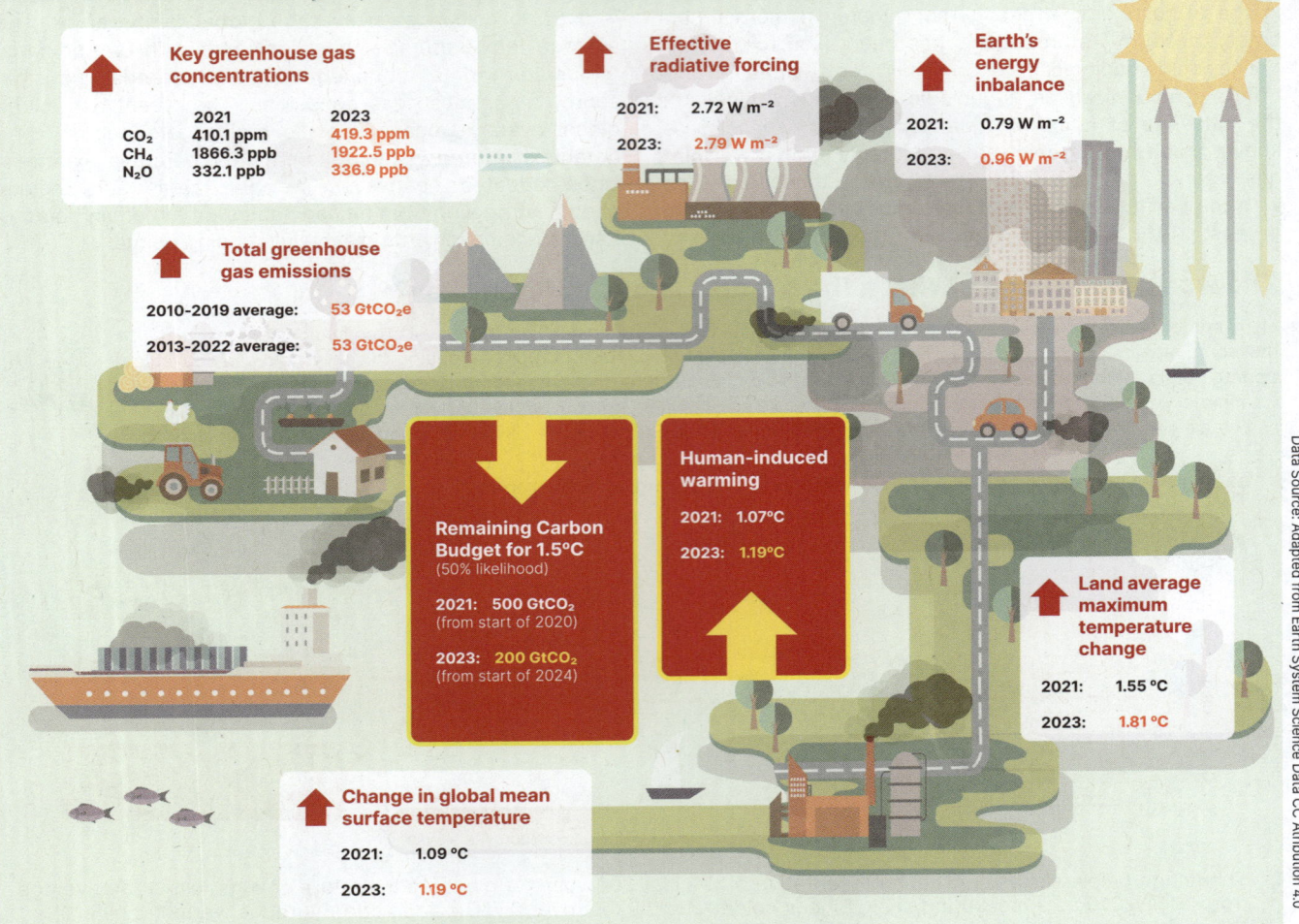

Key greenhouse gas concentrations

	2021	2023
CO_2	410.1 ppm	419.3 ppm
CH_4	1866.3 ppb	1922.5 ppb
N_2O	332.1 ppb	336.9 ppb

Effective radiative forcing
- 2021: 2.72 W m^{-2}
- 2023: 2.79 W m^{-2}

Earth's energy inbalance
- 2021: 0.79 W m^{-2}
- 2023: 0.96 W m^{-2}

Total greenhouse gas emissions
- 2010-2019 average: 53 GtCO$_2$e
- 2013-2022 average: 53 GtCO$_2$e

Remaining Carbon Budget for 1.5°C (50% likelihood)
- 2021: 500 GtCO$_2$ (from start of 2020)
- 2023: 200 GtCO$_2$ (from start of 2024)

Human-induced warming
- 2021: 1.07°C
- 2023: 1.19°C

Land average maximum temperature change
- 2021: 1.55 °C
- 2023: 1.81 °C

Change in global mean surface temperature
- 2021: 1.09 °C
- 2023: 1.19 °C

Data Source: Adapted from Earth System Science Data CC Attribution 4.0

3. Understanding the vocabulary associated with a branch of science such as climate change is crucial before any deeper dive into the topic. Use this book and online resources to define the following terms:

   (a) Greenhouse gases: _____

   (b) Greenhouse gas emissions: _____

   (c) Global mean surface temperature: _____

   (d) Carbon budget: _____

   (e) Human-induced (anthropogenic) warming: _____

   (f) Radiative forcing: _____

   (g) Earth's energy budget: _____

187 Finding the Evidence for Climate Change

Key Idea: Data on current and past indicators for climate change is collected using a wide variety of methods. Projections of future atmospheric CO_2 levels or temperature rises are extrapolated from current or past data. Accurate and consistent record keeping of temperatures on Earth only dates from the 1880's, while atmospheric CO_2 records are even more recent. The key scientific observation center, Mauna Loa, Hawaii, began collecting CO_2 data in 1958. Scientists use different methods to collect historic climate data including gas analysis from ice cores, fossilized pollen, sediments from glaciers and oceans, tree rings, and evidence in cave deposits. In recent times, atmospheric climate data has been collected by monitors positioned on space satellites, the **International Space Station (ISS)**, land observatories, ships, and ocean buoys. Near surface temperatures are collected from weather stations around the world by both individuals and large organizations such as the National Aeronautics and Space Administration (NASA).

Ice cores tell a story of the past

An ice core sample shows trapped bubbles of air captured at the time the snow was laid down.

▸ Ice cores are drilled samples of ice, laid down at the poles over a timespan of 800,000 years. They can be up to 3 km in length. The oldest samples are from Antarctic land ice.

▸ Annual snowfall captures small pockets of air from the atmosphere at the time. Each consecutive layer is compacted by further snowfall, compressing it into layers of ice. The deeper the layer of ice, the further back in time it was laid down.

▸ Samples are frozen on site and sent back to laboratories for analysis. Very thin slices are cut and the position determined to identify the date it was laid down. Temperature data can be determined by defrosting the slice and analysing the ratio of different oxygen isotopes. CO_2 and methane atmospheric concentrations can be ascertained from the trapped air bubbles.

▸ Examination of overlap in CO_2 data collected from different sites gives confidence in the accuracy of climate data.

Mapping greenhouse gases using the ISS

▸ The Orbiting Carbon Observatory-3 (OCO-3) was launched in 2019 and installed in the International Space Station (ISS). It will map CO_2 for approximately 10 years.

▸ The timing of the ISS orbit allows OCO-3 to see the same part of the Earth at different times of the day. An older satellite, the OCO-2, can only see a part of the Earth at the same time every day. The two sets of measurements should provide high precision data for changes in atmospheric CO_2.

▸ Satellites can also capture global surface temperatures, but indirectly. Brightness measurements can be translated into temperature data using computer climate models.

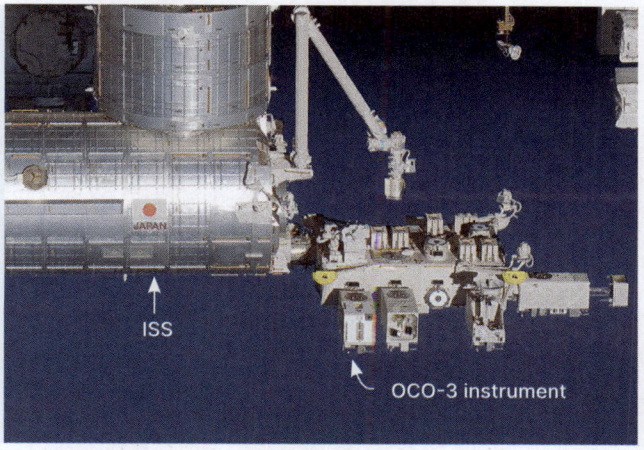

Climate scientists

Polar scientists setting up a temporary base

Investigating and collecting data to measure any **climate change** indicators and impacts can require scientists to spend periods of time in polar regions such as Greenland or Antarctica. Many different countries have their own polar research stations, such as the McMurdo station run by the USA Antarctic program. However, most climate scientists spend much of their time analysing data from models and writing reports.

1. How can climate data relating to periods of time before modern science be accepted as reliable?

_____
_____
_____
_____
_____
_____

2. What part of a climate scientist's job interests you the most?

_____
_____
_____

188 Climate Change Legislation

Key Idea: Climate change legislation is a set of internationally agreed policies and laws that govern activities relating to the impacts of climate change.

The 1992 Rio Earth Summit was the first major international conference recognizing that global cooperation was crucial to combat the potential dangers presented by **climate change**. Conference of the Parties (**COP**) gatherings began soon after. Five years after the Rio summit, the Kyoto protocol was released: a legally binding agreement between 'industrialized' governments on steps required to reduce **greenhouse gas emissions (GHG)**. The **Paris Agreement** was even more expansive. It involved every country, and provided a binding treaty on mitigation, **adaptation**, and also financing, particularly to support countries that were least resilient to current and projected impacts. The Paris Agreement is a landmark of climate legislation, with 195 of 198 Parties ratifying (formally consenting to) the treaty. The parties meet at each COP to present findings from Paris Agreement Global Stocktakes. They then release new or updated Nationally Determined Contributions (NDCs) to ensure the global temperature goal of <2°C (~ 1.5°C) is met.

IPCC - Intergovernmental Panel on Climate Change

▸ The **IPCC**, founded in 1988, is a body formed under the United Nations (UN) that brings together experts from every field and from many countries.

▸ IPCC publishes regular **Assessment Reports** which are the result of intensive research. AR6 (2021) is the most recent.

▸ Working Groups (WG) are also convened to focus on specifics: WGI, the physical science basis; WGII, impacts, adaptation, and vulnerabilities; and WGIII, mitigation of climate change.

▸ IPCC also publishes Special Reports (SP), such as the 2018 SR on **global warming** to support and inform the Paris Agreement.

▸ The reports inform policy makers, governments, industry, media, educators, and the general public.

COP - Conference of the Parties

▸ The COP are annual international climate change conferences, hosting delegates representing almost every country in the world.

▸ The conferences are convened under the United Nations Framework Convention on Climate Change (UNFCCC) and are where the main global climate change decisions, goals, and agreements are discussed, voted on, and assessed.

The **IPCC Assessment Reports** use a series of climate models based upon different possible scenarios. In **AR5** the scenarios are called **RCP** (Representative Concentration Pathways) and predict impacts due to different future greenhouse gas emissions.

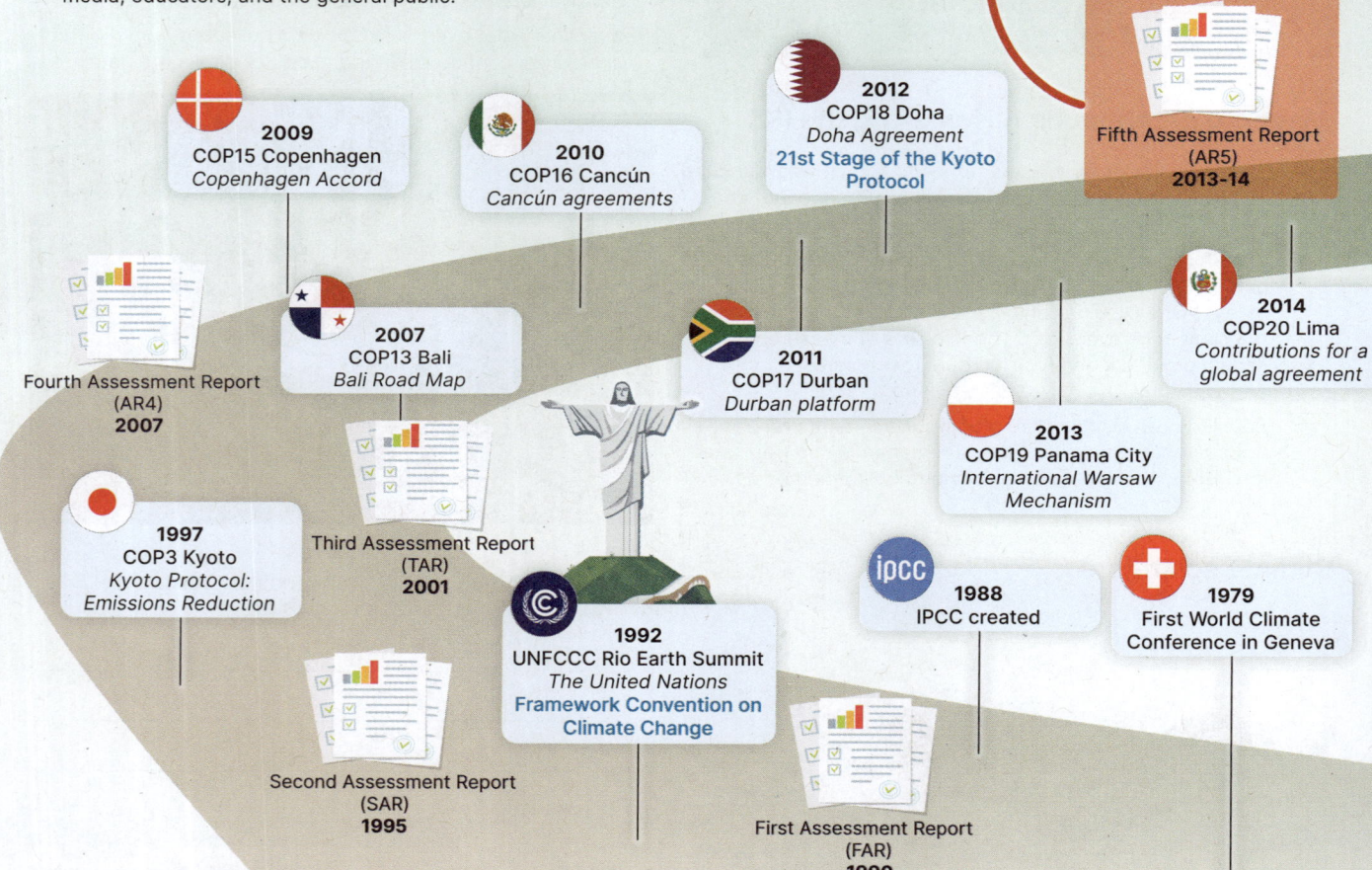

1. What role does the IPCC play in climate change legislation and how is that different from the COP?

___

___

___

___

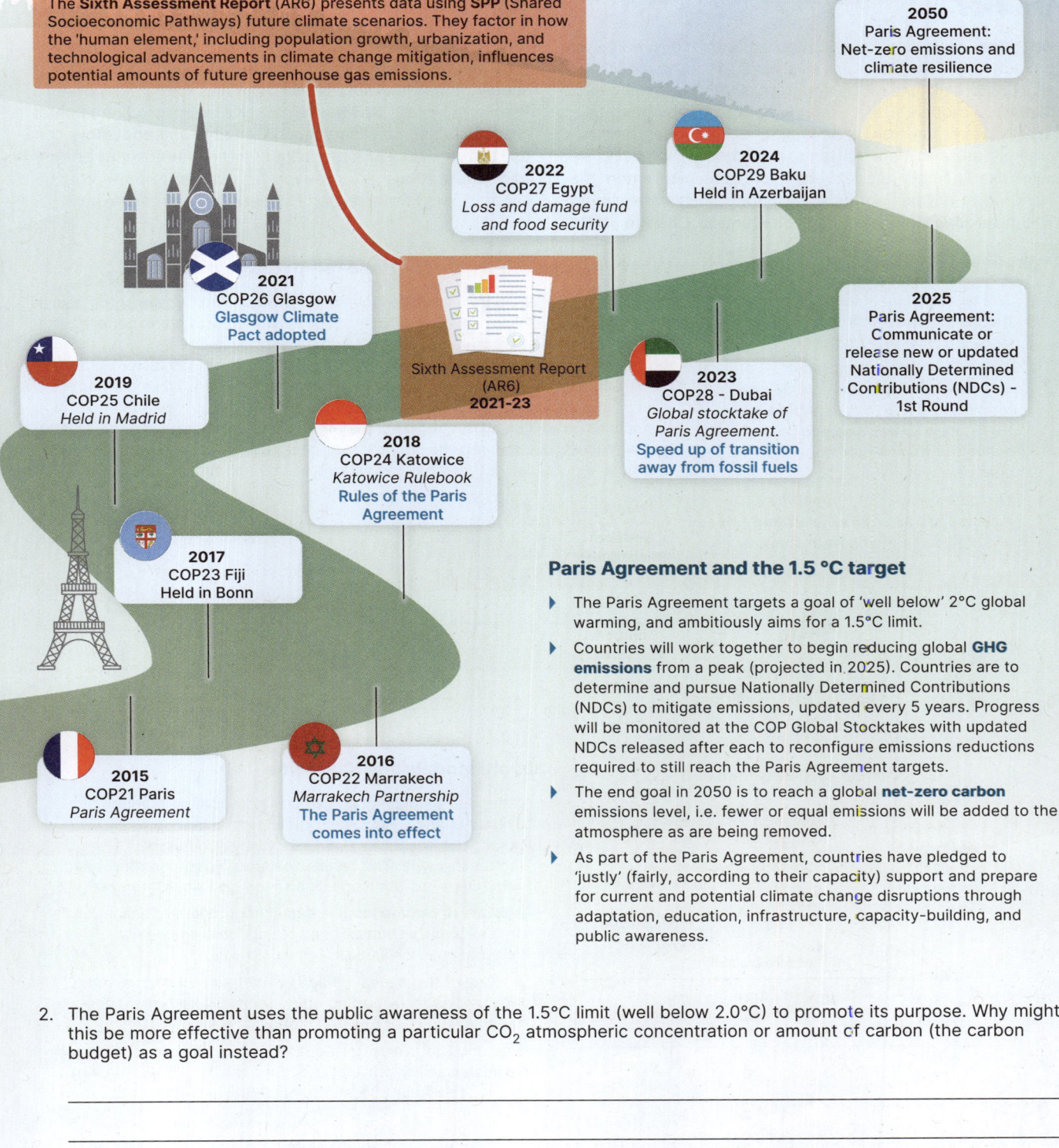

Paris Agreement and the 1.5 °C target

▸ The Paris Agreement targets a goal of 'well below' 2°C global warming, and ambitiously aims for a 1.5°C limit.

▸ Countries will work together to begin reducing global **GHG emissions** from a peak (projected in 2025). Countries are to determine and pursue Nationally Determined Contributions (NDCs) to mitigate emissions, updated every 5 years. Progress will be monitored at the COP Global Stocktakes with updated NDCs released after each to reconfigure emissions reductions required to still reach the Paris Agreement targets.

▸ The end goal in 2050 is to reach a global **net-zero carbon** emissions level, i.e. fewer or equal emissions will be added to the atmosphere as are being removed.

▸ As part of the Paris Agreement, countries have pledged to 'justly' (fairly, according to their capacity) support and prepare for current and potential climate change disruptions through adaptation, education, infrastructure, capacity-building, and public awareness.

2. The Paris Agreement uses the public awareness of the 1.5°C limit (well below 2.0°C) to promote its purpose. Why might this be more effective than promoting a particular CO_2 atmospheric concentration or amount of carbon (the carbon budget) as a goal instead?

___
___
___

3. Using a debate format, argue yay or nay for why or why not developed countries that have traditionally been large global contributors to GHG emissions should provide financial assistance and aid to countries that are most affected by climate change, such as small low-lying Pacific islands. Divide into two groups, plan your debate points, and summarize in the space below:

___
___
___
___
___

189 Models of Climate Change

Key Idea: Scientists use climate models to predict and project long term changes in climate patterns, which can also prepare us for a range of potential future climate events. Climate models are mathematical representations, usually very complex, that enable us to understand and predict future climate patterns and change, including sea level rise. Climate models are more accurate when they incorporate all the factors contributing to climate change. The **Intergovernmental Panel on Climate Change (IPCC)** regularly publishes climate change models based on different possible future atmospheric **carbon dioxide** concentration scenarios and the associated projected global temperature rise over time. Governments, NGOs (non-governmental organizations), scientists, action groups, and educators can use the models to inform mitigation (preventative measures) and **adaptations** to prepare for future climate change impacts, such as increased frequency of extreme events.

What is causing climate change?

- The average global temperature of the Earth has risen at least 1.19°C since the mid-19th century.
- An overwhelming proportion of scientists attribute this continuing temperature rise to **greenhouse gas** (GHG) from **anthropogenic** (human) activity, with high confidence (8/10 chance of being correct).
- The phenomenon of temperature increase is called **global warming**, and it is leading to **climate change**.

What does average global surface temperature rise mean?

The 1.19°C rise in temperature is determined by averaging across the entire planet. However, due to climate change, different regions of the Earth are experiencing varying rates of temperature increase. Over the past two decades, the Arctic region has warmed significantly more, at least two to three times faster than the global average, a phenomenon known as **Arctic amplification**. While the current level of global warming might appear minor, the substantial energy absorbed by the oceans, in particular, has 'masked' the otherwise rapid climate change. Due to global warming, certain regions are now facing temporary temperature anomalies that pose a potentially serious risk to human life and habitation.

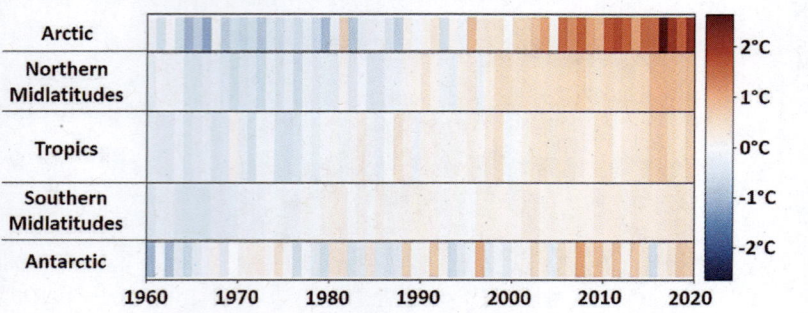

Climate models and scenarios

- Computer modeling can be a valuable tool for predicting climate change. However, predicting the outcomes can be difficult because of the complexity and number of factors involved. The result is often difficult to interpret and sometimes climate models can produce conflicting data.
- A series of climate models based upon different scenarios are called projections. These are updated when new information is available.
- A range of five scenarios (left) were frequently used in the IPCC AR6 when reporting on potential future climate change impacts, such as projected global temperature rise.
- Scenario SSP1 (blue, left) describes a global economy based on sustainability, while at the other extreme, SSP5 (brown) describes a future based on fossil-fuel developments.

Climate scenarios from AR6

Paris Agreement target of well below 2°C global warming, ideally 1.5°C.

- Very high GHG emissions: CO₂ emissions triple by 2075
- High GHG emissions: CO₂ emissions double by 2100
- Intermediate GHG emissions
- Low GHG emissions: net zero around 2075
- Very low GHG emissions: CO₂ emissions cut to net zero around 2050

Data source: IPCC assessment reports AR5 / AR6

1. (a) Why are different scenarios (future projected GHG emissions scenarios) useful to scientists, governments, and society when presenting future projections of global warming or related climate change impacts?

   _____

   _____

   _____

   (b) If global temperatures were to remain around or under the targeted 1.5°C rise, what type of changes would be required by the world's population (see assumptions), and how does the data show our current projection?

   _____

   _____

   _____

How are climate models tested?

- To see how well models work, scientists enter existing data from years gone by and see how accurately their models predict the climate changes that have already occurred. If the models recreate historical trends accurately, we can have confidence that they will also accurately predict future trends in climate change.

- The graph on the right shows an example of how climate models are tested. The data are collected from 41 historical models and multiple models for different RCP emission scenarios. The black line represents the average actual (observed) data for the same period. GMST = global mean surface temperature.

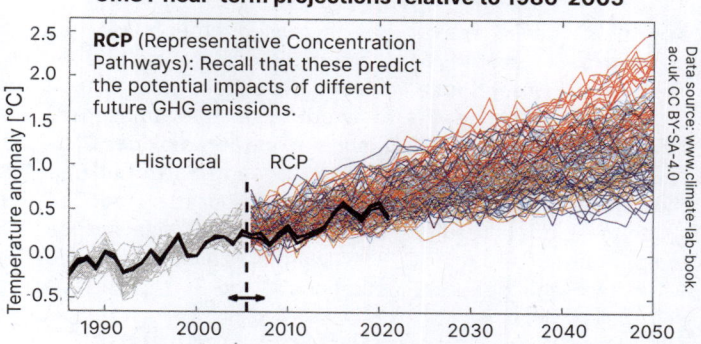

RCP (Representative Concentration Pathways): Recall that these predict the potential impacts of different future GHG emissions.

- Scientists also use models to predict the potential future impacts of climate change, such as sea level rise. (For an example, see QR code, right, for the NASA / IPCC sealevel rise tool).

- The graph (below right) shows modeled projections, based on IPCC report AR5, of future global sea level rise. The projections are based on potential future greenhouse gas emissions.

- Recorded data from IMBIE (Ice Sheet Mass Balance Inter-comparison Exercise) is gathered from 26 combined satellite surveys of the Greenland and Antarctica ice-sheets. Melt from these ice sheets is the single largest contributor to sea level rise.

2. (a) Data from the World Meteorological Organization (WMO) show global sea level rise has accelerated to +4.5mm a year in the Jan 2013 -Jan 2022 period (compared to +2.9mm/year from 2003-2012). How would this be reflected in the recorded data line? Explain your reasoning. (Hint: What AR5 projection would it trend towards, and why?)

_____

_____

_____

_____

(b) Suggest why the distance between the three projections of future sea level rise increases over time (graph right)?

_____

_____

_____

_____

(c) How could this model be used by governments when making climate change policies?

_____

_____

_____

3. (a) Would you expect the IPCC AR6 predictions of sea level rise to be higher than the AR5 predictions? Explain:

_____

_____

(b) What factors might cause changes in updated models of sea level rise?

_____

_____

190 The Enhanced Greenhouse Effect

Key Idea: Human activity has increased greenhouse gas concentrations, leading to an enhanced greenhouse effect. The term '**greenhouse effect**' describes the natural process by which heat is retained within the atmosphere by **greenhouse gases** (GHG) such as **carbon dioxide** (CO_2) and methane. These allow sunlight to warm Earth but trap the heat that would normally radiate back out to space. The greenhouse effect results in the Earth having a mean surface temperature of about 15°C, 33°C warmer than it would have without an atmosphere. Fluctuations in the Earth's surface temperature as a result of climate shifts are normal, and the current period of warming climate is partly explained by the recovery after the glacial period that finished 10,000 years ago. However, since the mid 19th century, the Earth's surface temperature has been increasing due to an enhanced greenhouse effect. **Anthropogenic** activity (changes made by humans) has rapidly released GHGs into the atmosphere, leading to a NET increase in most notably CO_2. Therefore, more radiated energy from the sun is retained in the Earth system than is lost from it, causing **global warming**.

Enhanced greenhouse effect mechanism

The incoming radiation is almost all in the form of short-wave light energy at the rate of **341 W/m²** (watts per meter square). Total outgoing radiation is mostly long-wave heat (IR) energy, at the rate of **239 W/m²**. The energy imbalance difference of **2 Wm²** retained on Earth is causing the enhanced greenhouse effect and global warming.

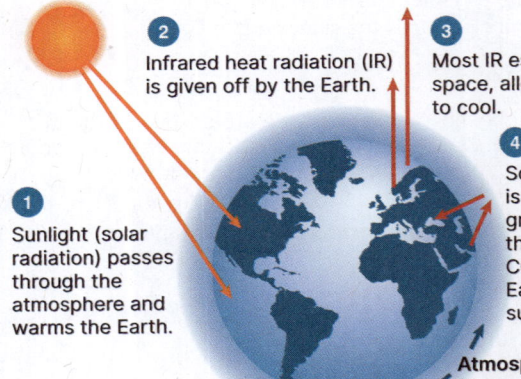

1. Sunlight (solar radiation) passes through the atmosphere and warms the Earth.
2. Infrared heat radiation (IR) is given off by the Earth.
3. Most IR escapes to outer space, allowing the Earth to cool.
4. Some IR (heat energy) is absorbed by greenhouse gases in the air (including CO_2), keeping the Earth warm enough to sustain life.

Atmosphere (Troposphere up to ~10km)

5. **Enhanced greenhouse effect**: Increasing levels of CO_2 increase the amount of heat retained, causing the atmosphere and Earth's surface to heat up (global warming).

Stratosphere (second layer of the atmosphere)
Troposphere (lowest layer of the atmosphere)
Buildup of CO_2 and other greenhouse gases
Earth's surface

Solar energy (mostly light) is absorbed as heat by Earth, where it is radiated back into the atmosphere.

Heat energy (IR) is absorbed by CO_2 in the troposphere and radiated back to Earth.

Greenhouse gas	Tropospheric conc.		GWP¶	Atmospheric lifetime (years)§
	Pre-industrial 1750	Present day (2023-2024)		
Carbon dioxide	278 ppm	425 ppm	1	120
Methane	729 ppb	1934 ppb	27-30	12
Nitrous oxide	270 ppb	337 ppb	273	109
Tropospheric ozone	25 ppb	34 ppb	17	Hours

Data source: CO2.earth, NOAA, IPCC, WMO

ppm = parts per million; ppb = parts per billion; ¶ GWP: Global warming potential: Figures contrast the radiative effect of different GHGs relative to CO_2 over 100 years, e.g. over 100 years, methane is 25 times more potent as a GHG than CO_2 § How long the gas persists in the atmosphere.

Greenhouse gases sources

Carbon dioxide
- Exhaust fumes from cars
- Combustion of fossil fuels and wood
- Burning rainforests

Methane
- Plant debris and growing vegetation
- Belching and flatus of cows

Nitrous oxide
- Car exhausts

Tropospheric ozone*
- Triggered by car exhausts (smog)

*Tropospheric ozone is found in the lower atmosphere (not to be confused with ozone in the stratosphere).

Water and the greenhouse effect

Water vapor plays an important part in keeping Earth's temperature stable. However, it does not directly influence the temperature, rather it is influenced by the temperature. An increase in temperature causes more water to evaporate and this can enhance the warming effect of other GHGs. Water constantly cycles from the atmosphere and back and so its effect is short-lived, unlike other GHGs which can remain in the atmosphere for many years.

1. Calculate the increase (as a %) in the 'greenhouse gases' between the pre-industrial era and the 2023-24 measurements (use the data from the table above). HINT: The calculation for carbon dioxide is: (425 - 278) ÷ 278 x 100 =

   (a) Carbon dioxide: _____   (b) Methane: _____   (c) Nitrous oxide: _____

Changes in global atmospheric CO_2

A network of observatories around the world are constantly measuring the **global concentration of CO_2** in the atmosphere. Below, right, are the measurements from Mauna Loa, Hawaii. These match readings from many different observatories. Note that the concentration rises and falls on an annual basis due to natural seasonal changes.

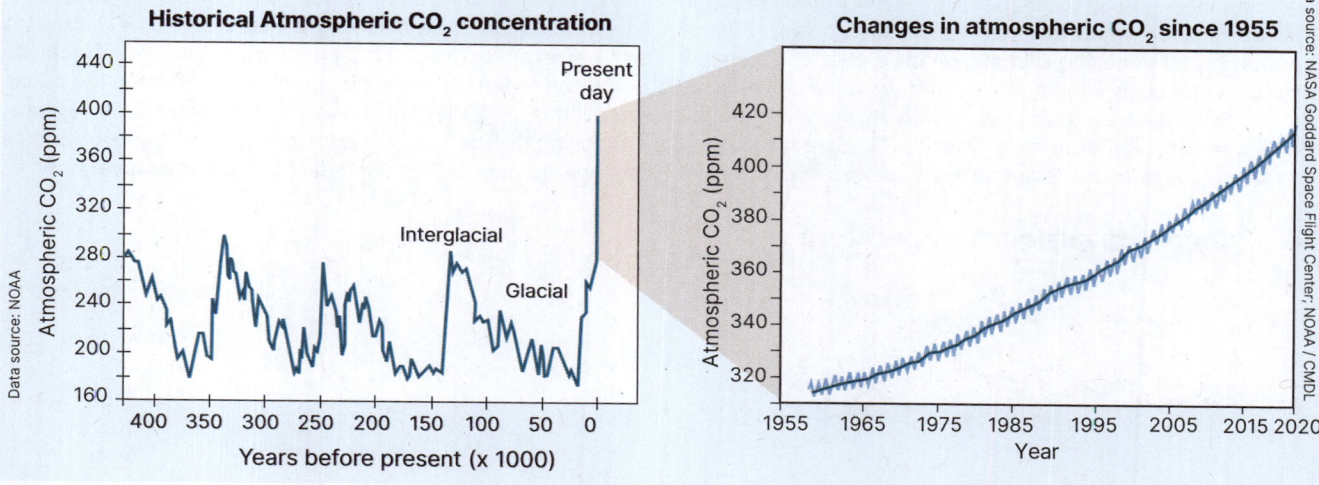

Changes in global, near-surface temperature

The enhanced greenhouse effect causes the Earth's surface temperature to rise. Measurements from around the globe show a steady increase in near-surface temperatures over the last half century.

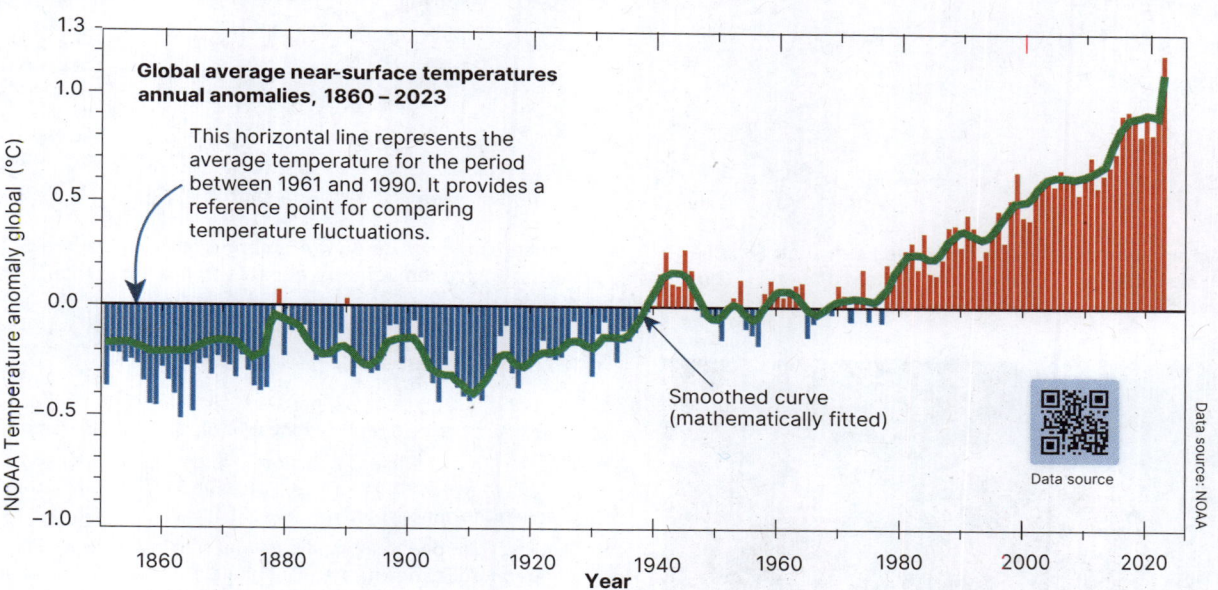

2. What is the difference between the greenhouse effect and the enhanced greenhouse effect?

___

3. Explain the relationship between the rise in concentrations of atmospheric CO_2, temperature, and the enhanced greenhouse effect:

___

191 Warming Oceans

Key Idea: Much of the extra energy retained by the Earth's system is absorbed by the oceans, leading to warming.

The enhanced greenhouse effect is causing an increase in retained heat energy in the Earth's system. In fact, 90% of the energy gain causing **climate change** over the last few decades has been absorbed by the oceans, specifically the top few surface meters. Measurements show that the average ocean temperature is rising. In 2023-24, recorded data showed significant temperature increases over previous years. This rise in temperature is concerning for two primary reasons. The first is that rising temperatures will affect marine communities adapted to live at certain temperatures. The second is that above 4°C, water volume increases as temperature rises. This could have serious effects on sea levels and coastal communities. Additionally, ocean warming is magnified by up to twice the rate in polar regions, increasing the amount of the Greenland and West-Antarctica **ice-sheets** melting, significantly adding to **sea level rise**.

Daily global sea surface temperature

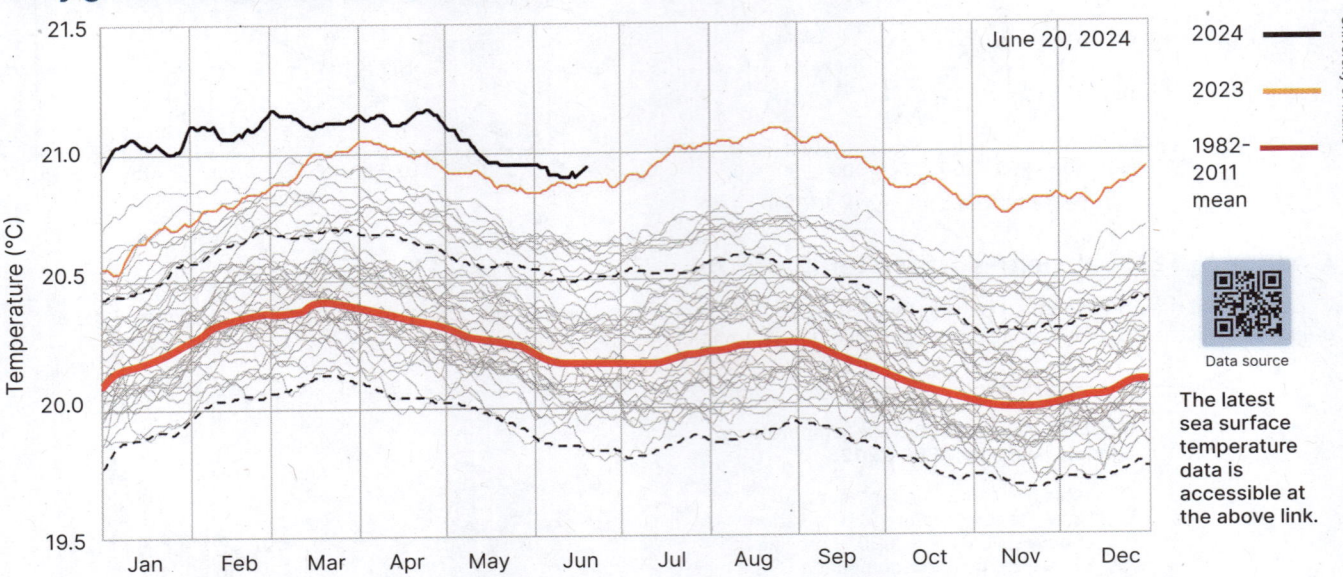

▸ Average ocean temperatures have risen sharply since 1970. Water absorbs a **large amount of energy** for every degree Celsius it rises (4.2 joules per milliliter or gram of water). Thus, even a small rise in sea temperature equates to the absorption of an enormous amount of energy when considering the entire ocean volume.

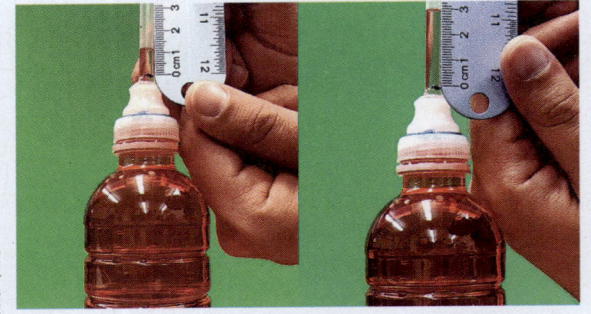

Investigation showing thermal expansion of (dyed) water when heated on right, compared to cooler water on the left.

Thermal expansion

▸ Heat energy in the atmosphere is transferred to the water molecules: the warmer the atmosphere, the more is transferred.

▸ Warm water molecules have more energy than cold molecules, so they move around more, and push apart from each other. This is called **thermal expansion**.

▸ Sea level rise due to expanding water from increasing temperature is known as steric height. The current rate of steric height change is an increase of 1.3mm/year, of which one third is thought to result from climate change.

▸ Data on both the ocean temperature and thermosteric change causing global mean sea level rise is gathered by free-floating meters called Argo floats.

1. Why has the ocean's water masked the global warming impacts of climate change?

2. How does thermal expansion contribute to sea level rise due to global warming?

3. Daily sea surface temperature fluctuates throughout the year, and between years. Why are the data from the 23/24 seasons so concerning?

192 Disappearing Islands

Key Idea: Sea level rise due to climate change is threatening the viability of low lying island nations in the near future.

Even under the most conservative projections of **climate change**, rising sea levels will place many coastal and low lying regions of the world at risk of inundation. Many of these at-risk island nations are located in the Pacific and Indian Oceans and, for many populations, permanent relocation is the only viable option for the future. The global mean sea level rose by about 15 cm during the 20th century. In the past decade it has continued to rise at close to 5mm a year, twice the rate of the two decades previous. A rise in global mean sea level of 1 m would inundate many island groups and coastal communities. Some low lying islands, such as Kiribati, are projected to be fully submerged by 2100. The **IPCC** AR6 report identified five island groups that are at imminent risk of being overcome by climate change induced **sea level rise** in the next 80 years: Kiribati, the Maldives, Tuvalu, the Marshall Islands, and Nauru.

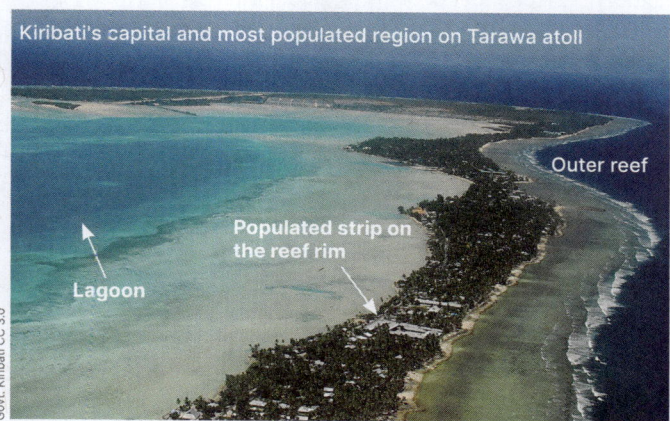

The island nation of Kiribati is made up of 33 atolls and reef islands and one raised coral island. More than 33% of its 100,000+ inhabitants live in an area of 16 km^2. Although atolls and reef islands can respond to sea level rise by increasing in surface area (through greater coral growth), there is no increase in height, so they are still vulnerable to inundation and salt water polluting drinking water.

Some 2800 km south of Kiribati, the tiny island nation of Tuvalu (maximum elevation 4.6 m) is also under threat from climate change, being vulnerable to tropical cyclones, storm surges, and king tide events. A sea level rise of 20-40 cm will make Tuvalu uninhabitable for its population of around 11,000 and already its leaders are making plans for evacuation, probably to nearby Fiji.

Adapting to sea level rise in the Maldives

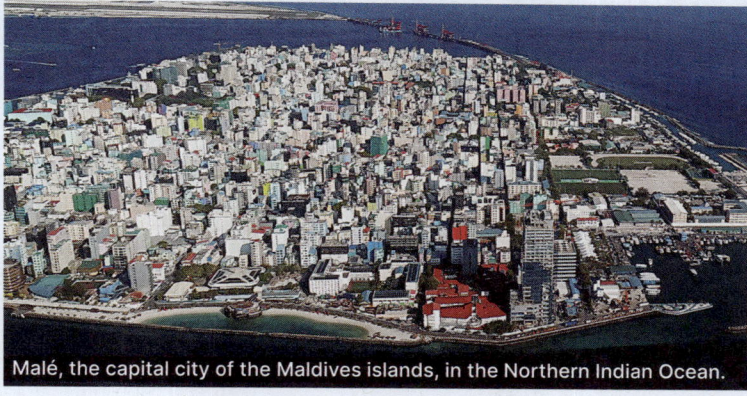

Malé, the capital city of the Maldives islands, in the Northern Indian Ocean.

- Building sea walls from waste construction material to prevent sea surge and flooding.
- Building up and reclaiming land in sheltered areas, such as Hulhumale built in a lagoon near to Malé. This was achieved by pumping sand from the sea floor and is anticipated to accommodate 250,000 when construction is completed.
- Placing high stilts and supports up to 8m high on buildings, .
- Future concepts to build floating buildings that interconnect, dubbed the 'Maldives Floating City', surrounded by protective reefs.

1. What common features do the nations identified as at risk from sea level rise in the IPCC report have?

2. Coral atolls can be relatively resilient to sea level rise by increasing in surface area over time. Why is this unlikely to help the people of island nations threatened by sea level rise due to climate change?

3. Aside from land inundation (flooding), what other issues might arise in these islands, due to sea level rise?

193 Ocean Acidification

Key Idea: Ocean acidification is occurring because higher CO_2 concentrations are lowering the pH of the water.

More CO_2 in the atmosphere is resulting in increased quantities of it dissolving in the ocean surface waters. This lowers the pH of the water through the formation of carbonic acid and H^+ ions and is known as **ocean acidification**. The ocean pH is the lowest than it has been in nearly 25 million years, becoming 30% more acidic in the past 200 years (although it is still above pH 7). Changes in ocean pH reduce the calcification rate of many species, including coral reefs and shelled molluscs. This limits their ability to form and maintain skeletal structures such as shells.

- The pH of the oceans has fluctuated throughout geological history but has usually remained at around pH 8.1 - 8.2. Recent studies have measured current ocean pH at around 8.1. A drop in 0.1 pH represents a 25% increase in acidity of the water.
- The oceans act as a **carbon sink**, absorbing much of the CO_2 produced from burning fossil fuels on Earth. When CO_2 reacts with water, it forms carbonic acid (H_2CO_3), which decreases the pH of the oceans.
- H_2CO_3 dissociates into HCO_3^- and H^+ ions. CO_3^{2-} ions from the ocean waters react with the extra H^+ ions to form more HCO_3^- ions. This process lowers the CO_3^{2-} ions available to shell-making organisms, leading to thinner and deformed shells.

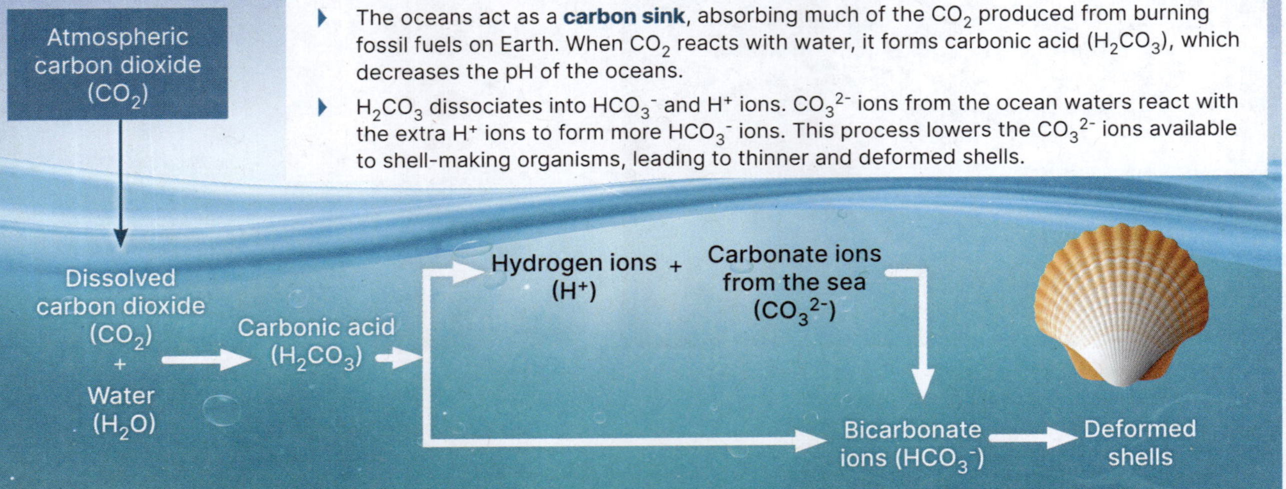

Atmospheric carbon dioxide (CO_2) → Dissolved carbon dioxide (CO_2) + Water (H_2O) → Carbonic acid (H_2CO_3) → Hydrogen ions (H^+) + Carbonate ions from the sea (CO_3^{2-}) → Bicarbonate ions (HCO_3^-) → Deformed shells

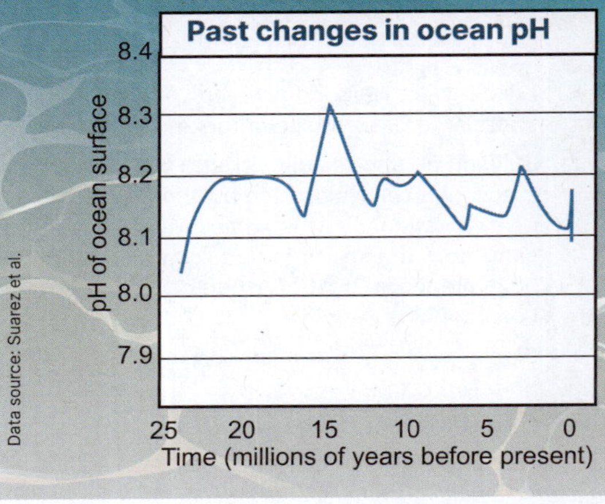

Past changes in ocean pH
(pH of ocean surface over 25 million years before present)
Data source: Suarez et al.

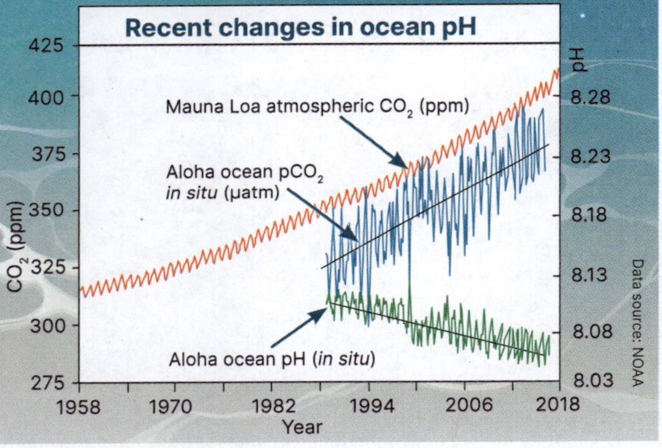

Recent changes in ocean pH
Mauna Loa atmospheric CO_2 (ppm), Aloha ocean pCO_2 in situ (µatm), Aloha ocean pH (in situ), 1958–2018
Data source: NOAA

The link between ocean temperature and CO_2 absorption

- Since pre-industrial times, the oceans have absorbed up to 30% of total **anthropogenic** CO_2 **emissions**, reducing the impacts of **climate change** but resulting in their increased acidification.
- Warming oceans hold less CO_2 and become net CO_2 emitters rather than net carbon sinks that the cold oceans are - the warmer surface water can 'hold' less dissolved CO_2, so the excess gas is released into the air above.
- Additionally, the increasing temperature reduces the mixing of ocean waters, so acidified water remains trapped under a warmer band of water, reducing nutrient and oxygen mixing.

1. Data indicates the oceans were less alkaline (more acidic) than they are today, such as 25 mya. Why is the current acidification so concerning?

   _____

   _____

   _____

   _____

©2025 **BIOZONE** International
ISBN: 978-1-99-101409-2
Photocopying prohibited

194 Albedo Effect

Key Idea: Light surfaces reflect more solar radiation and therefore stay cooler than dark surfaces. This is known as the albedo effect.

Albedo is a measure of reflection of light energy, ranging on a scale from 1, where all light is reflected, to 0, where all light is absorbed. The Earth has an average albedo of 0.3, meaning that around 70% of the solar energy is absorbed. Dark oceans and dark bare rock have a low albedo and therefore absorb almost all light (visible) energy which they then transform into heat (infrared) energy. Importantly, white ice has a high albedo of 0.7 so only 30% of the solar energy is absorbed. The surface temperature of the Earth is, in part, regulated by the amount of ice on its surface, which reflects a large amount of solar radiation back into space. The area and thickness of polar sea-ice is rapidly decreasing due to warming oceans and atmosphere driven by **climate change**.

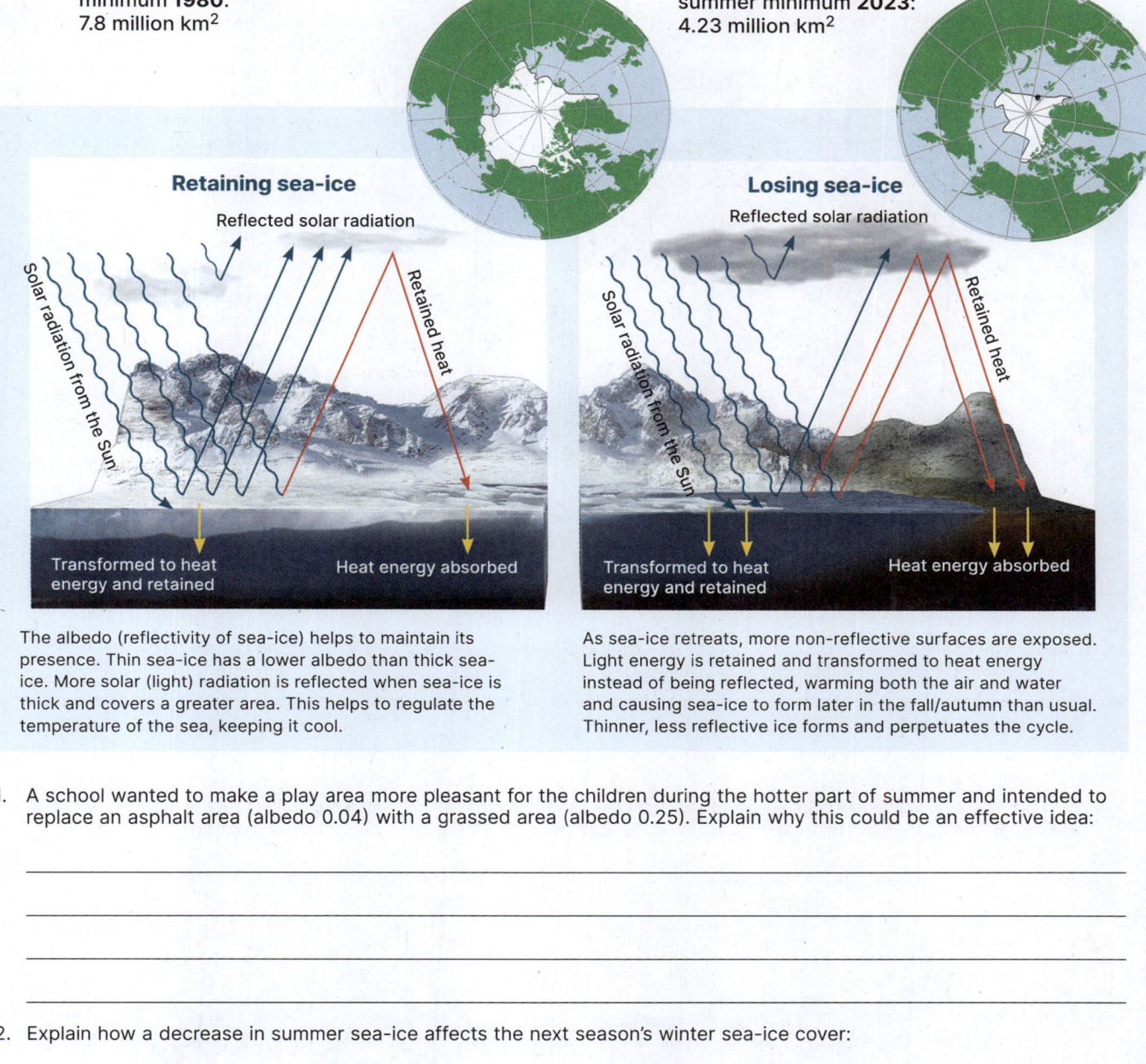

Arctic sea-ice summer minimum **1980**: 7.8 million km^2

Arctic sea-ice summer minimum **2023**: 4.23 million km^2

Retaining sea-ice

The albedo (reflectivity of sea-ice) helps to maintain its presence. Thin sea-ice has a lower albedo than thick sea-ice. More solar (light) radiation is reflected when sea-ice is thick and covers a greater area. This helps to regulate the temperature of the sea, keeping it cool.

Losing sea-ice

As sea-ice retreats, more non-reflective surfaces are exposed. Light energy is retained and transformed to heat energy instead of being reflected, warming both the air and water and causing sea-ice to form later in the fall/autumn than usual. Thinner, less reflective ice forms and perpetuates the cycle.

1. A school wanted to make a play area more pleasant for the children during the hotter part of summer and intended to replace an asphalt area (albedo 0.04) with a grassed area (albedo 0.25). Explain why this could be an effective idea:

2. Explain how a decrease in summer sea-ice affects the next season's winter sea-ice cover:

195 Extreme Weather Events

Key Idea: Climate change is causing an increase in frequency and intensity of extreme weather events.

Variation in weather is caused by complex, but naturally occurring interrelations between different components on Earth. They include precipitation, wind and storms, and drought. However, **extreme weather events** are more severe, and are often unexpected or unusual. They are characterized by systems containing more energy due to **global warming.** These can be category 5 hurricanes that form rapidly over warming waters, 'rainbombs' or atmospheric rivers that dump large amounts of rain leading to flash flooding, heatwaves that are higher in temperature and last longer, severe superdroughts that last longer and are spread wider, and extreme cold snaps due to destabilization of the polar jet stream. Scientists link the increase in extreme weather severity and frequency to **climate change**, as a warming planet has hotter nights and days, warmer oceans, and a higher atmospheric moisture load. However, the confidence of attribution and understanding of the exact mechanism is not clearly understood for every type of extreme event.

Hurricanes

Cyclone systems over the Gulf of Mexico, October 2023

- In 2023, the Atlantic hurricane season was predicted to produce only a moderate to below average number of hurricanes due to the dampening effect of an El Niño climate pattern.
- Instead, the season stood out as one with a high ACE (Accumulated Cyclone Energy) due to above average energy in the hurricanes and their frequency.
- A key factor in the formation of a hurricane is a sea surface temperature of at least 27°C. The July 2023 temperatures in the Gulf of Mexico were unusually high, producing an anomaly of over 1.09 °C hotter than normal.
- The energy in warmer seas allows hurricanes to form into larger categories faster. Each degree can add around 5% of wind speed, and 50% more storm damage on landfall.
- **Hurricane strength:** Hurricane categories 1-5 are determined by the Saffir-Simpson Hurricane Wind Scale, with each category representing a range of sustained wind speeds and potential damage. Warmer sea surface temperatures can fuel hurricanes to intensify more rapidly and reach higher categories due to the increased energy available for storm development.

Polar vortex and extreme cold snaps

- Climate change is normally associated with warm weather. Although appearing counter-intuitive, extreme cold weather events can also be attributed to a warming world.
- While climate change research would suggest that extreme cold events will decrease in **frequency** as global temperature rises, this appears not to affect the **intensity** those events.
- A strong polar jet stream, formed by convergence of cold Arctic air and warm tropical air, contains the colder air to the northern regions of the northern hemisphere.
- High latitude warming (Arctic amplification) can cause the jet stream to weaken and distort, allowing polar air to move down into typically warmer regions.
- In January 2024, a weakened polar vortex brought cold air down through the USA. One region in Montana recorded a -34°C weather station record low, and windchill (feels like) temperatures of -51°C in Montana and the Dakotas.

Stable polar vortex
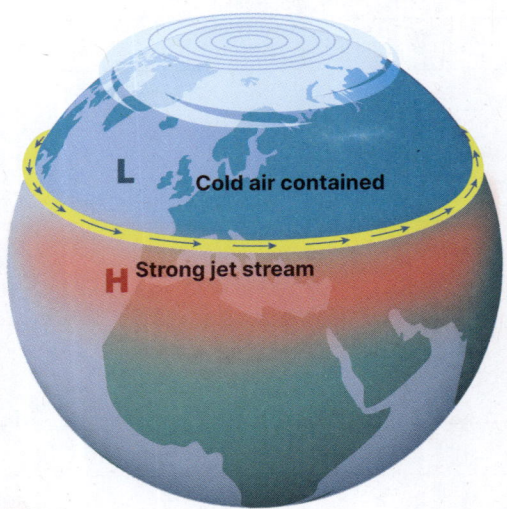

Unstable polar vortex
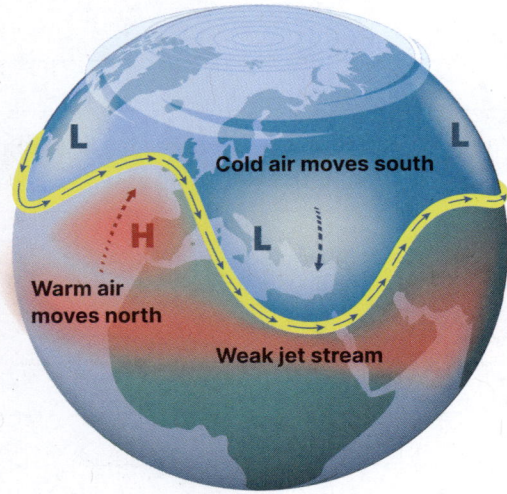

Heatwaves

- Climate change is increasing the **frequency** and **severity** of heat waves.
- Extreme events can be identified by how often they usually occur, i.e. a 10-year heat wave is likely to occur on average once every 10 years.
- Different future temperature scenarios have a very different impact on the likelihood of extreme events, including heatwaves.
- The **+1°C world** (on graph, right) of 2024 projects a 5x greater likelihood that 50-year heatwaves will occur in any given year. +4.0°C projects a 10-year and 50-year heatwave nearly every year.
- July 2023 saw 'historic' heat waves impacting southern USA/Texas, China, and Southern Europe.

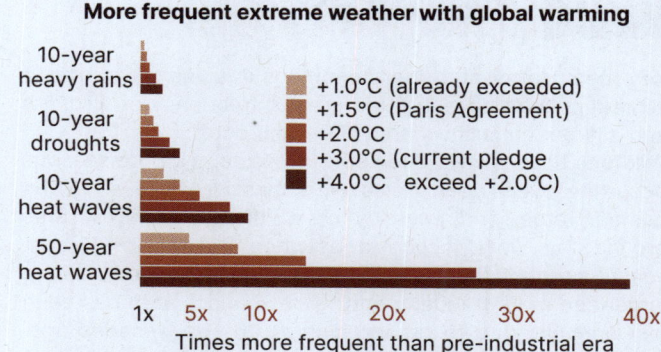

The IPCC AR6 projected increases in extreme event frequency of heat waves, droughts, and heavy precipitation events.

1. Predict the effect of increasing surface sea temperature on the frequency and strength of hurricanes:

   _____
   _____
   _____

2. Why is there more likelihood of both extreme cold snaps and heatwaves when the global temperature is rising?

   _____
   _____
   _____
   _____

3. The graph (right) shows various extreme weather events, and the confidence with which we can attribute their occurrence to climate change.

   (a) Which types of events can be most confidently assigned to climate change?

   _____
   _____

   (b) Suggest why there is such a range in the ability to assign events to climate change:

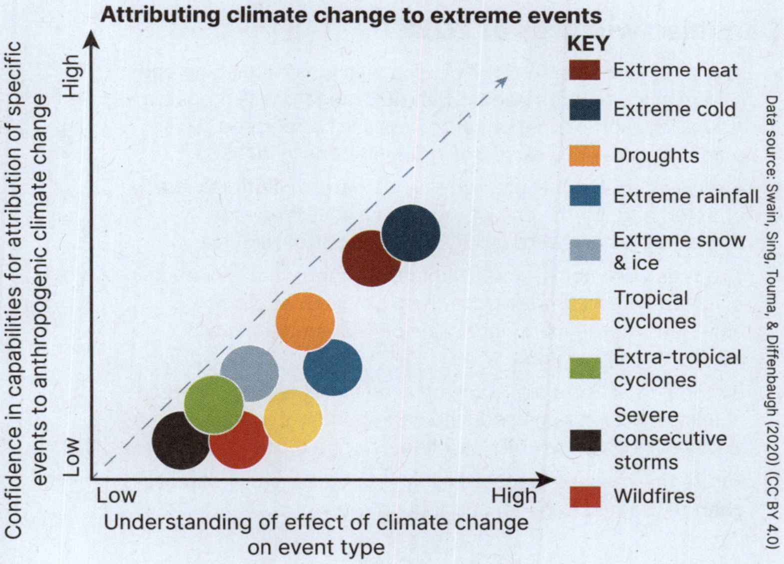

   _____
   _____
   _____
   _____

4. Research an recent extreme weather event, close to your location if possible, from the list in the graph above. (The **BIOZONE Resource Hub** is a possible starting source for data and study links). Record the year and event type. Reference the study and the study conclusion if applicable. This activity could be completed in pairs.

   Year: _____ Event Description: _____
   _____
   _____

   Reference and study summary: _____
   _____

196 Wildfires

Key Idea: Hotter and drier conditions in some areas due to climate change have led to a greater frequency of wildfires and a larger cumulative area of ground burnt.

Wildfires have always been part of nature, with fire seasons occurring every year. However, a massive increase in the number, area, and intensity of wildfires and forest fires around the world occurred between 2010-2020. Recent years have seen fires begin earlier in the season and become larger and more frequent, and twice as much area has been lost annually due to fires compared to two decades ago. Some of these fires are deliberately lit, while some arise naturally from lightning strikes. However, the warming world is increasing the fuel load and amount of dry vegetative fuel. This makes the consequences of the fire ignition far more severe, especially after droughts, which themselves are becoming more frequent. Moreover, the extra energy from global warming creates stronger winds. In 2019, more than 3 million hectares of tundra was affected by fire. The fires can be typical large surface fires, but they can also form slow smoldering fires underground. These fires can persist through cold and wet conditions, and relight as 'zombie fires' months later. Because they burn longer, these fires can transfer heat deeper into the soil and permafrost, melting and burning it.

Arctic tundra fires

- In recent years, fires in the Alaskan and Siberian tundra have threatened **permafrost**, which could fundamentally change the Arctic landscape.
- The Arctic region is warming at twice the rate of the rest of the world (polar amplification). This heating is melting permafrost and drying out the tundra, making it extremely susceptible to wildfires, both in ignition and burning intensity.
- Because of the freezing temperatures, there is normally little decay of plant material on the tundra. Huge quantities of organic material has built up over the centuries, storing vast amounts of carbon.
- Now that the Arctic is warming, this stored carbon is under threat of decaying and burning, both of which release **carbon dioxide**. The more warming there is, the more carbon dioxide (and trapped methane) could be released. Tundra fires in 2023 released at least 183 million tonnes of carbon dioxide.

Tundra fires, seen from space

Canadian wildfires of 2023

- The 2023 Canadian wildfire season was most damaging and severe since records began, with 400 megatons (1 megaton = 1 million metric tons) of carbon emissions released. This outdid the previous record of 138 megatons in 2014.
- Areas of Canada that are normally safe from wildfires were affected and, by the end of the typical season in mid-September, there were still 529 out of control wildfires.
- The fires were constrained within the country but the smoke pollution spread down across the USA (seen as carbon emissions on the map, right). The carbon emissions will impact the entire globe.
- By the end of February 2024, after an extremely dry winter, at least 100 fires remained smoldering underground. These 'zombie fires' can reignite as wildfires once more.
- Future fire seasons are increasingly exacerbated by **climate change**, both in Canada, and around the world.

Carbon monoxide concentration (1/5/23-15/6/23 average)
Lower — Higher

1. (a) The panhandle region of Texas had record setting wildfires in January-February, 2024. Over 400,000 ha was burnt in the Smokehouse Creek Fire alone. What was unusual about this wildfire, aside from the size?

   _____

   _____

   (b) Explain the link between increased wildfires, fuelled by dry and windy conditions, and climate change:

   _____

   _____

   _____

197 Megadroughts

Key Idea: Particularly severe, long-lasting, and widespread doughts are termed megadroughts. Climate change is projected to increase their occurrence.

Megadrought, as a term, was first coined in 1998. Colorado scientists have collected data from sediments, documents, and tree rings to identify historical megadroughts dating back over 2000 years, occurring on every continent except Antarctica. Some climate models are projecting that regions experiencing megadroughts in the past are more vulnerable to **climate change**-induced megadroughts or severe droughts in the future as the planet warms. Research suggests that the Western United States 2000-2023 megadrought was amplified by ~40-72% due to climate change, making what could have been a typical drought into a record breaker.

The Western US 23-year drought

▸ The megadrought experienced by the Western United States from 2000-2023 was the longest recorded dry spell in the area for least 1200 years. Data from annual tree rings, thinner in dry seasons, was able to verify the historical drought claims.

▸ The two largest lakes used as reservoirs on the Colorado River system, Lake Mead (right) and Lake Powell, reached their lowest water level in recorded history, at around 30% capacity. Authorities managing the Colorado River had to place almost total restrictions on water extraction from its remaining small flow during this time.

▸ Although heavy rainfall in 2023-24 alleviated most of the megadrought, 'patches' of severe (D3) drought are still present in areas such as western Texas.

What caused the megadrought?

▸ Scientists have partly linked the severity and length of this drought to environmental conditions caused by climate change. The drought was made more severe from higher-than-normal water loss from plants and early snowpack melting, both increased by warmer temperatures.

▸ The impacts of this exceptional drought included dead or dying crops, dry wells, harm to wildlife, loss of hydropower, and increased risk of intense wildfires.

At Lake Mead on the Colorado river, the white 'bathtub' ring (arrowed) on the cliff shows how high the water once sat. View from Hoover Dam at Nevada and Arizona border, USA.

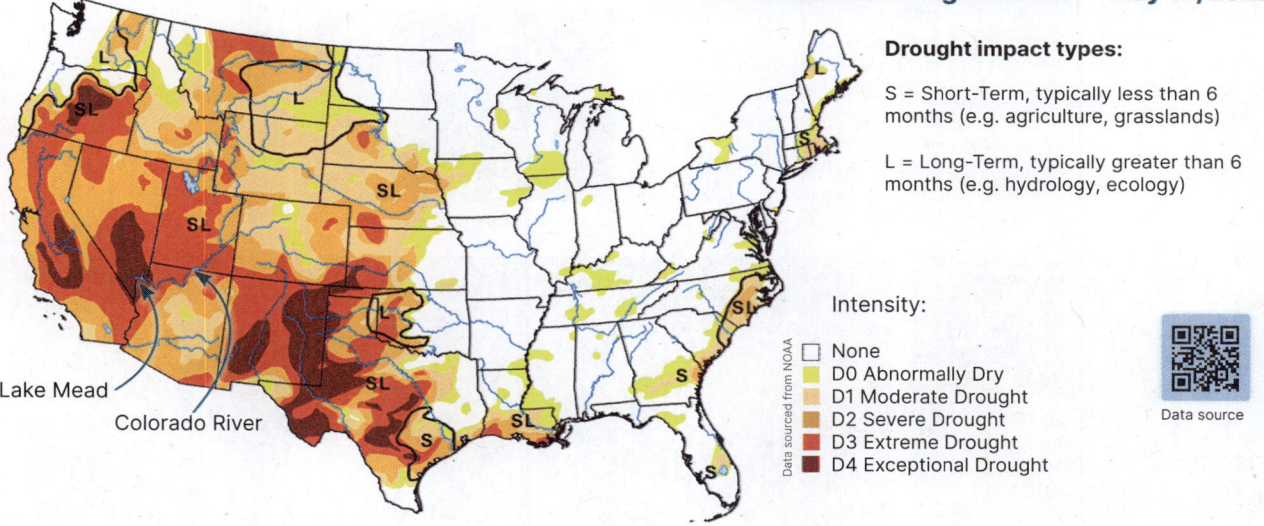

United States drought monitor - May 31, 2022

Drought impact types:

S = Short-Term, typically less than 6 months (e.g. agriculture, grasslands)

L = Long-Term, typically greater than 6 months (e.g. hydrology, ecology)

Intensity:
- None
- D0 Abnormally Dry
- D1 Moderate Drought
- D2 Severe Drought
- D3 Extreme Drought
- D4 Exceptional Drought

1. In late 2023, the El Niño Southern Oscillation (ENSO) caused higher than normal precipitation for the Western US region, relieving the megadrought. However, regions of the US are still experiencing drought.

   (a) Define a D3 and D4 drought: _____

   (b) What is the link between megadroughts and climate change? _____

198 Climate Change and Range Shift

Key Idea: Many habitats will change in a warmer climate. Climatic zones will shift, and some species will need to relocate to a more suitable area in order to survive.

As temperatures rise, organisms may be forced to move to areas better suited to their temperature tolerances. This may include moving to higher altitude or moving poleward to cooler areas. Those species that cannot move or tolerate the temperature change may face local or global extinction. Likewise, changes in precipitation, especially lack of water, may make their current habitat non-viable for previously adapted plants and animals. Long term changes in climate will ultimately result in a shift in vegetation zones as some habitats contract and others expand. Temperature shifts may force similar species into competition within overlapping ranges as newly arriving species are forced to share the same resources. Disease will increase as vector insects, such as mosquitoes, are able to spread into regions once too cold for them. **Climate change** may also impact migration timing of animals. They will have to leave their wintering location earlier to reach their summer feeding grounds further poleward.

Distribution and breeding of *Lithobates* frogs in North America

Water temperature during breeding and embryo development of *Lithobates* species

Distribution of *Lithobates* frog species in North America

The frog genus *Lithobates* is relatively common and widely distributed in North America (map, top right). The graph above shows the preferred water temperature for breeding in four common species of *Lithobates*, as well as the temperature tolerance range for embryonic development. Outside these ranges, embryonic development rate decreases or the embryos die. Increases in temperature could reduce the available breeding habitat for some species, e.g. *Lithobates sylvaticus* requires low temperatures to breed. Another likely outcome would be a change to the timing of breeding or a shift in the distribution patterns of populations.

L. pipiens, northern leopard frog

- *L. sylvaticus*
- *L. pipiens*
- *L. palustris*
- *L. clamintans*

The spread of a pathogenic fungal species *Batrachochytrium dendrobatidis* has been linked with global warming and the eventual extinction of the golden toad *Incilius periglenes*.

Distribution range changes of various tree species can be seen in the US. Overall, there has been a 17% decrease in forest cover from 2001-2022. Oak/pine and oak/hickory are predicted to increase in range while spruce/fir and maple/beech/birch communities will decrease.

Studies of the distributions of butterfly species in many countries show their populations are shifting. Surveys of Edith's checkerspot butterfly (*Euphydryas editha*) in Western North America have shown it is moving North and to higher altitudes.

1. Discuss the potential effects of a rise in global temperature on the US distribution of the frog *Lithobates* genus:

Effects of temperature increases on populations

A number of studies indicate that animals are beginning to be affected by increases in global temperatures. Data sets from around the world show that birds are migrating up to two weeks earlier to summer feeding grounds further poleward to compensate for temperature increase. This is also reflected in their wintering range which is proportionally more poleward as well. Collated research on a wide range of North American birds demonstrates movement poleward.

Animals living at altitude are also affected by warming climates and are being forced to shift their normal range. As temperatures increase, the snow line increases in altitude, pushing alpine animals to higher altitudes. In some areas of North America, this has resulted in the local extinction of the North American pika (*Ochotona princeps*).

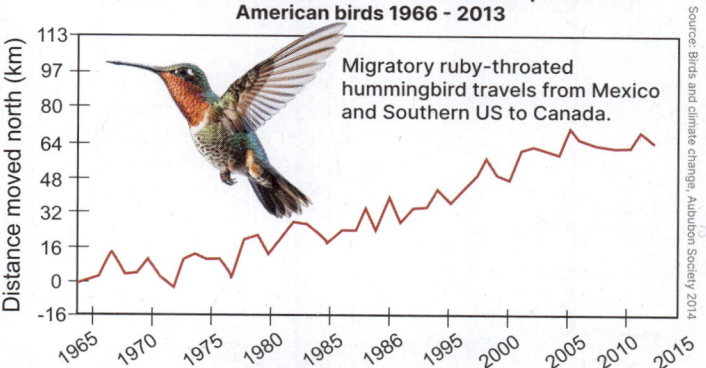

Migratory ruby-throated hummingbird travels from Mexico and Southern US to Canada.

Source: Birds and climate change, Audubon Society 2014

Higher-altitude range shifts in tropical-zone montane bird species in New Guinea

▶ The montane (temperate mountain with vegetation) regions of New Guinea form a group of unique, but unconnected, habitats. Due to their altitude-induced, cooler microclimates within a tropical region, the montane forests contain endemic species that are specifically adapted, and restricted to altitudes that correspond with their tolerance of temperature.

▶ Bird species in the montane regions of New Guinea appear to be particularly sensitive to long-term climate change and temperature increase in their habitats. Research has measured a shift upslope (to higher altitude) of 147m on Karkar Island, and 107m on Mt Karimui over the past 42 years, corresponding to around a 0.4 °C rise over the same period. Scientists estimate that just a further 1 °C rise will likely result in the extinction of at least four endemic bird species; scientists also predict another 2.5 °C rise is likely by 2100 in these habitats.

The Eastern crested berrypecker is a New Guinean montane species at risk of extinction from climate change.

2. Why are montane and alpine species particularly at risk from temperature increases, attributable to climate change?

___

3. North American tree species range movement has been documented over the past 30 years. The average shift for trees is around 32km north and 40km west. Deciduous trees tend to move westward and precipitation patterns are demonstrating less rainfall in the east. However, conifer species are moving poleward as temperatures rise beyond the plants' tolerance levels. What might be some barriers to current and future movement of habitats for tree species?

___

4. What advantage might there be for the Ruby-throated hummingbird to begin the spring migration earlier and what is the link to climate change?

___

199 Biodiversity and Climate Change

Key Idea: Climate change is creating rapid change in habitats, affecting the viability of plant and animal species.

Plants and animals have adapted to survive in their current habitats since the last significant period of **climate change** at the end of the last glaciation, 10,000 years ago. However, more recent **anthropogenic** climate change is contributing to an increase in extinction rates of species that is above the natural level, and to local and global reductions in biodiversity. Changes in precipitation and temperature as a result of climate change will affect where organisms can live. A decline in biodiversity, including loss of keystone species (essential species for stability of an ecosystem), reduces the ability of ecosystems to resist change and recover from disturbance. Humans depend both directly and indirectly on healthy ecosystems, so a loss of biodiversity affects us too.

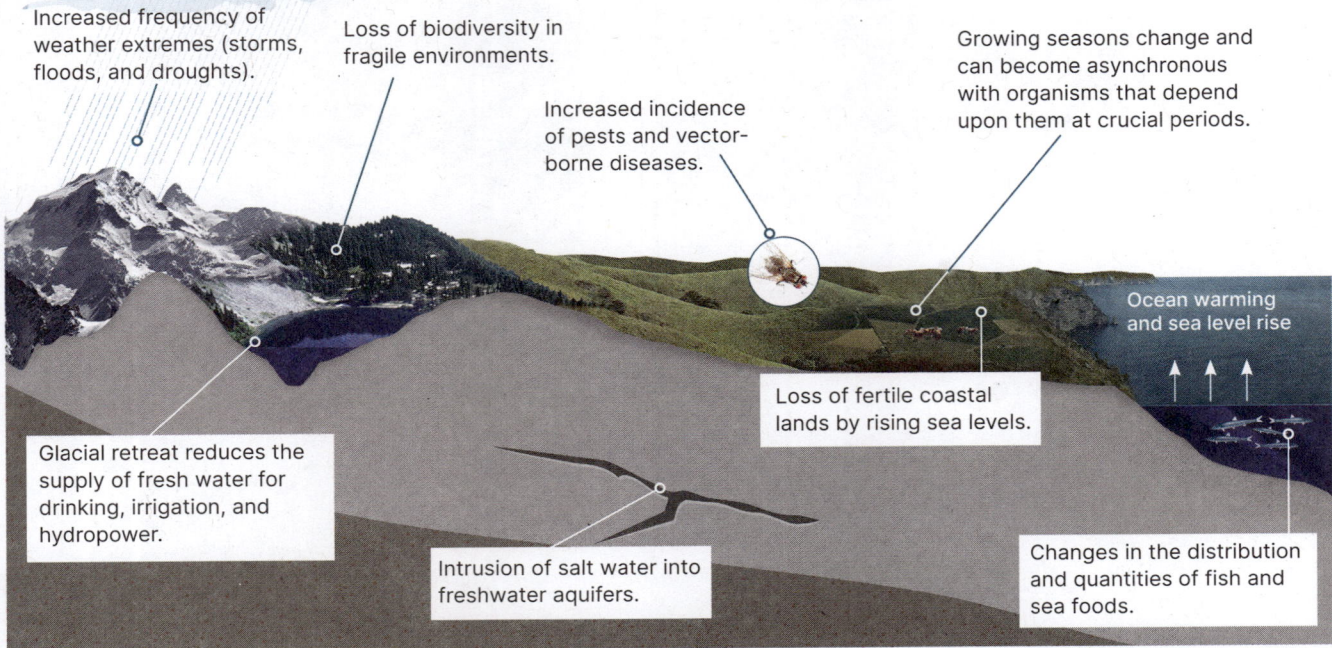

- Increased frequency of weather extremes (storms, floods, and droughts).
- Loss of biodiversity in fragile environments.
- Increased incidence of pests and vector-borne diseases.
- Growing seasons change and can become asynchronous with organisms that depend upon them at crucial periods.
- Ocean warming and sea level rise
- Loss of fertile coastal lands by rising sea levels.
- Glacial retreat reduces the supply of fresh water for drinking, irrigation, and hydropower.
- Intrusion of salt water into freshwater aquifers.
- Changes in the distribution and quantities of fish and sea foods.

Reduced rainfall and increased drought conditions are though to have contributed to the decrease in quaking aspen trees across the US. Quaking aspen is a keystone species (critically important to stability), so its loss in some regions has a significant effect on North American biodiversity. Moose, elk, deer, black bear, and snowshoe hare browse its bark, and aspen groves support to up to 34 species of birds.

Sex ratios of reptiles, including the sea turtle above, are affected by temperature. In turtles, males are produced at low incubation temperatures while females are produced at higher temperatures. Any rise in global temperature could significantly affect the composition of the reptile populations.

Reduced rainfall and increased drought...

The decreasing Arctic sea ice is having a detrimental effect on polar bears who rely on it to hunt their prey (seals), mainly in winter. In Canada's Hudson Bay, the sea ice is melting earlier and forming later. As a result, the bears must swim further to hunt and their hunting time is cut short. Survival and breeding success is reduced because they put on less weight during their main feeding time of year.

Changes in the climate over the past 200 million years are reflected in biodiversity changes in Greenland. Fossilized plants from 200 mya indicate a much warmer climate (and higher CO_2 atmospheric level). Scientists have also uncovered more recent fossils from ice cores, indicating at least two ice-free periods, after which biodiversity collapsed due to climate change.

©2025 **BIOZONE** International
ISBN: 978-1-99-101409-2
Photocopying prohibited

Walrus and sea ice

- Biodiversity in the Arctic region is being affected by climate change.
- Walrus are large, carnivorous marine mammals, distributed around oceans that surround the Arctic and sub-Arctic polar regions.
- On average, adults weigh around 1000 kg and require an extensive and continuous intake of mollusks, a wide variety of ocean invertebrates, and the occasional polar cod fish, if available.
- The walrus dive for their food but need to rest between dives on floating sea ice, usually in depths less than 80m. The sea ice is essential for their young calves who tire easily. From April-May, female walrus give birth and nurse their young on sea ice. The floating sea ice allows the walrus to reduce density of populations and therefore competition for food.
- Rising ocean temperatures and other impacts of climate change are reducing sea ice and in some regions, such as the Chukchi Sea along the Russian coast, it completely disappears through summer. Current trends have led to models predicting total sea ice melt during summer by 2040.
- Walrus can migrate up to 3000 km a year but when the sea ice disappears they are forced to occupy coastal rocky shorelines, far away from their ocean feeding grounds. This is called haulout. These areas tend to be crowded, with colonies of up to 35,000 walrus, as they look for space that is close enough to feeding areas. Walrus have to make round trips of up to 250 km in order to feed, as food availability decreases when competition rises. Young walrus are easily squashed to death without the space afforded by sea ice.

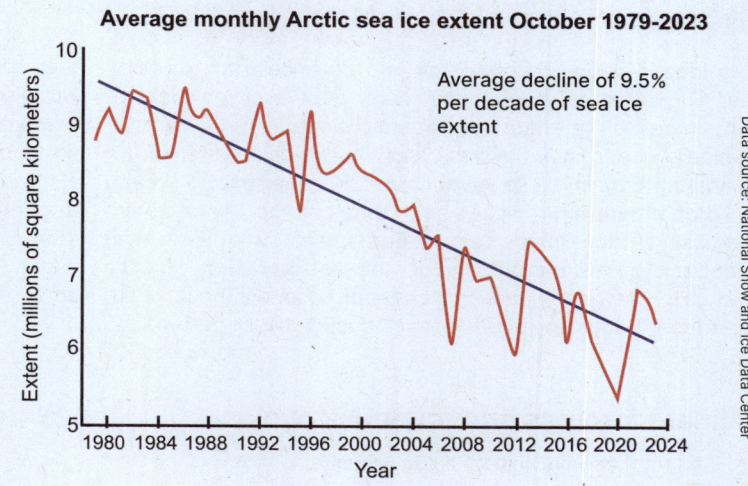

Female walrus and pup resting on sea ice between feeding trips, Norway.

1. Organisms are adapted to a wide range of habitats, from desert to polar environmental conditions. So why is anthropogenic climate change placing the biodiversity of species at risk?

2. Why is climate change impacting the biodiversity of species in the Arctic habitats, including polar bear and walrus?

3. Quaking aspen, sea turtles, polar bear, and walrus are all keystone species in their ecosystems. Why does climate change impact to those species have a wider implication for the biodiversity in their ecosystems?

4. The cane toad is an invasive species introduced into Australia where it competes with many native species for a wide variety of food and is continually spreading across the country. The toad has a physiology that allows it to adapt to warming temperatures more effectively than other species. How might climate change impact biodiversity in this case?

200 Positive Feedback Cycles

Key Idea: Positive feedback cycles are accelerating global warming, emitting greenhouse gas or heat energy which, in turn, increase the effects of climate change.

Typically, feedback cycles allow physical systems to maintain stability. However, **positive feedback cycles** in **global warming** accelerate the change, leading to increasing, non-linear temperature rise. Many positive feedback cycles contributing to the net **emission** of CO_2 and CH_4 (methane) gases into the atmosphere have been identified. Some positive feedback cycles affect others, i.e. they are interconnected: an increase in effects in one cycle will also cause an increase in the effects of another. For example, the **albedo effect** increases heat absorbed, both on land, increasing methane release from **permafrost**; and also under the oceans, increasing release of CH_4 and CO_2 from previously frozen clathrates and hydrates, respectively. Moreover, warmer temperatures increase the amount of water vapor evaporated from water surfaces, and the amount of water vapor the atmosphere can hold. An increase in water vapor then affects many different feedback cycles.

Climate forcings and feedback cycles

- Factors that drive climate are called climate forcings. Some are natural and cyclical, such as radiation fluctuations from the Sun, and volcanic activity.
- Human-induced climate forcings include increases in **greenhouse gases** and land changes affecting albedo.
- Human-induced forcings are now dominant over natural climate forcings in driving **climate change**.

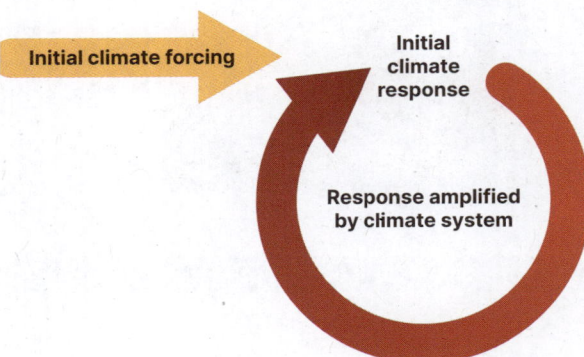

Water vapor release

Water evaporation from lake surface

Water vapor adds to the **greenhouse effect** in the atmosphere, trapping heat energy, although it has a shorter lifespan than other gases. Warmer temperatures result in higher rates of evaporation from water on Earth's surface, thereby increasing the amount of atmospheric water vapor. This energized water cycle can lead to intensified weather events.

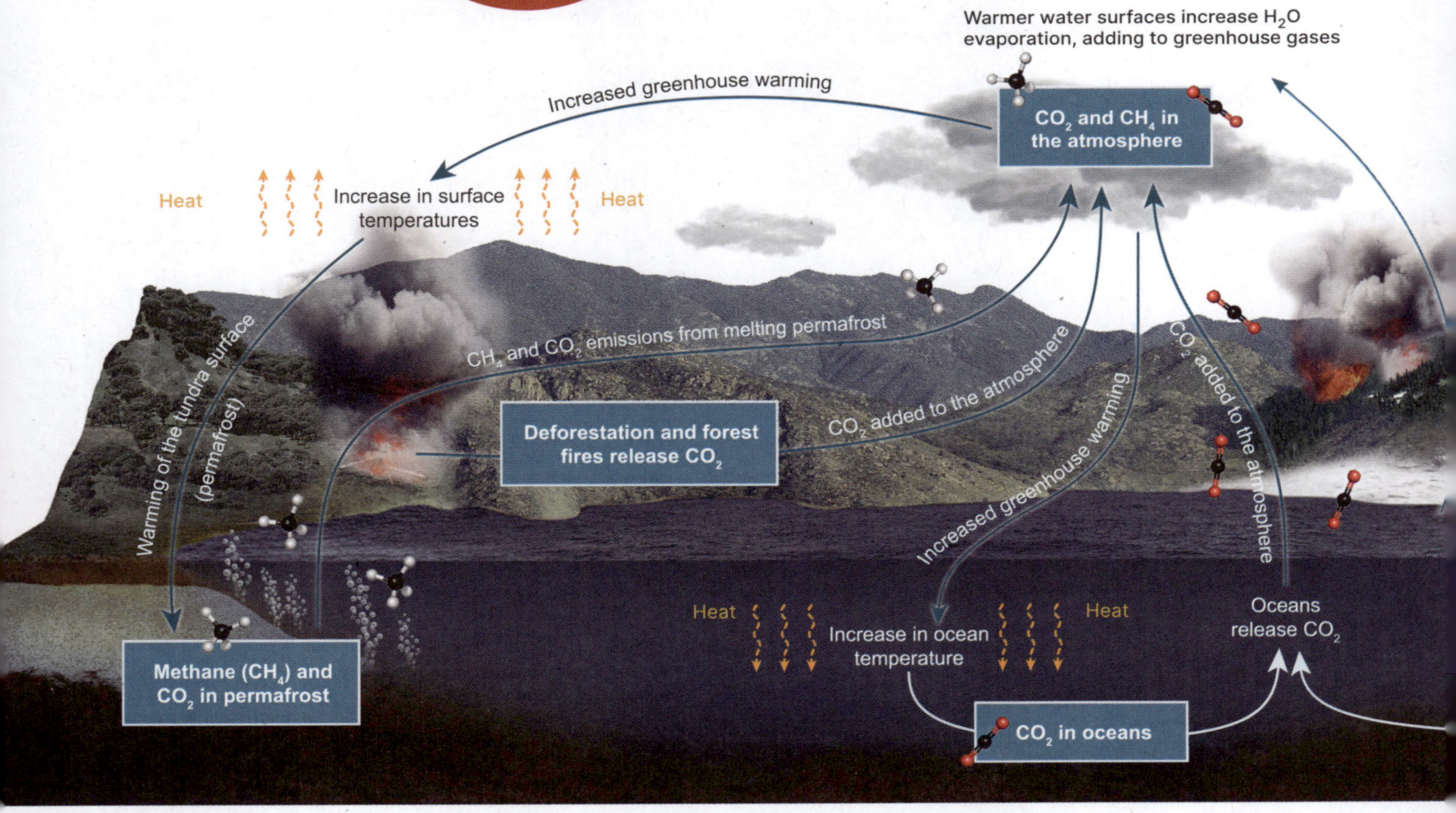

1. Explain what a positive feedback cycle is and why these are so significant for climate change and global warming:

   _____
   _____
   _____
   _____
   _____

2. What are the consequences of interconnected positive feedback cycles? _____
   _____
   _____
   _____
   _____

3. How does a decrease in albedo from melting sea ice contribute to an increase of CO_2 released from the ocean, and add to global warming? (Hint: reread information on CO_2 absorption in the acidification activity).

   _____
   _____
   _____
   _____
   _____
   _____
   _____

Feedback in the Albedo effect

Icebreaker in melting sea ice

The high albedo (reflectivity) of sea-ice helps to maintain its presence. As sea-ice retreats, more non-reflective surfaces (darker sea) are exposed. Heat is absorbed instead of reflected, warming the air and water and causing sea-ice to form later in the fall/autumn than usual. Thinner and less reflective ice forms, continuing the positive feedback cycle.

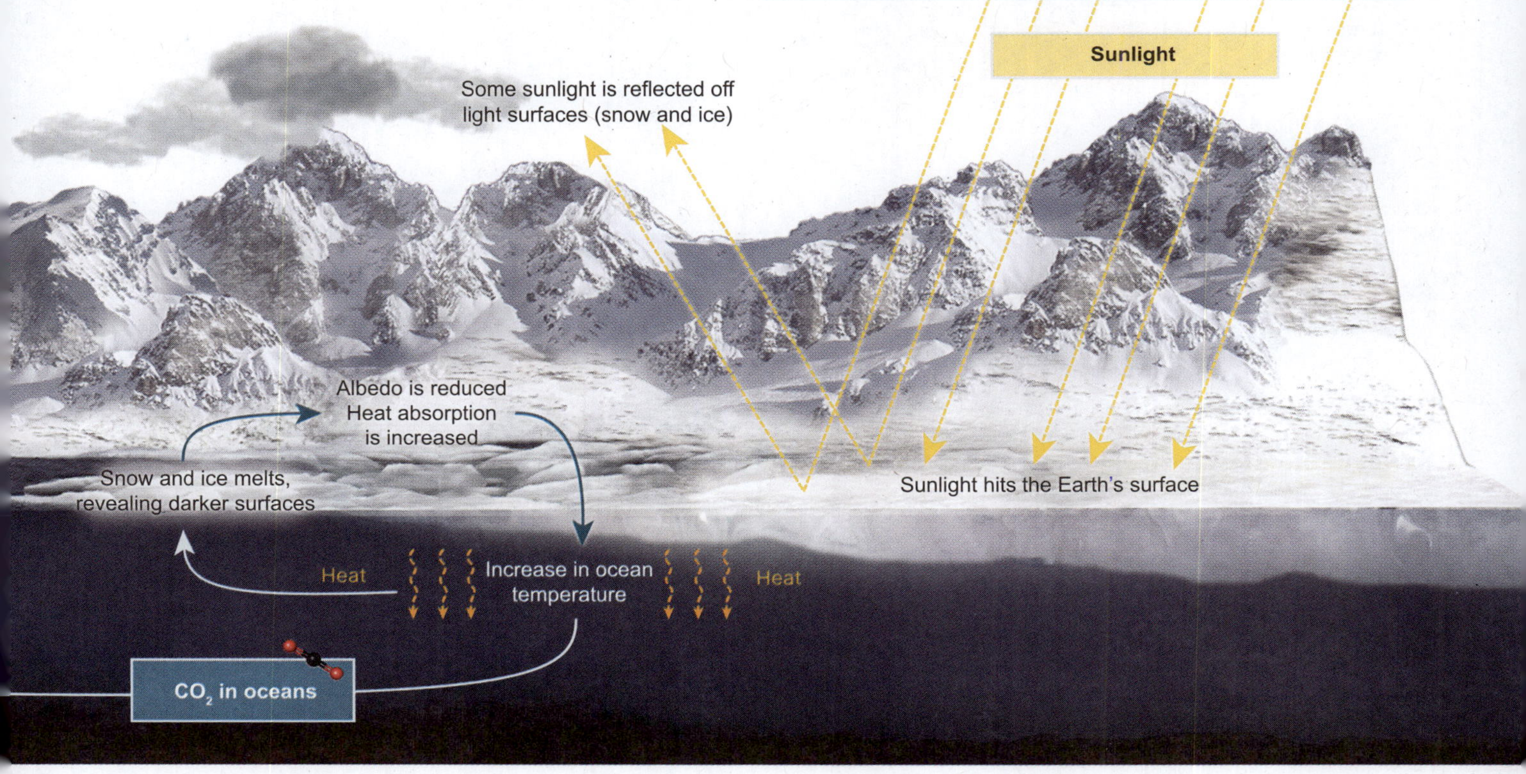

201 Tipping Points

Key Idea: Climate change tipping points are reached at certain critical threshold temperatures and must include a positive feedback cycle that becomes self-perpetuating. When changes become too great in Earth systems, they may cause the systems to rapidly shift from one state to another. This phenomenon is known as a **tipping point**. The changes become self-reinforcing through positive feedback cycles. **Climate change** effects are already noticeable, with a worldwide average global temperature rise of 1.19°C since pre-industrial times. The **IPCC** publishes regular, in-depth reports on the causes, mitigation, and **adaptation** responses to climate change and recommends that the worst of the effects can be prevented by limiting **global warming** to 1.5 °C. Governments meet at yearly United Nations climate change conferences to discuss and sign accords (agreements or letters of intent). They agree on how they can reduce emissions to contribute to global reductions and meet the **Paris Agreement**: an international climate change treaty. Different tipping points are projected to occur at different temperature rises in different systems.

Significant climate change tipping points

Greenland ice sheet

The **albedo effect**, exposing dark rock of Greenland's bare surface, and the loss of ice sheet elevation are causing positive feedback cycles. Scientists estimate a rise of 1.5 °C will shift Greenland into an irreversible tipping point of collapse. Total loss of ice from Greenland will cause a **sea level rise** of over 7m over the entire Earth, however this will occur over 100s to 1000s of years.

AMOC (and Gulf Stream)

The **AMOC** (Atlantic Meridional Overturning Circulation) circulates warm water North up the Eastern coast of the USA as the Gulf Stream System. This current maintains the temperate climate of Britain, Ireland and Western European countries. An increase in fresh water from melting Greenland **ice sheets** slows this ocean current. A collapsing Gulf Stream current will result in a colder climate in Northern Europe.

Sahel / West African Monsoon

The West Africa Monsoon (WAM) brings regular heavy rain to areas in a band south of the Sahara called the Sahel. Warm sea surface temperatures (SSTS) influence the variability of timing of the occurrence of the WAM and any disruption can result in widespread drought in the North as rainfall shifts southward (causing greening in that area). The WAM system is sensitive and a change in the Atlantic current (AMOC) can add to its destabilization.

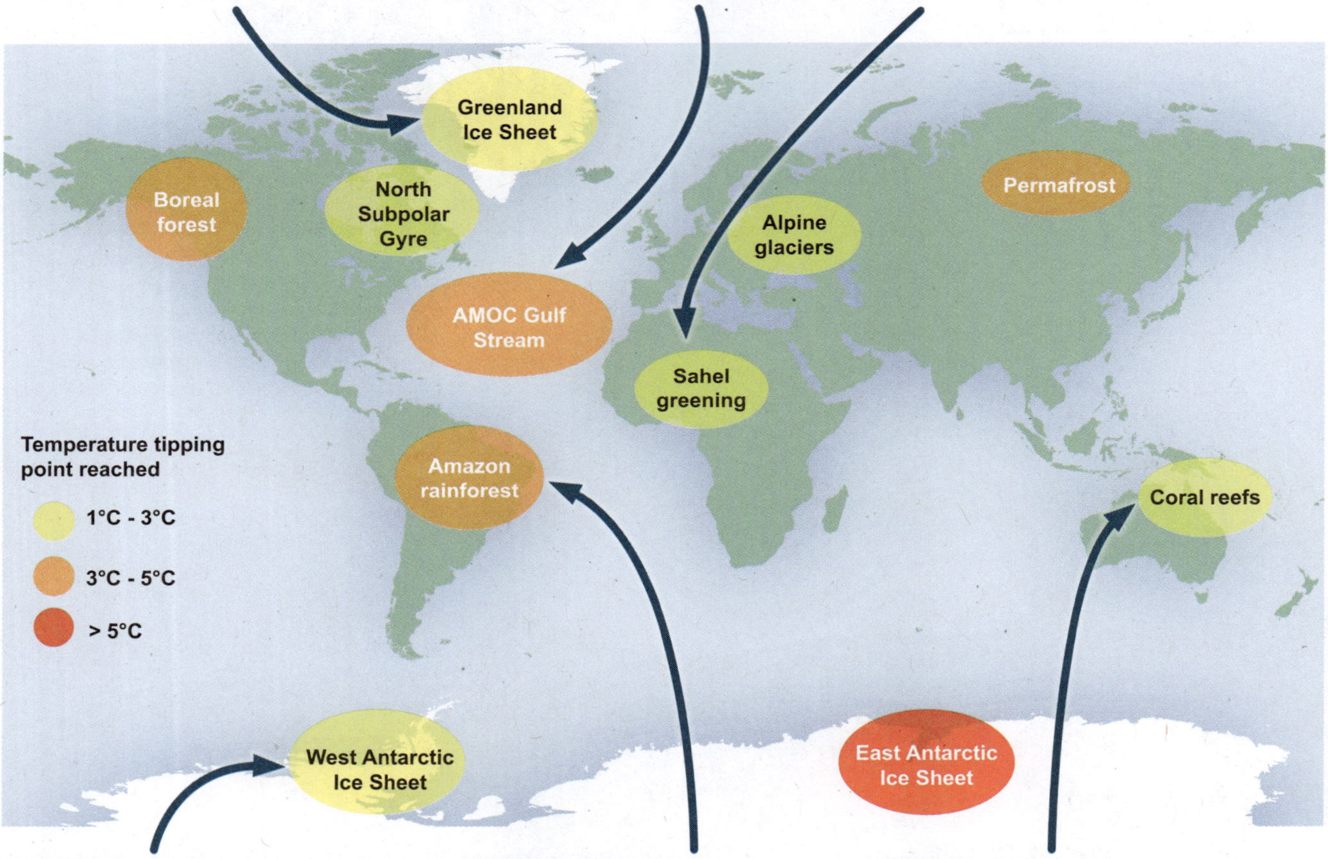

Temperature tipping point reached
- 🟡 1°C - 3°C
- 🟠 3°C - 5°C
- 🔴 > 5°C

Map labels: Greenland Ice Sheet, Boreal forest, North Subpolar Gyre, Permafrost, Alpine glaciers, AMOC Gulf Stream, Sahel greening, Amazon rainforest, Coral reefs, West Antarctic Ice Sheet, East Antarctic Ice Sheet

Adapted from Steffen et al (2018), PNAS

West Antarctic ice sheet

This ice sheet holds enough water to raise sea level by around 3.3 meters if it is completely lost. The bedrock of this ice sheet sits below sea level, making it especially vulnerable to rising ocean temperatures and warming sea breezes. The thinning ice sheet causes a retreating grounding line (where the ice attaches to bedrock) and further loss is accelerated as parts of the ice sheet break off.

Amazon rainforest

Scientists predict that a deforestation of around 25% of the Amazon rainforest will lead to a dieback tipping point. By 2023, around 15-17% had been lost. Removal of timber for agriculture and export is the main cause of deforestation currently, but rising global temperatures are leading to less rainfall and death of parts of the forest due to drought. This results in lower rainfall in other areas of the forest.

Coral reefs

Many coral reefs depend on a very narrow temperature range in order to sustain their mutualistic relationship with photosynthetic protists. Scientists have estimated that coral reefs take on average 10 years to recover from an ocean heat wave, but heat wave events are occurring more than once every 8 years. A 2 °C rise would cause up to 99% of coral reefs to become extinct from **bleaching** and death.

Confidence and timelines in tipping points

▸ Tipping points are non-linear consequences of climate change forcing, leading to self-perpetuating and often sudden shifts to another state.

▸ However, not all systems impacted by climate change have tipping points, and they will continue with a response that rises at an equivalent rate as the climate forcing. For example, although global warming leads to less Arctic sea ice, melting Arctic sea ice systems also incorporate negative feedback cycles. This is because open water loses more heat and is less insulating, therefore counteracting the impact of albedo loss.

▸ Confidence in whether a system is a climate tipping point or not is the result of multiple, concordant (in agreement) peer reviewed research findings. For example, evidence provides high confidence that the West-Antarctic and Greenland ice sheet systems have tipping points, and high confidence that the Arctic sea ice systems do not.

▸ Scientists are still unclear about whether some systems have climate tipping points. For example, East-Antarctica ice sheets, temperate forests, and the South American monsoon.

▸ Additionally, research provides clarity as to whether the impact is likely to be local, regional, or global when a tipping point is reached in a system. For example, the Greenland ice sheet melt will have global consequences, while the kelp forest loss will be restricted to local-scale impacts.

▸ The timeline for different tipping points is not the same for different systems. For example, total melt of the Greenland ice sheet would take at least 1000 - 10,000 years, depending on the global temperature reached, whereas complete death of all warm water coral could occur over just a few years if the temperature of ocean water remained high enough.

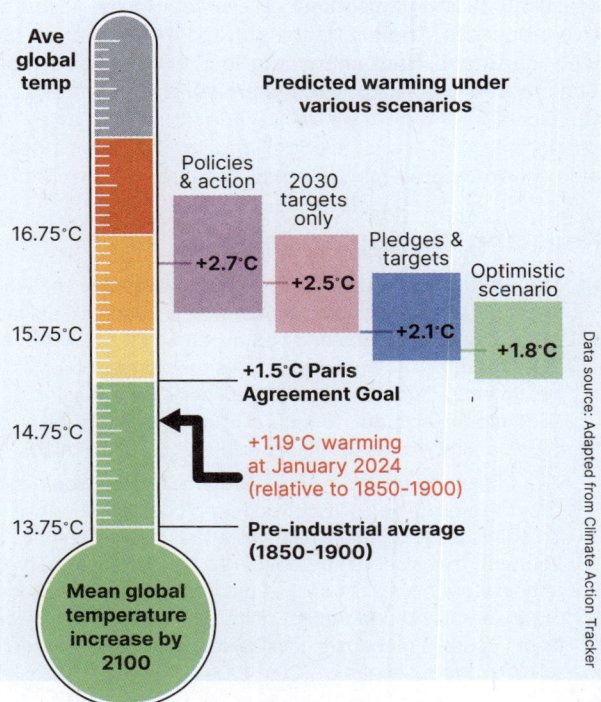

Global warming projections

Current climate change policies from governments around the world will not prevent a number of tipping points being reached before 2100. Full implementation of net zero targets can bring the warming closer to the 1.5 °C Paris Agreement goal (see Activity 188).

1. What distinguishes systems that have climate tipping points from those that do not?

2. What factors differentiate between different climate tipping point systems? Provide examples from the map (left):

3. Lowering carbon dioxide and other greenhouse gas emissions is called mitigation and can help us avoid some of the worse predicted consequences of climate change by limiting global warming. Why does this approach have less impact once a climate change tipping point is reached?

4. What role do updated data, new research, and climate models play in shifting current government polices towards more assertive pledges and targets, especially regarding climate tipping points?

202 Tipping Point: Greenland Ice Sheet

Key Idea: Continued global warming may push the Greenland ice sheet system towards a tipping point of total ice loss.

Ice sheets cover land and hold most of Earth's freshwater reserves. The Greenland ice sheet covers 1.71 million km² of land, and is situated northeast of Canada, close to the Arctic. The fluctuation of ice present on a sheet can be measured by two processes. The first is the surface mass balance (SMB), with ice added from snow, and lost from surface melting. The second is called ice dynamics and relates to both the acceleration of iceberg calving (breaking away) from glaciers around the sea edge, and the retreat of glaciers, also referred to as marine mass balance, (MMB). Together, both comprise the total mass balance (TMB). The global temperature threshold for the southern end of the Greenland ice sheet to enter a melting tipping point is a sustained 1.5°C. Higher **global warming** temperatures would lead to total ice sheet collapse. Aside from the consequences of a significant **sea level rise**, the loss of the ice sheet could interact with other systems, such as the disruption and slowing of the **AMOC**, and induce a tipping cascade. Observations indicate a positive feedback cycle is already occurring (see right), suggesting the Greenland ice sheet will become the first system to reach **tipping point**.

Reaching the tipping point

- The Greenland ice sheet is the second largest in the world after the Antarctic ice sheet. It contains about 2.8 million cubic kilometers of ice and has a mean thickness of over 2000 meters. The continuous ice sheet formed over 2.6 million years ago during the Pleistocene ice ages. It has fluctuated in size due to natural climate forcings (factors that cause an imbalance of energy in=energy out of Earth).

- Two tipping points have been identified in the Greenland ice sheet system. The current trending global temperature rise is soon likely to initiate the smaller, first tipping point of the Southern ice sheet melt. Reaching a further sustained global temperature over 2.0°C would push the system into total collapse: the second tipping point. The added meltwater would cause a global sea level rise of around 6.9 meters, albeit over a timespan of 100s-1000s of years.

- Once the second tipping point of the ice sheet has been reached, the Greenland sheet would not reform, even if the CO_2 levels in the atmosphere returned to levels occurring in pre-industrial times.

In the past three decades, Greenland has lost around 28,700 km² of ice, 1.6% of its entire mass. However, the ice loss is not linear and will accelerate as temperatures continue to increase.

Measuring Greenland ice sheet loss

- Greenland surface ice melt resulting in mass balance (TMB) losses can be measured by satellites, such as the GRACE from 2002-2017, and the GRACE-FO from 2018 onwards. The data is gathered indirectly by measuring gravity anomalies.

- The melt duration and extent are measured by microwave tools, also from satellites, combined with land-based monitors to form an integrated model of changes occurring on the ice sheet. Another method, called the input-output (IO) method, combines data and models and can identify if ice loss is due to SMB or MMB. NASA MODIS satellites are used to monitor **albedo** changes.

- Data collection involves international cooperation by scientists and sharing of research findings.

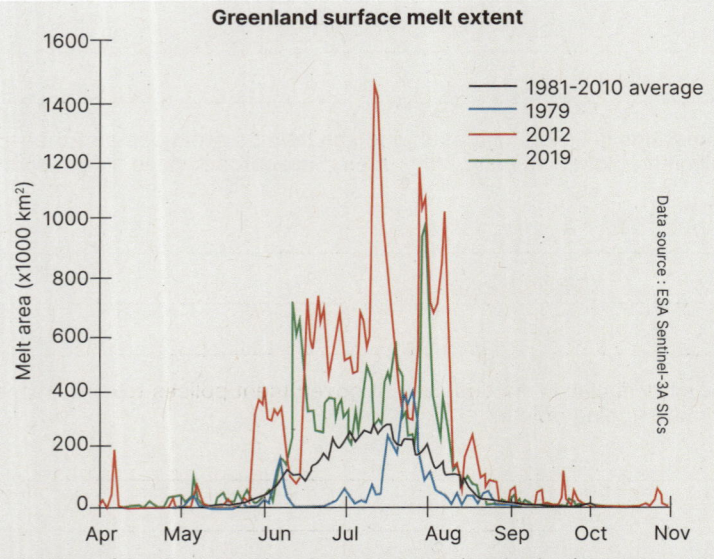

Melting of the Greenland ice sheet occurs during the Arctic summer. Since 1979, the area of ice melting and the length of time melting occurs has increased. 2012, 2019, and 2023 have been record years of ice loss.

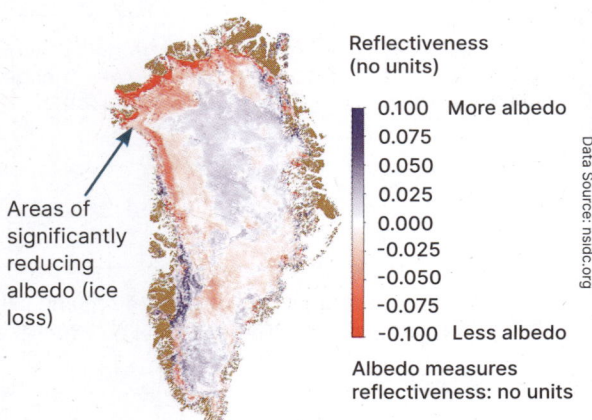

The map above shows the difference in sunlight reflected during summer (July) 2022 compared to the average (2017-20) reflection. Less reflectiveness (red) is caused by lower albedo, and thinner ice.

Positive feedback cycle and tipping point

- The height or elevation of the Greenland ice sheet reaches over 2km, a much cooler high altitude zone where falling snow remains frozen as ice.
- As the global temperature warms and ice is lost as surface melt, the absolute height of the ice surface is lower and therefore exposed to a warmer environment, thus melting quicker. This is called **positive melt-elevation feedback**.
- This cycle is accelerated, as the air temperature in the polar regions, including Greenland, is warming **twice as fast** as other areas around the globe.
- Scientists predict that, once the ice sheet melts to a certain point, an irreversible tipping point will be reached that will continue until Greenland moves into an ice-free state.

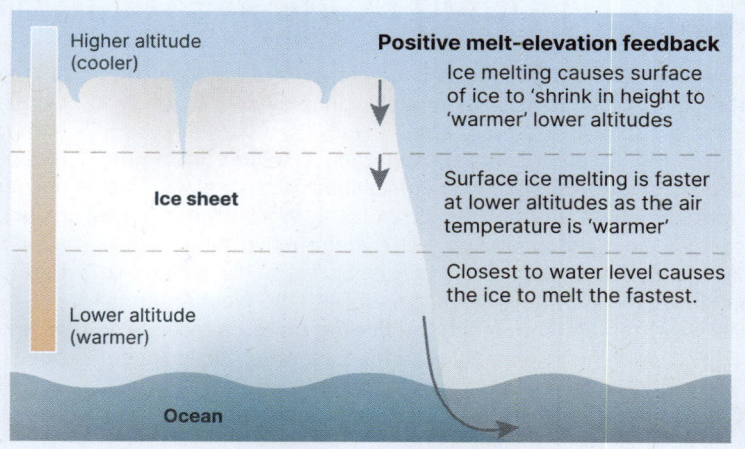

1. The Greenland ice sheet loses most of its ice by a combination of SMB and MMB processes. Explain these terms and how they differ from each other:

   _____

2. Greenland ice sheet loss can be measured by adding together SMB and MMB to obtain the total mass balance (mass is also lost at the base of the ice sheet in very small amounts, BMB). Ice loss for previous time periods is: ~41 Gt/yr (1990-2000), ~187 Gt/yr (2000-2010), ~286 Gt/yr (2010-2018). (Hint: draw a simple line plot to see trend).

   (a) Is this a linear (geometric) rate of ice loss? Explain your answer: _____

   (b) How would reaching the threshold of the tipping point effect this rate, and therefore length of time to reach 100% ice sheet collapse?

   _____

3. An average loss of 269 Gt/yr from the Greenland ice sheet adds about 0.69mm of sea level rise per year, globally. Why then does the melt of sea ice, particularly that covering the Arctic sea, only negligibly contribute to sea level rise?

   _____

4. The average albedo of the Greenland ice sheet in summer fluctuates, as shown in the graph below. Which years have record summer ice loss? Justify your answer (hint: see albedo effect, Activity 194):

   _____

5. How do albedo loss and positive melt-elevation contribute towards the tipping point?

   _____

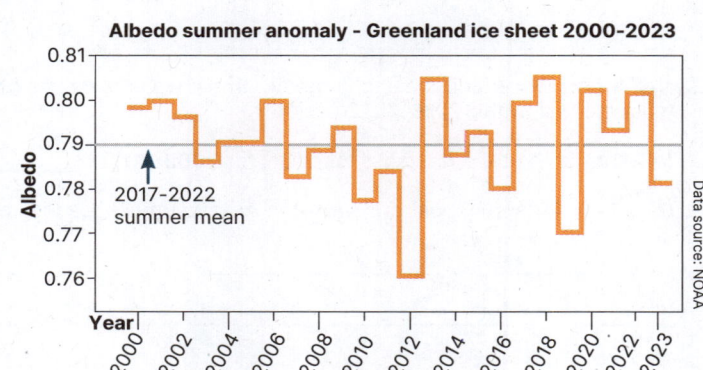

203 Tipping Point: West Antarctic Ice Sheet

Key Idea: Climate change is accelerating the ice loss from the West Antarctic ice sheet, pushing the system towards a climate tipping point of collapse.

The ice in Antarctica sits on top of a continent and is the largest single **ice sheet** on the planet. Whereas the East Antarctic sheet sits on top of land, the West Antarctic ice sheet is mostly grounded on rock that is submerged to nearly 2.5km in some places, in marine basins. This difference influences the different rates of melt, with the West Antarctic ice sheet projected to reach the **tipping point** of collapse at a 1.5°C warming threshold, while the East Antarctic ice sheet is projected to maintain its current state until around a 7.5°C warming threshold. Antarctica sits directly over the South Pole, so temperatures tend to be lower than Greenland, reducing surface melt. Due to greater contact between ocean and ice at the ice sheet margins, ice shelves are formed, also called fast ice, which thin and melt rapidly. The ice that is supported behind will then accelerate forward to replace that lost in the ice shelves. The retreat of the ice grounding line due to warming oceans undercutting ice shelves acts as a destabilizing **positive feedback cycle**. After the tipping point, the ice sheet would continue to melt over 1000 years.

Collapse of Thwaites Glacier

Thwaites Glacier, situated in a marine basin in West Antarctica, is a large, fast retreating glacier. It has been termed the 'Doomsday Glacier,' due to its vulnerability and risk of collapse. Together with the neighboring and equally fast retreating Pine Island Glacier, a collapse could add up to a meter of **sea level rise**.

- Ice shelves extend from the base of the Thwaites Glacier into the ocean. The outer edge of the glacier that touches the sea floor is known as the grounding line (see right).
- Water can flow under the ice shelf and cause melting: the warmer the water, the faster the melt, and the more rapid the retreat of the glacier. Ice loss in Thwaites Glacier has been significant for several decades.
- As more of the ice shelf disappears, the warm water can access deeper under the ice, shifting the grounding line back. The cliff calving front (break off) above the grounding line becomes higher due to the downward, or retrograde, tilt of the sea floor further back. It is therefore more unstable: taller ice cliffs break off (calve) more easily.
- The loss of ice from Thwaites Glacier creates a positive feedback cycle: the more the glacier retreats, the faster the ice is lost. This type of feedback is called **marine ice sheet instability (MISI)** and evidence points to this process already starting at the Thwaites and Pine Island Glaciers.

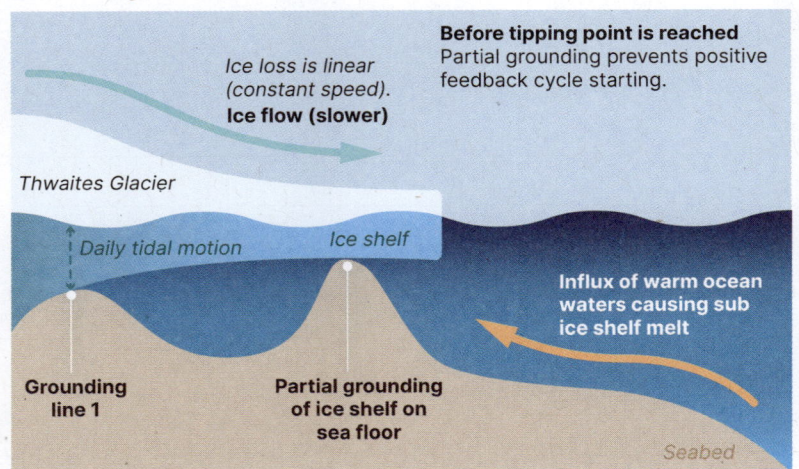

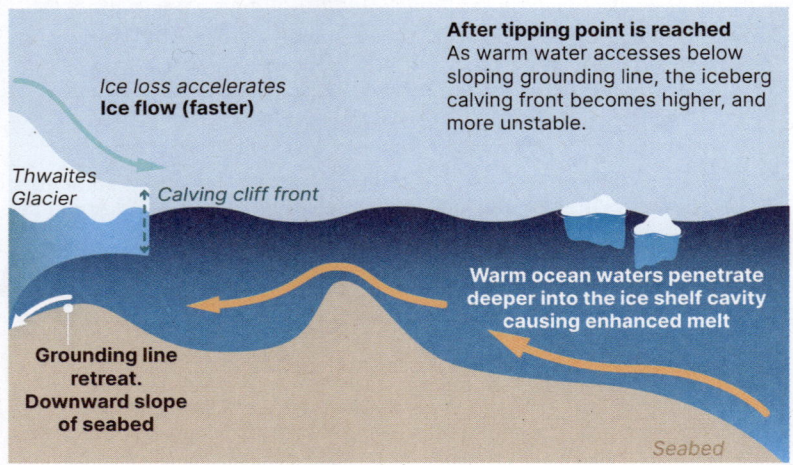

1. The seabed under the Thwaites Glacier is sloping backwards. How does this contribute to a ice loss feedback cycle?

   _____
   _____
   _____

2. NASA Antarctica land mass changes data shows an average of 142 GT per year from the period of 2002-2022, with the most ice lost in the 2018-2022 period.

   (a) What is the total ice mass loss (Gt) for 2002-2022? _____

   (b) Which Antarctic locations have the greatest mass ice loss and why? _____
   _____
   _____
   _____

204 Tipping Point: Boreal Permafrost

Key Idea: Global warming could make large boreal ecosystems too warm and dry to maintain carbon-storing permafrost. Cold temperate regions between northern polar regions and forested temperate regions are called boreal biomes; they include Siberia and Northern Europe. The soil is cold enough in most areas to maintain a permanently frozen ground called **permafrost** that locks away large quantities of carbon in peat. Increasingly warming temperatures, contributing to **climate change**, are thawing the permafrost and activating microbes that decompose the peat. **Greenhouse gases**, including CO_2 and CH_4 (methane), are products of peat decomposition and are being emitted at higher levels, leading to even more warming. Thus, a **positive feedback cycle** is being activated. Climate models have projected that specific regions such as Norway, Sweden, and Finland may have passed a **tipping point** already. Unless the world rapidly lowers greenhouse gas emissions, an inevitable collapse of the **boreal permafrost** regions, acting as net **carbon sinks** (taking in more carbon than releasing), is almost certain. However, recent research indicates the collapse will more likely occur at a regional or local level, and aggregation into runaway climate change is unlikely.

Exploding craters and climate change

▸ By April 2024, eight large craters up to 20 meters wide and nearly 50 meters deep had been identified in the Siberian permafrost. Local residents reported hearing loud explosions and seeing debris thrown across a large distance when craters formed.

▸ Scientists originally hypothesized that the built-up gas was due to decomposition of material in a trapped underground prehistoric lake. Microbial processes were 'reawakened' by climate change-induced ground warming, leading to the gas bursting through when enough pressure had built up.

▸ Another hypothesis for crater formation proposes that hot gases, built-up underground via geological processes, were able to escape rapidly through the permafrost which has thawed and weakened due to climate change.

Permafrost crater, Yamal peninsular, Russia

Feedback cycle and permafrost thawing

The melting of permafrost has the potential to produce a positive feedback cycle, producing even more heating.

1. Why does permafrost prevent greenhouse gases from being released?

2. Why might there be more than one hypothesis to explain the formation of the permafrost craters?

3. What is the link between the positive feedback cycle of permafrost melting and passing the tipping point of regional permafrost collapse?

205 Tipping Point: Boreal Forests

Key Idea: Climate change may push boreal forests from carbon sinks into carbon emitters due to drought-fuelled wildfires, forming a positive feedback cycle of continued loss. **Boreal forests**, or taiga, make up 30% of all global forests and surround polar regions. The forests consist mainly of cold-tolerant conifer species that require reasonable precipitation. Boreal forests typically act as **carbon sinks**, storing one third of global carbon, mostly underground, while only accounting for 11% land coverage of the Earth's surface. Boreal forests are at risk of reaching a **tipping point** due to **global warming** leading to **climate change**.

Warmer temperatures and subsequent loss of snowfall are creating drought stress, browning of forests, and increased vulnerability to insect pests. Additionally, fire risk, increasing in both intensity and frequency, lowers primary production, where less photosynthesis leads to less carbon being stored. The carbon released comes primarily from soil organic matter as it is combusted. Models have predicted that regional tipping points will begin at around 1.5°C global warming. A higher 4.0°C warming will shift the boreal forest ecosystem into a treeless state. Projected reduced precipitation will prevent regeneration after fire disturbances.

Increases in droughts and boreal forest fires

- Boreal forests act as carbon sinks, storing huge amounts of their plant material.
- Much of the carbon is sequestered underground, forming peaty organic layers. It is contained in roots and root microorganisms, where colder temperatures slow decomposition.
- Evidence shows that global warming, leading to climate change, is creating more prolonged and severe droughts in areas around the world.
- Under drought conditions, vegetation and dead matter on the boreal forest floor become drier which increases the risk of wildfires.
- Boreal fires are increasing in intensity, the fire seasons are becoming longer, and there is less time between fires.
- When wildfires occur, combustion of the organic material releases large amounts of CO_2 into the atmosphere.

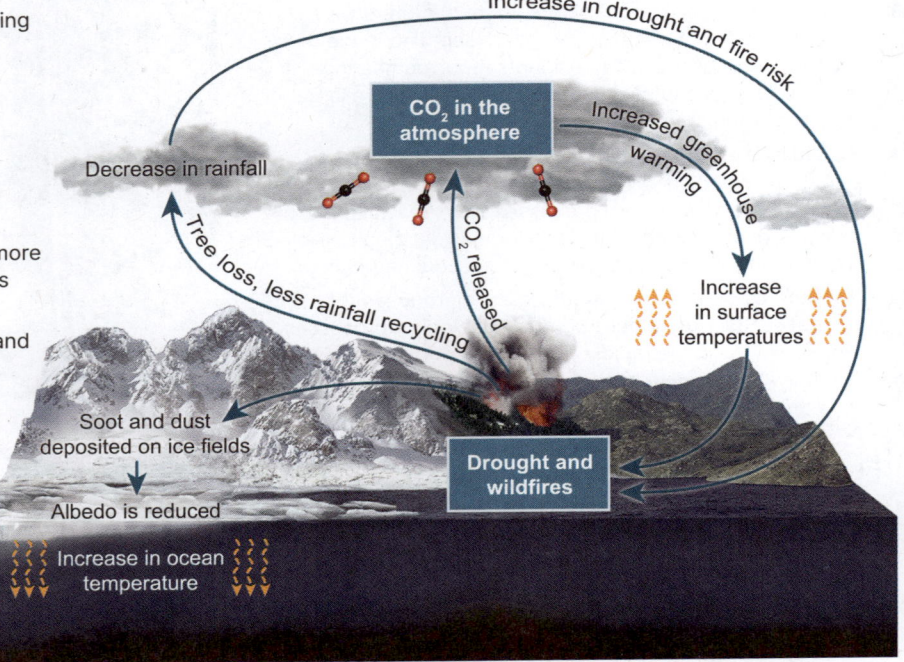

- Old boreal forests that have not experienced recent forest fires have a protective damp layer over 'legacy carbon' stored underground. The more frequent the forest fires, the shallower and less protective the layer, putting the organic material more at risk from burning in forest fires.

1. (a) Why might the southern regions of boreal forest be more at risk due to climate change?

_____

(b) What might you observe as the boreal forest reach its tipping point? _____

2. Arctic amplification, occurring mainly due to sea ice loss, is warming the region nearly two to three times faster than other regions. Boreal forest loss is one of several systems with potential tipping points in this region. Some scientists have proposed a tipping point cascade may occur between the ice sheets, boreal forests, and permafrost systems. How are the three systems connected by a tipping cascade and how might this impact the boreal forests?

_____

206 Tipping Point: Amazon Rainforest

Key Idea: The Amazon rainforest could reach a tipping point if enough is lost through climate change induced deforestation. The Amazon rainforest is located in tropical South America. It covers 5.5 million km² of land and is the world's largest continuous rainforest by far. A healthy rainforest draws water from the soil. Warming temperatures and shifting rain patterns due to **climate change** can increase the risk of forest fires, leading to deforestation in the Amazon rainforest.

Deforested land does not have enough plants to return water to the atmosphere through transpiration and most of the rain is lost in runoff. The resulting lower humidity affects forested areas downwind, such that less rainfall occurs, leading to degradation (continuing rainforest death). As a certain quantity of healthy rainforest is required to ensure the movement of precipitation, when a **tipping point** is reached, even healthy forest begins to be affected.

Feedback cycles and deforestation tipping points in the Amazon rainforest

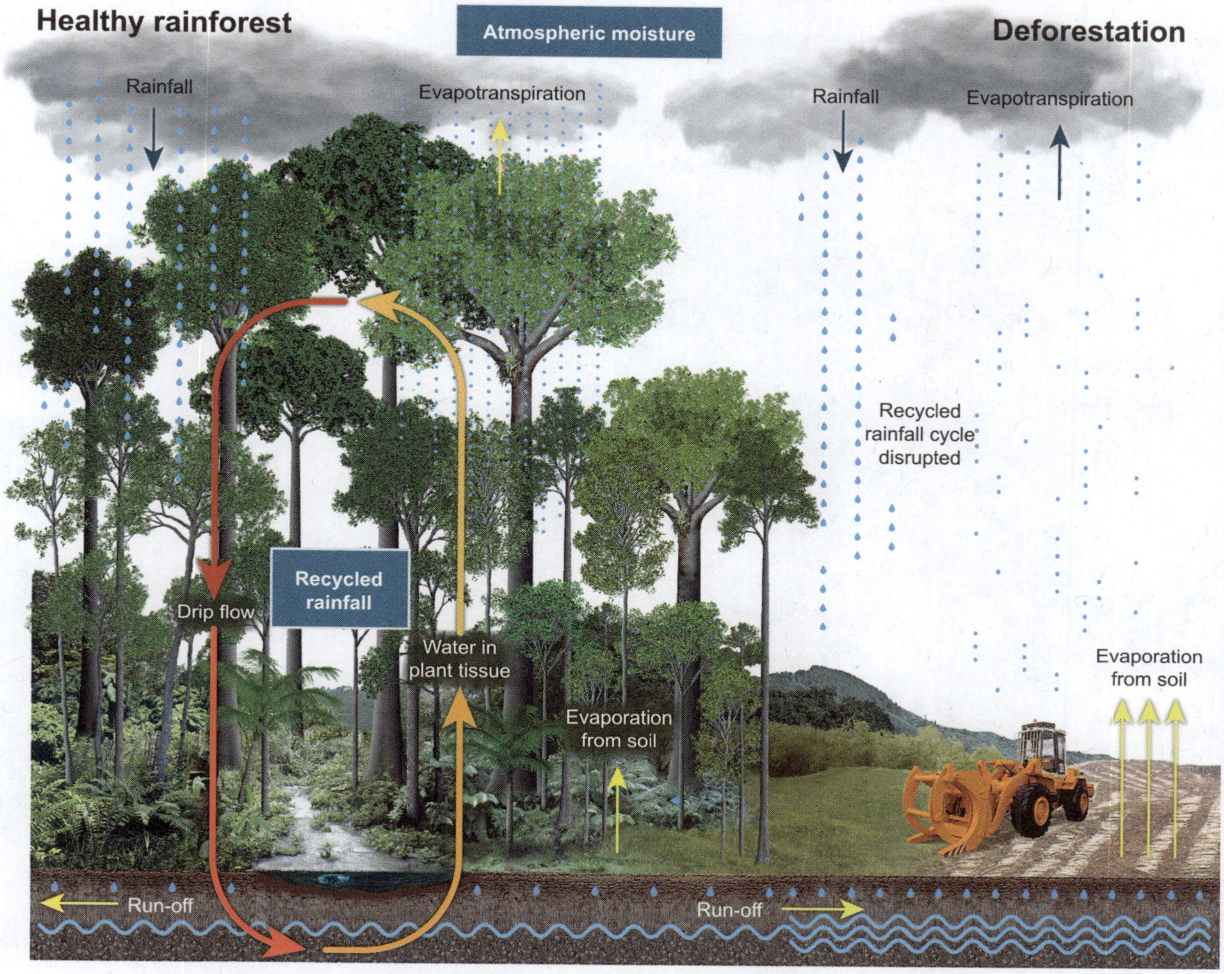

- The water circulates through the plants and returns to the atmosphere via transpiration, causing humidity. This humid air continues to circulate further through the forest in airflows.
- Estimates of upwards of 30% of the water falling on the Amazon forest has already been circulated through tree transpiration processes at least once.

1. Reaching a climate tipping point would shift the Amazon forest and boreal forests from a carbon sink to a carbon source. Explain the difference between the two terms, in the context of the forests:

   _____
   _____
   _____

2. How does climate change lead to an increase in forest fires in the Amazon rainforest and how does that process lead to a feedback mechanism at a regional level?

   _____
   _____
   _____

207 Tipping Point: Warm Water Coral Reefs

Key Idea: Warmer oceans are causing thermal stress to coral ecosystems, leading to wide-scale mortality.

An increase in sea temperatures could mean the death of coral reefs. Healthy coral reefs depend on the symbiotic relationship between a coral polyp that builds the reef and photosynthetic organisms called zooxanthellae. Zooxanthellae live within the polyp tissues and provide coral with most of its energy. A 1-2°C temperature increase maintained for weeks is enough to disrupt its photosynthetic enzymes. The zooxanthellae either die or are expelled from the coral. The result is **coral bleaching**. Some coral bleaching is reversible if water temperature cools once more to tolerance levels but this process is much slower than the original bleaching event. If the temperature remains outside tolerable levels, the coral dies. Coral bleaching events can affect more than just the organisms themselves. The coral is essential for transfer of energy through the food chain as it provides habitats for other species to live, breed, and be protected. Although the **tipping point** of each coral reef system is regional and dependent on the temperature of the surrounding ocean, the extent of coral bleaching world-wide has made it a global phenomenon.

Coral bleaching in the Great Barrier Reef

- Great Barrier Reef, off the coast of Eastern Australia, is the largest warm water coral system in the world. It is formed from a patchwork of over 2,900 individual reefs.
- The Great Barrier Reef has experienced five mass bleaching events since 2016, the latest in 2024.
- In some regions of the reef, only a few small areas of coral in deeper water were unaffected by bleaching.
- Record, warm ocean temperatures around the reef, linked to **anthropogenic climate change** by scientists, have coincided with past bleaching events.

Healthy corals, Great Barrier Reef, Australia.

Bleaching and dead coral on the Great Barrier Reef, Australia.

1. Summarize the link between ocean warming, and coral bleaching:

2. Coral ecosystems provide a habitat for around 25% of all marine organisms, including photosynthesizing plankton and bacteria, while only occupying 1% of the ocean. They act as a nursery for many open ocean species of fish. What would be some likely consequences of the coral reefs reaching their climate tipping point?

3. Ocean heatwaves are occurring more frequently. How does this affect a coral reef's ability to recover?

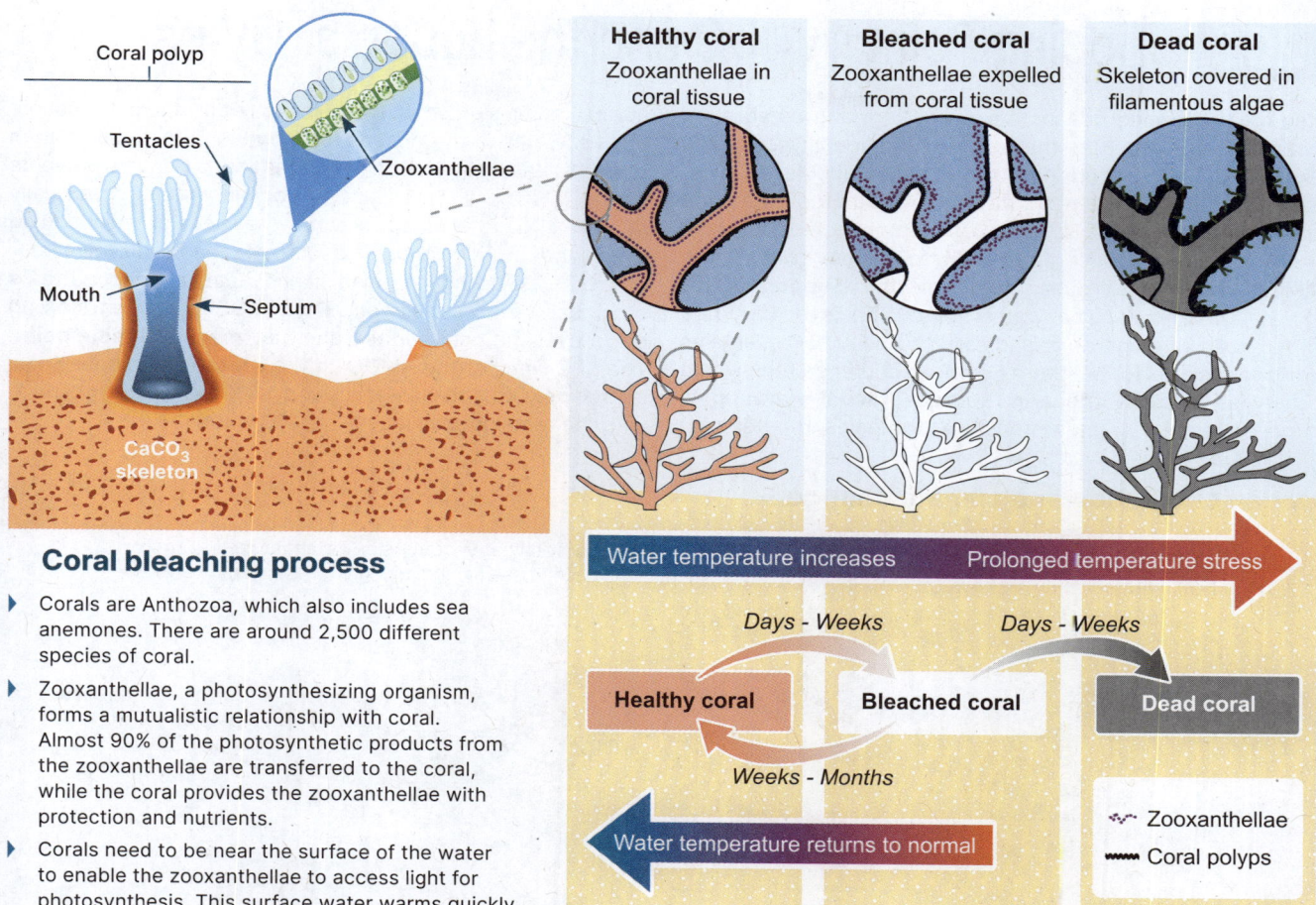

Coral bleaching process

▶ Corals are Anthozoa, which also includes sea anemones. There are around 2,500 different species of coral.

▶ Zooxanthellae, a photosynthesizing organism, forms a mutualistic relationship with coral. Almost 90% of the photosynthetic products from the zooxanthellae are transferred to the coral, while the coral provides the zooxanthellae with protection and nutrients.

▶ Corals need to be near the surface of the water to enable the zooxanthellae to access light for photosynthesis. This surface water warms quickly.

▶ Increased light and warmth increase the photosynthetic rate of zooxanthellae, overwhelming the coral polyps with waste material. The expulsion of zooxanthellae by the coral occurs as a stress mechanism to avoid tissue damage.

▶ The removal of zooxanthellae is called bleaching and the coral appears as a distinctive white color. This is because the zooxanthellae pigments give the coral their bright colors.

▶ Corals can survive in a bleached state for only a limited number of weeks and starve without the zooxanthellae.

Coral bleaching as a tipping point

Climate scientists project that around 70-90% of warm water corals will be lost once the global temperature threshold reaches 1.5°C for a sustained period. Around 99% of corals will disappear at just half a degree more.

Some corals are resilient and the zooxanthellae can return once ocean waters cool. However, once enough coral has died because of prolonged temperature stress, a tipping point will be reached where the coral ecosystem will fail to recover. Coral will not reproduce and spawn, therefore there will be no larvae to regenerate new coral colonies. The system will typically tip into a different algae dominated ecosystem.

4. Why does the death of the coral in an area often lead to a tipping point, while this is not necessarily the case with bleached coral?

_____
_____
_____

5. Why are warm water corals particularly vulnerable to ocean temperature rise?

_____
_____

6. Observe the images of coral reefs on the previous pages. What are some observable differences between bleached and healthy coral? Discuss in pairs and note your ideas below:

_____
_____

208 Tipping Point: AMOC and the Subpolar Gyres

Key Idea: Warming oceans are disrupting long established ocean currents such as the Atlantic Meridional Overturning Circulation (AMOC) and the North Atlantic Subpolar Gyre. The oceans transport heat, moisture, and nutrients around the globe through a system of interconnected currents. Surface currents move from cold to warmer water. The **AMOC** deep-water ocean currents, the **Subpolar Gyre** (SPG), and the surface water are distinguished by their differences in temperature and salinity. The AMOC maintains warmer climates for Northern European countries than would otherwise occur without it. Because **global warming** is experienced at a greater intensity at the poles, greater than normal melting of ice at the poles is releasing vast amounts of freshwater to the oceans. Freshwater is less dense than seawater and stratification of different density layers prevents mixing. This slows the sinking of ocean waters, specifically at the SPG, that loop around the oceans surrounding Greenland and Iceland. The currents carry dissolved O_2 and CO_2, decaying organisms, and heat, altering patterns of global ocean circulation. The AMOC has been slowing down over the past century and may reach a **tipping point**. The SPG 'limb' of the AMOC, in the Labrador-Irminger Seas, is predicted to collapse at around 1.8°C warming threshold, with the remainder collapsing at 4°C warming.

Ocean circulation and climate change

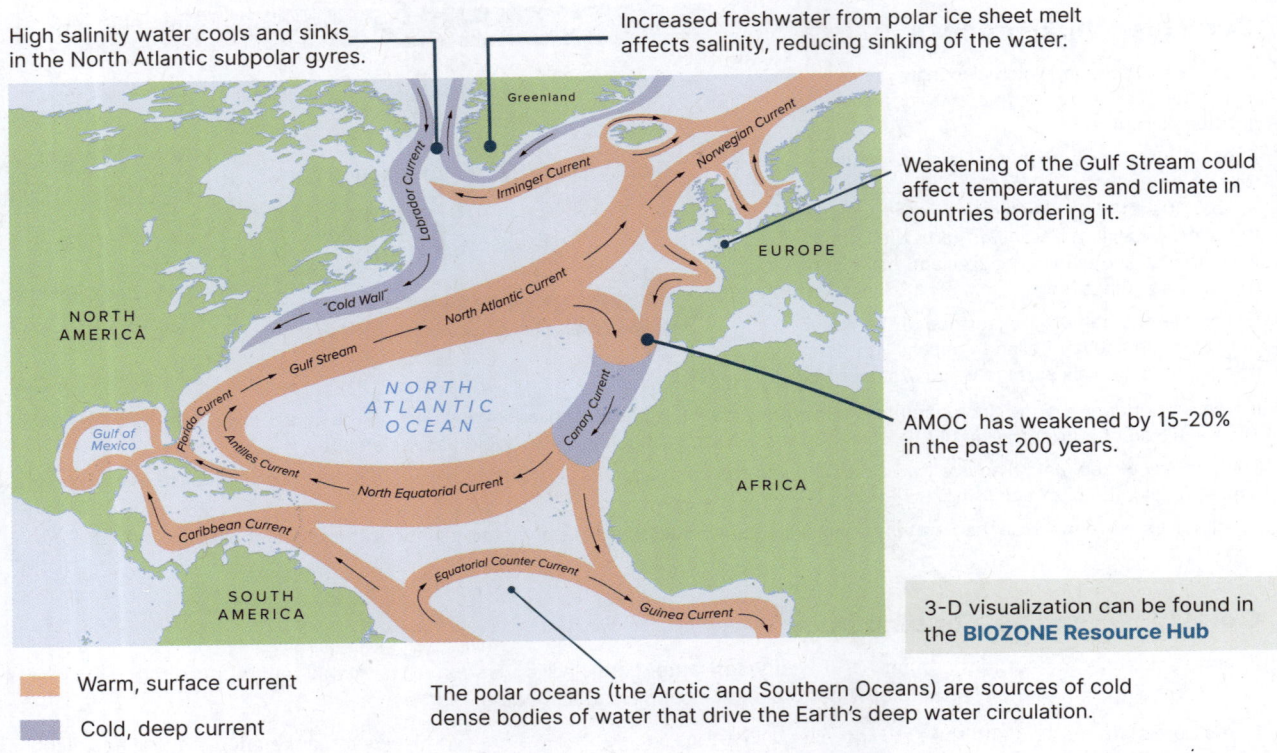

1. Higher global temperatures increase polar ice melt. Freshwater mixing into the North Atlantic oceans lowers salinity. Lower salinity water slows sinking, and slows the AMOC. A slower AMOC delivers less salty waters up to the subpolar gyres. Draw a flow chart showing this positive feedback mechanism. Share with other students and amend if needed:

The Little Ice Age

- The **Little Ice Age** (LIA) began in the early 1300s and lasted until 1850. It was a period of cooling restricted to Northern Europe, not experienced globally. It caused an average temperature drop of around 2°C.
- The LIA was responsible for crop failure, famine, and high death rates, due to the temperature lows in Europe being the coldest in over 10,000 years.
- Scientists have hypothesized that the AMOC underwent significant strengthening at the end of the **Medieval warm period** prior, delivering warm ocean water North and causing excessive ice melt in the polar region. Lowered salinity led to a collapsing AMOC and/or SPG 50 years later.
- Evidence of volcanic activity and wind system changes also occurred during this time which may have magnified the disruption. However, past disruptions of the AMOC and SPG highlight how vulnerable these systems are to change.

Abraham Hondius: Museum of London

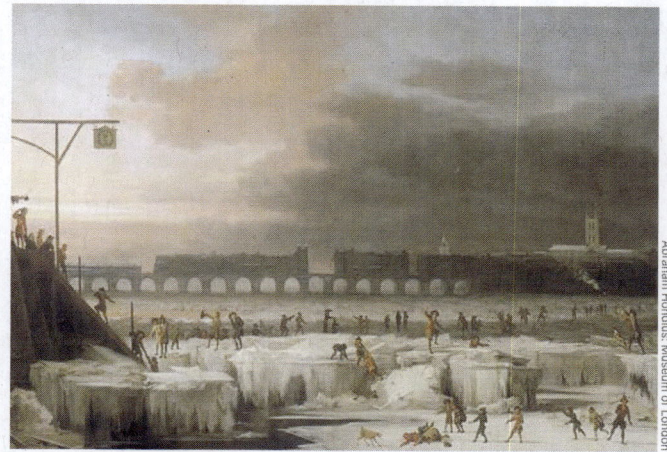

The frozen Thames during the Little Ice Age. It was cold enough for this English river to freeze solid for months.

- Analysis of tree ring data suggests that high solar radiation and a clear atmosphere from low volcanic activity led to the Medieval warm period prior to AMOC collapse in the Little Ice Age. This phenomenon was due to natural causes.
- In contrast, **anthropogenic** causes leading to global warming will be the trigger that leads to AMOC collapse once it reaches its tipping point. However, more research will be required for better understanding of how the AMOC and related SPG systems act as a tipping element.

Data source: Ed Hawkins CC BY-SA 4.0

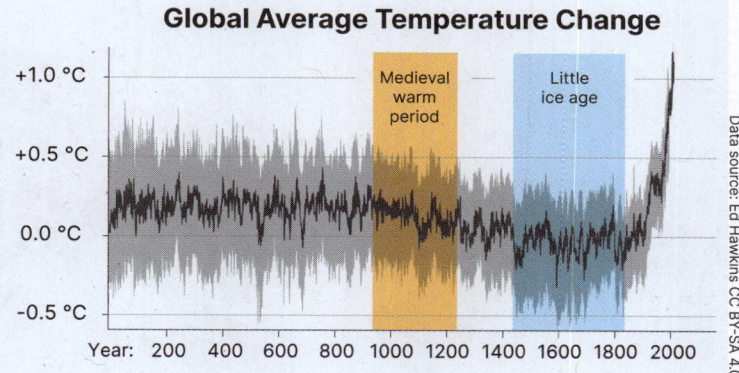

2. Summarize how the AMOC allows northern European countries, such as England and northern France, to have warmer climates than other countries at a similar latitude, such as Canada:

3. Why is salinity important in the Subpolar Gyre (SPG) for the AMOC to remain strong?

4. Climate change is associated with warming by most of the public who assume that places will become hotter. Yet some regions would instead become colder if the AMOC tipping point is reached. Write a persuasive paragraph targeted to your peer group to provide clarity:

ISBN: 978-1-99-101409-2

209 Climate Risk

Key Idea: Some countries are more at risk from climate change than others.

Countries are not evenly exposed to risk from current and projected **climate change** impacts. Some countries face more risk than others. The **IPCC** defines **climate risk** as a combination of three factors: hazard, exposure, and vulnerability. Some of the most at-risk countries are some of the lowest carbon emitters, such as the Pacific or Caribbean Islands, and many African countries. Some countries are facing increased likelihood of drought, fire, and heatwaves, e.g. USA and Australia, but have the financial means to adapt, so are less at risk. Countries facing high risk will need financial assistance from less vulnerable countries. Climate change has already created 'climate refugees', such as in Kiribati, where residents need to migrate to another country permanently. Climate change will also cause internal migration of people within a country in response to an extreme weather event or sea level rise around the coasts. Movement to cooler areas for growing food, or areas with a more secure water supply may be required.

Factors determining climate change risk across countries

Hazard is the potential for higher frequency and severity of physical phenomena, e.g. **sea level rise**, flooding, and **extreme weather events**, such as heat waves. **Exposure** relates to the proportion of population, ecosystems, and infrastructure that could be impacted by climate change. **Vulnerability** measures how well a country could adjust to climate change. It can be further divided into sensitivity affected by social and economic factors, including war, levels of poverty, and historical famine, and the coping capacity, or the resilience of a population to adapt to the impacts.

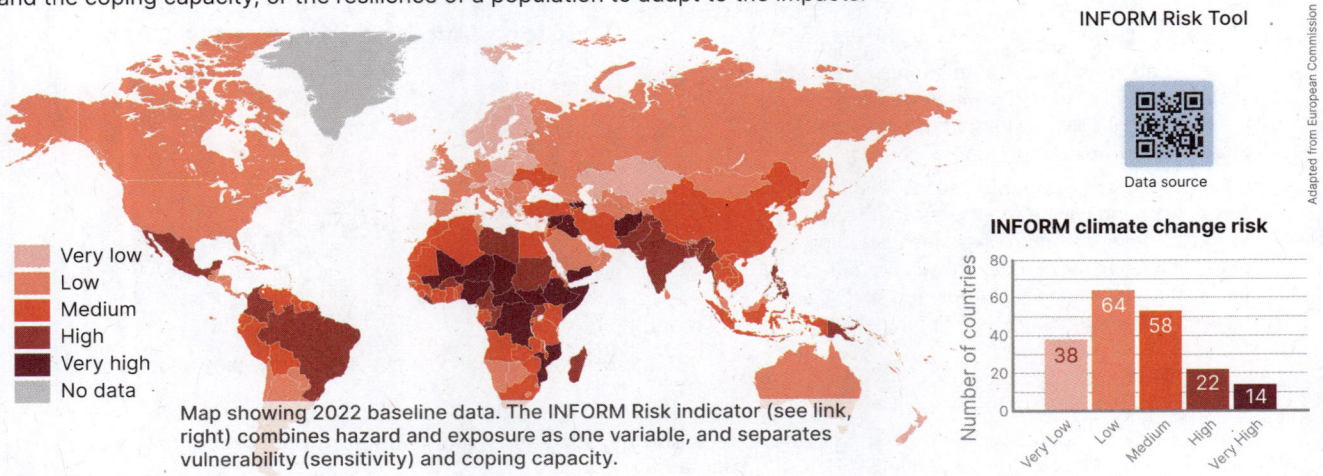

Map showing 2022 baseline data. The INFORM Risk indicator (see link, right) combines hazard and exposure as one variable, and separates vulnerability (sensitivity) and coping capacity.

INFORM Risk Tool

Data source

INFORM climate change risk: Very Low 38, Low 64, Medium 58, High 22, Very High 14

High climate risk countries:

▸ Low lying island nations are particularly vulnerable. Sea level rise is causing water to cover land used both for housing and agriculture. Fresh water supplies are becoming polluted with salt water, and disease rates are increasing.

▸ Equatorial African countries are more vulnerable due to their reliance on rain-fed agriculture. Increasing heat and drought is likely to result in lower agricultural productivity, leading to a lack of food and water security.

▸ Afghanistan faces increasing desertification, deforestation, and soil degradation. Poverty and other effects from wars limit the country's ability to respond.

▸ **Comparing climate risk in Arabian Peninsula countries:** Despite sharing similar arid environments and extreme weather risk factors, Yemen, Syria, and Iraq are all at high risk of climate change vulnerability compared to Saudi Arabia and Oman. The former three countries have been involved in numerous wars, leading to poverty and food insecurity. Saudi Arabia and Oman are wealthy countries, mostly due to fossil fuel reserves; they will be better able to cope with risks.

1. What three factors combine to make a country 'high-risk' to climate change?

   _____

   _____

2. What are climate change refugees? _____

   _____

3. (a) Select a country at high risk from the map above (updated data and risk information for each country can be accessed from **QR link / bit.ly** above). Record the country name and levels of overall risk. Research the factors impacting hazard & exposure (listing areas of higher risk), vulnerability, and lack of coping capacity. Record below:

   _____

   _____

   (b) Compare the climate risk values with another low risk country. Discuss the key differences with a classmate:

   _____

210 Climate Change and Agriculture

Key Idea: Productivity in agriculture is projected to decrease as the effects of climate change become more pronounced.

The effects of **climate change** will disrupt the productivity of agriculture and the location of where our food is grown. As the global temperature rises, some cereal crops such as wheat may be able to expand their range into areas that were once too cold. Growing seasons will begin earlier in the year and may be extended, especially so in northern America, Europe, and Asia. However, warmer and drier conditions in southern and equatorial regions will result in crop declines for corn and rice. **Extreme weather events** affect the available land for crop, dairy, and meat production. Global food production is responsible for around 35% of **anthropogenic GHG emissions**. However, some agricultural sectors, such as beef production, emit many more times the amount of greenhouse gas (GHG) than most plant crops to produce the equivalent quantity of protein for human consumption. While some agricultural changes may be required to compensate for shifts in growing conditions, an increased move to low carbon or 'green' practices in agriculture is also required to mitigate (action to reduce) GHG emissions and reduce future climate change severity.

Agricultural productivity changes due to climate change

Although some crops can counter climate shifts with a range extension to traditionally colder areas, other crops will be affected by reduced access to water and failure of reliable precipitation, and temperature levels above a tolerable level. Stock will also be affected by heat, loss of food supply due to extreme events and shortages, and water restrictions. Farmers may have to switch to alternative crops and implement water saving practices.

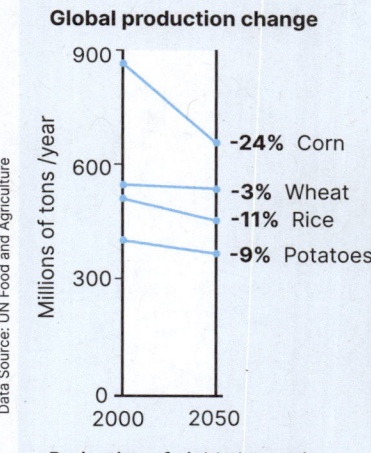

Projection of yield change in crops due to climate change.
- Corn: -24%
- Wheat: -3%
- Rice: -11%
- Potatoes: -9%

Grain crops (such as wheat, above) are at higher risk of crop failure if levels of precipitation decrease.

Crops grown near to their climate temperature threshold may suffer reductions in yield/quality, due to excessive heat.

Citrus production will shift slightly North with reduced yields in Texas and Florida.

Agricultural practices and the need for change

- According to the UN, the human population is projected to grow from 7.6 billion to a projected 9.7 billion by 2050. Demand for more animal-based protein from increasing prosperity in countries such as India and China will require production of nearly twice the current food crops as are grown currently to feed everyone.
- Currently, 45% of all habitable land on Earth is used for agriculture. Only 16% of this is used to directly grow crops eaten by humans; 4% is non-food crops, such as biofuels; and 80% is used to raise livestock (grazing and animal feed crops).
- A shift from livestock to crops would reduce GHG but would require an equivalent lifestyle shift to a more plant-based diet, especially in developed countries.

GHG emissions for agricultural productivity	
Protein source	Kg of CO_2 equivalents emitted per 100 grams of protein produced.
Beef	49.89kg
Lamb and mutton	19.85kg
Poultry meat	9.87kg
Pork	7.61kg
Farmed fish	5.98kg
Eggs	4.21kg
Grain	2.7kg
Nuts	0.26kg

1. Why will climate change benefit some agricultural crops, while disadvantaging other crops?

2. How can a change in utilizing more agricultural land for crop production, and less livestock, mitigate climate change?

211 Mitigation and Adaptation

Key Idea: Mitigation action aims to reduce the greenhouse gas emissions responsible for climate change. Adaptation prepares populations and ecosystems for projected impacts. The most recent projections of **global warming** show that increasingly significant and immediate mitigation efforts to reduce **GHG emissions** will be required to stay below the 1.5°C **Paris Agreement** target. For this to occur, **IPCC** states with high confidence that global **net carbon zero** (global emissions equal to or less than total carbon removal) will need to be achieved by the early 2050s. Once the global temperature rise continues over this 1.5°C target, the effects of **climate change** will become increasingly severe and many tipping points will be reached. Alongside mitigation, **adaptation** enables populations to cope with and avoid displacement when projected and present events occur as a result of climate change caused.

Mitigation strategies

- **Energy systems:** Rapid reduction in fossil fuel use. Replace with low-emission energy sources. Increase energy efficiency and conservation of appliances, transport, and machinery.
- **Industry:** Reduction in industry emissions by changes in production processes, circular material flow (more recycled or reused), recapture and storage of GHGs produced, and switching to renewable energy wherever possible.
- **Urban systems:** Electrification of urban areas, recycling and reduction of carbon-based resources, and incorporation of carbon uptake and storage options. Increased infrastructure to increase public transport, walking, and cycling in urban areas.
- **Buildings:** Design of buildings that capture and store carbon, including building materials and planting. Reduction of energy to construct and maintain buildings.
- **Transport:** Continued shift to plug-in electric vehicles (PEVs) and other transport means away from fossil fuels.
- **Agriculture and Forestry:** Shift to more sustainable and lower GHG methods in agro-forestry. Carbon sequestration (natural processes for carbon removal) by restoration of forests and peatlands acting as **carbon sinks**.
- **Policy and global cooperation:** Proposal and implementation of governmental laws, policy, and institutions focused on GHG emission reduction.
- **Technology:** Research and deployment at scale of **carbon capture and storage** (CCS) technology (Activity 214).

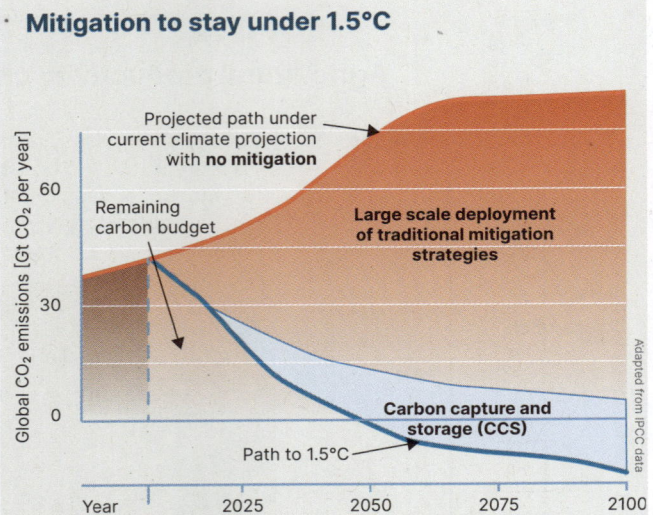

In order to remain under the 1.5°C Paris Agreement target, the combined global total of emissions (the carbon budget) must not exceed an already specified amount. The carbon budget also requires a carbon-zero global target by 2050. Based on our current emissions rate, we will quickly use up the last of the carbon budget. Traditional mitigation strategies will compensate for the majority of the projected emissions, thus preventing greenhouse gases from being emitted in the first place. However, it is likely that carbon removal solutions (CCS) will also be required to allow the 1.5°C target to be reached, capturing and sequestering carbon already in the atmosphere in the form of CO_2.

The new public library in Nord Odal, Norway has been built to zero heating standards, requiring no external source of energy to heat it aside from natural solar radiation. The building is designed to remove the need for emissions-releasing energy supply during winter months (when solar power is less effective). It uses heat retaining materials, quadruple glazing, and flexible shading, while retaining a comfortable climate inside for library users.

1. Why is mitigation action an essential component of response to climate change?

2. Why are 'zero heating' buildings considered mitigation?

Adaptation strategies

- **Coastal systems:** Building of seawalls and water barriers to counter **sea level rise**. Stabilizing coastal land (including planting).
- **Food security:** Efficient meat/livestock production and crop management, including alternative, heat resistant variants.
- **Ecosystem protection:** Sustainable fishery/aquaculture and agroforestry, management of biodiversity, and linking of conservation areas.
- **Water security:** Improvement of water use efficiency by shifting to lower water use processes. Adaptation to cope with future projected water shortfalls and droughts, both for agriculture and populations.
- **Infrastructure:** Building of housing and commercial spaces to naturally allow for cooling and/or release of heat. Shift to reliable and resilient energy systems.
- **Populations:** Preparation for health and disease epidemics, and response to weather disasters. Human migration and resettlement on a large scale may also be required when in-place adaptation is no longer feasible.

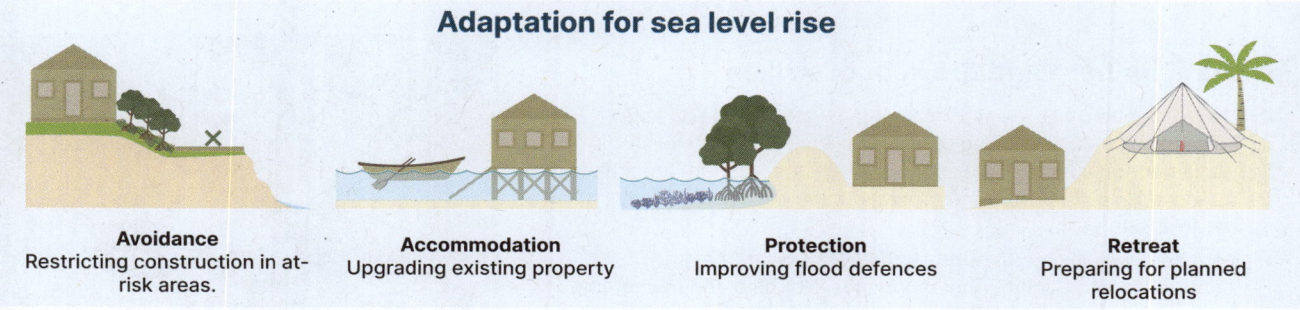

Adaptation for sea level rise

- **Avoidance** — Restricting construction in at-risk areas.
- **Accommodation** — Upgrading existing property
- **Protection** — Improving flood defences
- **Retreat** — Preparing for planned relocations

3. Why are climate change adaptation strategies still required if we are implementing mitigation strategies?

_____
_____
_____

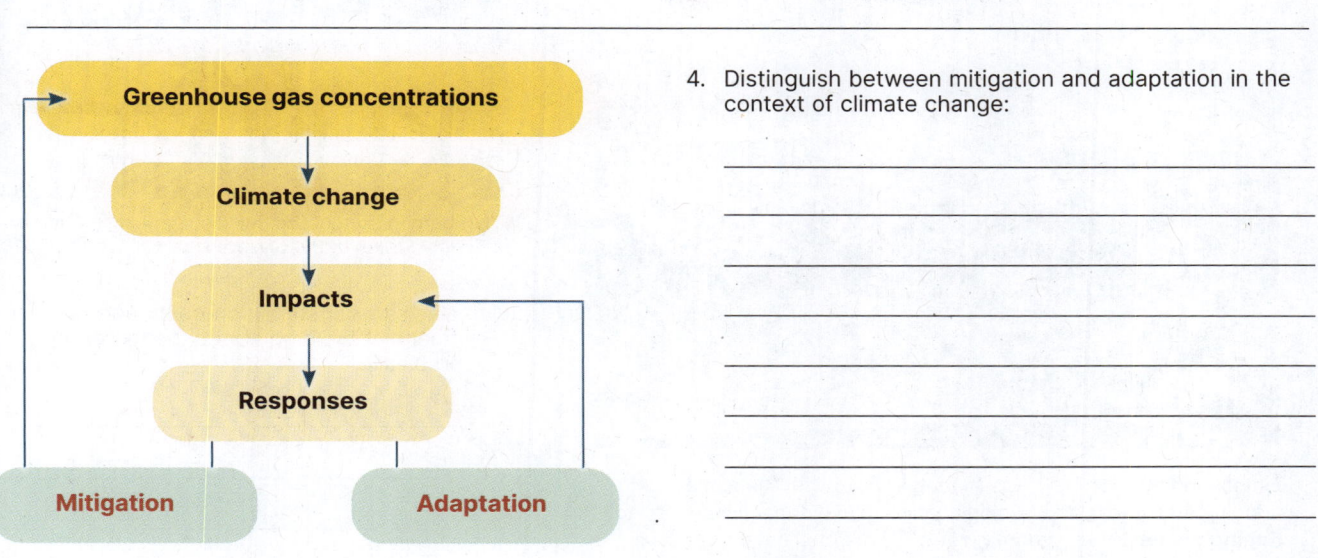

4. Distinguish between mitigation and adaptation in the context of climate change:

_____
_____
_____
_____
_____
_____
_____
_____
_____

5. Work in small groups to brainstorm any mitigation or adaptation strategies or policies that are being considered or currently implemented in your local region or country. You may need to research government climate change web pages to locate the information. Complete the chart below to summarize your findings.

Mitigation current	Adaptation current
Mitigation planned	**Adaptation planned**

212 The Climate Action Movement

Key Idea: Individuals and organized groups are participating in climate action to bring awareness to and force changes in government policy and business practices.

Climate action is a form of civil protest. It can involve just one individual, such as Greta Thunberg holding up a sign in her early school strikes as a teenager in Sweden. Large protests can also be successful at raising **climate change** awareness: the climate strike, led by Greta Thunberg, coincided with the 2019 UN Climate Summit and involved up to 6 million protesters across 150 countries. Climate action originated in the 1970s with the first Earth Day and continued into the 1990s with Climate Action Network, a coalition of over 1,300 environmental groups around the world, including Greenpeace and World Wide Fund for Nature. Groups like Extinction Rebellion have acted through non-violent civil disobedience. Terms like climate emergency and climate crisis have become synonymous with the protest groups, as they push their message of urgency to address the rising atmospheric CO_2 concentration and global temperature rise.

Greta Thunberg and the school strikes

▸ Greta Thunberg began her climate school strike protests as a 15 year-old in 2018, holding up a sign in front of the Swedish parliament on a Friday. She soon inspired other students to strike: up to a million students were protesting and striking on Fridays the following year.

▸ Thunberg was invited to address the UN Climate Change Conference and Climate Action Summits. Her bold and memorable speeches included the 2019, 'How dare you!' scold to world leaders that was reported widely in media. Thunberg continues to lead climate youth protests, write books, and speak up for climate change action, having been arrested a number of times for pushing her views.

Extinction Rebellion

Extinction Rebellion (XR) was formed in 2018 with the aim of bringing awareness and forcing change to avoid the climate change **tipping points**. XR uses non-violent civil disobedience protests as a means of gaining publicity for their cause. Past actions include occupation and blockages of buildings and infrastructure, and the use of mass arrests at protests. Like many climate action movements, XR includes youth in its protests to underscore how this group will disproportionately be impacted in future.

Thunberg in front of the Swedish parliament, holding a *School Strike for the Climate* sign, Stockholm, 2018.

1. Why can climate action be an effective means of bringing about change to the problem?

   _____
   _____
   _____
   _____
   _____
   _____

2. Why have 'school strikes for climate' become a significant way for youth to have a voice in climate action?

   _____
   _____

3. In 2023, Extinction Rebellion pivoted away from law-breaking action, such as seen by 'Just Stop Oil' when throwing paint at public art, into more 'lawful' peaceful protests. Suggest why they might they have made that decision:

   _____
   _____
   _____

213 Carbon Trading

Key Idea: Carbon credits can financially offset GHG emissions. Reducing **carbon dioxide emissions** is a challenge in a world mostly powered by carbon-based fossil fuels. One of the proposed strategies to compensate for CO_2 emissions is the **carbon trading** scheme based on carbon credits. Certain parts of industry, such as forestry and farming, can produce carbon credits by growing plant material to act as a **carbon sink**. Other companies can buy these carbon credits to offset the CO_2 they produce. Each 1000 kg of CO_2 is given a credit and these may be bought or sold on an exchange. The value of the carbon credit depends on the demand for and the quality of the credit.

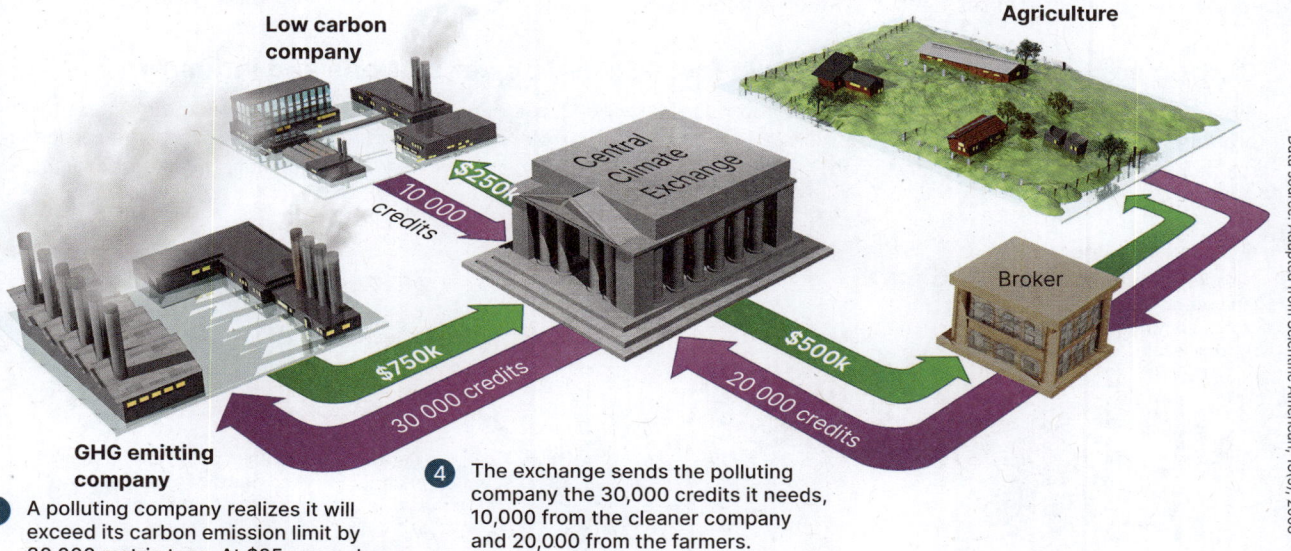

2 A cleaner company joins the exchange and produces 10,000 tonnes less CO_2 than its allowable limit. It sells this as carbon credits to the exchange for $25 a credit, receiving $250,000.

3 Farmers manage grassland so that they absorb more carbon than normal, qualifying for credits. Their broker sells 20,000 credits to the exchange for $500,000.

1 A polluting company realizes it will exceed its carbon emission limit by 30,000 metric tons. At $25 per carbon credit, it pays the exchange $750,000.

4 The exchange sends the polluting company the 30,000 credits it needs, 10,000 from the cleaner company and 20,000 from the farmers.

Data source: Adapted from Scientific American, 18(5), 2008

Advantages of trading carbon credits	Disadvantages of trading carbon credits
• Caps the amount of carbon dioxide emissions produced. • Makes polluting companies pay a penalty. • May force companies to become more efficient. • Companies that are less polluting or more efficient can sell their extra credits for financial benefit. • Projects, such as reforestation and peatland regeneration that act as carbon sinks, can be funded. • Renewable energy projects can be subsidized.	• As with any stock market, prices will fluctuate. Very high or low prices will remove incentives to use credits. • Many credits are not regulated and may not actually be producing any offsets at all. • The scheme may give large, highly profitable companies a free rein to emit **GHG** as they can easily afford credits and expenses being passed on to the consumer.

1. Explain how farmlands are able to generate carbon credits: _____

_____

2. Some large industries can invest in carbon offset projects, such as renewable energy, reforestation, or methane capture technology, often outside their own country. If the offset is more than the GHGs emitted, the company can receive carbon credits. How might carbon offset be a more effective way in reducing CO_2 emissions than carbon trading only?

_____

_____

3. Discuss the advantages and disadvantages of a carbon trading system: _____

_____

_____

_____

214 Carbon Capture and Storage

Key Idea: Carbon capture and storage prevent CO_2 emissions from being released into the atmosphere.

Coal, oil, and gas, all fossil fuels, are used by power stations to generate electricity. They are a major source of **greenhouse gas emissions**. The **IPCC** states that, in order to remain at or below the 1.5°C **Paris Agreement** limit, we must utilize **carbon capture and storage** (CCS) technology in addition to reducing emissions. CCS can be classified into three main groups: post-combustion technology, mainly used by fossil fuel power stations; pre-combustion technology, mainly used by industrial processes; and oxy-fuel combustion.

Captured CO_2 needs to be transported either via roads, by shipping, or through pipelines. It must then be stored long-term in carbon pools. Carbon storage is a type of technology-facilitated **sequestration**. Geologic sequestration can use deep underground rock formations, empty oil and gas reservoirs, coal beds, and saline (salt water) formations. Mineral sequestration uses a chemical reaction to bond carbon with minerals to form carbonates. Recent technology has also seen success in injecting CO_2 into deep sea geologic formations to react and form carbon compounds within the basalt rock.

Carbon capture and storage (CCS) in fossil fueled power stations and industry

Even power stations using high quality coal and oil release huge volumes of CO_2. Worldwide, electricity use and heat production release close to 50% of global CO_2 emissions each year. Manufacturing, construction, and industrial processes release an additional 20%. Generating 70% of total emissions, energy and industry are critical areas to target for CCS.

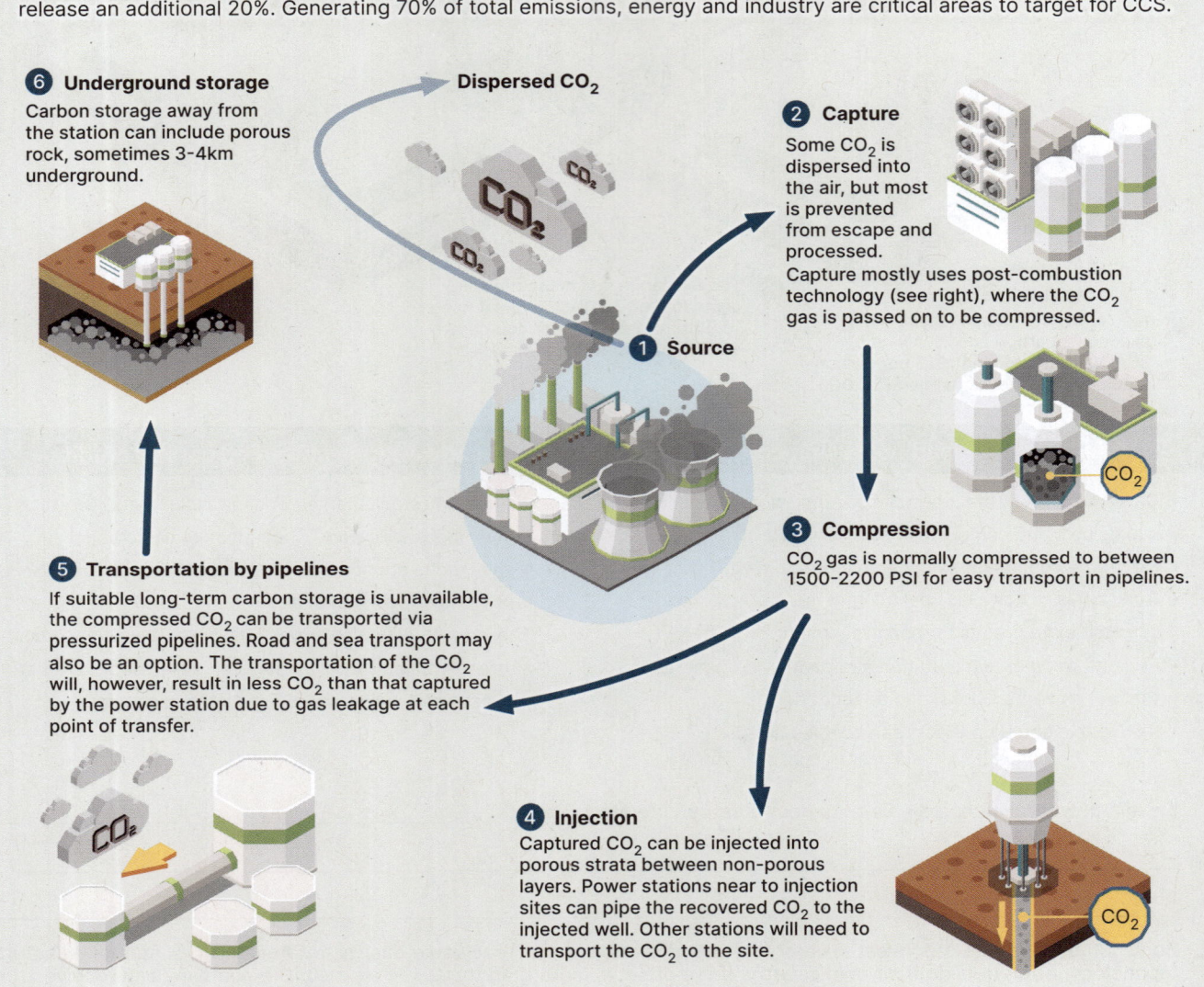

⑥ Underground storage
Carbon storage away from the station can include porous rock, sometimes 3-4km underground.

Dispersed CO_2

② Capture
Some CO_2 is dispersed into the air, but most is prevented from escape and processed. Capture mostly uses post-combustion technology (see right), where the CO_2 gas is passed on to be compressed.

① Source

③ Compression
CO_2 gas is normally compressed to between 1500-2200 PSI for easy transport in pipelines.

⑤ Transportation by pipelines
If suitable long-term carbon storage is unavailable, the compressed CO_2 can be transported via pressurized pipelines. Road and sea transport may also be an option. The transportation of the CO_2 will, however, result in less CO_2 than that captured by the power station due to gas leakage at each point of transfer.

④ Injection
Captured CO_2 can be injected into porous strata between non-porous layers. Power stations near to injection sites can pipe the recovered CO_2 to the injected well. Other stations will need to transport the CO_2 to the site.

1. Why is carbon capture and storage technology a vital component of reaching the Paris Agreement temperature targets, and ultimately carbon-zero (emissions are equal or less than carbon taken out of atmosphere) in 2050?

Three main carbon capture and storage (CCS) systems

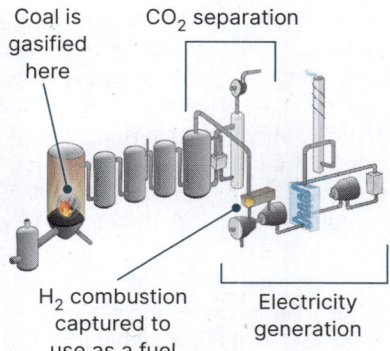
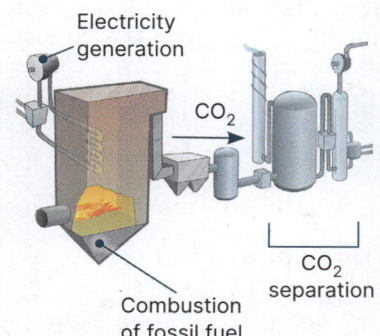
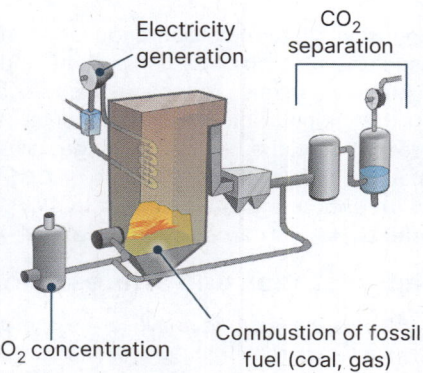

Pre-combustion capture:
The coal is converted to CO_2 and H_2 using a gasification process. The CO_2 is recovered while the H_2 gas is combusted. This process is more efficient than post-combustion, but equipment can be more expensive.

Post combustion capture:
CO_2 is washed from the flue gas after combustion. It is then passed to a desorber to re-gasify the CO_2, where it is then compressed for storage. This equipment can usually be retro-fitted to older power stations.

Oxyfuel combustion:
Concentrated O_2 is used in the furnace, producing only CO_2 in the flue gas. This is then compressed for storage. Compressed CO_2 is useful as an inexpensive, nonflammable, pressurized gas, e.g. for inflation and for carbonated water.

Storing captured CO_2

- Underground storage capacity in the United States can hold approximately 1.7 trillion metric tons of CO_2.
- Ideal storage locations include porous layers between non-porous rock, where the layers act as a seal to prevent gas from escaping.
- Basalt rock, a volcanic rock forming most of the bedrock of the oceans, is an ideal substance for carbon storage. The CO_2 chemically reacts with the rock minerals and is then stored indefinitely. This is a naturally occurring process that can take millions of years without technological assistance.
- The limitations of carbon storage result from location of sources, convenience for storage, scale of operation, and financial costs to inject and stabilize the carbon.
- Risks associated with CO_2 storage include potential ocean pH reduction with deep ocean storage and the risk of sudden CO_2 release from geological formations if the rock is unstable.

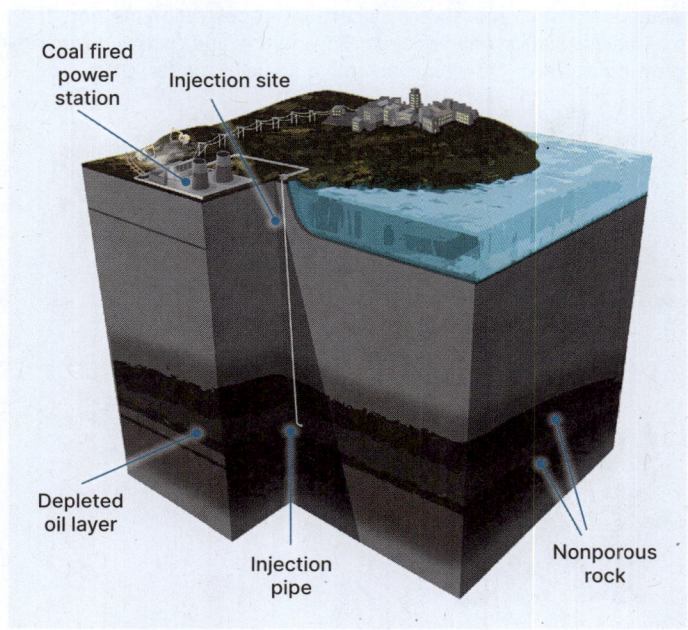

2. Describe the similarities and differences between the three types of carbon dioxide capture systems described above:

3. Some CCS are being developed to draw in carbon dioxide gas directly from the air, separating and storing the carbon dioxide, and then releasing the 'CO_2 free' air. Explain what natural process this is attempting to replicate?

215 Carbon Sequestration

Key Idea: Natural carbon sequestration in forestry can be used as a method of carbon capture and storage (CCS).

If greenhouse gas production was to completely stop today, Earth's temperature would continue to rise slightly. To prevent continued **climate change**, climate scientists agree that we need multiple methods of **CCS** to reduce our level of **greenhouse gas emissions** to slow the most damaging effects. Natural **carbon sequestration** by newly planted trees is a viable means of reducing atmospheric concentrations of CO_2. Technology can be used to develop a wide range of possible sequestration solutions, but natural ecosystem development approaches can be effective. They include reforesting bare areas (afforestation), regenerating depleted forest, and restoring peat wet-lands. Scientific research is still ongoing to compare the effectiveness of natural rewilding methods, and native and non-native plantations.

Reforestation and afforestation carbon sequestration

▸ For carbon storage to take place, there needs to be a mechanism to capture atmospheric CO_2 from the air. Plants naturally capture carbon through photosynthesis. They convert CO_2 into glucose which is further processed into other, carbon-based organic substances.

▸ Some carbon is stored in the soil when the plant decomposes, other carbon passes through the food chain when plants are eaten, and carbon is released through aerobic respiration as CO_2. The longer a plant lives, the longer it sequesters carbon.

▸ Plants grown in cooler areas, such as **boreal forests**, tend to grow much more slowly and are therefore more effective at sequestering carbon, compared with faster growing tropical plants.

▸ Damaged forest can be restored to tree covered ecosystems (reforestation). Bare land can be afforested with either native or plantation (non-native) crops as a method of increasing carbon sequestration. Rewilding occurs with little human input.

What trees should be planted for afforestation-driven carbon sequestration?

When afforestation is used as a carbon sequestration method, there are a number of planting options. These are: planting and maintaining native tree species; allowing the land to rewild with no intervention; growing a non-native short-rotation plantation; or growing a slower growing, non-native (usually conifer) plantation. The four different methods are represented below:

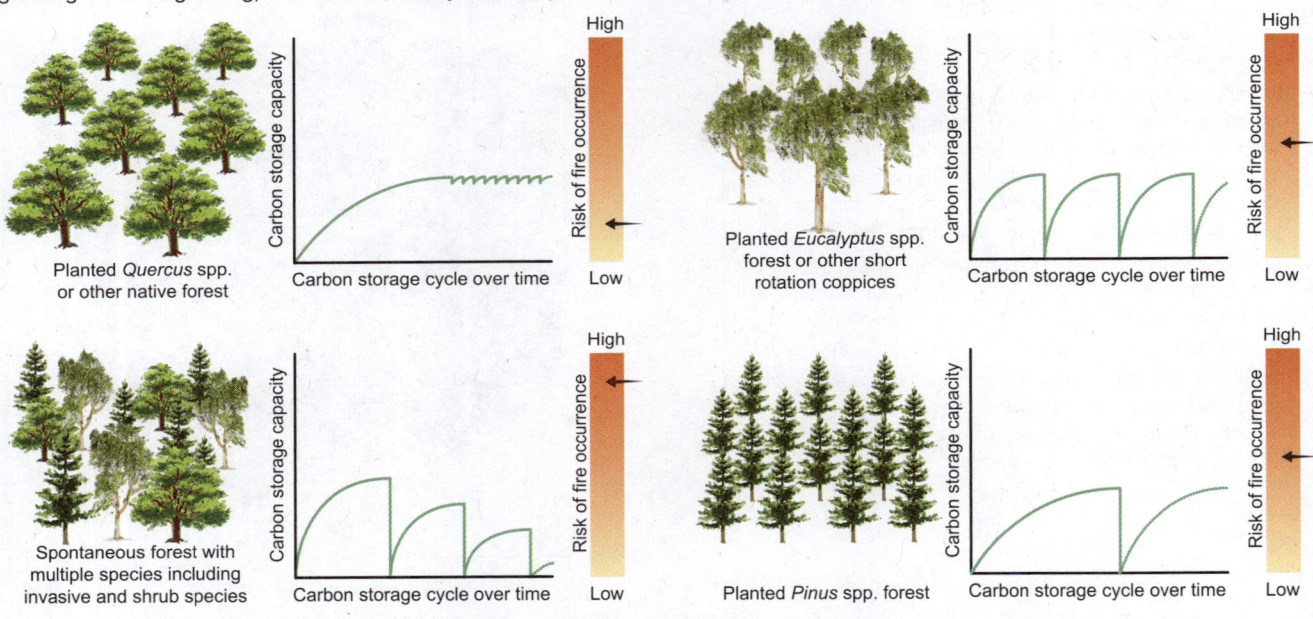

1. Which form of carbon sequestration appears to be the most effective and why? _____
_____
_____

2. Why is natural sequestration considered a complementary method of CCS with technology, rather than an alternative? _____
_____

3. How is natural sequestration linked to carbon trading (see Activity 213)? _____
_____

216 Carbon Footprints

Key Idea: A carbon footprint is a measure of greenhouse gases added by human activities. It can be measured personally, by household, by industry, or by country.

The carbon footprint per country can be compared by multiplying the footprint per person (or per capita) by the total population. High per capita footprint but relatively low population countries, such as Saudi Arabia, will therefore have a much lower footprint than China or India, both of which have lower per capita footprint but much larger populations. Households in high footprint per capita countries tend to have higher incomes, leading to larger houses, greater energy needs for heating and electricity, a diet richer in meat, frequent powered transport and travel, and increased purchase of goods. Wealthy countries also have the greatest capacity to reduce their footprint, by both personal and industrial change.

Carbon footprint across countries

- Carbon footprints are measured in tCO_2e, which is metric tons of CO_2 equivalent, enabling other **GHGs** to be accounted for.
- The current average carbon footprint across the world is around 4 tons equivalent of CO_2 per person per year. To stay below the **Paris Agreement** (1.5°C), this will need to halve to 2 tCO_2e by 2050.
- India has a very low per capita carbon footprint, correlating to a low average household income.
- An ethical dilemma occurs when large population /low income countries are encouraged to keep living standards low to enable global emissions targets to be reached, while households in wealthy countries expect a much higher living standard.

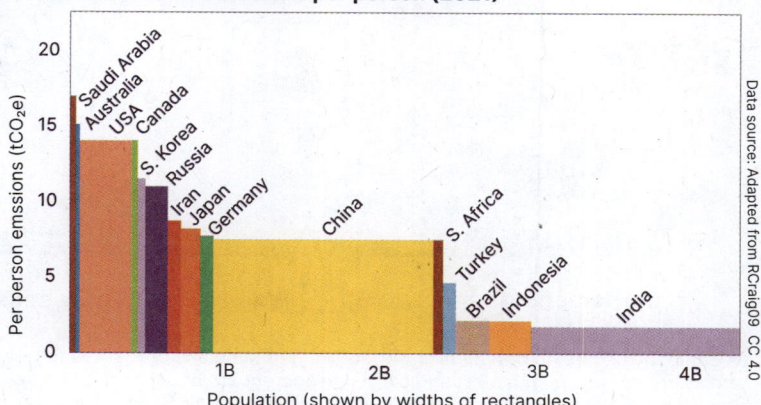

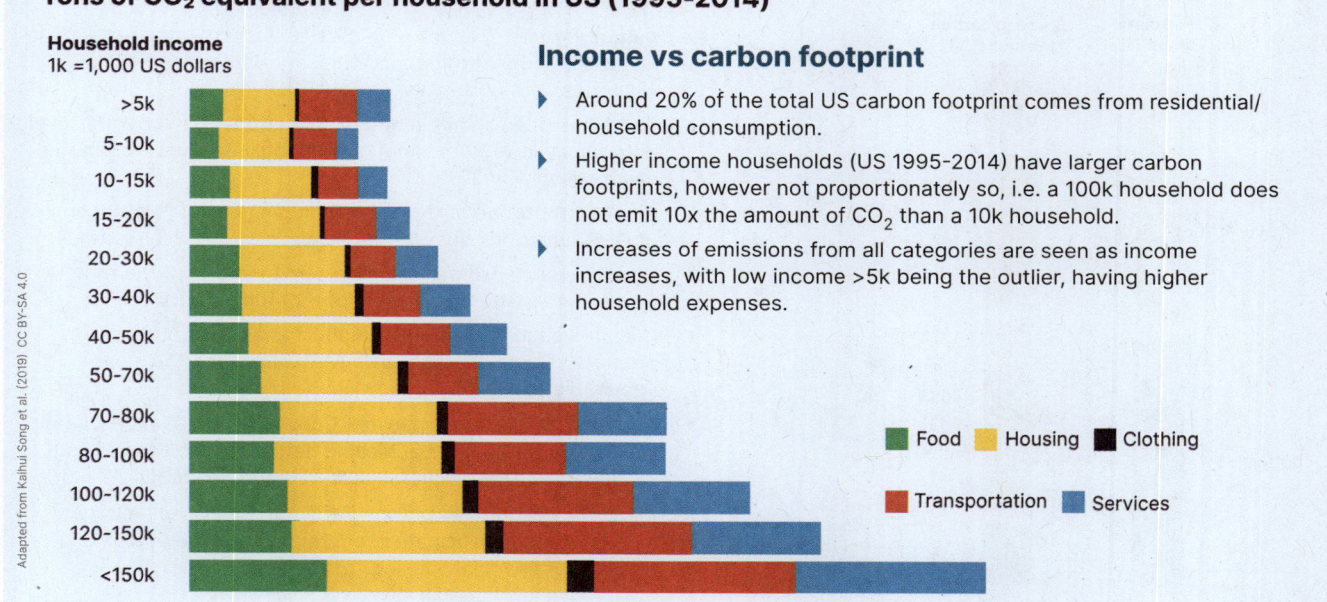

Income vs carbon footprint

- Around 20% of the total US carbon footprint comes from residential/household consumption.
- Higher income households (US 1995-2014) have larger carbon footprints, however not proportionately so, i.e. a 100k household does not emit 10x the amount of CO_2 than a 10k household.
- Increases of emissions from all categories are seen as income increases, with low income >5k being the outlier, having higher household expenses.

1. What are some barriers to reducing the carbon footprint of countries and households?

___
___
___

2. Brainstorm in small groups to suggest some solutions, including personal, to reducing the global carbon footprint:

___
___
___
___

217 Moving to Net Zero Carbon

Key Idea: Net zero carbon (emissions are less than or equal to carbon captured from the atmosphere) is a 2050 goal.

The **Paris Agreement** pathway to **net zero carbon** involves each country working towards decarbonization by reducing their CO_2 **emissions** each year. The decarbonization rates are pre-determined by each country's Nationally Determined Contributions (NDCs) (Activity 188), i.e. mitigation goals/obligations agreed to by the country itself. Fossil fuel emissions dropped in 2020 during the COVID-19 pandemic but rebounded the year after. However, this rapid reduction indicates that the world is able to pivot to significant decarbonization when required. Many countries have signed up to reaching carbon zero by or before 2050, and two countries, Bhutan and Suriname, have already achieved net zero goals. As up to 50% of global emissions are released from around 10% of the countries, **climate justice** advocates for a fair and equitable financial contribution from wealthier, high emitting countries to ensure carbon net-zero is reached.

A company can be carbon neutral without transforming any processes to reduce emissions at the source. Although carbon in and out are equal, carbon neutral status in itself will not be sufficient to limit **global warming** to the 1.5°C limit

What's the difference between carbon neutral or net zero?

- **Carbon neutral:** can be achieved by a business through purchasing carbon credits or investing in carbon offset projects, focusing solely on CO_2 emissions without emphasizing long-term reduction. It allows for a 'business as usual' approach to continue.

- **Net zero:** encompasses all greenhouse gases (GHG) and aims to reduce emissions at the source to residual levels. Achieving net zero requires significant changes in industrial processes but is considered a more effective strategy for limiting global warming to 1.5°C.

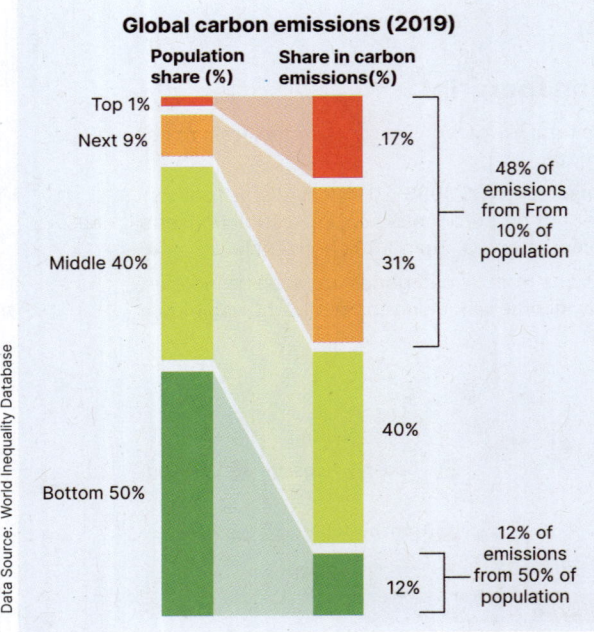

Addressing emission inequity

- **Inequity** refers to a lack of fairness or justice, often resulting in unequal treatment or opportunities for different individuals or groups.

- There has been a shift in **emission inequity** from country-level differences in 1990 to variations among individuals within the same country by 2019.

- A small proportion of the population emits a much larger share of GHG emissions due to lifestyle choices and wealth status.

- Wealthy people from all countries emit more GHG than the poorest, no matter the wealth level of the country.

- Climate action and financial support at a global scale can promote climate justice, particularly benefiting the most vulnerable countries.

- The United Nations Framework Convention on Climate Change (UNFCCC) emphasizes achieving net zero while considering climate justice, ensuring fairness for countries at risk.

- Equitable net zero, as per the Paris Agreement, encourages all countries to participate in emission reduction efforts and receive fair financial backing.

1. How does climate justice lead to a fair pathway to carbon zero for all countries? _____
_____
_____

2. Research the progress of your country towards reaching their 2030 goals carbon zero goals (Data source link provided in **BIOZONE Resource Hub** if required). Summarize your country's current pathway (policies and action), NDC target, and gap to 1.5°C:
_____
_____
_____
_____

Net zero carbon cities

▸ Currently, around half of Earth's population lives in cities that are spread over only 3% of the total available land surface. Scientists project that city or urban occupation will increase to 68% of the global population by 2050.

▸ Of the world's GHG emissions, 70% (including energy use) are generated to supply the people and industries in cites. Therefore, significant changes in how urban areas are designed and built, supplied with and use energy, and how people are transported around, will be required to reach the 2050 global net zero carbon goal.

▸ Some cities, such as Oslo in Norway, are making good progress towards becoming emissions free. Over 98% of all electricity in Norway is generated by renewable sources, mostly hydropower. Public transport is shifting to becoming 100% electrified and almost all new car sales are plug-in electric vehicles (PEVs), both in the city and the country. The government eliminates many car and fuel taxes on PEV owners in addition to removing parking fees and road tolls, while making numerous free charging stations available in the city and around the country.

3. Suggest some barriers to implementing the 'Norway' system in other countries?

___

4. Although the PEV share in some countries, such as the USA, is much lower than Norway, why might that be an advantage when considering a shift to net zero?

___

PEV in use as a share of all passenger vehicles (Dec, 2022)

Top markets by PEV market share
Top markets by PEV sales volume

- USA: 1.3%
- World: 2.1%
- Europe: 2.4%
- Germany: 4.0%
- China: 4.9%
- Norway: 27.0%
- Iceland: 16.0%
- Sweden: 8.8%
- Denmark: 7.8%
- Netherlands: 5.6%
- Finland: 5.4%

Data source: Adapted from Mariordo CC 4.0

The concept of the 20 minute city

▸ The 20 minute city is a concept in urban planning that aims to reduce dependency on personal vehicle use. All daily activities including shopping, school, and work should be done by walking, cycling, or public transport and take no more than 20 minutes.

▸ Residential areas should have a range of affordable housing types to suit everyone.

▸ Healthcare could be offered via video consultations, where appropriate, with actual trips to healthcare facilities only occurring when necessary.

▸ Transportation via private, fossil fuel powered vehicles is a large contributor to GHG emissions. Cutting down on a need for private vehicles would help mitigate the issue.

▸ Having schools within a 20 minute walk for children would remove the congestion caused by the school drop off/pickup.

▸ The urban heat-island effect could be reduced by inclusion of green spaces. These would also ease flood risk as green spaces absorb rainwater. Accessible green spaces that also act as community areas contribute towards people's mental wellbeing.

5. Sustainable energy cities with low emissions transport are key to moving towards net zero carbon, but what other considerations need to be made when designing and building in future?

___

6. What changes have you, or will you, make in your own personal capacity to lower your carbon footprint?

___

218 Possibilities of Solar Radiation Modification

Key Idea: Solar radiation modification (SRM) is technology that reduces the amount of incoming solar radiation.

Solar radiation modification (SRM) or solar geoengineering, is a group of technologies that reflect or prevent incoming solar radiation, mostly light, from reaching Earth's surface. Although not a substitute for **climate change mitigation**, the appropriate use of this technology may temporarily reduce global temperature overshoot of the 1.5°C target until **CCS** and other **GHG emission** reductions are deployed at larger scale. Much of this technology is still in the initial development phase or at small scale, and would require strong cooperation between countries to begin use. The SRM that 'seeds' clouds with aerosols to increase reflectivity replicates a naturally occurring phenomenon during volcanic eruptions but the environmental impact of artificial SRM is still uncertain. SRM does not reduce GHG emissions, therefore, effects related to climate change such as **ocean acidification** would not be reduced. Only technologies that remove GHG from the atmosphere are now classified as mitigation strategies by **IPCC**. Scientists propose that careful use of SRM, in tandem with CCS technology and **sequestration**, will allow peak **global warming** to be reduced in the near future. Research estimates each 1.0°C of global cooling from SRM could cost tens of billions (US dollars) each year. However, the technology may be an important last resort, where the need to temporarily reduce temperature outweighs risks and cost.

Stratosphere aerosol injection

Stratospheric aerosol injection (SAI) technology uses particles which combine with water and reflect solar energy. Planes and hot air balloons are used to seed clouds in the stratosphere.

Space mirrors

This SRM is still at the theoretical phase, and uses the reflective ability of the mirror surface (with very high **albedo**) to prevent solar energy from entering Earth's atmosphere. The space mirrors would be huge structures, many kilometers long. They could be positioned over vulnerable regions, like the poles, but would be very expensive to deploy at any scale.

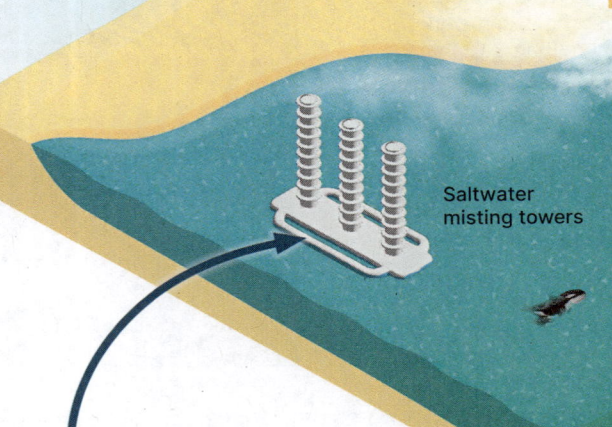

Drones

Saltwater misting towers

Marine cloud brightening

This SRM is a localized process where harmless sea salt is made into an aerosol and 'shot up' into low lying marine clouds. The US conducted its first trials of this technology in April 2024. The clouds become temporarily brighter, increasing their albedo and enabling more solar energy to be reflected back into space. The equipment needs to be close to a source of salt water, so is likely to be fixed in place, and is limited to the height the sea salt can be shot up. Although considered a safer form of SRM, there are still concerns about unintended impacts to the climate.

Cirrus cloud thinning

Cirrus clouds are thin and wispy and form above 6km. They retain heat and prevent it from escaping higher into the atmosphere or space. The SRM process would 'seed' the clouds with aerosol particles to form ice crystals. Drones could be used to deliver the particles. The clouds would thin-out as a consequence and allow more heat to escape. This process would also allow more light to enter through them, but on balance they would induce more cooling than heating. Like all SRM, scientists are still unsure how deployment might impact the ozone layer and weather patterns, or if it might cause a harmful cooling overshoot.

©2025 BIOZONE International
ISBN: 978-1-99-101409-2
Photocopying prohibited

SAI deployment

- Volcanic eruptions release aerosol sulfates into the atmosphere, causing temporary pollution that leads to an albedo effect, reducing incoming solar radiation. Historical records and ice core data link major eruptions to rapid cooling events, such as the significant cooling in 536 AD, associated with death, famine, and a dusty haze. Similarly, the Pinatubo eruption in 1991 cooled the Earth by approximately 0.5°C in the subsequent months.

- Stratospheric aerosol injection (SAI) technology aims to mimic the natural dimming effect caused by volcanic particles.

- This SAI would need to be deployed at a large scale, and specifically in a direction that has the least impact on precipitation or swings in extreme temperature, and for an extended period of decades.

- The technology is still largely untested and faces scientific and public backlash due to uncertainties around the potential for **extreme weather events**, acid rain, and ozone layer damage.

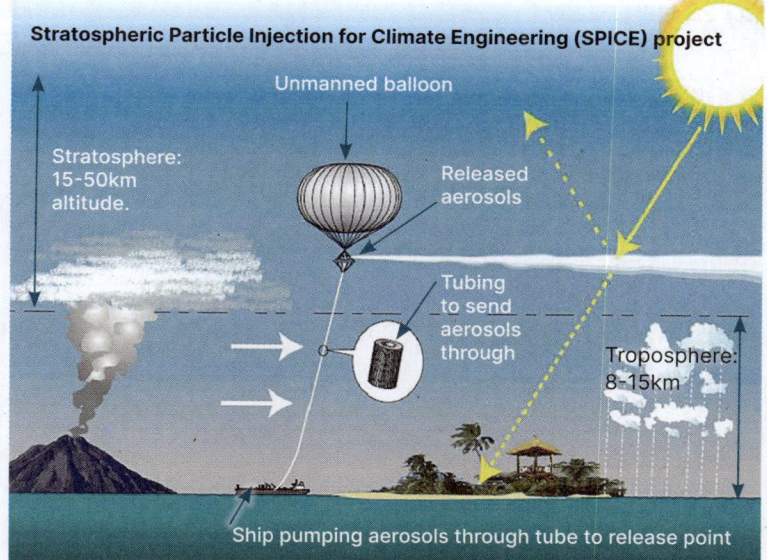

The Stratospheric Particle Injection for Climate Engineering (SPICE) was an early adopter UK project investigating how particles released into the stratosphere could replicate volcanic aerosols.

1. How is SRM technology different from Carbon Capture and Storage in the way it reduces global warming?

2. Reduction to net zero carbon will require both GHG emissions cut and CO_2 removal through CCS and sequestration. How does SRM 'bridge' the gap to reduce the projected potential climate impacts?

3. Discuss the advantages and disadvantages of using SRM to reduce the impact of climate change:

4. Why might replicating a natural phenomenon in SRM be better than a more conceptual technology, e.g. space mirrors?

219 Did You Get It?

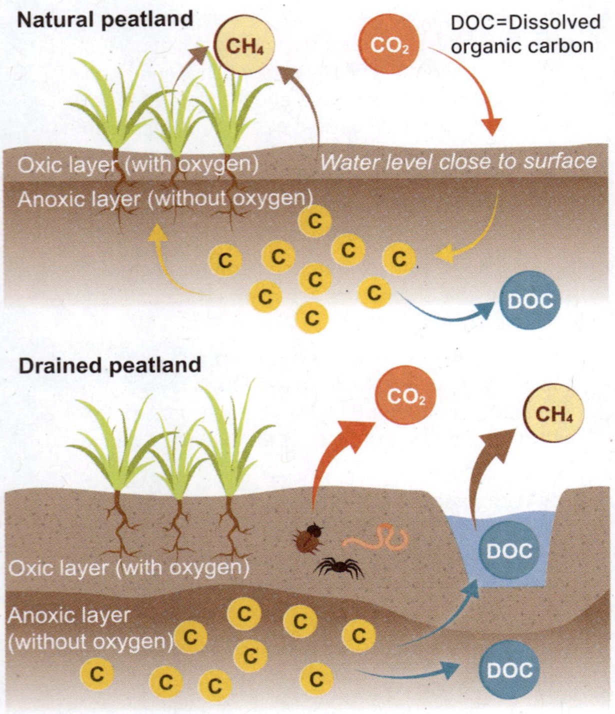

1. (a) The image to the right shows drained peatland for agriculture (bottom) and restored peatland (top). Drained peatland is a net carbon emitter. What does this imply and how does it differ from natural peatland?

___

(b) Why is CO_2 considered greenhouse gas (GHG)?

___

(c) What is the link between GHG and climate change?

___

2. (a) Restoring peatlands is one form of carbon sequestration. What does this term mean and why is it considered a climate change mitigation rather than an adaptation?

___

(b) Restoring peatlands could be used as a carbon offset project. Explain this process:

___

3. (a) Frozen peatland, or permafrost, is thawing due to global warming. This creates a positive feedback cycle. What is a positive feedback cycle in climate change, and why is it an issue?

___

(b) Permafrost is one system that could reach a climate tipping point. What is a climate tipping point?

___

4. (a) Climate change legislation is being used to reduce GHG emissions to limit global warming and reaching tipping points. What is the key agreement called, and what are the goals of it?

___

(b) Climate justice is an important aspect of the legislation. What is climate justice and why is it required?

___

Chapter 10
Science Practices

Resource Hub
bit.ly/3S18k3w

Key Terms
- continuous data
- data
- discontinuous data
- mean
- median
- mode
- model
- qualitative data
- quantitative data

Key Concepts
- Models allow parts of a system to be represented in a simpler form.
- The graphical presentation of data depends on the type of data collected.
- Transforming numbers can make them easier to work with.
- When the changes in two variables correlate, it does not necessarily mean that the change in one variable is causing the change in the other.

Science Practices

Learning Outcomes:

			Activity Number
☐	1	Explain the purpose and use of scientific models. Describe the different types of models.	220
☐	2	Classify data as quantitative, ranked, or qualitative.	221
☐	3	Evaluate the suitability of collecting qualitative or quantitative data in different types of investigations.	221-222
☐	4	Calculate mean, median, and mode, from provided data.	222
☐	5	Identify which type of graph to use when displaying specific information.	223
☐	6	Analyze information in tables and graphs.	224
☐	7	Plot a line graph from provided data.	224
☐	8	Convert between decimal and standard form in given numerical values.	225
☐	9	Use exponential and logarithmic functions to analyze exponential and logarithmic data.	225
☐	10	Convert between different rates and units to analyze data and information.	226
☐	11	Distinguish between correlation and causation in data.	227

220 Models and Modeling

Key Idea: Models are a way of representing data and ideas in a way that can help understanding.

Scientists often used diagrams and **models** to learn about environmental systems. A model is a representation of a system and is useful for breaking a complex system down into smaller parts that can be studied more easily. Often, only part of a system is modeled. As more information is gathered, more **data** can be put into the model so that eventually it represents the real system more closely. How the data is represented depends on the concept being expressed.

Modeling data

There are many different ways to model data. Often, seeing data presented in different ways can help to understand it better. Some common examples of models are shown here.

Mathematical models

Displaying data in a graph or as a mathematical equation, as shown below for logistic growth, often helps us to see relationships between different parts of a system.

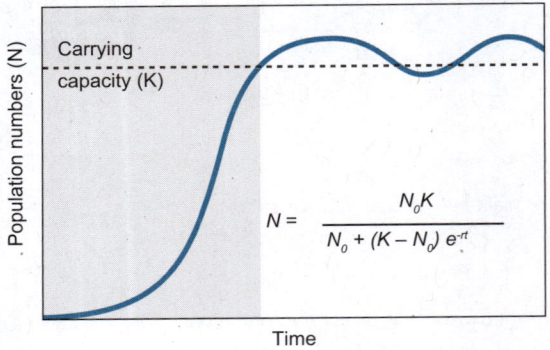

$$N = \frac{N_0 K}{N_0 + (K - N_0) e^{-rt}}$$

Analogy

An analogy is a comparison between two things. Sometimes, comparing an environmental system with a simpler object enables us to better understand it: the effect of Earth's atmosphere on the temperature is sometimes compared to the way a greenhouse produces a warm environment.

Visual models

Visual models can include drawings. These might be diagrams showing the movement of energy or matter, or a simplified picture of an object.

Diagram of nutrient cycling in an ecosystem

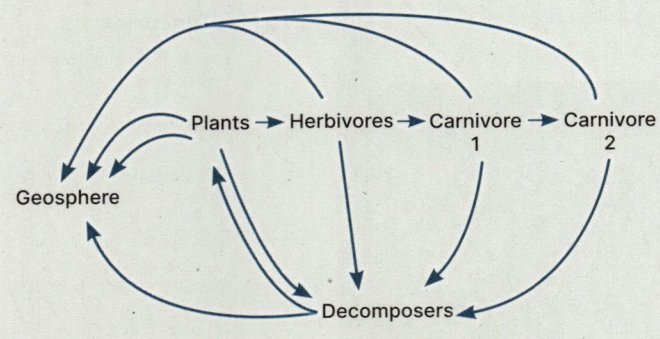

3D model of the Earth's layers

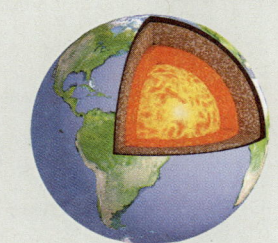

Three dimensional models can help us understand the spatial arrangement of layers and parts of the system.

1. What is a model? _____

2. Explain why a model is never a 100% accurate representation of the system being studied: _____

3. What are the advantages and disadvantages of using models to explain a system? _____

4. Why is it easier to use a series of simple models to explain a system that using one large complex model? _____

221 Types of Data

Key Idea: Different types of data are recorded and displayed in different ways.

Data is information collected during an investigation. Data may be **quantitative**, **qualitative**, or **ranked**. When planning a biological investigation, it is important to consider the type of data that will be collected. It is best to collect quantitative or numerical data, because it is easier to analyze it objectively (without bias).

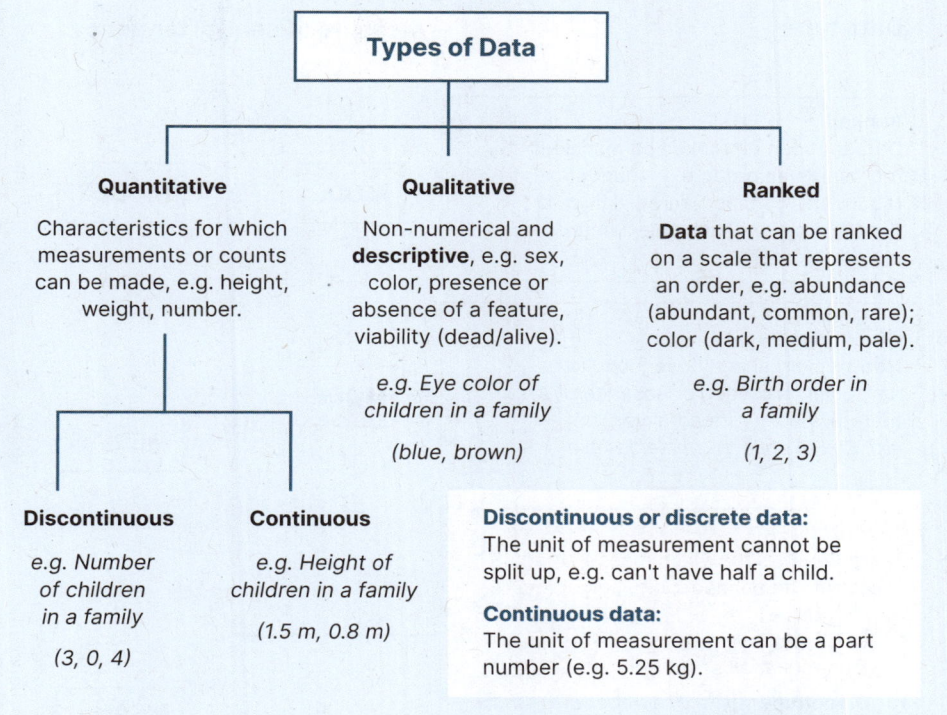

A: Flower color

B: Wind turbines per site

C: Tree trunk diameter

1. For each of the photographic examples A–C above, classify the data as quantitative, ranked, or qualitative:

    (a) Flower color: _____

    (b) Wind turbines per site: _____

    (c) Bacterial colony diameter: _____

2. Why is it best to collect quantitative data in biological studies, where possible? _____
_____
_____
_____

3. Give an example of data that could not be collected quantitatively, and explain your answer:
_____
_____
_____
_____

©2025 **BIOZONE** International
ISBN: 978-1-99-101409-2
Photocopying prohibited

222 Mean, Median, Mode

Key Idea: Descriptive statistics such as the mean, median, and mode can be used to describe and analyze data.

When describing a set of **data**, it is usual to give a measure of central tendency. This is a single value identifying the central position within that set of data. Descriptive statistics, such as **mean**, **median**, and **mode**, are all valid measures of central tendency, depending of the type of data and its distribution. They help to summarize features of the data, so are often called summary statistics. The appropriate use of these statistics is shown below:

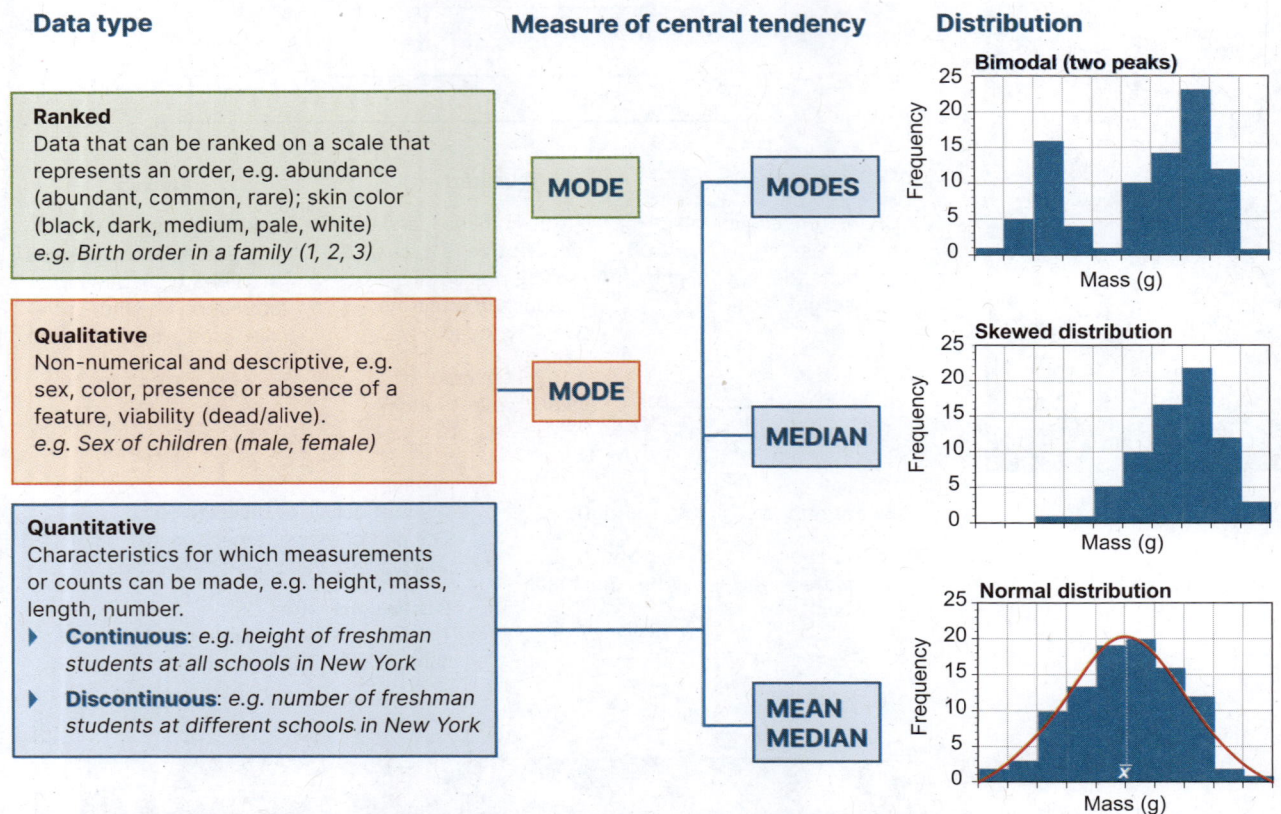

Statistic	Definition and when to use it	How to calculate it
Mean	• The average of all data entries. • Measure of central tendency for normally distributed data.	• Add up all the data entries. • Divide by the total number of data entries.
Median	• The middle value when data entries are placed in rank order. • A good measure of central tendency for skewed distributions.	• Arrange the data in increasing rank order. • Identify the middle value. • For an even number of entries, find the mid point of the two middle values.
Mode	• The most common data value. • Suitable for bimodal distributions and qualitative data.	• Identify the category with the highest number of data entries using a tally chart or a bar graph.

1. Match the data distribution shown above (Bimodal, Skewed, Normal) with their description below:

   (a) Data is spread symmetrically about the mean. It has a classic bell shape when plotted: _____

   (b) Data which has two peaks: _____

   (c) Data is not centered around the middle but has a "tail" to the left or right: _____

2. In a class of 20 students, the individual heights of the students in cm are: 135, 139, 141, 146, 147, 149, 156, 151, 158, 155, 156, 159, 161, 167, 162, 163, 161, 172, 171, 170.

   (a) Calculate the mean height of the students: _____

   (b) A person takes a sample of five of the students: 139, 151, 162, 172, 170. Calculate the mean of the sample and comment on its accuracy (how close it is to the true value):

   _____

223 Which Graph to Use?

Key Idea: The type of data collected will affect the type of graph chosen to display your data.
Different types of graph display different types of information.

It is important that the type of information you want to display is matched to the correct type of graph so that information is clearly communicated.

What type of data have you collected?

- One variable is a category
- One variable is a count

Use a pie graph

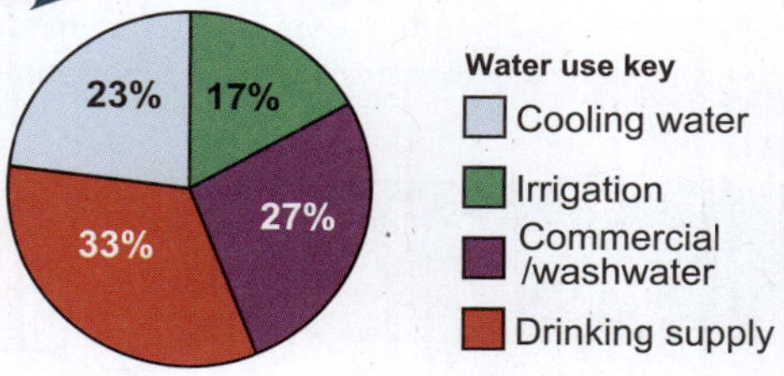
23%
17%
27%
33%
Water use key
Cooling water
Irrigation
Commercial /washwater
Drinking supply

Use to compare proportions in different categories.

- One variable is a category
- One variable is **continuous data** (measurements)

Use a bar or column graph

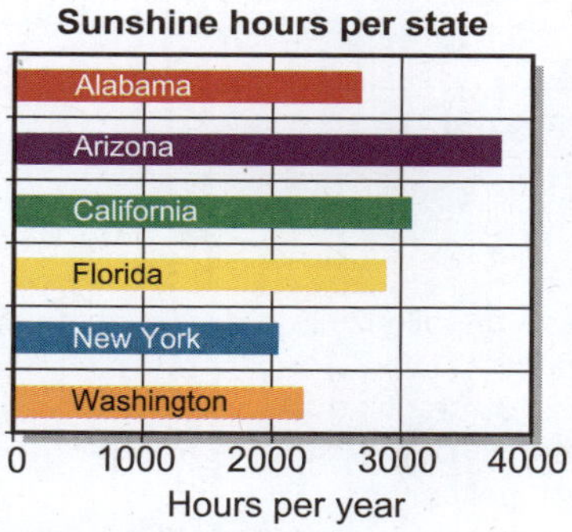
Sunshine hours per state
Alabama
Arizona
California
Florida
New York
Washington
0
1000
2000
3000
4000
Hours per year

Use to compare different categories (or treatments) for a continuous variable.

- One variable is continuous data (measurements)
- One variable is a count

Use a histogram

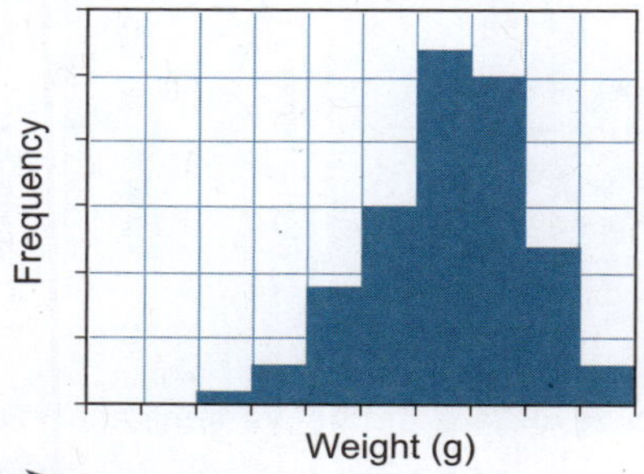
Frequency
Weight (g)

Use to show a frequency distribution for a continuous variable.

- Both variables are continuous
- The response variable is dependent on the independent (manipulated) variable

Use a line graph

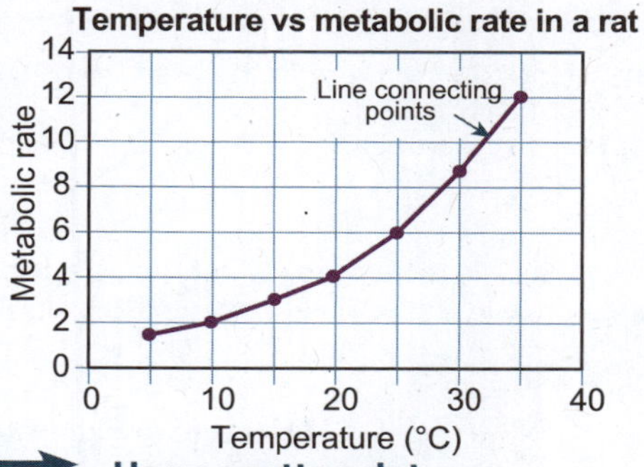
Temperature vs metabolic rate in a rat
Metabolic rate
14
12
10
8
6
4
2
0
Line connecting points
0
10
20
30
40
Temperature (°C)

Use to illustrate the response to a manipulated variable.

- Both variables are continuous
- The two variables are inter-dependent but there is no manipulated variable

Use a scatter plot

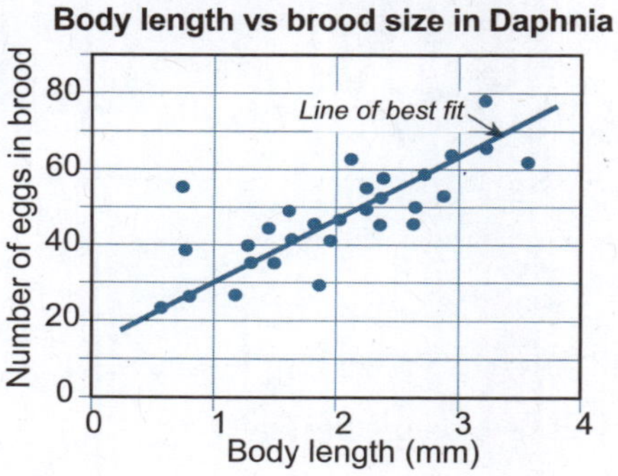
Body length vs brood size in Daphnia
Number of eggs in brood
80
60
40
20
0
Line of best fit
0
1
2
3
4
Body length (mm)

Use to illustrate the relationship between two correlated variables.

ISBN: 978-1-99-101409-2

224 Analyzing and Interpreting Data

Key Idea: Collected data needs to be processed and interpreted to provide meaningful information.

Processed **data** is usually displayed in a table or graph. Tables, and especially graphs, can help make any trends in the data easier to see. Graphs can be used to predict values that do not appear in the original data. Sometimes, many different data points may be plotted on the one graph so that a large amount of information is shown in a compact way.

Analyzing population density

Tables provide a way to systematically record and condense a large amount of information. They can also make calculations simpler because relevant data is presented in close proximity. The data below can be used to calculate the population density of various US states.

State	Population	Total area (km²)	Population density (people /km)
Alabama	5,024,279	135,765	
Florida	22,610,726	170,312	
Montana	1,122,867	380,800	
Texas	30,503,301	695,662	

1. Which of the four states in the table shown left has the greatest population?

   _____

2. Calculate the population density for each of the four states. Write your answers on the table (left):

3. Which of the four states in the table shown left has the greatest population density?

   _____

Pollen in sediments

The data right shows the percentage of pollen in sediments from a region in northeastern United States, laid down over 15,000 years. By graphing the percentages of pollen from different plants beside each other and matching the pollen type to known trees and landscapes, we can develop a picture of what the land looked like at any particular time in the last 15,000 years.

We can see from the graph that about 5000 years ago, the land was mostly hardwood forest, including oak and beech trees. However, 15,000 years ago, we can see that the land was covered mostly by tundra. This is consistent with what is known about the climate conditions 15,000 years ago.

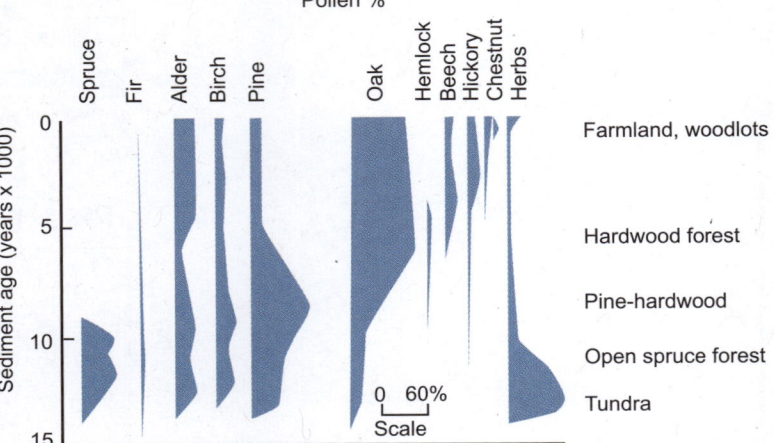

4. (a) Using the pollen graph above, determine the predominant type of plant cover in the northeastern United States 10,000 years ago:

   _____

   (b) What appears to be the most common tree in the area from 10,000 years ago to present? _____

   (c) Which type of tree went into steep decline between 10,000 and 5000 years ago? _____

Stalactite formation

Stalactites grow down from the top of cave roofs as rainwater deposits minerals onto them. The more rain there is, the faster the stalactites grow. They also tend to grow in annual rings as there is usually more rain in winter than in summer. Analyze the stalactite data below:

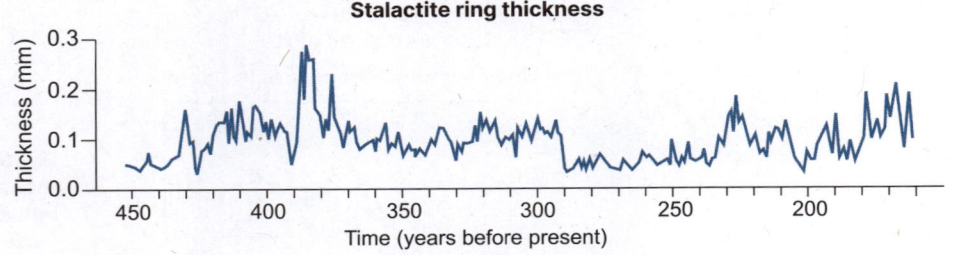

5. (a) Identify the wettest period of time before present: _____

   (b) Identify the driest period of time before present: _____

Analyzing body length in a population of fish

6. (a) Using the data of body length of a population of smelt presented below, create a tally chart, size classes of 10 mm:

Length of smelt from the Red River
Cloudy, 16°C   2- July-2021

81	71	42	61
47	47	78	41
57	57	47	61
31	63	46	41
63	71	67	53
66	43	33	43
36	54	68	49
54	48	36	37
29	28	56	34
26	68	56	51
51	62	38	38
27	25	62	25
52	51	64	52
11	18	58	6
74	66	57	58

All fish netted and returned after measuring (n = 60)

Size class (mm)	Tally	Total
0 – 10	I	1
11 – 20	II	2
21 – 30	IIII I	6
31 – 40	IIII III	8
41 – 50	IIII IIII I	11
51 – 60	IIII IIII IIII	15
61 – 70	IIII IIII II	12
71 – 80	IIII	4
81 – 90	I	1

(b) What kind of distribution does body length follow? _____ Normal (bell-shaped) distribution _____

(c) Calculate the body length mean, median, and mode for the population:

Mean: _____ 48.9 mm _____

Median: _____ 51.5 mm _____

Mode: _____ 47, 51, and 57 mm (multimodal, each appearing 3 times) _____

Analyzing tree ring data and rainfall

Trees produce annual rings. In dry years, the rings are narrow and closer together. In wet years, the rings are larger and further apart. Using specific formulae, the width of the tree rings can be used to reconstruct rainfall. The data for an area of eucalyptus forest in Australia is shown in the table below:

Year	Tree ring (mm)	Actual rainfall (mm)	Reconstructed rainfall (mm)
1910	1.672	115.20	301.36
1920	0.378	193.80	165.53
1930	0.432	163.80	171.20
1940	0.600	121.80	188.83
1950	0.433	148.40	171.30
1960	1.288	320.80	261.06
1970	0.726	160.30	202.06
1980	2.147	420.40	351.23
1990	0.197	199.20	146.53
2000	3.902	627.60	535.45
2010	0.122	102.30	138.66

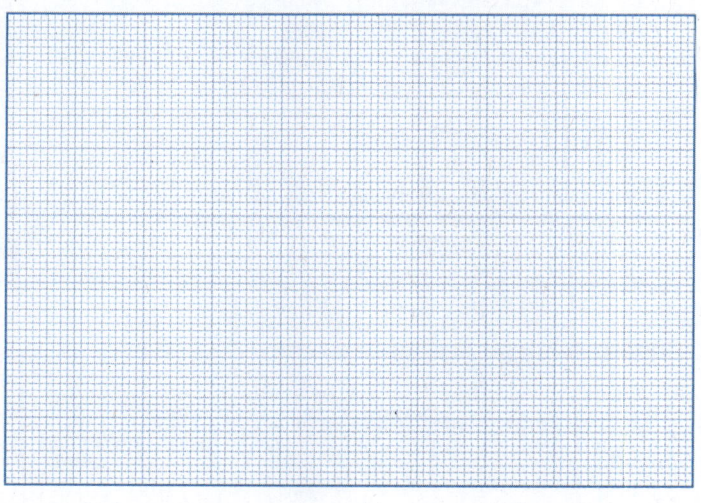

7. (a) Plot the actual rainfall data and the reconstructed rainfall data on the graph above. Remember a key:

(b) Does the reconstructed data accurately represent the actual data? What does this mean for reconstructing rainfall when we do not know the actual rainfall?

_____

_____

225 Working with Numbers

Key Idea: Sometimes the numbers we have to deal with in science are very large or are in a notation we are not familiar with. Using mathematical rules we can transform these numbers into ones that are simpler to work with.

Mathematics is used in environmental sciences to analyze, interpret, and compare **data**. It is important that you are familiar with mathematical notation (the language of mathematics) and can confidently apply some basic mathematical principles and calculations to your data. Much of our understanding of the sciences is based on our ability to use mathematics to interpret the patterns seen in collected data, and express the laws of the universe in simple notation.

Commonly used mathematical symbols

In mathematics, universal symbols are used to represent mathematical concepts. They save time and space when writing. Some commonly used symbols are shown below:

= Equal to

< The value on the left is **less than** the value on the right

> The value on the left is **greater than** the value on the right

∝ Proportional to. A ∝ B means that A = a constant × B

~ Approximately equal to

∞ Infinity

$\sqrt{b}$ The square root of b

b^2 b squared (b × b)

b^n b to the power of n (b × b... n times)

Δ The change in. For example $\Delta T / \Delta d$ = the change in T ÷ the change in d (see rates, below right).

Length
Kilometer (km)	1000 m
Meter (m)	1000 mm

Volume
Liter (L)	1000 mL
Milliliter (mL)	1 mm³

Area
Square kilometer	1,000,000 m²
Hectare	10,000 m²
Square meter	1,000,000 mm²

Temperature
0°C = freezing point of pure water
100°C = boiling point of pure water

Kelvin scale (K) and °C have the same magnitude.
Kelvin scale starts at absolute zero (−273.15 °C).

Decimal and standard form

▶ Decimal form (also called ordinary form) is the longhand way of writing a number, e.g. 15,000,000. Very large or very small numbers can take up too much space if written in decimal form and are often expressed in a condensed standard form. For example, 15,000,000 is written as 1.5×10^7 in standard form.

▶ In standard form, a number is always written as $A \times 10^n$, where A is a number between 1 and 10, and n (the exponent) indicates how many places to move the decimal point. n can be positive or negative.

▶ For the example above, A = 1.5 and n = 7 because the decimal point moved seven places (see below).

$$15{,}000{,}000 = 1.5 \times 10^7$$

▶ Small numbers can also be written in standard form. The exponent (n) will be negative. For example, 0.00101 is written as 1.01×10^{-3}.

$$0.00101 = 1.01 \times 10^{-3}$$

Adding numbers in standard form

▶ Numbers in standard form can be added together so long as they are both raised to the same power of ten. E.g. $1 \times 10^4 + 2 \times 10^3 = 1 \times 10^4 + 0.2 \times 10^4 = 1.2 \times 10^4$

Rates

▶ Rates are expressed as a measure per unit of time and show how a variable changes over time. Rates are used to provide meaningful comparisons of data that may have been recorded over different time periods.

▶ Often, rates are expressed as a mean rate over the duration of the measurement period, but it is also useful to calculate the rate at various times to understand how rate changes over time. The table below shows the distance traveled by a rolling ball. A worked example for the rate at 4 seconds is provided below.

Time (s)	Distance traveled (m)	Rate of movement (speed) (m/s)
0	0	0
2	34	17
4	42	4*
6	48	3
8	50	1
10	50	0

* meters moved between 2-4 seconds: 42 m − 34 m = 8 m

Rate of movement (speed) between 2-4 seconds
8 m ÷ 2 seconds = 4 m/s

1. Use the information above to complete the following calculations:

   (a) $\sqrt{9}$: _____

   (b) 4^3: _____

   (c) Write 6,340,000 in standard form: _____

   (d) Write 0.00103 in standard form: _____

   (e) Convert 10 cm to millimeters: _____

   (f) Convert 4 liters to milliliters: _____

   (g) Write 7.82×10^7 as a number: _____

   (h) $4.5 \times 10^4 + 6.45 \times 10^5$: _____

Dealing with large numbers

▶ Earth and space sciences often deal with very large numbers or scales. Numerical data, indicating scale, can often increase or decrease exponentially. Large scale changes in numerical data can be made more manageable by transforming the data using logarithms.

Exponential function

▶ Exponential growth or decay occurs at an increasingly rapid rate in proportion to the increasing or decreasing total number or size.

▶ In an exponential function, the base number is fixed (constant) and the exponent is variable.

▶ The equation for an exponential function is $y = c^x$ (where c = constant).

▶ An example of exponential decay is radioactive decay. Any radioactive element has a half-life, the amount of time required for its radioactivity to fall to half its original value.

▶ Here, the amount of radioactivity (y) is dependent on a constant (c) to the power of the time passed (x).

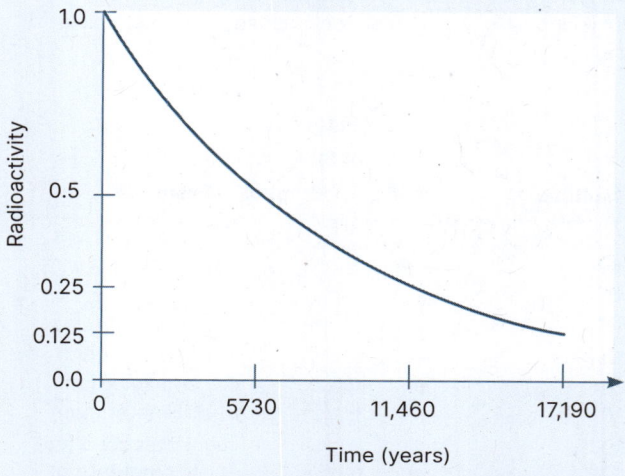

Example above: Carbon-14 (^{14}C) has a half life of 5730 years. If a sample with a mass of 10 g was left for 5730 years half the sample will have decayed, leaving 5 g of radioactive material. After another 5730 years, 2.5 g of radioactive carbon will be left.

Log transformations

▶ A log transformation can make very large numbers easier to work with.

▶ The log of a number is the exponent to which a fixed value (the base) is raised to get that number. So $\log_{10}(1000) = 3$ because $10^3 = 1000$.

▶ Both $\log_{10}$ and $\log_e$ (natural logs or $\ln$) are commonly used.

▶ Log transformations are useful for data where there is an exponential increase or decrease in numbers. In this case, the transformation will produce a straight line plot.

▶ To find the log10 of a number, e.g. 32, using a calculator, key in log 32 =   The answer should be 1.51.

▶ An example of a log scale is the Moment Magnitude scale used to measure the energy released during earthquakes. Each step of the scale is approximately $10^{1.5}$ times greater than the step below it. Calculating the difference in energy released between earthquakes can be done by finding the inverse $\log_{10}$ of the difference in magnitude ($10^{(1.5 \times (m1-m2))}$).

▶ Also, the number of earthquakes around the world at each magnitude follows a negative logarithmic spread (below).

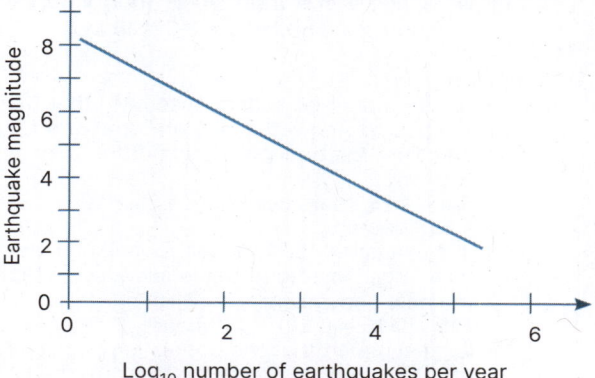

2. The Moment Magnitude scale is a measure of the energy released during an earthquake.

   (a) How many times more energy is released by the magnitude 6 earthquake than a magnitude 4 earthquake?

   _____

   (b) How many times more energy is released by the magnitude 7.5 earthquake than a magnitude 4.3 earthquake?

   _____

3. The pH scale measures the acidity of a substance. It is a negative logarithmic scale. A pH of 3 has a hydrogen ion concentration (which is responsible for acidity) ten times greater than a pH of 4.

   How many times greater is the hydrogen ion concentration of a pH 2 solution than a pH 6 solution?

   _____

4. Carbon-14 (^{14}C) is found in living organisms. It has a half life of 5730 years. When an organism dies, it stops taking in ^{14}C and the ratio of ^{14}C to ^{12}C changes.
   Using these pieces of information, explain how we can calculate how long ago an organism died:

   _____

   _____

   _____

   _____

226 Calculations, Conversions, and Multiples

Key Idea: Being able to use international units and convert between different types of measurement is a useful skill. When answering computational questions, it is important to show all the working associated with calculating the answer. Some examples of the calculations and conversions you may encounter are described below.

1. **Converting between multiples:**
   (a) Convert the following to kilometers:
   (i) 5 mm: 5 mm ÷ 1,000,000 = 5 ×10^{-6} km.
   (ii) 10,000 cm: 10,000 cm ÷ 100,000 = 0.1 km.
   (iii) 8000 m: 8000 m ÷ 1000 = 8 km.

   (b) Convert 12 m/s to km/h: 12 × 60 × 60 = 43,200 m/h. 43,200/1000 = 43.2 km/h.

2. **Energy calculations:**
   (a) Calculate the amount of electricity (in joules) produced by a 5 MW generator over three hours:
   1 W = 1 J/s. 5 MW = 5,000,000 J/s. Seconds in 3 hours = 60 × 60 × 3 = 10,800 s.
   10,800 s x 5,000,000 J/s = 54,000,000,000 J = 54,000 MJ.

   (b) Calculate the energy (in joules) used by a 2000 W heater over two hours:
   2000 W = 2000 J/s. 2000 × 60 × 60 × 2 = 14,400,000 J = 14.4 MJ.

   (c) Calculate the amount of energy (in joules) expressed by the notation 5 kWh:
   1 kWh = 1000 W/h = 1000 J/s/hr. 1000 × 60 × 60 = 3,600,000 J = 3.6 MJ. 5 × 3.6 = 18 MJ.

3. **Sampling:**
   (a) A study of a bear population discovers that there were 5 bears living within a 4 km^2 area of a forest:
   (i) Calculate the total population if the forest is 100 km^2:
   5/4 = 1.25 bears per 1 km^2. 1.25 × 100 = 125 bears.
   (ii) It is estimated that the bear population may be 20% larger than the sample suggests. Calculate the new population:
   125 + (125 × 0.20) = 150 bears.
   (iii) It is estimated that the population has an annual growth rate of 1.3 percent. Calculate the bear population in a further five years:
   $Pop_{future\ bears}$ = $Pop_{present\ bears}$ x (1 + $0.013)^5$ = 150 x $(1.013)^5$ = 160 bears in 5 years time.

   (b) A coal deposit is estimated at 2000 tonnes (t). If the coal is extracted at a rate of 30 t/h calculate how long the deposit will last: 2000/30 = 66.67 hours.

4. **Reading off a graph:**

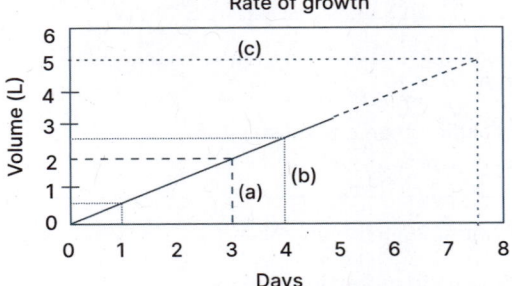

   (a) State the volume after 3 days: 1.95 L

   (b) Calculate the rate of growth per day:
   Day 1 = 0.5 L. Day 4 = 2.5 L. Change = 2.5-0.5 = 2.0 over 3 days. 2.0/3 = 0.667 L per day.

   (c) Extrapolate the graph to determine how long it will take to reach a volume of 5 L: 7.55-7.65 days. (confirm from calculation 5/0.667 = 7.5).

International system of units (si)
Examples of SI derived units

Derived Quantity	Name	Symbol
area	square meter	m^2
volume	cubic meter	m^3
speed, velocity	meter per second	m/s
acceleration	meter per second squared	m/s^2
mass density	kilogram per cubic meter	kg/m^3
specific volume	cubic meter per kilogram	m^3/kg
amount-of-substance	mole per cubic meter	mol/m^3
concentration	mole per liter	mol/L
luminance	candela per square meter	cd/m^2

Multiples
Examples of SI derived units

Multiple	Prefix	Symbol	Example
10^9	giga	G	gigawatt (GW)
10^6	mega	M	megawatt (MW)
10^3	kilo	k	kilogram (kg)
10^2	hecto	h	hectare (ha)
10^{-1}	deci	d	decimeter (dm)
10^{-2}	centi	c	centimeter (cm)
10^{-3}	milli	m	milliimeter (mm)
10^{-6}	micro	μ	microsecond (μs)
10^{-9}	nano	n	nanometer (nm)
10^{-12}	pico	p	picosecond (ps)

Conversion factors for common units of measure
For all conversions multiply by the factor shown

Length	
Centimeters to inches:	0.393
Meters to feet:	3.280
Kilometers to miles:	0.621

Volume	
Milliliters to fluid ounces:	0.034
Liters to gallons:	0.264
Cubic meters to gallons:	264.1

Area	
Square meters to square feet:	10.76
Hectares to acres:	2.471
Square kilometers to square miles:	0.386

Temperature	
°C to °F:	0 °C = 32 °F
	100 °C = 212 °F
Formula °C to °F:	°F = °C x 1.8 + 32

Energy	
BTU to joules	1055.0558

©2025 BIOZONE International
ISBN: 978-1-99-101409-2
Photocopying prohibited

227 Correlation or Causation?

Key Idea: Correlation does not mean causation.
Researchers often want to know if two variables have any correlation (relationship) to each other. This can be achieved by plotting the **data** as a scatter graph and drawing a line of best fit through the data, or by testing for correlation using a statistical test. The strength of a correlation is indicated by the correlation coefficient (r or R), which varies between 1 and -1. A value of 1 indicates a perfect (1:1) relationship between the variables. A value of -1 indicates a 1:1 negative relationship, and 0 indicates no relationship.

Correlation does not imply causation

You may come across the phrase "correlation does not necessarily imply causation". This means that even when there is a strong correlation between variables, i.e. they vary together in a predictable way, you cannot assume that change in one variable caused change in the other.

Example: When data from the organic food association and the office of special education programs is plotted (below), there is a strong correlation between the increase in organic food and rates of diagnosed autism. However, it is unlikely that eating organic food causes autism, so we cannot assume a causative effect here.

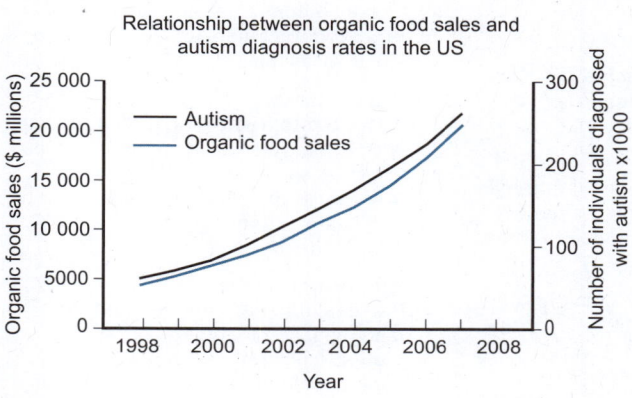

Drawing the line of best fit

Some simple guidelines need to be followed when drawing a line of best fit on your scatter plot.
- Your line should follow the trend of the data points.
- Roughly half of your data points should be above the line of best fit, and half below.
- The line of best fit does not necessarily pass through any particular point.
- The line of best fit should pivot around the point which represents the mean of the x and the mean of the y variables.

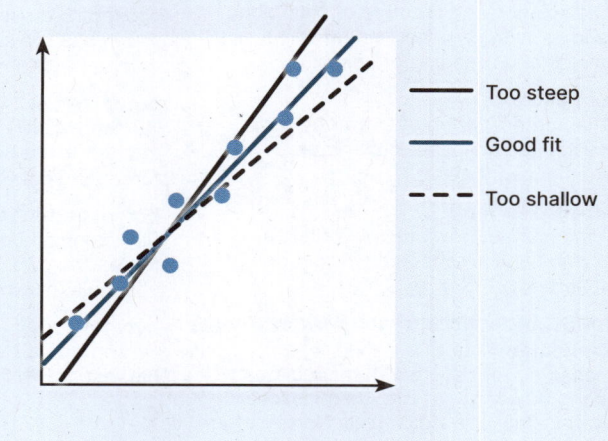

1. What is meant by the phrase, 'correlation does not imply causation'?

   _____
   _____
   _____
   _____

2. A student measured the length of eruptions of Old Faithful and the time between the eruptions for a day, and plotted a scatter graph of the results (right).

   (a) Draw a line of best fit through the data:

   (b) Using your line of best fit as a guide, comment on the correlation between eruption length and the time between eruptions:

   _____
   _____
   _____
   _____

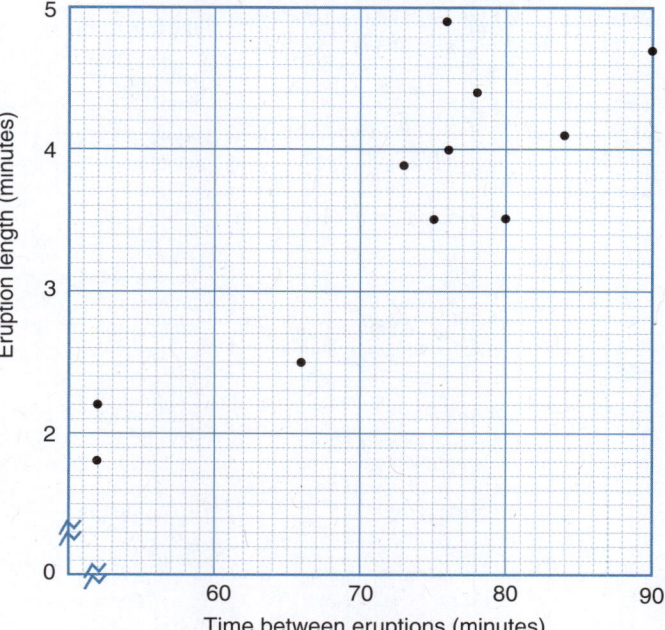

Glossary

A

abiotic factors
Non-living, physical features in an ecosystem, including temperature, humidity, and rainfall).

abundance
The relative representation of a species in an ecosystem or region.

acid rain
Rain or other precipitation that has elevated H+ levels or pH below about 5.2, usually produced from emissions of sulfur.

adaptation (climate change)
The taking of action or building of structures that prepare for or are able to adjust to projected changes in climate.

agriculture
The science and practice of cultivating land, growing and harvesting crops, and raising and breeding of animals.

air pollution
The contamination of the air by particulate matter or hazardous substances (gases).

albedo (-effect)
A measure of how much light hits a surface and is reflected without being absorbed.

AMOC (Atlantic Meridional Overturning Circulation)
Ocean current system that circulates along the length of the Atlantic Ocean. Warm shallow water travels North before cooling, sinking and traveling back South at depth.

anthropogenic
Relating to humans or human influences. Anthropogenic changes relate to changes in the environment as a result of direct or indirect human actions.

asthenosphere
The upper layer of the Earth's mantle that lies below the lithosphere.

atmosphere
The envelope of gases surrounding Earth

B

biodiversity
The amount of biological variation present in a region (includes genetic, species, and habitat diversity).

biodiversity hotspots
Threatened biogeographic regions that contain high species diversity with at least 1500 endemic plants and have 30% or less of the original habitat left.

biofuel
A fuel derived from biomass, such as plant or algal material or animal waste.

biomagnification
The process by which pesticides and other substances become more concentrated in each link of the food chain.

biome
Major regional ecological community of plants and animals.

biosphere
A combination of all parts of the Earth that support life, from the depth of the ocean to a few km into the atmosphere. Also called the ecosphere.

biotic factors
Relating to the living factors in an ecosystem, including distribution and abundance.

biotic potential
The capacity of a population of organisms to increase in numbers under optimal environmental conditions.

bleaching (coral reef)
The loss of symbiotic algae from zooxanthellae (coral polyps), usually caused by a rise in sea temperature which has the effect of the coral losing color and turning white.

boreal forest
Northern hemisphere biome characterized by continuous conifer forest and cold temperatures.

boreal permafrost
Sediments in boreal forests that remain frozen (below 0°C) for more than two years, i.e. not melting between winters.

bycatch
Unwanted fish or animals caught during a harvesting/fishing operation.

C

carbon capture and storage (CCS)
The process of separating, removing, and storing greenhouse gases from waste emissions of industrial processes.

carbon cycle
The process by which carbon is exchanged between living organisms, the Earth and its atmosphere.

carbon dioxide (CO_2)
Molecule that makes up about 0.04% of the atmosphere. It is an important greenhouse gas, helping to trap heat in the atmosphere.

carbon footprint
A calculated value that allows the comparison of greenhouse gases, as carbon dioxide equivalent, generated by people and or industry.

carbon sequestration
The process of capturing and storing atmospheric carbon dioxide, and preventing it from re-entering the carbon cycle (in order to reduce greenhouse gases).

carbon sink
Any reservoir, process, or mechanism that removes more carbon dioxide (or other greenhouse gas) from the atmosphere than it releases.

carbon trading
The buying and selling credits within a limited market to allow specific amounts of carbon dioxide equivalent emissions by the purchasing company while maintaining a limit on the total greenhouse gas emissions.

carrying capacity (K)
Number of individual organisms the resources of a given area can support, usually through the most unfavorable period of the year.

cellular respiration
The series of metabolic reactions that oxidize organic molecules to produce ATP.

classification keys
A series of dichotomous steps or questions designed to identify and/or separate organisms into groups based on the observable characteristics of the organisms being classified.

climate change
A change in the patterns of climate such as temperature and rainfall that is attributed to human activity, including the use of fossil fuels.

climate justice
Term used to frame global climate change as an ethical and political issue instead of just an environmental issue, and endeavours to spread the responsibility for climate action fairly across all interested parties.

climate legislation
The acts, decrees, laws, and policies used by governments to address actions related to mitigation, adaptation, and management of climate related issues.

climate risk
The potential for the adverse consequences of climate change to affect human activities and infrastructure.

climax community
A community that has reached an equilibrium with the environment and no longer appears to be changing in composition.

coal
A brown/black sedimentary rock formed by compression of plant material over a long period. Rich in carbon and can be burned as a fuel source

commensalism
A relationship between two organisms in which one benefits, and one is unaffected.

community
The living component of an ecosystem.

Conference of Parties (COP)
Main decision making body of the UNFCCC. Meets annually to negotiate measures and review progress against the goal of limiting climate change.

conservation
The active management of natural populations in order to rebuild numbers and ensure species survival.

consumer
An organism that feeds on producers,

other consumers, or non-living organic material

convergent boundary
A tectonic boundary where two plates are moving toward each other and colliding.

continuous data
Quantitative data representing a scale of measurement that can consist of numbers other than whole numbers.

core
The innermost layer of the Earth; a solid mass of mostly iron and nickel.

Coriolis effect
Physical consequence of the law of conservation of angular momentum; as a result of the Earth's rotation, a moving object veers to the right in the northern hemisphere and to the left in the southern hemisphere relative to the Earth's surface.

crust
The outermost layer of the Earth, composed of a great variety of igneous, metamorphic, and sedimentary rocks.

D

data
Facts and statistics collected for analysis.

dataloggers
Devices that are able to record data as it changes in real time.

decomposer
Organism that obtains energy from the breakdown of dead organic matter to simpler substances; most precisely refers to bacteria and fungi.

deforestation
The removal of forests by cutting, burning or other large scale activity by humans, usually to make way for crops or monoculture plantations.

demographic transition model (DTM)
A theoretical framework that explains how populations transition overtime from small populations with high birth and death rates to larger populations with low birth and death rates as a result of economic and social development (from pre to post industrial societies).

density dependent factor
A factor that regulates the size of a population in proportion to the density of the population.

density independent factor
A factor that affects population size but acts in the same way regardless of population density.

detritivore
An organism that feeds on decaying matter (detritus).

developed country
A sovereign state with a high quality of life, developed stable economy, and technically advanced infrastructure.

developing country
A sovereign state with underdeveloped industry, low standard of living, and an underdeveloped or limited economy with low gross domestic product.

discontinuous data
A quantitative data set representing a scale of measurement that can consist only of whole numbers (*cf. continuous data*).

divergent boundary
A tectonic boundary where two plates are moving away from each other and new crust is forming from magma rising to the surface between the two plates.

DNA barcode
A short section of standardized DNA that can be used to identify organisms to the species level.

E

ecological niche
The functional role in the environment performed by an organism and the conditions it requires.

ecosystem
All the organisms in a given area and the abiotic factors with which they interact.

ecotourism
A form of tourism taking people to ecologically interesting or important areas while considering and promoting environmental, social and ethical issues.

emigration
The leaving of a place or country of residence to permanently settle elsewhere.

emissions
Gases, heat, light etc. released into the atmosphere as a result of chemical or industrial processes.

endangered species
A species of animal or plant that is facing a very high risk of extinction in the wild.

endocrine disrupting chemicals (EDCs)
Natural or man made chemicals that can mimic or block the body's hormones, affecting the functioning of the endocrine system.

energy conservation
The deliberate reduction in the use of energy and electricity via the use of devices and designs that use less energy to produce the same output (as a former device or design), thus lowering the amount of primary energy resources required per unit of output.

ENSO
El Niño Southern Oscillation. A periodic global climate pattern resulting from changes in the air pressure over and surface temperature of the equatorial Pacific Ocean.

environmental disaster
A catastrophic event occurring in the natural environment resulting from some form of human activity (as opposed to a natural disaster).

environmental legislation
The acts, decrees, laws, and policies used and enforced by governments to regulate human activities in and treatment of the natural world.

environmental remediation
The clean up, treatment, and removal of hazardous substances, pollutants, and contaminants from the environment and restoration of the natural parts of the environment within the clean up area.

eutrophication
Nutrient enrichment of a body of water.

ex situ (conservation)
The conservation of species outside of their natural environment, usually in enclosed environments e.g. zoos and botanical gardens.

exponential growth
Instantaneous rate of population growth, expressed as a proportional increase per unit of time.

extinction
The dying out or extermination of a species.

extreme weather event
Weather events that are at the extremes of the historical distribution of weather events. They are relatively rare and tend to have severe consequences.

F

fault
A fracture of the Earth's crust, between different rock masses.

fertility rate
The average number of offspring born to a woman over her lifetime. Rates can be specific for given groups, e.g. age, country etc.

fertilizer
A natural or artificial substance added to soil that contains nutrients and supplies them to plants to enhance their growth and productivity.

food chain
A model that is used to demonstrate the feeding relationships between organisms in an ecosystem.

food web
The combination of all the food chains in a particular ecosystem.

forever chemicals
Informal name for the class of chemicals called polyfluoroalkyl substances (PFAs). They can persist for an extremely long length of time in the environment.

fossil fuel
A natural fuel such as coal or gas, formed in the geological past from the remains of living organisms.

G

geothermal energy
Energy derived from geothermal activity, normally in the form of hot water for heating or steam that is used to drive turbines to produce electricity.

globalization
The global inter-connection of people, trade, technology, and culture.

global warming
The gradual heating of the Earth, attributed to human activities such as the burning of fossil fuels.

GPS tracking
The use of global positioning system (GPS) devices attached to animals to

track movements over time.

greenhouse effect
The retention of solar energy in the Earth's atmosphere by gases that absorb heat and prevent it being released back into space.

greenhouse gas (GHG)
Any gas in the atmosphere that causes the retention of heat in the Earth's atmosphere. Major gases are water vapor and carbon dioxide.

green revolution
The global spread of agricultural techniques including selective crossbreeding and genetic modification that dramatically increased the production and productivity of global food crops.

H

habitat
The natural environment in which an organism lives, including all of the biotic and abiotic factors

habitat fragmentation
The fragmentation of larger continuous tracts of habitat into a mosaic of smaller, often isolated areas.

heavy metals
A metal with either a high atomic number, high density, or high toxicity. Heavy metals generally include those that are not normally found in the environment in detectable or toxic levels and have been leached into the environment due to human activities.

human activity
Humans actions and operations that may affect the environment.

hydroelectric power
The production of electricity via the use of water moving through turbines connected to electrical generators.

hydrogen fuel cell
An electrochemical cell that uses hydrogen as a fuel source and oxygen as an oxidizer. Electricity is produced when electrons are striped from hydrogen gas and forced to travel around a circuit before uniting with oxygen and hydrogen ions to form water.

I

ice sheet
A geological period when there are substantial ice sheets and alpine glaciers on the planet.

igneous
Rock type produced from volcanic or plutonic activity, e.g. extrusion of lava.

immigration
The arrival in a new place or country to permanently live after leaving another country or place.

insecticide
A substance specifically designed to kill insects.

in situ (conservation)
The conservation of species inside their natural environment, usually by protective measures such as trapping pests, monitoring, or providing extra nests, food, etc.

indirect sampling
Sampling method that uses recordings of the signs of a species, rather than the species itself, e.g. scat, calls, tracks, and markings on vegetation. These are used to assess population abundance.

integrated pest management
A method of pest control that relies on a range of strategies to control plant pests.

interspecific
Interactions that occur between members of different species.

intertropical convergence
A planetary band of heavy precipitation close to the equator.

intraspecific
Interactions that occur between members of the same species.

introduced (invasive) species
A species that is not native to that ecosystem.

IPCC (Intergovernmental Panel on Climate Change)
An intergovernmental body of the United Nations formed to research, report on, and advance scientific knowledge on human caused climate change.

IUCN (International Union for Conservation of Nature)
An international organization for the UN working to enhance conservation and sustainability of natural resources. It is involved in the gathering and analysis of data, research, and education.

J-K

K-selected species
Species that maintains its population at or around the carrying capacity, and favors the production of a few, highly competitive offspring.

keystone species
A species that occupies an essential role in an ecosystem and on which most or all of the other species in an ecosystem depend, directly or indirectly.

kite graphs
Graph that shows the observed data, e.g. counts, relative abundance (y axis), as a shaded area along the length of a (centrally placed) transect line (x axis).

L

legislation (conservation)
The acts, decrees, laws, and policies used and enforced by governments to protect, conserve, and enhance the environment and its biodiversity.

life expectancy
A statistical measure of the estimate of a person's remaining years of life at a given age. This is typically given as the life expectancy at birth, which is based on the mean length of life of a cohort from birth. Life expectancy is different for different ages and can change as medical technology and quality of life improves).

light pollution
The presence of unwanted light produced by artificial outdoor lights. This can disrupt the natural patterns of wildlife and humans, and can affect astronomical observations.

Lincoln index
A statistical measure of a population's size based on mark, release, and recapture of members of the population.

lithosphere
The rigid outer part of the Earth, consisting of the crust and upper mantle.

loam
A mixture of sand, silt, and clay. Specific types can be defined, but loam itself is a 40:40:20 mix of silt, sand, and clay.

logistic growth
Population growth that follows a sigmoidal curve, plateauing at carrying capacity.

M

mantle
The mostly solid bulk of Earth's interior, which lies between the dense, super-heated core and the crust.

mark and recapture
Sampling procedure that captures, marks and counts a sample of the population, releases them back into the environment to mix into the population, then captures a new sample and counts the marked individuals in the new sample.

mean
The sum of the data divided by the number of data entries.

median
The number that occurs in the middle of a set of sorted numbers. It divides the upper half of the data set from the lower half.

metamorphic
Rocks that change from one format to another, e.g. as a result of heat and/or pressure.

mineral
A naturally occurring substance that can be in the form of a crystalline solid, ore, element, or compound.

mitigation
A way of reducing the impact of a negative event.

mode
The value that occurs most often in a data set.

model
A conceptual, mathematical or physical representation of a real-world phenomenon.

monoculture
Consisting of only one primary agricultural crop (or organism).

mortality
The death rate; the ratio of total number of death to the total population. The ratio of deaths in an area to the population of that area, expressed per 1000 per year.

mutualism
Biological interaction between (usually two) species that benefits both parties.

N

nanoplastic (microplastic)
Plastic particles less than 1 μm across that form from the breakdown of larger plastic pieces. Nano and micro plastics can be small enough to enter the tissues of even very small organisms.

natality
Production of new individuals in a population.

natural gas
A naturally occurring fossil fuel that is a mixture of short chain hydrocarbons (primarily methane) and forms as part of the natural production of oil.

net zero carbon
An emissions regime in which greenhouse gas emissions are matched by the removal of the equivalent amount of greenhouse gases from the atmosphere (net greenhouse gas emissions are zero).

nitrogen cycle
The processes by which nitrogen, in different forms, is cycled between living and non-living things in marine, terrestrial and atmospheric ecosystems.

noise pollution
The non natural and unwanted production of outdoor noise that can detrimentally affect humans and wildlife.

non-renewable energy
Energy production using fuels that can not be replaced or renewed in feasible time periods. Usually refers to the production of energy using fossil fuels or radioactive materials.

nuclear accidents (pollution)
Accidents involving nuclear or radioactive material, usually as a result of breach of containment.

nuclear fission
The splitting of a large atomic nucleus into two smaller atomic nuclei, normally by the introduction of a neutron.

nutrient cycling
The transfer of matter, minerals, and energy between biotic and abiotic parts of the environment.

O

ocean acidification
A reduction in the pH of the ocean over time, caused primarily by uptake of carbon dioxide from the atmosphere.

oil
A liquid fossil fuel formed by intense heat and pressure on the remains of ancient, dead marine organisms.

oil spill
Loss of containment of a volume of oil into the environment. This may be from the breach of a transport vessel, pipeline, or well head.

overfishing
The catching of so many fishing at one time or at such a rate that the targeted population can not replace its losses fast enough to maintain the population.

ozone depletion
The loss of stratospheric ozone via chemical interactions with chemicals (normally containing chlorine) that convert O3 into O2.

P

parasitism
Biological interaction in which one organism, the parasite, benefits at the expense of the other, the host.

Paris Agreement
International agreement on climate change adopted in 2015. Its goals are to limit global warming to 1.5°C, strengthen the ability of governments to deal with climate change, and support low carbon and climate resilient economies.

permafrost
Permanently frozen soil.

persistent organic pollutants (POPs)
Organic compounds that are resistant to environmental degradation through chemical, biological, and photolytic processes

phosphorus cycle
The process by which phosphorous is exchanged between living organisms and non living reservoirs.

photochemical smog
Hazardous brownish-gray haze produced when sunlight interacts with nitrogen and carbon oxides and aromatic hydrocarbons near the ground.

photosynthesis
A process used by green plants, algae, and some bacteria to convert light energy into chemical energy (carbohydrate).

photovoltaic cell
The conversion of light into electricity using semiconducting materials that exhibit the photovoltaic effect.

pioneer species
A species that is first to colonize areas of bare ground or soil, usually providing shelter for seedlings or other plants that arrive later.

plastic pollution
The accumulation of waste plastic material in the environment and the negative environmental effects of this.

plate boundary
The location where the edges of two or more tectonic plates meet.

plate tectonics
The scientific theory in which the Earth's crust is divided into separate plates which sit on top of a layer of hot plastic rock, the mantle, and are moved about due to convection currents.

point sampling
A sampling method used to determine vegetation cover in a sample area. Points are placed within the sample area and the proportion of points that intercept vegetation are recorded.

polar
Relating to the North or South pole and the area around them.

pollutants
Chemical or biological substance that causes harm to the environment.

pollution
The introduction of a contaminant into the natural environment that has harmful or poisonous effects.

population
A group of individuals of the same species living in a given area at a given time

population density
The number of individuals per unit area. For humans this is normally given as people per square kilometer.

population distribution
Description of how a population is spread across a habitat; the pattern of where individuals live.

population growth
The increase in the number of individuals in a population over a specific period of time. Population growth may be exponential or logistic.

positive feedback cycle
A mechanism in which the output of a system acts to oppose changes to the input of the system. The net effect is to stabilize the system and dampen fluctuations.

primary succession
Vegetational development starting on a new site never before colonized by life.

producer
An organism that produces its own food using materials from inorganic sources (also known as an autotroph).

Q

quadrats
A measured and marked region used to isolate a sample area for study.

qualitative data
Data described in descriptors of terms rather than by numbers

quantitative data
Data that can be expressed in numbers. Numerical values derived from counts or measurements.

R

r-selected species
Species that favors rapid growth rates and population increases, and tends to live in transient environments.

radio tracking
A method of tracking individuals of a species using transmitters.

radioactive waste
Any material that is itself radioactive or has been contaminated with radioactive material and is no longer deemed to be of any further use.

radiometric dating
A method for determining the age of an object based on the proportion of a radioactive isotope within it and the half-life of that isotope.

random sampling
A technique to ensure that every possible sample of a given size has the same chance of selection.

rangeland
Any extensive area of land usually

growing scrub or sporadic grasses which is grazed by native or introduced animals.

recycling
The process of converting waste material into its constituent components that can then be used to remake new material.

renewable energy
Useful energy (normally as electricity) produced using energy sources that can be replenished by the environment, e.g. wind turbines, photovoltaic cells.

resilience
The property of ecosystems or populations to recover from disturbances.

rewilding
A form of ecological restoration aimed at restoring biodiversity, ecosystem health, and ecosystem services.

rock cycle
Series of processes that create rocks in the Earth's crust and change them from one form into another.

S

sample
A sub-set of a whole used to estimate the values that might have been obtained if every individual or response was measured.

sampling
The act of selecting representative parts of a population for the purpose of determining useful parameters or characteristics for the whole population.

sampling error
A statistical error that occurs when sampling a population. The difference between the sample value and the real population value.

satellite tracking
The use of satellites and transmitters to track the movement of animals across their environment.

sea level rise
The rise in the level of the sea over time. This may be absolute (compared to the center of the Earth) or relative (absolute change plus any movement, up or down, of coastal land).

secondary succession
Development of vegetation after a disturbance.

sedimentary
Composed of layers of buried and compressed sediments.

sewage
Waste and wastewater containing high levels of organic matter, nutrients, and pollutants discharged from homes and industry.

sewerage
The infrastructure designed to carry and process sewage, including pipes, pumping stations and treatment facilities.

shield volcano
A broad based, shallow sloped, dome shaped volcano that erupts fluid basaltic lava.

Simpson's index of diversity
A measure of biodiversity that takes into account the number of species present and their relative abundance.

Sixth Mass Extinction
The apparent ongoing loss of many modern species far above and beyond the background extinction rate as a result of human activities.

soil
Loose mixture of materials on Earth's surface in which plants grow; composed of rock particles, organic matter and clay

soil horizon
A layer of soil with distinct physical, chemical, and biological characteristics. Soil horizons are classified based on their properties.

solar energy
Useful energy (usually as electricity) derived from sunlight. System may include photovoltaic cells, heliostat and power tower designs, or solar based water heating.

solar radiation modification (SRM)
A geoengineering technique in which sunlight arriving at Earth is reflected or radiated back into space in an attempt to lower the Earth's surface temperature to counteract global warming.

species
A group of organisms that can successfully interbreed and produce fertile offspring, and is reproductively isolated from other such groups.

stability
Ability of an ecosystem to resist change, i.e. remain unchanged over time.

strato-volcano
A steep sided volcano made up of layers of andesitic lava from many, generally explosive, eruptions.

stratospheric ozone
Ozone (O_3) that exists in the stratosphere at about 20-25 kilometers above the Earth's surface.

subpolar gyre (SPG)
A region of ocean circulation that sits under a region of low atmospheric pressure in polar regions.

survivorship curve
A curve showing the age specific mortality of a population.

sustainability
The ability of the earth to maintain itself without depletion of non-renewable resources.

T

thermohaline circulation
Global deep ocean currents driven by differences in water temperature and salinity.

tipping point
A critical climate threshold that, once exceeded, will result in significant and possibly irreversible change to the climate and/or ecosystems.

transect
A line placed across a community of organisms to provide information on their distribution.

transform boundary
A region where two tectonic plates, side by side, slide past each other.

tricellular model
A model of atmospheric circulation that uses three cells in each hemisphere to explain weather and climatic conditions at certain latitudes.

trophic level
A level or position in a food chain, food web or ecological pyramid. An organism's trophic level is determined by its feeding behavior.

U-V

urbanization
The movement of people out of small areas into cities

vertical farming
The process of growing crops in vertical layers or stacked racks in interior environments designed to optimize growth, normally using hydroponics. It includes the use of tall buildings providing a massive floor/growing surface while minimizing the building's ground footprint.

W

waste management
The processes involved in monitoring, collecting, handling, transport, and disposal of waste material to minimize its effects on the environment.

water cycle
Biogeochemical cycle that collects, purifies, and distributes the planet's water from the abiotic to the biotic environment and back.

water footprint
A measure of the volume of water used to grow or produce goods, including food.

water pollution
Any physical or chemical change in a body of water that can result in the water being unfit for use by living organisms.

water quality
The degree of purity in a sample of water or suitability of a sample water for a particular purpose, e.g. drinking.

watt
The SI unit of power, measured in joules per second (J/s).

wind energy
Form of renewable energy derived from harnessing the movement of air currents in the atmosphere using equipment such as wind mills and wind turbines.

wind turbine
Device designed to transform the energy in the wind into electrical energy via a generator connected to large, wind driven, propeller-like blades.

WWF (Worldwide Fund for Nature Organization)
Non governmental organization that works to help local and global communities to conserve and restore natural resources and habitats.

Questioning Terms

The following terms are often used when asking questions in examinations and assessments.

Analyse:	Interpret data to reach stated conclusions.
Annotate:	Add brief notes to a diagram, drawing or graph.
Apply:	Use an idea, equation, principle, theory, or law in a new situation.
Calculate:	Find an answer using mathematical methods. Show the working unless instructed not to.
Compare:	Show similarities between two or more items, referring to both (or all) of them throughout.
Construct:	Represent or develop in graphical form.
Contrast:	Show differences. Set in opposition.
Define:	Give the precise meaning of a word or phrase as concisely as possible.
Derive:	Manipulate a mathematical equation to give a new equation or result.
Describe:	Define, name, draw annotated diagrams, give characteristics of, or an account of.
Design:	Produce a plan, object, simulation or model.
Determine:	Find the only possible answer.
Discuss:	Show understanding by linking ideas. Where necessary, justify, relate, evaluate, compare and contrast, or analyze.
Distinguish:	Give the difference(s) between two or more items.
Draw:	Represent by means of pencil lines. Add labels unless told not to do so.
Estimate:	Find an approximate value for an unknown quantity, based on the information provided and application of scientific knowledge.
Evaluate:	Assess the implications and limitations.
Explain:	Provide a reason as to how or why something occurs.
Identify:	Find an answer from a number of possibilities.
Illustrate:	Give concrete examples. Explain clearly by using comparisons or examples.
Interpret:	Comment upon, give examples, describe relationships. Describe, then evaluate.
List:	Give a sequence of answers with no elaboration.
Outline:	Give a brief account or summary. Include essential information only.
Predict:	Give an expected result.
Solve:	Obtain an answer using numerical methods.
State:	Give a specific name, value, or other answer. No supporting argument or calculation is necessary.
Suggest:	Propose a hypothesis or other possible explanation.
Summarize:	Give a brief, condensed account. Include conclusions and avoid unnecessary details.

Image Credits

We acknowledge the generosity of those who have provided photographs for this edition:
• PASCO for photographs of probeware • Otago • University Environment Waikato for sampling photos • Sam Banks for the Wombat scat photo • Stephen Moore for his photos of aquatic invertebrates • US Coast Guard • Jane Ussher for her photograph og the albatross • Precession Seafood Harvesting • Kurchatov Institute for the photograph of Chernobyl

We also acknowledge the photographers that have made their images available through Wikimedia Commons under Creative Commons Licences 1.0, 2.0, 2.5, 3.0, or 4.0:
Gunnar Ries • Taiwankengo • Brokeninaglory • Gemini Observatory/AURA Joy Pollard • Archives New Zealand • Jon Zander • Javier Rubilar • P.Trusler • Tuxyso • Friedrich Kircher • StateStreet • Franco Folini • Svdmolen • "Mike" Michael L. Baird • JJ Harrison • Marco vinc • Luc Viatour www.Lucnix.be • JB-BU • Dustin M Ramsey • Janke • Mikrolit • Rasbak • Daderot • Bob Blaylock • Benjamint444 • Tapesh Yadav • Ubiquinoid • Tapesh Yada • Ian Lambot • CMG Lee • Tim Mossholder • Wawny • Beyond My Ken • Sam Fentress • Stahlkoche • Ravedave • W. Oelen • Hullwarren • BS Thurner Hof • David Monniaux • Harro5 • QFSE Media • Fundy • Erik Friis-Madsen • Tetris L • MisterRichValentine • IGV Biotech • Thzorro77 • Rufino Uribe • Mario Roberto Duran Ortiz • Karlis Dambrans • F1jmm • Geni • Tokyo Electric Power Co., TEPCO • Alexandr Trubetskoy • American • Johantheghost • Tomskyhaha • Fidel Gonzalez • Andrew Dunn • Dave Giles • Julian Nyča • Ibama • Nick Carson • USDA Natural Resources Conservation Service • Paul Hanly • Z22 • SuSanASecretariat • Richard Webb • Raul654 • DAVID ILIFF. • Schwede66 • Govt. Kiribati • Davidarfonjones • Shahee Ilyas • Pomfoto • Walter Siegmund • Famartin • Richard Ling • RCraig09 • Ellen Levy Finch • Alek14 • Anders Hellberg • Julia Hawkins • Hughhun • Sandia National Laboratories • F1jm

Contributors identified by coded credits are:
NASA: National Aeronautics and Space Administration • **USGS:** United States Geological Survey • **NOAA:** National Ocean and Atmosphere Administration • **USDA:** United States Department of Agriculture • **KP:** Kent Pryor • **USFWS:** US Fish and Wildlife Service • **WBS:** Warwick Silvester • **EW:** Environment Waikato: **CDC:** Centers for Disease Control and Prevention • **COD:** Colin O'Donnell • **USDE:** United States Department of Energy • **USAF** United States Air Force • **RA:** Richard Allan • **USCG:** United States Coast Guard • **DoC:** Department of Conservation (NZ) **NPS:** National Park Service

Royalty free images, purchased by BIOZONE International Ltd, are used throughout this workbook and have been obtained from the following sources:
Corel Corporation from their Professional Photos CD-ROM collection; IMSI (Intl Microcomputer Software Inc.) images from IMSI's MasterClips® and MasterPhotos™ Collection, 1895 Francisco Blvd. East, San Rafael, CA 94901-5506, USA; ©1996 Digital Stock, Medicine and Health Care collection; © 2005 JupiterImages Corporation www.clipart.com; ©Hemera Technologies Inc, 1997-2001; ©Click Art, ©T/Maker Company; ©1994., ©Digital Vision; Gazelle Technologies Inc.; PhotoDisc®, Inc. USA, www.photodisc.com. • TechPool Studios, for their clipart collection of human anatomy: Copyright ©1994, TechPool Studios Corp. USA (some of these images were modified by Biozone) • Totem Graphics, for their clipart collection • Corel Corporation, for use of their clipart from the Corel MEGAGALLERY collection • 3D images created using Poser and Pymol • iStock images • Art Today • Adobestock • Image stills from Sketchfab

©2025 BIOZONE International
ISBN: 978-1-99-101409-2
Photocopying prohibited

Index

10% rule 64

A
Abiotic factor 41, 48-52
ACFOR abundance scale 122
Acid rain 252, 256
Agricultural revolution 145-146
Agriculture
 - and climate change 349
 - increasing productivity 149
 - intensive 148-149
 - land use 143
 - productivity of 67
 - sustainable 150-151
 - water use 181
Air pollution 231, 252-257
 - monitoring 253
 - reducing 257
Albedo 325, 335
Amazon rainforest
 - tipping point 336, 343
AMOC, tipping point 336, 346-347
Animal populations, sampling 120
Anthropogenic climate change 318
Aquaculture 187
Asthenosphere 16, 17
Atlantic Meridional Overturning Circulation (AMOC) 336, 346-347
Atmosphere 6
Atmospheric circulation 28-29
Atmospheric CO_2,
 - and climate change 320-321
Autotroph 56-58, 66
Axial tilt 2

B
Batteries, rechargeable 225
Biodiversity 279-287
 - and climate change 332-333
 - and introduced species 291-292
 - loss of 284-287
 - of insects 286-287
Biodiversity hotspot 283, 288
Biofuel 205, 215-216
Biogeochemical cycles 76-83
Biological controls 160
Biological oxygen demand 237
Biomagnification 233, 236, 238-240
Biomass, as fuel 215-216
Biome 41-47
 - aquatic 47
 - distribution 42
 - terrestrial 44-45
Biomes, and temperature 46
Bioremediation 270
Biosphere 41, 42
Biotic factor 41
Biotic potential 101
Bird flu 293
Birth rate 94
Bophal disaster 274
Boreal forest, tipping point 342
Bycatch, of fishing 183

C
Caldera 19
Carbon capture 354-355
Carbon credits 353
Carbon cycle 77-78
Carbon emissions 358-359
Carbon footprint 357
 - and agriculture 151
Carbon mitigation 353-356
Carbon sequestration 356
Carbon storage 354-355
Carbon trading 353
Carbon, net zero 358-359
Carrying capacity (K) 91, 98-101
Causation, and correlation 373
Cellular respiration 59-60, 77, 81
Central tendency, measures 366
Cereal crops 146, 152-153
CFCs, and ozone depletion 258, 260
Chemical spill 274
Chernobyl disaster 273
Circulation, oceanic 32-33
CITES legislation 309
City planning 170-171
Classification key 135-137
Climate action 352
Climate adaptation 350-351
Climate change, defined 313-314
 - and agriculture 349
 - and biodiversity 295, 332-333
 - evidence for 315
 - extreme weather 326-327
 - legislation 316-317
 - mitigation 350-351
 - models 318-319
 - species distribution 330-331
Climate models 318-319
Climate pattern 30
Climate risk 348
Climate, past 34-35
Coal 192-194
Colorado River
 - electricity generation 209
Commensalism 71
Community 41
Competition 71
 - effect on niche 54-55
Competitive exclusion 55
Conference of the Parties (COP) 316
Conservation legislation 309
Conservation methods 304-310
Conservation sustainability 310
Consumer 56, 61, 66
Continental drift 18
Continuous data 365, 366
Convergent boundary 14-15
Coral bleaching 345
Coral reef, tipping point 336, 344-345
Core, Earth's 8, 13
Coriolis effect 28-29, 33
Correlation vs causation 373
Cretaceous climate 35
Crops, cereal 146, 152-153
Crust, Earth's 8, 13, 16
Currents, oceanic 32-33
Cycles, celestial bodies 11-12

D
Dams, power generation 208
Data analysis 368-369
Data interpretation 368-369
Data, types of 365
Datalogger 118
Dating methods 5
Dating, relative 10
DDT, and biomagnification 238
Death rate 94
Decay, of isotopes 5
Decimal form 370
Decomposer 56
Deep Water Horizon oil spill 267
Deforestation 288-289
 - and ozone hole 260
Demographic transition model 109-110
Demography, human 109-110
Density dependent factors 93
Density independent factors 93
Density, of populations 91-92
Desert, abiotic factors 48
Detritivore 56
Developed country 107
Developing country 107
Dichotomous key 135-137
Discontinuous data 365, 366
Disease, effects on populations 293-294
Distribution patterns, in populations 92
Divergent boundary 14-15
Diversity index 133-134
DNA barcode 132
Drought 329
Dutch Elm disease 293

E
Earth, cycles 11
Earth
 - formation of 6
 - structure of 8
Earth's history 6-7
Earthquake 21
Ecological niche 54-55
Ecological pyramids, types 69-70
Ecological succession 84-88
Ecosystem diversity 133-134
Ecosystem stability 72-73
Ecosystem
 - components of 41
 - energy flow in 64
 - scales of 42
Ecotourism 302-303
Efficiency, of transportation 218-220
El Niño Southern Oscillation 30
Electric vehicles 218
Electricity production 190
Electron transport chain 60
Elephants, evolution of 10
Emigration 94
Endangered species 295-297
Endocrine disruptor 240
Energy calculations 372
Energy conservation 219-220
Energy efficiency 218-220
Energy flow 56, 64
Energy security 221-222
Energy storage 223-225
Energy transformation 190
Energy types, use of 191
Energy use, global 191
Energy
 - in the home 219-220
 - non-renewable 192-204
 - renewable 205-217
Enhanced greenhouse effect 320
Environmental accidents 272-275
Environmental change, scales 74
Environmental DNA 132
Environmental issues
 - of oil 201
 - of wind turbines 207
Environmental legislation 279-280
Environmental remediation 276-277
Environmental resources, managing 300-303
Erosion, soil 161, 163
Ethanol, as fuel 216
Eutrophication 79, 88, 232, 234, 236
Evolution, elephants 10
Ex-situ conservation 306-307
Exponential functions 371
Exponential growth 93, 98
Extinction 298-299
Extreme oil 197
Extreme weather 326-327
Exxon Valdez oil spill 266

F
Farming, impacts of 147
Farming
 - intensive 148-149
 - sustainable 150-151
Fast fashion 182
Fault 14
Feedback cycle 334-335
Fertility rate 107-108
Fertilizer production 147
Fertilizer
 - and nitrogen cycle 80
 - and pollution 232
 - use of 149
Fishing
 - and biodiversity 295
 - impacts of 183-184
 - management of 185-187
Food chain 61-62
Food security 155
Food web 56, 61-63
Forest
 - abiotic factors 49
 - stratification 49
Forestry 164-165
 - detrimental effects 165
Forever chemicals 250, 263
Fossil fuel 192-201, 205
 - health effects of 271
Fossil
 - formation of 9-10
 - record 10
Fracking 199, 201
Frontal weather systems 29
Fuel efficiency, comparisons 218

Fuel types 192-201
Fukushima Daiichi disaster 272-273

G
Galápagos Islands, ecotourism 302-303
Geothermal power 205, 212-213
Glaciation, Ordovician 35
Global warming, projections 337
Globalization 175
Glycolysis 60
GPS tracking 131
Graph
 - types 367
 - reading 372
Grazing, effects of 166-167
Great Pacific Garbage Patch 247
Green Revolution 145-146
Greenhouse effect 320-321
Greenhouse gas 320-321

H
Haber process 147, 235
Habitat 53
Habitat fragmentation 290
Heatwaves 327
Heterotroph 61, 66
Hotspots, for biodiversity 283
Humans
 - age structure 102
 - and resource use 111-112
 - demography 109-110
 - effects on species 295-297
 - global distribution 104-106
 - population growth 107-108
 - survivorship curve 96-97
Hurricanes 326
Hydraulic fracturing 199, 201
Hydroelectric power 190, 205, 208-209
 - issues 209
Hydrogen fuel cells 217
Hydrologic cycle 82

I
Ice age, little 347
Ice sheet
 - melting 322
 - tipping point 336, 338-340
Igneous rock 13, 22-23
Illegal fishing 186
Immigration 94
In-situ conservation 304-305
Indicator species 133-134
Indirect sampling 129
Industrial metals 174
Insects, and biodiversity 286-287
Integrated pest management 159-160
Intensive farming 148-149
Intergovernmental Panel on Climate Change 316, 318
International units 372
Intertropical Convergence Zone 28
Introduced species
 - and biodiversity 291-292, 295
IPCC 316, 318
Isotope decay, role in dating 5

J
J shaped curve 98

K
K-selected species 101
Keystone species 73
Kinetic energy 190
Kinetic energy recovery system 224
Kite graph 127
Krebs cycle 60

L
La Niña 30
Lake, abiotic factor 52
Land, and agriculture 143
Legislation
 - climate change 316-317
 - conservation 309
 - environmental 279-280
Life expectancy, humans 96-97
Light pollution 231, 264
Lighting, efficiency 220
Lincoln Index 128
Link reaction 60
Lithosphere 16
Log transformations 371
Logging methods 164-165
Logistic growth 98-100

M
Maasai Mara 301
Mantle, Earth's 8, 13, 16
Mark and recapture sampling 121, 128
Mathematical symbols 370
Mean, calculating 366
Meat, production of 154
Median, calculating 366
Megadrought 329
Mercury, biomagnification 239
Metamorphic rock 13, 22-23
Methane production 215
Microclimate 48, 51
Microhabitat 48
Microplastic 250-251
Migration 94
Minerals
 - mining of 173-174
 - types 23
Mining
 - disaster 275
 - of coal 193-194
 - of minerals 173-174
Mode, calculating 366
Modeling 364
Models 364
Monoculture 165
Moon, cycle 11
Mortality 94
Mutualism 71

N
Nanoplastic 250-251
Natality 94
Natural gas 192, 195-196
Natural gas, extraction of 198
Net zero carbon 358-359
Niche, ecological 54-55
Nitrogen cycle 79-80
Nitrogen fixing 80
Nitrogen pollution 232, 234-237
Noise pollution 231, 261
Non-random sampling 117
Non-renewable energy 192-204
Nuclear accidents 272-273

Nuclear fission 192, 202-203
Nuclear power 192, 202-204
Numbers, converting 372

O
Ocean acidification 324
Ocean circulation 32-33
Ocean currents 32-33
Ocean warming 322
Oil 192, 195-197
 - environmental effects of 201
 - extraction of 198-200, 201
Oil sand 197-198, 201
Oil shale 197-198, 201
Oil spills 266-270
 - mitigation 269-270
Orbital cycles 11-12
Orbital eccentricity 12
Overfishing 183-184
Oxygen cycle 81
Ozone depletion 231, 258-260
Ozone hole 259

P
P-wave 21
Pacific ring of fire 20
Parasitism 71
Paris Agreement 316-317, 350
Parks 168-169
Passive solar heating 210
PCPs 263
Peak oil 195
Permafrost, tipping point 341
Persistent organic pollutants (POP) 238-239
Pesticide resistance 158
Pesticides
 - problems with 157-158
 - resistance to 158
 - use of 149, 156-157
Pest control 156-157
Phosphorus cycle 83
Photochemical smog 254-255
Photosynthesis 57-59, 77, 81
Photovoltaic cells 190, 205, 210
Physical factor 41, 48-52
Pioneer species 84
Plants, importance of 144
Plastic pollution 246-251
 - solutions to 249
Plate boundaries, types 14-15
Plate tectonics 17
Point sampling 121
Polar vortex 326
Pollination, role of insects 287
Pollutant 231
Pollution 231-240, 246-251, 252-257, 261, 262-263, 264, 266-270
 - air 231, 252-257
 - and biodiversity 295
 - economic effects 278
 - health effects 265
 - in the home 262-263
 - light 231, 264
 - nitrogen 232, 234-237
 - noise 231, 261
 - plastic 246-251
 - water 231-233, 266-268
Polyculture 150
POPs (persistent organic pollutants) 238-239
Population 116
 - density 91-92

 - factors affecting 93
 - features of 91
Population age structure 102-103
Population density 104-105
Population distribution 92
 - human 104-106
Population growth curves 98-100
Population growth
 - calculating 94
 - human population 107-108
Population sampling 116-117
Populations dynamics 91
Positive feedback 339
 - and climate 334-335
Predation 71
Primary productivity 59, 66-67
Primary succession 84-85
Producer 56, 57-58, 61, 66
Productivity, measuring 59, 66-68
Pyramids
 - age structure 102-103
 - ecological 69-70

Q
Quadrat sampling 121-125
Qualitative data 365, 366
Quantitative data 365, 366

R
r-selected species 101
Radio tracking 131
Radiometric dating 5
Rain shadow 42
Random sampling 117
Rangeland management 166-167
Ranked data 365, 366
Rates, calculating 370
Realized niche 54
Rechargeable battery 225
Recycling 243-244
Relative dating 5, 10
Remediation, environmental 276-277
Renewable energy 205-217
Reserve lands 168-169
Residential pollution 262-263
Resources, human use of 111-112
Rewilding 308
Rice crops 146
Rock cycle 22-23
Rock formations 22
Rock profile, and fossil record 10
Rock types 13, 22-23
Rocky shore, abiotic factor 51

S
S-wave 21
Salinization, of soil 161
Sampling 118
Sampling error 116
Sampling techniques 21-132
Sampling
 - animal populations 120
 - of populations 116-117, 120
 - reasons for 115-116
 - methods 117
Satellite tracking 131
Sea level rise 322-323
Secondary productivity 68
Secondary succession 86-87

Sedimentary rock 13, 22-23
Seismic wave 21
Sensors, for sampling 118
Sewage treatment 241
Shield volcano 19
SI units 372
Simpson's Index of diversity 133-134
Sixth extinction 298
Smog, photochemical 254-255
Soil 24-26
Soil degradation 161-162
Soil erosion 161
- limiting 163
Soil
- composition 24
- development of 25
- dynamics 25-26
- horizons 25-26
- texture 26
Solar power 205, 210-211
Solar radiation modification 360-361
Solar storms, and energy supply 222
Species distribution, response to climate change 330-331
Species interactions 71
Standard form 370
Storage, of energy 223-225
Stratification, forest 49
Strato-volcano 19
Sun, cycle 11
Survivorship curve
- humans 96-97
- types 95
Sustainability
- in farming 150-151
- in fishing 184
- in planning 171

T
Tectonic plate 17
Temperature, effect on biome 46
Thermal expansion 322
Thermohaline circulation 32
Thorium reactor 204
Thunberg, Greta 352
Tidal power 205, 214
Tipping points 336-347
Tolerance range 53
Tragedy of the Commons 175
Transect sampling 121, 126-127
Transform boundary 15
Transportation methods 172
- efficiency of 218-220
Tricellular model 28
Trophic efficiency 66-68
Trophic level 56, 61, 64, 69
Tundra fires 328

U
Uranium, and nuclear power 202-203
Urbanization 106

V
Valeriepieris circle 105
Vehicles, efficiency comparison 218
Volcanic dome 19
Volcano 14-15
Volcanoes, types 19-20

WXYZ
Waste management 242
Waste reduction 243-245
Water conflicts 180
Water cycle 82
Water pollution 231-233, 266-268
Water quality, measuring 119
Water resources 176-178
Water use 179-182
Water
- industrial use 181-182
- locations 31
Wave power 205, 214
Weather fronts 29
Weather, extreme 326-327
Wetland succession 88
Wheat production 146
Wildfires 328
Wind power 205-207
Wind turbine 190, 205-207

Zika virus 294